# APPLICATIONS

*Introduction to the Practice of Statistics* presents a wide variety of applications from diverse disciplines. The following list indicates a number of Examples and Exercises related to different fields. Note that some items appear in more than one category.

## Examples by Application

### Biology and Environmental Science
**Ch 1:** 1.12, 1.23, 1.34, 1.35. **Ch 2:** 2.1, 2.3, 2.7, 2.8, 2.10, 2.18, 2.20. **Ch 3:** 3.15, 3.19, 3.34. **Ch 4:** 4.19. **Ch 6:** 6.1, 6.28. **Ch 14:** 14.7, 14.8. **Ch 15:** 15.1, 15.2, 15.3, 15.4, 15.5, 15.13, 15.14.

### Business and Consumer Behavior
**Ch 1:** 1.6, 1.8, 1.18. **Ch 2:** 2.2, 2.7, 2.10, 2.20, 2.35, 2.36. **Ch 3:** 3.8, 3.13, 3.16, 3.26, 3.31. **Ch 4:** 4.12, 4.13, 4.14, 4.27, 4.29, 4.32, 4.33, 4.34, 4.35, 4.36. **Ch 5:** 5.6, 5.7, 5.8, 5.11, 5.12, 5.13, 5.14, 5.16, 5.17, 5.18, 5.19. **Ch 8:** 8.3, 8.4, 8.5. **Ch 9:** 9.8, 9.9, 9.10, 9.11. **Ch 12:** 12.1, 12.2, 12.3, 12.4, 12.5, 12.6, 12.7, 12.8, 12.9, 12.10, 12.11, 12.12, 12.13, 12.14, 12.15, 12.16, 12.17, 12.18, 12.19, 12.20, 12.21, 12.22, 12.23, 12.24, 12.25. **Ch 13:** 13.1, 13.2, 13.7, 13.8. **Ch 14:** 14.9, 14.10. **Ch 16:** 16.1, 16.2, 16.3, 16.4, 16.5, 16.6, 16.8, 16.9, 16.13. **Ch 17:** 17.20, 17.21.

### College Life
**Ch 1:** 1.1, 1.2, 1.3, 1.4. **Ch 2:** 2.5, 2.23, 2.27, 2.28, 2.29, 2.31, 2.32, 2.33, 2.34. **Ch 3:** 3.24, 3.25. **Ch 4:** 4.11, 4.22, 4.26. **Ch 6:** 6.4, 6.5, 6.6, 6.7, 6.8, 6.9, 6.10, 6.11, 6.12, 6.13, 6.14, 6.20. **Ch 8:** 8.1, 8.6, 8.9, 8.11, 8.12. **Ch 9:** 9.1, 9.2. **Ch 14:** 14.1, 14.2, 14.3, 14.4, 14.6.

### Demographics and Characteristics of People
**Ch 1:** 1.7, 1.25, 1.26. **Ch 2:** 2.42. **Ch 4:** 4.30. **Ch 5:** 5.1, 5.5, 5.15. **Ch 7:** 7.13.

### Economics and Finance
**Ch 1:** 1.20. **Ch 2:** 2.10, 2.24, 2.37, 2.41, 2.42. **Ch 4:** 4.38. **Ch 6:** 6.31. **Ch 7:** 7.4, 7.5, 7.6, 7.16, 7.17.

### Education and Child Development
**Ch 1:** 1.9, 1.10, 1.22, 1.24, 1.27, 1.28, 1.29, 1.30, 1.31, 1.32. **Ch 2:** 2.6, 2.10, 2.23, 2.26, 2.37, 2.40. **Ch 3:** 3.2, 3.4, 3.5, 3.7, 3.20, 3.21. **Ch 4:** 4.17, 4.22, 4.26, 4.37, 4.42, 4.45. **Ch 6:** 6.3, 6.4, 6.5, 6.6, 6.7, 6.8, 6.9, 6.10, 6.11, 6.12, 6.13, 6.14, 6.16, 6.19, 6.20. **Ch 7:** 7.14, 7.15, 7.18. **Ch 11:** 11.1. **Ch 12:** 12.27, 12.28. **Ch 15:** 15.8, 15.9, 15.10. **Ch 16:** 16.11, 16.12. **Ch 17:** 17.19.

### Ethics
**Ch 3:** 3.35, 3.36, 3.37, 3.38, 3.39, 3.40, 3.41.

### Health and Nutrition
**Ch 1:** 1.19. **Ch 2:** 2.4, 2.12, 2.13, 2.14, 2.15, 2.16, 2.19, 2.21, 2.22, 2.25, 2.37, 2.38, 2.39, 2.44, 2.45. **Ch 3:** 3.1, 3.6, 3.9, 3.14, 3.18, 3.39. **Ch 4:** 4.12, 4.20. **Ch 5:** 5.10. **Ch 6:** 6.2, 6.15, 6.24, 6.29. **Ch 7:** 7.19, 7.20, 7.21, 7.22, 7.23. **Ch 8:** 8.1, 8.2, 8.9, 8.10, 8.11, 8.12. **Ch 9:** 9.1, 9.2, 9.7. **Ch 10:** 10.10, 10.13, 10.14, 10.15, 10.16, 10.17, 10.18, 10.19, 10.20, 10.21. **Ch 13:** 13.3, 13.4, 13,5, 13.6, 13.7, 13.8, 13.9, 13.11. **Ch 14:** 14.1, 14.2, 14.3, 14.4, 14.6. **Ch 15:** 15.6, 15.7.

### Humanities and Social Sciences
**Ch 2:** 2.37, 2.42. **Ch 3:** 3.3, 3.23, 3.27, 3.28, 3.29, 3.30, 3.40, 3.41. **Ch 5:** 5.2, 5.3, 5.4. **Ch 6:** 6.25, 6.27. **Ch 7:** 7.7, 7.8, 7.12. **Ch 9:** 9.3, 9.4, 9.5, 9.6, 9.12, 9.13. **Ch 13:** 13.9, 13.10. **Ch 16:** 16.14.

### Manufacturing, Products, and Processes
**Ch 1:** 1.11, 1.33. **Ch 3:** 3.22. **Ch 5:** 5.5, 5.22, 5.23, 5.25. **Ch 6:** 6.17, 6.18, 6.30, 6.32. **Ch 12:** 12.26. **Ch 17:** 17.1, 17.2, 17.3, 17.4, 17.5, 17.7,

17.8, 17.9, 17.11, 17.12, 17.13, 17.14, 17.15, 17.18, 17.19.

### International
**Ch 2:** 2.25. **Ch 7:** 7.1, 7.2, 7.3. **Ch 9:** 9.7, 9.8, 9.9, 9.10, 9.11. **Ch 13:** 13.10.

### Motor Vehicles and Fuel
**Ch 1:** 1.5, 1.13, 1.14, 1.15, 1.16, 1.17. **Ch 2:** 2.9, 2.10. **Ch 4:** 4.9, 4.10, 4.11, 4.32, 4.33. **Ch 9:** 9.14, 9.15. **Ch 10:** 10.1, 10.2, 10.3, 10.4, 10.5, 10.6, 10.7, 10.8, 10.9, 10.10, 10.12, 10.22, 10.23.

### Physical Sciences
**Ch 1:** 1.21. **Ch 2:** 2.17. **Ch 4:** 4.16.

### Sports and Leisure
**Ch 1:** 1.27, 1.28, 1.29, 1.30, 1.31, 1.32. **Ch 2:** 2.11. **Ch 4:** 4.2, 4.31, 4.34, 4.35, 4.36, 4.41, 4.44, 4.46. **Ch 5:** 5.9, 5.24. **Ch 7:** 7.10, 7.11. **Ch 8:** 8.7, 8.8. **Ch 15:** 15.11, 15.12. **Ch 16:** 16.7, 16.10.

### Technology and the Internet
**Ch 3:** 3.10, 3.11, 3.12, 3.17. **Ch 4:** 4.9, 4.10, 4.11, 4.43, 4.47, 4.48. **Ch 7:** 7.1, 7.2, 7.3, 7.10, 7.11. **Ch 8:** 8.1. **Ch 9:** 9.14, 9.15.

## Exercises by Application

### Biology and Environmental Science
**Ch 1:** 1.18, 1.28, 1.29, 1.30, 1.31, 1.32, 1.33, 1.34, 1.36, 1.42, 1.56, 1.63, 1.64, 1.71, 1.78, 1.79, 1.80, 1.91, 1.93, 1.95, 1.115, 1.119, 1.144, 1.148, 1.149, 1.152, 1.155, 1.164, 1.168. **Ch 2:** 2.2, 2.5, 2.15, 2.18, 2.26, 2.29, 2.34, 2.40, 2.50, 2.63, 2.64, 2.66, 2.80, 2.81, 2.85, 2.88, 2.96, 2.141, 2.143, 2.149. **Ch 3:** 3.21, 3.22, 3.23, 3.44, 3.49, 3.58, 3.67, 3.83. **Ch 5:** 5.46, 5.49, 5.67, 5.70. **Ch 6:** 6.66, 6.67, 6.97, 6.110, 6.111, 6.122. **Ch 7:** 7.10, 7.25, 7.44, 7.65, 7.66, 7.81, 7.82, 7.89, 7.91, 7.92, 7.93, 7.103, 7.104, 7.105, 7.106, 7.108, 7.109, 7.122, 7.123, 7.124, 7.134, 7.135. **Ch 8:** 8.48, 8.63, 8.80. **Ch 9:** 9.20,

9.25, 9.40. **Ch 10:** 10.14, 10.15, 10.16, 10.17, 10.18, 10.19, 10.23, 10.24, 10.31, 10.56, 10.57, 10.58, 10.59. **Ch 11:** 11.40, 11.41, 11.42, 11.43, 11.44, 11.45, 11.46, 11.47, 11.48, 11.49, 11.50. **Ch 12:** 12.8, 12.10, 12.12, 12.14, 12.25, 12.34, 12.35, 12.36, 12.37, 12.51, 12.52, 12.53. **Ch 13:** 13.6, 13.7, 13.25, 13.26, 13.27, 13.36, 13.37, 13.38, 13.39, 13.40, 13.41, 13.42, 13.43, 13.44, 13.45. **Ch 14:** 14.17. **Ch 15:** 15.10, 15.11, 15.13, 15.22, 15.34, 15.35, 15.36, 15.38, 15.42, 15.43, 15.47, 15.48. **Ch 16:** 16.5, 16.11, 16.12, 16.17, 16.20, 16.39, 16.41, 16.64, 16.74.

## Business and Consumer Behavior
**Ch 1:** 1.39, 1.66, 1.67, 1.69, 1.156. **Ch 2:** 2.3, 2.4, 2.19, 2.29, 2.37, 2.39, 2.62, 2.66, 2.84, 2.90, 2.124, 2.129, 2.134, 2.138, 2.161, 2.162, 2.163, 2.164. **Ch 3:** 3.3, 3.11, 3.15, 3.16, 3.25, 3.28, 3.30, 3.32, 3.35, 3.36, 3.41, 3.48, 3.54, 3.57, 3.60, 3.61, 3.70, 3.78, 3.83, 3.89, 3.107, 3.114, 3.117, 3.120, 3.125, 3.130. **Ch 4:** 4.5, 4.14, 4.15, 4.32, 4.53, 4.54, 4.55, 4.75, 4.77, 4.87, 4.89, 4.90, 4.106, 4.107, 4.108, 4.109, 4.111, 4.135, 4.144. **Ch 5:** 5.20, 5.55, 5.56, 5.60, 5.69, 5.71, 5.77, 5.78. **Ch 6:** 6.7, 6.8, 6.20, 6.21, 6.31, 6.34, 6.35, 6.42, 6.53, 6.54, 6.55, 6.72, 6.100, 6.112, 6.115, 6.117, 6.123, 6.124, 6.125, 6.132. **Ch 7:** 7.8, 7.9, 7.14, 7.34, 7.38, 7.42, 7.45, 7.63, 7.65, 7.66, 7.67. 7.68, 7.74, 7.75, 7.80, 7.83, 7.87, 7.93, 7.101, 7.103, 7.104, 7.110, 7.125, 7.141, 7.142. **Ch 8:** 8.22, 8.30, 8.31, 8.34, 8.35, 8.36, 8.37, 8.38, 8.51, 8.64, 8.65, 8.66, 8.67, 8.70, 8.71, 8.80. **Ch 9:** 9.30, 9.31, 9.36. **Ch 10:** 10.37, 10.38, 10.52. **Ch 11:** 11.17. **Ch 12:** 12.24, 12.34, 12.36, 12.40, 12.41, 12.63, 12.69. **Ch 13:** 13.14, 13.15, 13.17, 13.18, 13.19, 13.20, 13.33, 13.34. **Ch 14:** 14.3, 14.4, 14.5, 14.6, 14.7, 14.8, 14.9, 14.11, 14.12, 14.13, 14.14, 14.15, 14.16, 14.18, 14.19, 14.20, 14.21, 14.22, 14.23, 14.24, 14.31, 14.33, 14.34, 14.36, 14.37, 14.38. **Ch 15:** 15.17, 15.28, 15.40, 15.41. **Ch 16:** 16.1, 16.15, 16.18, 16.19, 16.22, 16.26, 16.28, 16.31, 16.40, 16.42, 16.43, 16.54, 16.55,
16.58, 16.67, 16.73, 16.79, 16.84, 16.85. **Ch 17:** 17.5, 17.14.

## College Life
**Ch 1:** 1.1, 1.2, 1.3, 1.5, 1.6, 1.7, 1.8, 1.9, 1.10, 1.11, 1.12, 1.13, 1.14, 1.27, 1.37, 1.38, 1.41, 1.47, 1.48, 1.49, 1.50, 1.51, 1.52, 1.53, 1.55, 1.61, 1.72, 1.116, 1.146, 1.154, 1.157, 1.160. **Ch 2:** 2.1, 2.6, 2.7, 2.8, 2.9, 2.30, 2.31, 2.31, 2.33, 2.49, 2.50, 2.58, 2.59, 2.60, 2.61, 2.78, 2.79, 2.105, 2.106, 2.107, 2.108, 2.109, 2.110, 2.120, 2.154. **Ch 3:** 3.1, 3.2, 3.20, 3.50, 3.51, 3.54, 3.68, 3.69, 3.79, 3.83, 3.123. **Ch 4:** 4.10, 4.46, 4.60, 4.65, 4.97, 4.99, 4.101, 4.102, 4.103, 4.104, 4.105, 4.112, 4.113, 4.114, 4.115. **Ch 5:** 5.2, 5.18, 5.25, 5.28, 5.73, 5.77. **Ch 6:** 6.1, 6.2, 6.3, 6.5, 6.6, 6.14, 6.15, 6.16, 6.19, 6.23, 6.24, 6.36, 6.52, 6.63, 6.69, 6.72. **Ch 7:** 7.1, 7.2, 7.3, 7.6, 7.38. **Ch 8:** 8.3, 8.5, 8.6, 8.11, 8.12, 8.13, 8.16, 8.19, 8.20, 8.23, 8.32, 8.33, 8.43, 8.46, 8.47, 8.52, 8.74, 8.76, 8.83. **Ch 9:** 9.4, 9.7, 9.8, 9.17, 9.21, 9.23, 9.24, 9.34, 9.35, 9.41. **Ch 10:** 10.10, 10.11, 10.42, 10.43. **Ch 12:** 12.9, 12.10, 12.11, 12.12, 12.14, 12.23, 12.26. **Ch 13:** 13.22, 13.23, 13.24. **Ch 16:** 16.83. **Ch 17:** 17.1, 17.2, 17.3, 17.7, 17.8, 17.11.

## Demographics and Characteristics of People
**Ch 1:** 1.4, 1.15, 1.16, 1.17, 1.20, 1.21, 1.22, 1.62, 1.65, 1.68, 1.82, 1.83, 1.84, 1.117, 1.159, 1.161, 1.169, 1.172. **Ch 2:** 2.10, 2.14, 2.92, 2.111, 2.112, 2.113, 2.114, 2.115, 2.116, 2.120, 2.126. **Ch 3:** 3.76, 3.112, 3.113. **Ch 4:** 4.11, 4.21, 4.22, 4.28, 4.29, 4.30, 4.34, 4.37, 4.42, 4.43, 4.44, 4.45, 4.52, 4.59, 4.84, 4.89, 4.90, 4.106, 4.107, 4.108, 4.116, 4.117, 4.118, 4.119, 4.120, 4.121, 4.124, 4.141, 4.143. **Ch 5:** 5.4, 5.21, 5.31, 5.68. **Ch 6:** 6.116. **Ch 7:** 7.130, 7.131. **Ch 8:** 8.26. **Ch 9:** 9.7, 9.8. **Ch 16:** 16.26.

## Economics and Finance
**Ch 1:** 1.170, 1.172. **Ch 2:** 2.12, 2.27, 2.28, 2.36, 2.38, 2.48, 2.70, 2.71, 2.92, 2.103, 2.142, 2.151, 2.152. **Ch 3:** 3.40, 3.49, 3.87. **Ch 4:** 4.39, 4.91, 4.92, 4.93, 4.110, 4.136. **Ch 5:** 5.57, 5.62.

**Ch 6:** 6.116. **Ch 7:** 7.59, 7.60, 7.90. **Ch 10:** 10.5, 10.20, 10.33, 10.45, 10.47, 10.48, 10.49. **Ch 11:** 11.22, 11.23, 11.34, 11.25, 11.26. **Ch 16:** 16.45, 16.49, 16.61, 16.80.

## Education and Child Development
**Ch 1:** 1.25, 1.43, 1.44, 1.45, 1.74, 1.99, 1.100, 1.101, 1.102, 1.103, 1.104, 1.105, 1.106, 1.126, 1.127, 1.128, 1.129, 1.130, 1.131, 1.132, 1.133, 1.134, 1.135, 1.146, 1.153, 1.157, 1.167, 1.173. **Ch 2:** 2.11, 2.13, 2.35, 2.51, 2.65, 2.74, 2.91, 2.122, 2.127, 2.128, 2.131, 2.135, 2.155, 2.156, 2.159, 2.160, 2.161. **Ch 3:** 3.9, 3.12, 3.13, 3.14, 3.17, 3.19, 3.93, 3.94, 3.127. **Ch 4:** 4.46, 4.73, 4.97, 4.99, 4.103, 4.104, 4.105, 4.122, 4.138. **Ch 5:** 5.28, 5.32, 5.48, 5.50, 5.66. **Ch 6:** 6.5, 6.6, 6.54, 6.64, 6.65, 6.69, 6.95, 6.96, 6.113. **Ch 7:** 7.111, 7.127, 7.128, 7.138, 7.139. **Ch 9:** 9.29, 9.35. **Ch 10:** 10.10, 10.11, 10.30, 10.32, 10.34, 10.53, 10.54, 10.55, 10.61. **Ch 11:** 11.1, 11.3, 11.5, 11.6, 11.27, 11.28, 11.29, 11.30. **Ch 12:** 12.26, 12.42, 12.43, 12.44, 12.61, 12.62, 12.64. **Ch 13:** 13.5, 13.7, 13.12, 13.22, 13.23, 13.24, 13.46, 13.47, 13.48, 13.49. **Ch 14:** 14.39, 14.40, 14.41, 14.42. **Ch 15:** 15.7, 15.8, 15.9, 15.12, 15.14, 15.26. **Ch 16:** 16.4, 16.9, 16.14, 16.32, 16.51, 16.53, 16.56, 16.59. **Ch 17:** 17.75.

## Ethics
**Ch 3:** 3.96, 3.97, 3.98, 3.99, 3.100, 3.101, 3.102, 3.103, 3.104, 3.105, 3.106, 3.107, 3.108, 3.109, 3.110, 3.111, 3.112, 3.129, 3.130, 3.131, 3.132, 3.133.

## Health and Nutrition
**Ch 1:** 1.23, 1.24, 1.35, 1.57, 1.58, 1.59, 1.70, 1.92, 1.112, 1.114, 1.136, 1.137, 1.138, 1.139, 1.140. **Ch 2:** 2.9, 2.20, 2.21, 2.53, 2.54, 2.55, 2.68, 2.73, 2.82, 2.83, 2.87, 2.89, 2.93, 2.121, 2.129, 2.130, 2.132, 2.133, 2.136, 2.137, 2.145, 2.146, 2.166. **Ch 3:** 3.7, 3.24, 3.26, 3.27, 3.29, 3.31, 3.37, 3.38, 3.39, 3.45, 3.88, 3.100, 3.101, 3.104, 3.106, 3.108, 3.109, 3.116, 3.124, 3.126, 3.131. **Ch 4:** 4.6, 4.21, 4.22, 4.29, 4.34, 4.42, 4.43, 4.44, 4.45,

4.52, 4.84, 4.101, 4.102, 4.125, 4.126, 4.127, 4.128, 4.129, 4.143. **Ch 5:** 5.47, 5.51, 5.53, 5.65. **Ch 6:** 6.17, 6.18, 6.22, 6.25, 6.29, 6.30, 6.37, 6.60, 6.63, 6.70, 6.73, 6.92, 6.121. **Ch 7:** 7.26, 7.27, 7.28, 7.30, 7.31, 7.32, 7.33, 7.37, 7.39, 7.40, 7.46, 7.47, 7.51, 7.61, 7.62, 7.63, 7.64, 7.67, 7.68, 7.76, 7.78, 7.79, 7.85, 7.86, 7.88, 7.94, 7.99, 7.100, 7.102, 7.107, 7.117, 7.118, 7.119, 7.120, 7.121, 7.126, 7.136, 7.137. **Ch 8:** 8.23, 8.24, 8.27, 8.28, 8.46, 8.47, 8.54. **Ch 9:** 9.4, 9.11, 9.21, 9.39. **Ch 10:** 10.13, 10.25, 10.26, 10.27, 10.28, 10.29, 10.35, 10.36, 10.50, 10.60, 10.62, 10.63. **Ch 11:** 11.13, 11.14, 11.15, 11.19, 11.20, 11.21, 11.34, 11.35, 11.36, 11.37, 11.38, 11.39. **Ch 12:** 12.9, 12.11, 12.13, 12.21, 12.22, 12.29, 12.30, 12.31, 12.32, 12.33, 12.39, 12.45, 12.46, 12.47, 12.48, 12.49, 12.50, 12.58, 12.59, 12.60, 12.68. **Ch 13:** 13.7, 13.8, 13.21, 13.28, 13.29, 13.30, 13.36. **Ch 14:** 14.25, 14.27, 14.29, 14.35. **Ch 15:** 15.15, 15.16, 15.23, 15.24, 15.27, 15.29, 15.30, 15.31, 15.32, 15.33, 15.37, 15.39, 15.44, 15.45, 15.46. **Ch 16:** 16.7, 16.10, 16.64, 16.65, 16.66, 16.68, 16.69, 16.70, 16.71, 16.72, 16.81, 16.82, 16.83. **Ch 17:** 17.50.

## Humanities and Social Sciences

**Ch 1:** 1.26, 1.60, 1.85, 1.124, 1.125. **Ch 2:** 2.17, 2.41, 2.47, 2.50, 2.67, 2.75, 2.123, 2.125, 2.134, 2.157, 2.158, 2.162, 2.163. **Ch 3:** 3.8, 3.55, 3.61, 3.63, 3.64, 3.66, 3.71, 3.72, 3.73, 3.74, 3.75, 3.80, 3.85, 3.86, 3.98, 3.100, 3.101, 3.105, 3.110, 3.111, 3.112, 3.128, 3.132, 3.133. **Ch 4:** 4.24, 4.66, 4.148. **Ch 5:** 5.18, 5.22, 5.24, 5.25, 5.26, 5.27, 5.29, 5.72, 5.76. **Ch 6:** 6.53, 6.62, 6.68, 6.93. **Ch 7:** 7.29, 7.41, 7.43, 7.48, 7.129. **Ch 8:** 8.17, 8.18, 8.25, 8.44, 8.45, 8.53, 8.55, 8.56, 8.69, 8.72, 8.73, 8.81. **Ch 9:** 9.6, 9.9, 9.10, 9.15, 9.16, 9.18, 9.19, 9.22, 9.27, 9.28, 9.33, 9.38. **Ch 10:** 10.46. **Ch 11:** 11.31, 11.32, 11.33. **Ch 12:** 12.27, 12.28, 12.38, 12.66, 12.67. **Ch 13:** 13.9, 13.10, 13.12. **Ch 14:** 14.26, 14.28, 14.30. **Ch 15:** 15.25. **Ch 16:** 16.23, 16.50, 16.77, 16.78, 16.86, 16.87.

## International

**Ch 2:** 2.14, 2.15, 2.23, 2.25, 2.29, 2.37, 2.87, 2.88, 2.89, 2.126, 2.146, 2.151, 2.152, 2.156. **Ch 3:** 3.28, 3.48, 3.78, 3.86, 3.88, 3.104, 3.116. **Ch 4:** 4.22, 4.24, 4.59. **Ch 6:** 6.60, 6.62, 6.70. **Ch 9:** 9.37. **Ch 10:** 10.46, 10.47, 10.48, 10.49, 10.56, 10.57, 10.58. **Ch 11:** 11.31, 11.32, 11.33. **Ch 12:** 12.27, 12.28, 12.35, 12.38, 12.49, 12.50, 12.66, 12.67. **Ch 13:** 13.12. **Ch 15:** 15.13, 15.30, 15.38, 15.43, 15.47. **Ch 16:** 16.64, 16.65, 16.66. **Ch 17:** 17.22.

## Manufacturing, Products, and Processes

**Ch 1:** 1.145. **Ch 2:** 2.119, 2.144, 2.147. **Ch 3:** 3.33. **Ch 4:** 4.81. **Ch 5:** 5.40, 5.43, 5.45, 5.54, 5.58, 5.59, 5.63, 5.74, 5.75. **Ch 6:** 6.123. **Ch 7:** 7.11, 7.50, 7.76, 7.78, 7.79, 7.94, 7.107. **Ch 8:** 8.68. **Ch 9:** 9.5, 9.32. **Ch 10:** 10.21, 10.22. **Ch 11:** 11.51, 11.52, 11.53, 11.54, 11.55, 11.56, 11.57, 11.58, 11.59, 11.60. **Ch 12:** 12.29, 12.30, 12.31, 12.32, 12.33, 12.54, 12.55, 12.56, 12.57, 12.58, 12.59, 12.60. **Ch 13:** 13.31, 13.32. **Ch 14:** 14.32. **Ch 17:** 17.9, 17.10, 17.12, 17.13, 17.15, 17.16, 17.17, 17.18, 17.19, 17.20, 17.26, 17.31, 17.32, 17.34, 17.35, 17.36, 17.37, 17.38, 17.39, 17.40, 17.41, 17.42, 17.43, 17.44, 17.45, 17.46, 17.53, 17.54, 17.55, 17.56, 17.59, 17.60, 17.61, 17.65, 17.66, 17.67, 17.68, 17.69, 17.70, 17.71, 17.72, 17.73, 17.74, 17.76, 17.77, 17.79, 17.83, 17.84, 17.85, 17.86, 17.87.

## Motor Vehicles and Fuel

**Ch 1:** 1.146. **Ch 2:** 2.16, 2.22, 2.44, 2.45, 2.52, 2.86, 2.97, 2.99. **Ch 3:** 3.4. **Ch 4:** 4.12, 4.13, 4.109, 4.111. **Ch 6:** 6.26, 6.27, 6.71. **Ch 7:** 7.24, 7.35, 7.49. **Ch 8:** 8.14, 8.15, 8.58, 8.59. **Ch 10:** 10.2, 10.3, 10.4, 10.12. **Ch 16:** 16.24, 16.60.

## Physical Sciences

**Ch 1:** 1.40, 1.73, 1.94, 1.147. **Ch 2:** 2.24, 2.40, 2.69, 2.72, 2.77, 2.94. **Ch 6:** 6.32, 6.61. **Ch 10:** 10.39, 10.40, 10.41. **Ch 16:** 16.44, 16.48.

## Sports and Leisure

**Ch 1:** 1.46, 1.81, 1.143, 1.165, 1.166. **Ch 2:** 2.19, 2.23, 2.25, 2.39, 2.62, 2.95, 2.98, 2.100, 2.139, 2.140, 2.148, 2.150, 2.165. **Ch 3:** 3.43, 3.47, 3.121. **Ch 4:** 4.2, 4.3, 4.4, 4.7, 4.9, 4.18, 4.27, 4.31, 4.32, 4.38, 4.50, 4.56, 4.58, 4.72, 4.80, 4.83, 4.96, 4.98, 4.100, 4.133, 4.134, 4.137, 4.139, 4.140, 4.142. **Ch 5:** 5.23, 5.30, 5.42, 5.44, 5.52. **Ch 6:** 6.13, 6.14, 6.16, 6.128. **Ch 8:** 8.11, 8.12, 8.43, 8.54, 8.61, 8.82. **Ch 9:** 9.23, 9.24. **Ch 10:** 10.62, 10.63. **Ch 11:** 11.18. **Ch 12:** 12.23, 12.47, 12.48. **Ch 13:** 13.8. **Ch 14:** 14.1, 14.35. **Ch 15:** 15.1, 15.2, 15.3, 15.4, 15.5, 15.6, 15.18, 15.19, 15.20, 15.21. **Ch 16:** 16.6, 16.9, 16.13, 16.16, 16.37, 16.38, 16.47, 16.62, 16.76. **Ch 17:** 17.4, 17.49.

## Technology and the Internet

**Ch 1:** 1.19, 1.158, 1.162. **Ch 3:** 3.5, 3.6, 3.7, 3.10, 3.12, 3.14, 3.34, 3.42, 3.50, 3.51. **Ch 4:** 4.19, 4.20, 4.26, 4.33, 4.35, 4.36, 4.51, 4.74, 4.78, 4.123, 4.145. **Ch 5:** 5.1, 5.13, 5.14, 5.15, 5.16, 5.17, 5.42, 5.44. **Ch 6:** 6.9, 6.23, 6.24. **Ch 7:** 7.54, 7.55, 7.63, 7.74, 7.75, 7.87, 7.101. **Ch 8:** 8.1, 8.2, 8.4, 8.13, 8.41, 8.49, 8.50, 8.58, 8.59, 8.60, 8.62, 8.65, 8.66, 8.67. **Ch 9:** 9.26. **Ch 10:** 10.44. **Ch 11:** 11.17, 11.22, 11.23, 11.34, 11.25, 11.26. **Ch 12:** 12.10, 12.12, 12.13, 12.14, 12.65. **Ch 13:** 13.50. **Ch 14:** 14.21, 14.22, 14.23, 14.24, 14.33, 14.34. **Ch 15:** 15.12. **Ch 16:** 16.6, 16.9, 16.13, 16.16, 16.38.

# Introduction to the
# Practice of Statistics

## Authors' note about the cover

*Introduction to the Practice of Statistics* emphasizes the use of graphical and numerical summaries to understand data. The front cover shows a painting entitled *0 to 9,* by the American artist Jasper Johns in 1961. In this work, the structure of the painting is determined by number sequence, just as our graphical summaries are determined by the numerical calculations that we perform when we analyze data. Can you find all of the digits in the painting?

# Introduction to the
# Practice of Statistics

**SIXTH EDITION**

**DAVID S. MOORE**

**GEORGE P. McCABE**

**BRUCE A. CRAIG**

*Purdue University*

**W. H. Freeman and Company**
**New York**

| | |
|---|---|
| *Senior Publisher:* | CRAIG BLEYER |
| *Publisher:* | RUTH BARUTH |
| *Development Editors:* | SHONA BURKE, ANNE SCANLAN-ROHRER |
| *Senior Media Editor:* | ROLAND CHEYNEY |
| *Assistant Editor:* | BRIAN TEDESCO |
| *Editorial Assistant:* | KATRINA WILHELM |
| *Marketing Coordinator:* | DAVE QUINN |
| *Photo Editor:* | CECILIA VARAS |
| *Photo Researcher:* | ELYSE RIEDER |
| *Cover and Text Designer:* | VICKI TOMASELLI |
| *Senior Project Editor:* | MARY LOUISE BYRD |
| *Illustrations:* | INTEGRE TECHNICAL PUBLISHING CO. |
| *Production Manager:* | JULIA DE ROSA |
| *Composition:* | INTEGRE TECHNICAL PUBLISHING CO. |
| *Printing and Binding:* | RR DONNELLEY |

TI-83™ screen shots are used with permission of the publisher: ©1996, Texas Instruments Incorporated. TI-83™ Graphic Calculator is a registered trademark of Texas Instruments Incorporated. Minitab is a registered trademark of Minitab, Inc. Microsoft © and Windows © are registered trademarks of the Microsoft Corporation in the United States and other countries. Excel screen shots are reprinted with permission from the Microsoft Corporation. S-PLUS is a registered trademark of the Insightful Corporation. SAS© is a registered trademark of SAS Institute, Inc. *CrunchIt!* is a trademark of Integrated Analytics LLC.

Library of Congress Control Number: 2007938575

ISBN-13: 978-1-4292-1623-4
ISBN-10: 1-4292-1623-9 (Extended Version, Casebound)

ISBN-13: 978-1-4292-1622-7
ISBN-10: 1-4292-1622-0 (Casebound)

ISBN-13: 978-1-4292-1621-0
ISBN-10: 1-4292-1621-2 (Paperback)

Printed in the United States of America

Second printing

W. H. Freeman and Company
41 Madison Avenue
New York, NY 10010
Houndmills, Basingstoke RG21 6XS, England
www.whfreeman.com

# Brief Contents

# Contents

## PART I    Looking at Data

### CHAPTER 1
### Looking at Data—Distributions                1

### CHAPTER 2
### Looking at Data—Relationships                83

Sections marked with an asterisk are optional.

## CHAPTER 3
## Producing Data 171

## PART II  Probability and Inference

## CHAPTER 4
## Probability: The Study of
## Randomness 237

Companion Chapters (on the IPS Web site
**www.whfreeman.com/ips6e** and CD-ROM)

## CHAPTER 17
## Statistics for Quality: Control and Capability ... 17-1

Statistics is the science of data. *Introduction to the Practice of Statistics* (*IPS*) is an introductory text based on this principle. We present the most-used methods of basic statistics in a way that emphasizes working with data and mastering statistical reasoning. *IPS* is elementary in mathematical level but conceptually rich in statistical ideas and serious in its aim to help students think about data and use statistical methods with understanding.

Some schematic history will help place *IPS* in the universe of texts for a first course in statistics for students from a variety of disciplines. Traditional texts were almost entirely devoted to methods of inference, with quick coverage of means, medians, and histograms as a preliminary. No doubt this reflected the fact that inference is the only part of statistics that has a mathematical theory behind it. Several innovative books aimed at nontraditional audiences pioneered a broader approach that paid more attention to design of samples and experiments, the messiness of real data, and discussion of real-world statistical studies and controversies. All were written by widely known statisticians whose main business was not writing textbooks. *The Nature of Statistics* (Wallis and Roberts) has passed away, but *Statistics* (Freedman and collaborators) and *Statistics: Concepts and Controversies* (Moore) remain alive and well. None of these books tried to meet the needs of a typical first course because their audiences did not need full coverage of standard statistical methods.

*IPS* was the first book to successfully combine attention to broader content and reasoning with comprehensive presentation of the most-used statistical methods. It reflects the consensus among statisticians—even stronger now than when the first edition appeared—concerning the content of an introduction to our discipline. This consensus is expressed in a report from the joint curriculum committee of the American Statistical Association and the Mathematical Association of America[1] and in discussions in leading journals.[2] *IPS* has been successful for several reasons:

1. *IPS* examines the nature of modern statistical practice at a level suitable for beginners. Attention to data analysis and data production as well as to probability and inference is "new" only in the world of textbooks. Users of statistical methods have always paid attention to all of these. Contemporary research in statistics, driven by advances in computing, puts more emphasis on sophisticated "looking at data" and on data-analytic ways of thinking. Formal inference remains important and receives careful treatment, but it appears as part of a larger picture.

2. *IPS* has a logical overall progression, so data analysis and data production strengthen the presentation of inference rather than stand apart from it. We stress that data analysis is an essential preliminary to inference because inference requires clean data. The most useful "goodness of fit" procedure, for example, is the normal quantile plot presented in Chapter 1 and used frequently in the inference chapters. We emphasize that when you do formal statistical inference, you are acting as if your data come from properly randomized data production. We use random samples and experimental randomization to motivate the need for probability as a language for inference.

3. *IPS* presents data analysis as more than a collection of techniques for exploring data. We integrate techniques with discussion of systematic ways of thinking about data. We also work hard to make data-analytic thinking accessible to beginners by presenting a series of simple principles: always plot your data; look for overall patterns and deviations from them; when looking at the overall pattern of a distribution for one variable, consider shape, center, and spread; for relations between two variables, consider form, direction, and strength; always ask whether a relationship between variables is influenced by other variables lurking in the background. Inference is similarly treated as more than a collection of methods. We warn students about pitfalls in clear cautionary discussions—about regression and correlation, experiments, sample surveys, confidence intervals, and significance tests. Our goal throughout *IPS* is to present principles and techniques together in a way that is accessible to beginners and lays a foundation for students who will go on to more advanced study.

4. *IPS* integrates discussion of techniques, reasoning, and practice using real examples to drive the exposition. Students learn the technique of least-squares regression and how to interpret the regression slope. But they also learn the conceptual ties between regression and correlation, the importance of looking for influential observations (always plot your data), and to beware of averaged data and the restricted-range effect.

5. *IPS* is aware of current developments both in statistical science and in teaching statistics. For example, the first edition already favored the version of the two-sample $t$ procedures that does not assume equal population variances and discussed the great difference in robustness between standard tests for means and for variances. In the fourth edition, we introduced the modified ("plus four") confidence intervals for proportions that are shown by both computational studies[3] and theory[4] to be superior to the standard intervals for all but very large samples. Brief optional "Beyond the Basics" sections give quick overviews of topics such as density estimation, scatterplot smoothers, nonlinear regression, and data mining. Chapter 16 on resampling methods offers an extended introduction to one of the most important recent advances in statistical methodology.

The title of the book expresses our intent to introduce readers to statistics as it is used in practice. Statistics in practice is concerned with gaining understanding from data; it focuses on problem solving rather than on methods that may be useful in specific settings. A text cannot fully imitate practice because it must teach specific methods in a logical order and must use data that are not the reader's own. Nonetheless, our interest and experience in applying statistics have influenced the nature of *IPS* in several ways.

## Statistical Thinking

Statistics is interesting and useful because it provides strategies and tools for using data to gain insight into real problems. As the continuing revolution in computing automates the details of doing calculations and making graphs, an emphasis on statistical concepts and on insight from data becomes both more practical for students and teachers and more important for users who must supply what is not automated. No student should complete a first statistics course, for example, without a firm grasp of the distinction between observational studies and experiments and of why randomized comparative experiments are the gold standard for evidence of causation.

We have seen many statistical mistakes, but few have involved simply getting a calculation wrong. We therefore ask students to learn to explore data, always starting with plots, to think about the context of the data and the design of the study that produced the data, the possible influence of wild observations on conclusions, and the reasoning that lies behind standard methods of inference. Users of statistics who form these habits from the beginning are well prepared to learn and use more advanced methods.

**Data** Data are numbers with a context, as we say in "To Students: What Is Statistics?" A newborn who weighs 10.3 pounds is a big baby, and the birth weight could not plausibly be 10.3 ounces or 10.3 kilograms. Because context makes numbers meaningful, our examples and exercises use real data with real contexts that we briefly describe. Calculating the mean of five numbers is arithmetic, not statistics. We hope that the presence of background information, even in exercises intended for routine drill, will encourage students to always consider the meaning of their calculations as well as the calculations themselves. Note in this connection that a calculation or a graph or "reject $H_0$" is rarely a full answer to a statistical problem. We strongly encourage requiring students always to state a brief conclusion in the context of the problem. This helps build data sense as well as the communication skills that employers value.

**Mathematics** Although statistics is a mathematical science, it is not a field of mathematics and should not be taught as if it were. A fruitful mathematical theory (based on probability, which *is* a field of mathematics) underlies some parts of basic statistics, but by no means all. The distinction between observation and experiment, for example, is a core statistical idea that is ignored by the theory.[5] Mathematically trained teachers, rightly resisting a formula-based approach, sometimes identify conceptual understanding with mathematical understanding. When teaching statistics, we must emphasize statistical ideas and recognize that mathematics is not the only vehicle for conceptual understanding. *IPS* requires only the ability to read and use equations without having each step parsed. We require no algebraic derivations, let alone calculus. Because this is a *statistics* text, it is richer in ideas and requires more thought than the low mathematical level suggests.

**Calculators and Computers** Statistical calculations and graphics are in practice automated by software. We encourage instructors to use software of their choice or a graphing calculator that includes functions for both data analysis and basic inference. *IPS* includes some topics that reflect the dominance of software in practice, such as normal quantile plots and the version of the two-sample $t$ procedures that does not require equal variances. Several times we display the output of multiple software systems for the same problem. The point is that a student who knows the basics can interpret almost any output. Students like this reassurance, and it helps focus their attention on understanding rather than reading output.

**Judgment** Statistics in practice requires judgment. It is easy to list the mathematical assumptions that justify use of a particular procedure, but not so easy to decide when the procedure can be safely used in practice. Because judgment develops through experience, an introductory course should present clear guidelines and not make unreasonable demands on the judgment of students. We have given guidelines—for example, on using the *t* procedures for comparing two means but avoiding the *F* procedures for comparing two variances—that we follow ourselves. Similarly, many exercises require students to use some judgment and (equally important) to explain their choices in words. Many students would prefer to stick to calculating, and many statistics texts allow them to. Requiring more will do them much good in the long run.

**Teaching Experiences** We have successfully used *IPS* in courses taught to quite diverse student audiences. For general undergraduates from mixed disciplines, we cover Chapters 1 to 8 and Chapter 9, 10, or 12, omitting all optional material. For a quantitatively strong audience—sophomores planning to major in actuarial science or statistics—we move more quickly. We add Chapters 10 and 11 to the core material in Chapters 1 to 8 and include most optional content. We de-emphasize Chapter 4 (probability) because these students will take a probability course later in their program, though we make intensive use of software for simulating probabilities as well as for statistical analysis. The third group we teach contains beginning graduate students in such fields as education, family studies, and retailing. These mature but sometimes quantitatively unprepared students read the entire text (Chapters 11 and 13 lightly), again with reduced emphasis on Chapter 4 and some parts of Chapter 5. In all cases, beginning with data analysis and data production (Part I) helps students overcome their fear of statistics and builds a sound base for studying inference. We find that *IPS* can be flexibly adapted to quite varied audiences by paying attention to our clear designation of some material as optional and by varying the chapters assigned.

## The Sixth Edition: *What's New?*

- **Co-author** We are delighted to welcome Professor Bruce Craig to the *Introduction to the Practice of Statistics* author team. Bruce is currently Director of the Statistical Consulting Service at Purdue University and is an outstanding teacher. His vast experience consulting and collaborating with individuals who use statistical methods in their work provides him with perspective on the field of statistics that resonates with the approach of this text.

- **Ethics** Chapter 3 now contains a new section (3.4) on ethics. We believe that this topic is a very important part of the undergraduate curriculum and that a course in statistics is an ideal forum to stimulate thought and discussion about ethical issues.

- **Text Organization** Logistic Regression, previously treated in Chapter 16, now appears in Chapter 14. Similarly, Bootstrap Methods and Permutation Tests has moved to Chapter 16. This change is in line with the increasing importance of logistic regression in statistical practice. In response to suggestions from current *IPS* users, we have moved the material on data analysis for two-way tables from Chapter 9 back to Chapter 2 (Section 2.5). In addition, the large sample confidence procedures are now the featured methods for one and two proportions in Chapter 9, and the plus-four have been moved to Beyond the Basics sections, a more appropriate location. The table of contents follows what we consider to be the best ordering of the topics from a pedagogical point of view. However, the text chapters are generally written to enable instructors to teach the material in the order they prefer.

- **Design** A new design incorporates colorful, revised figures throughout to aid students' understanding of text material. Photographs related to chapter examples and exercises make connections to real-life applications and provide a visual context for topics.

- **Exercises and Examples** Exercises and examples are labeled to help instructors and students easily identify key topics and application areas. The number of total exercises has increased by 15%. Approximately half the total exercises are new or revised to reflect current data and a variety of topics. *IPS* examples and exercises cover a wide range of application areas. An application index is provided for instructors to easily select and assign content related to specific fields.

- **Use Your Knowledge Exercises** Short exercises designed to reinforce key concepts now appear throughout each chapter. These exercises are listed, with page numbers, at the end of each section for easy reference.

- **Look Back** At key points in the text Look Back margin notes direct the reader to the first explanation of a topic, providing page numbers for easy reference.

In addition to the new Sixth Edition enhancements, *IPS* has retained the successful pedagogical features from previous editions:

- **Caution** Warnings in the text, signaled by a caution icon, help students avoid common errors and misconceptions.

- **Challenge Exercises** More challenging exercises are signaled with an icon. Challenge exercises are varied: some are mathematical, some require open-ended investigation, and so on.

- **Applets** Applet icons are used throughout the text to signal where related, interactive statistical applets can be found on the text Web site (www.whfreeman.com/ips6e) and CD-ROM.

- **Statistics in Practice** Formerly found at the opening of each chapter, these accounts by professionals who use statistics on the job are now located on the *IPS* Web site and CD-ROM.

- **CrunchIt! Statistical Software** Developed by Webster West of Texas A&M University, CrunchIt! is an easy-to-use program for students and offers capabilities well beyond those needed for a first course. CrunchIt! output, along with other statistical software output, is integrated throughout the text. Access to CrunchIt! is available online through an access-code–protected Web site. Access codes are available in every new copy of *IPS* 6e or can be purchased online.

## Acknowledgments

We are pleased that the first five editions of *Introduction to the Practice of Statistics* have helped move the teaching of introductory statistics in a direction supported by most statisticians. We are grateful to the many colleagues and students who have provided helpful comments, and we hope that they will find this new edition another step forward. In particular, we would like to thank the following colleagues who offered specific comments on the new edition:

Mary A. Bergs
*Mercy College of Northwest Ohio*
John F. Brewster
*University of Manitoba*
Karen Buro
*Grant MacEwan College*
Smiley W. Cheng
*University of Manitoba*
Gerarda Darlington
*University of Guelph*
Linda Dawson
*University of Washington, Tacoma*
Michael Evans
*University of Toronto*
Mary Gray
*American University*
Rick Gumina
*Colorado State University*
Patricia Humphrey
*Georgia Southern University*
Mohammad Kazemi
*The University of North Carolina
at Charlotte*
Jeff Kollath
*Oregon State University*
William J. Kubik
*Hanover College*
Charles Liberty
*Keene State College*
Brian Macpherson
*University of Manitoba*
Henry Mesa
*Portland Community College*
Helen Noble
*San Diego State University*

Richard Numrich
*College of Southern Nevada*
Becky Parker
*Montana State University*
Maria Consuelo C. Pickle
*St. Petersburg College*
German J. Pliego
*University of St. Thomas*
Philip Protter
*Cornell University*
John G. Reid
*Mount Saint Vincent University*
Diann Reischman
*Grand Valley State University*
Shane Rollans
*Thompson Rivers University*
Nancy Roper
*Portland Community College*
Teri Rysz
*University of Cincinnati*
Engin Sungur
*University of Minnesota, Morris*
Todd Swanson
*Hope College*
Anthony L. Truog
*University of Wisconsin,
Whitewater*
Augustin Vukov
*University of Toronto*
Erwin Walker
*Clemson University*
Nathan Wetzel
*University of Wisconsin,
Stevens Point*

The professionals at W. H. Freeman and Company, in particular Mary Louise Byrd, Ruth Baruth, and Shona Burke, have contributed greatly to the success of *IPS*. Additionally, we would like to thank Anne Scanlan-Rohrer, Pam Bruton, Jackie Miller, and Darryl Nester for their valuable contributions to the Sixth Edition. Most of all, we are grateful to the many people in varied disciplines and occupations with whom we have worked to gain understanding from data. They have provided both material for this book and the experience that enabled us to write it. What the eminent statistician John Tukey called "the real problems experience and the real data experience" have shaped our view of statistics, convincing us of the need for beginning instruction to focus on data and concepts, building intellectual skills that transfer to more elaborate settings and remain essential when all details are automated. We hope that users and potential users of statistical techniques will find this emphasis helpful.

# Media and Supplements

## For Students

NEW! STATS P◭RTAL

**portals.bfwpub.com/ips6e** (Access code required. Available packaged with *Introduction to the Practice of Statistics*, Sixth Edition, or for purchase online.) StatsPortal is the digital gateway to *IPS* 6e, designed to enrich the course and enhance students' study skills through a collection of Web-based tools. StatsPortal integrates a rich suite of diagnostic, assessment, tutorial, and enrichment features, enabling students to master statistics at their own pace. It is organized around three main teaching and learning components:

- **Interactive eBook** offers a complete and customizable online version of the text, fully integrated with all the media resources available with *IPS* 6e. The eBook allows students to quickly search the text, highlight key areas, and add notes about what they're reading. Similarly, instructors can customize the eBook to add, hide, and reorder content, add their own material, and highlight key text for students.

- **Resources** organizes all the resources for *IPS* 6e into one location for students' ease of use. These resources include the following:

  - **StatTutor Tutorials** offer over 150 audio-multimedia tutorials tied directly to the textbook, including videos, applets, and animations.

  - **Statistical Applets** are 16 interactive applets to help students master key statistical concepts.

  - **CrunchIt! Statistical Software** allows users to analyze data from any Internet location. Designed with the novice user in mind, the software is not only easily accessible but also easy to use. CrunchIt! offers all the basic statistical routines covered in the introductory statistics courses. **CrunchIt!** statistical software is available via an access-code protected Web site. Access codes are available in every new copy of *IPS* 6e or can be purchased online.

  - **Stats@Work Simulations** put students in the role of statistical consultants, helping them better understand statistics interactively within the context of real-life scenarios. Students are asked to interpret and analyze data presented to them in report form, as well as to interpret current event news stories. All tutorials are graded and offer helpful hints and feedback.

  - **EESEE Case Studies** developed by The Ohio State University Statistics Department provide students with a wide variety of timely, real examples with real data. Each case study is built around several thought-provoking questions that make students think carefully about the statistical issues raised by the stories. **EESEE** case studies are available via an access-code-protected Web site. Access codes are available in every new copy of *IPS* 6e or can be purchased online.

- **Podcast Chapter Summary** provides students with an audio version of chapter summaries to download and review on an mp3 player.

- **Data Sets** are available in ASCII, Excel, JMP, Minitab, TI, SPSS, and S-Plus formats.

- **Online Tutoring with SMARTHINKING** is available for homework help from specially trained, professional educators.

- **Student Study Guide with Selected Solutions** includes explanations of crucial concepts and detailed solutions to key text problems with step-by-step models of important statistical techniques.

- **Statistical Software Manuals** for TI-83/84, Minitab, Excel, JMP, and SPSS provide instruction, examples, and exercises using specific statistical software packages.

- **Interactive Table Reader** allows students to use statistical tables interactively to seek the information they need.

**Resources (instructors only)**

- **Instructor's Guide with Full Solutions** includes worked-out solutions to all exercises, teaching suggestions, and chapter comments.

- **Test Bank** contains complete solutions for textbook exercises.

- **Lecture PowerPoint slides** offer a detailed lecture presentation of statistical concepts covered in each chapter of *IPS*.

- **Assignments** organizes assignments and guides instructors through an easy-to-create assignment process providing access to questions from the Test Bank, Web Quizzes, and Exercises from *IPS* 6e. The Assignment Center enables instructors to create their own assignments from a variety of question types for self-graded assignments. This powerful assignment manager allows instructors to select their preferred policies in regard to scheduling, maximum attempts, time limitations, feedback, and more!

## Online Study Center: www.whfreeman.com/osc/ips6e

(Access code required. Available for purchase online.) In addition to all the offerings available on the Companion Web site, the OSC offers:

- **StatTutor Tutorials**

- **Stats@Work Simulations**

- **Study Guide with Selected Solutions**

- **Statistical Software Manuals**

## Companion Web site: www.whfreeman.com/ips6e

Seamlessly integrates topics from the text. On this open-access Web site, students can find the following:

- **Interactive Statistical Applets** that allow students to manipulate data and see the corresponding results graphically.

- **Data Sets** in ASCII, Excel, JMP, Minitab, TI, SPSS, and S-Plus formats.

- **Interactive Exercises and Self-Quizzes** to help students prepare for tests.

- **Optional Companion Chapters 14, 15, 16, and 17,** covering logistic regression, nonparametric tests, bootstrap methods and permutation tests, and statistics for quality control and capability.

- **Supplementary Exercises** for every chapter.

**Interactive Student CD-ROM**   Included with every new copy of *IPS*, the CD contains access to the companion chapters, applets, and data sets also found on the Companion Web site.

**Special Software Packages**   Student versions of JMP, Minitab, S-PLUS, and SPSS are available on a CD-ROM packaged with the textbook. This software is not sold separately and must be packaged with a text or a manual. Contact your W. H. Freeman representative for information or visit www.whfreeman.com.

**NEW! SMARTHINKING Online Tutoring** (access code required) W. H. Freeman and Company is partnering with SMARTHINKING to provide students with free online tutoring and homework help from specially trained, professional educators. Twelve-month subscriptions are available for packaging with *IPS*.

**Printed Study Guide**   prepared by Michael A. Fligner of The Ohio State University offers students explanations of crucial concepts in each section of *IPS*, plus detailed solutions to key text problems and stepped-through models of important statistical techniques. ISBN 1-4292-1473-2

## For Instructors

The **Instructor's Web site** www.whfreeman.com/ips6e requires user registration as an instructor and features all the student Web materials plus:

- **Instructor version of EESEE** (Electronic Encyclopedia of Statistical Examples and Exercises), with solutions to the exercises in the student version and **CrunchIt!** statistical software.

- **Instructor's Guide,** including full solutions to all exercises in .pdf format.

- **PowerPoint slides** containing all textbook figures and tables.

- **Lecture PowerPoint slides** offering a detailed lecture presentation of statistical concepts covered in each chapter of *IPS*.

- **Full answers to the Supplementary Exercises** on the student Web site.

**Instructor's Guide with Solutions**   by Darryl Nester, Bluffton University.   This printed guide includes full solutions to all exercises and provides video and Internet resources and sample examinations. It also contains brief discussions of the *IPS* approach for each chapter. ISBN 1-4292-1472-4

**Test Bank**  by Brian Macpherson, University of Manitoba. The test bank contains hundreds of multiple-choice questions to generate quizzes and tests. Available in print as well as electronically on CD-ROM (for Windows and Mac), where questions can be downloaded, edited, and resequenced to suit the instructor's needs.

Printed Version, ISBN 1-4292-1471-6
Computerized (CD) Version, ISBN 1-4292-1859-2

**Enhanced Instructor's Resource CD-ROM**  Allows instructors to search and export (by key term or chapter) all the material from the student CD, plus:

- All text images and tables

- Statistical applets and data sets

- Instructor's Guide with full solutions

- PowerPoint files and lecture slides

- Test bank files

ISBN 1-4292-1503-8

**Course Management Systems**  W. H. Freeman and Company provides courses for Blackboard, WebCT (Campus Edition and Vista), and Angel course management systems. They are completely integrated courses that you can easily customize and adapt to meet your teaching goals and course objectives. On request, Freeman also provides courses for users of Desire2Learn and Moodle. Visit www.bfwpub.com/lmc for more information.

**i-clicker**  is a new two-way radio-frequency classroom response solution developed by educators for educators. University of Illinois physicists Tim Stelzer, Gary Gladding, Mats Selen, and Benny Brown created the i-clicker system after using competing classroom response solutions and discovering they were neither classroom-appropriate nor student-friendly. Each step of i-clicker's development has been informed by teaching and learning. i-clicker is superior to other systems from both pedagogical and technical standpoints. To learn more about packaging i-clicker with this textbook, please contact your local sales rep or visit www.iclicker.com.

S tatistics is the science of collecting, organizing, and interpreting numerical facts, which we call *data*. We are bombarded by data in our everyday lives. The news mentions imported car sales, the latest poll of the president's popularity, and the average high temperature for today's date. Advertisements claim that data show the superiority of the advertiser's product. All sides in public debates about economics, education, and social policy argue from data. A knowledge of statistics helps separate sense from nonsense in the flood of data.

The study and collection of data are also important in the work of many professions, so training in the science of statistics is valuable preparation for a variety of careers. Each month, for example, government statistical offices release the latest numerical information on unemployment and inflation. Economists and financial advisors, as well as policymakers in government and business, study these data in order to make informed decisions. Doctors must understand the origin and trustworthiness of the data that appear in medical journals. Politicians rely on data from polls of public opinion. Business decisions are based on market research data that reveal consumer tastes. Engineers gather data on the quality and reliability of manufactured products. Most areas of academic study make use of numbers, and therefore also make use of the methods of statistics.

## Understanding from Data

*The goal of statistics is to gain understanding from data.* To gain understanding, we often operate on a set of numbers—we average or graph them, for example. But we must do more, because data are not just numbers; they are numbers that have some context that helps us understand them.

You read that low birth weight is a major reason why infant mortality in the United States is higher than in most other advanced nations. The report goes on to say that 7.8% of children born in the United States have low birth weight, and that 13.4% of black infants have low birth weight.[1] To make sense of these numbers you must know what counts as low birth weight (less than 2500 grams, or 5.5 pounds) and have some feeling for the weights of babies. You probably recognize that 5.5 pounds is small, that 7.5 pounds (3400 grams) is about average, and that 10 pounds (4500 grams) is a big baby.

Another part of the context is the source of the data. How do we know that 7.8% of American babies have low birth weight or that the average weight of newborns is about 3400 grams? The data come from the National Center for Health Statistics, a government office to which the states report information from all birth certificates issued each month. These are the most complete data available about births in the United States.

When you do statistical problems—even straightforward textbook problems—don't just graph or calculate. Think about the context and state your conclusions in the specific setting of the problem. As you are learning how to do statistical calculations and graphs, remember that the goal of statistics is not calculation for its own sake but gaining understanding from numbers. The

# Looking at Data—
# Distributions

Students planning a referendum on college fees. See Example 1.1.

## Introduction

*Statistics is the science of learning from data.* Data are numerical facts. Here is an example of a situation where students used the results of a referendum to convince their university Board of Trustees to make a decision.

**EXAMPLE**

**1.1 Students vote for service learning scholarships.** According to the National Service-Learning Clearinghouse: "Service-learning is a teaching and learning strategy that integrates meaningful community service with instruction and reflection to enrich the learning experience, teach civic responsibility, and strengthen communities."[1] University of Illinois at Urbana–Champaign students decided that they wanted to become involved in this national movement. They proposed a $15.00 per semester Legacy of Service and Learning Scholarship fee. Each year, $10.00 would be invested in an endowment and $5.00 would be used to fund current-use scholarships. In a referendum, students voted 3785 to 2977 in favor of the proposal. On April 11, 2006, the university Board of Trustees approved the proposal. Approximately $370,000 in current-use scholarship funds will be generated each year, and with the endowment, it is expected that in 20 years there will be more than a million dollars per year for these scholarships.

To learn from data, we need more than just the numbers. The numbers in a medical study, for example, mean little without some knowledge of the goals of the study and of what blood pressure, heart rate, and other measurements contribute to those goals. That is, *data are numbers with a context*, and we need to understand the context if we are to make sense of the numbers. On the other hand, measurements from the study's several hundred subjects are of little value to even the most knowledgeable medical expert until the tools of statistics organize, display, and summarize them. We begin our study of statistics by mastering the art of examining data.

## Variables

Any set of data contains information about some group of *individuals*. The information is organized in *variables*.

### INDIVIDUALS AND VARIABLES

**Individuals** are the objects described in a set of data. Individuals are sometimes people. When the objects that we want to study are not people, we often call them **cases.**

A **variable** is any characteristic of an individual. A variable can take different values for different individuals.

**EXAMPLE**

**1.2 Data for students in a statistics class.**   Figure 1.1 shows part of a data set for students enrolled in an introductory statistics class. Each row gives the data on one student. The values for the different variables are in the columns. This data set has eight variables. ID is an identifier for each student. Exam1, Exam2, Homework, Final, and Project give the points earned, out of a total of 100 possible, for each of these course requirements. Final grades are based on a possible 200 points for each exam and the final, 300 points for Homework, and 100 points for Project. TotalPoints is the variable that gives the composite score. It is computed by adding 2 times Exam1, Exam2, and Final, 3 times Homework plus 1 times Project. Grade is the grade earned in the course. This instructor used cut-offs of 900, 800, 700, etc. for the letter grades.

**FIGURE 1.1** Spreadsheet for Example 1.2.

| | A | B | C | D | E | F | G | H |
|---|---|---|---|---|---|---|---|---|
| 1 | ID | Exam1 | Exam2 | Homework | Final | Project | TotalPoints | Grade |
| 2 | 101 | 89 | 94 | 88 | 87 | 95 | 899 | B |
| 3 | 102 | 78 | 84 | 90 | 89 | 94 | 866 | B |
| 4 | 103 | 71 | 80 | 75 | 79 | 95 | 780 | C |
| 5 | 104 | 95 | 98 | 97 | 96 | 93 | 962 | A |
| 6 | 105 | 79 | 88 | 85 | 88 | 96 | 861 | B |

Microsoft Excel

spreadsheet    The display in Figure 1.1 is from an Excel **spreadsheet.** Most statistical software packages use similar spreadsheets and many are able to import Excel spreadsheets.

---

## USE YOUR KNOWLEDGE

**1.1**    **Read the spreadsheet.** Refer to Figure 1.1. Give the values of the variables Exam1, Exam2, and Final for the student with ID equal to 104.

**1.2**    **Calculate the grade.** A student whose data do not appear on the spreadsheet scored 88 on Exam1, 85 on Exam2, 77 for Homework, 90 on the Final, and 80 on the Project. Find TotalPoints for this student and give the grade earned.

---

Spreadsheets are very useful for doing the kind of simple computations that you did in Exercise 1.2. You can type in a formula and have the same computation performed for each row.

Note that the names we have chosen for the variables in our spreadsheet do not have spaces. For example, we could have used the name "Exam 1" for the first exam score rather than Exam1. In many statistical software packages, however, spaces are not allowed in variable names. For this reason, when creating spreadsheets for eventual use with statistical software, it is best to avoid spaces in variable names. Another convention is to use an underscore (_) where you would normally use a space. For our data set, we could use Exam_1, Exam_2, and Final_Exam.

In practice, any set of data is accompanied by background information that helps us understand the data. When you plan a statistical study or explore data from someone else's work, ask yourself the following questions:

1. **Why? What purpose** do the data have? Do we hope to answer some specific questions? Do we want to draw conclusions about individuals other than those for whom we actually have data?

2. **Who?** What **individuals** do the data describe? **How many** individuals appear in the data?

3. **What?** How many **variables** do the data contain? What are the **exact definitions** of these variables? Some variables have units. Weights, for example, might be recorded in pounds, in thousands of pounds, or in kilograms. For these kinds of variables, you need to know the **unit of measurement.**

---

**EXAMPLE**

**1.3 Individuals and variables.**    The data set in Figure 1.1 was constructed to keep track of the grades for students in an introductory statistics course. The individuals are the students in the class. There are 8 variables in this data set. These include an identifier for each student and scores for the various course requirements. There are no units for ID and grade. The other variables all have "points" as the unit.

---

Some variables, like gender and college major, simply place individuals into categories. Others, like height and grade point average, take numerical values for which we can do arithmetic. It makes sense to give an average salary for a company's employees, but it does not make sense to give an "average" gender. We can, however, count the numbers of female and male employees and do arithmetic with these counts.

---

### CATEGORICAL AND QUANTITATIVE VARIABLES

A **categorical variable** places an individual into one of two or more groups or categories.

A **quantitative variable** takes numerical values for which arithmetic operations such as adding and averaging make sense.

The **distribution** of a variable tells us what values it takes and how often it takes these values.

---

**EXAMPLE**

**1.4 Variables for students in a statistics course.** Suppose the data for the students in the introductory statistics class were also to be used to study relationships between student characteristics and success in the course. For this purpose, we might want to use a data set like the spreadsheet in Figure 1.2. Here, we have decided to focus on the TotalPoints and Grade as the outcomes of interest. Other variables of interest have been included: Gender, PrevStat (whether or not the student has taken a statistics course previously), and Year (student classification as first, second, third, or fourth year). ID is a categorical variable, total points is a quantitative variable, and the remaining variables are all categorical.

| | A | B | C | D | E | F |
|---|---|---|---|---|---|---|
| | ID | TotalPoints | Grade | Gender | PrevStat | Year |
| 1 | | | | | | |
| 2 | 101 | 899 | A | F | Yes | 4 |
| 3 | 102 | 866 | B | M | Yes | 3 |
| 4 | 103 | 780 | C | M | No | 3 |
| 5 | 104 | 962 | A | M | No | 1 |
| 6 | 105 | 861 | B | F | No | 4 |

**FIGURE 1.2** Spreadsheet for Example 1.4.

In our example, the possible values for the grade variable are A, B, C, D, and F. When computing grade point averages, many colleges and universities translate these letter grades into numbers using A = 4, B = 3, C = 2, D = 1, and F = 0. The transformed variable with numeric values is considered to be quantitative because we can average the numerical values across different courses to obtain a grade point average.

Sometimes, experts argue about numerical scales such as this. They ask whether or not the difference between an A and a B is the same as the difference between a D and an F. Similarly, many questionnaires ask people to respond on a 1 to 5 scale with 1 representing strongly agree, 2 representing agree, etc. Again we could ask about whether or not the five possible values for this scale are equally spaced in some sense. From a practical point of view, the averages that can be computed when we convert categorical scales such as these to numerical values frequently provide a very useful way to summarize data.

## USE YOUR KNOWLEDGE

1.3 **Apartment rentals.** A data set lists apartments available for students to rent. Information provided includes the monthly rent, whether or not cable is included free of charge, whether or not pets are allowed, the number of bedrooms, and the distance to the campus. Describe the individuals or cases in the data set, give the number of variables, and specify whether each variable is categorical or quantitative.

## Measurement: know your variables

The context of data includes an understanding of the variables that are recorded. Often the variables in a statistical study are easy to understand: height in centimeters, study time in minutes, and so on. But each area of work also has its own special variables. A psychologist uses the Minnesota Multiphasic Personality Inventory (MMPI), and a physical fitness expert measures "VO2 max," the volume of oxygen consumed per minute while exercising at your maximum capacity. Both of these variables are measured with special **instruments**. VO2 max is measured by exercising while breathing into a mouthpiece connected to an apparatus that measures oxygen consumed. Scores on the MMPI are based on a long questionnaire, which is also an instrument. Part of mastering your field of work is learning what variables are important and how they are best measured. Because details of particular measurements usually require knowledge of the particular field of study, we will say little about them.

*instrument*

*Be sure that each variable really does measure what you want it to. A poor choice of variables can lead to misleading conclusions.* Often, for example, the **rate** at which something occurs is a more meaningful measure than a simple count of occurrences.

*rate*

**1.5 Accidents for passenger cars and motorcycles.** The government's Fatal Accident Reporting System says that 27,102 passenger cars were involved in fatal accidents in 2002. Only 3339 motorcycles had fatal accidents that year.[2] Does this mean that motorcycles are safer than cars? Not at all—there are many more cars than motorcycles, so we expect cars to have a higher *count* of fatal accidents.

A better measure of the dangers of driving is a *rate*, the number of fatal accidents divided by the number of vehicles on the road. In 2002, passenger cars had about 21 fatal accidents for each 100,000 vehicles registered. There were about 67 fatal accidents for each 100,000 motorcycles registered. The rate for motorcycles is more than three times the rate for cars. Motorcycles are, as we might guess, much more dangerous than cars.

# 1.1 Displaying Distributions with Graphs

Statistical tools and ideas help us examine data in order to describe their main features. This examination is called **exploratory data analysis.** Like an explorer crossing unknown lands, we want first to simply describe what we see. Here are two basic strategies that help us organize our exploration of a set of data:

- Begin by examining each variable by itself. Then move on to study the relationships among the variables.

- Begin with a graph or graphs. Then add numerical summaries of specific aspects of the data.

We will follow these principles in organizing our learning. This chapter presents methods for describing a single variable. We will study relationships among several variables in Chapter 2. Within each chapter, we will begin with graphical displays, then add numerical summaries for more complete description.

## Graphs for categorical variables

The values of a categorical variable are labels for the categories, such as "female" and "male." The **distribution** of a categorical variable lists the categories and gives either the **count** or the **percent** of individuals who fall in each category. For example, how well educated are 30-something young adults? Here is the distribution of the highest level of education for people aged 25 to 34 years:[3]

| Education | Count (millions) | Percent |
|---|---|---|
| Less than high school | 4.6 | 12.1 |
| High school graduate | 11.6 | 30.5 |
| Some college | 7.4 | 19.5 |
| Associate degree | 3.3 | 8.7 |
| Bachelor's degree | 8.6 | 22.6 |
| Advanced degree | 2.5 | 6.6 |

Are you surprised that only 29.2% of young adults have at least a bachelor's degree?

bar graph    The graphs in Figure 1.3 display these data. The **bar graph** in Figure 1.3(a) quickly compares the sizes of six education groups. The heights of the bars show the percents in the six categories. The **pie chart** in Figure 1.3(b) helps

pie chart    us see what part of the whole each group forms. For example, the "Bachelor's" slice makes up 22.6% of the pie because 22.6% of young adults have a bachelor's degree but no higher degree. We have moved that slice out to call attention to it. Because pie charts lack a scale, we have added the percents to the labels for the slices. *Pie charts require that you include all the categories that make up a whole. Use them only when you want to emphasize each category's relation to the whole.* Bar graphs are easier to read and are also more flexible. For example, you can use a bar graph to compare the numbers of students at your college majoring in biology, business, and political science. A pie chart cannot make this comparison because not all students fall into one of these three majors.

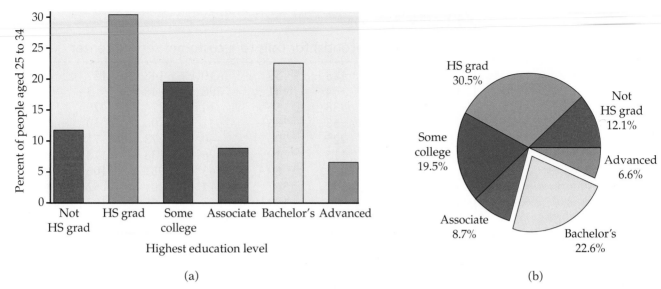

FIGURE 1.3 (a) Bar graph of the educational attainment of people aged 25 to 34 years. (b) Pie chart of the education data, with bachelor's degree holders emphasized.

**USE YOUR KNOWLEDGE**

1.4 **Read the pie chart.** Refer to Figure 1.3(b). What percent of young adults have either an associate degree or a bachelor's degree?

Bar graphs and pie charts help an audience grasp a distribution quickly. They are, however, of limited use for data analysis because it is easy to understand data on a single categorical variable, such as highest level of education, without a graph. We will move on to quantitative variables, where graphs are essential tools.

## Data analysis in action: don't hang up on me

Many businesses operate call centers to serve customers who want to place an order or make an inquiry. Customers want their requests handled thoroughly. Businesses want to treat customers well, but they also want to avoid wasted time on the phone. They therefore monitor the length of calls and encourage their representatives to keep calls short. Here is an example of the difficulties this policy can cause.

**EXAMPLE**

**1.6 Individuals and variables for the customer service center.** We have data on the length of all 31,492 calls made to the customer service center of a small bank in a month. Table 1.1 displays the lengths of the first 80 calls. The file for the complete data set is *eg01-004*, which you can find on the text CD and Web site.[4]

Take a look at the data in Table 1.1. The numbers are meaningless without some background information. The *individuals* are calls made to the bank's call center. The *variable* recorded is the length of each call. The *units* are

**TABLE 1.1**

Service times (seconds) for calls to a customer service center

| | | | | | | | |
|---|---|---|---|---|---|---|---|
| 77 | 289 | 128 | 59 | 19 | 148 | 157 | 203 |
| 126 | 118 | 104 | 141 | 290 | 48 | 3 | 2 |
| 372 | 140 | 438 | 56 | 44 | 274 | 479 | 211 |
| 179 | 1 | 68 | 386 | 2631 | 90 | 30 | 57 |
| 89 | 116 | 225 | 700 | 40 | 73 | 75 | 51 |
| 148 | 9 | 115 | 19 | 76 | 138 | 178 | 76 |
| 67 | 102 | 35 | 80 | 143 | 951 | 106 | 55 |
| 4 | 54 | 137 | 367 | 277 | 201 | 52 | 9 |
| 700 | 182 | 73 | 199 | 325 | 75 | 103 | 64 |
| 121 | 11 | 9 | 88 | 1148 | 2 | 465 | 25 |

seconds. We see that the call lengths vary a great deal. The longest call lasted 2631 seconds, almost 44 minutes. More striking is that 8 of these 80 calls lasted less than 10 seconds. What's going on?

Figure 1.4 is a histogram of the lengths of all 31,492 calls. We did not plot the few lengths greater than 1200 seconds (20 minutes). As expected, the graph shows that most calls last between about a minute and 5 minutes, with some lasting much longer when customers have complicated problems. More striking is the fact that 7.6% of all calls are no more than 10 seconds long. It turned out that the bank penalized representatives whose average call length was too long—so some representatives just hung up on customers in order to bring their average length down. Neither the customers nor the bank were happy about this. The bank changed its policy, and later data showed that calls under 10 seconds had almost disappeared.

**FIGURE 1.4** The distribution of call lengths for 31,492 calls to a bank's customer service center, for Example 1.6. The data show a surprising number of very short calls. These are mostly due to representatives deliberately hanging up in order to bring down their average call length.

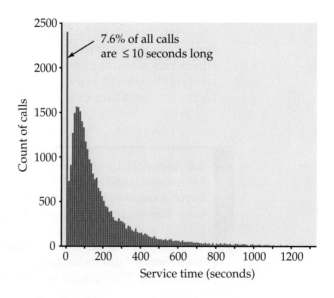

tails      The extreme values of a distribution are in the **tails** of the distribution. The high values are in the upper, or right, tail and the low values are in the lower, or left, tail. The overall pattern in Figure 1.4 is made up of the many moderate call lengths and the long right tail of more lengthy calls. The striking departure from the overall pattern is the surprising number of very short calls in the left tail.

Our examination of the call center data illustrates some important principles:

- After you understand the background of your data (individuals, variables, units of measurement), the first thing to do is almost always **plot your data.**

- When you look at a plot, look for an **overall pattern** and also for any **striking departures** from the pattern.

We now turn to the kinds of graphs that are used to describe the distribution of a quantitative variable. We will explain how to make the graphs by hand, because knowing this helps you understand what the graphs show. However, making graphs by hand is so tedious that software is almost essential for effective data analysis unless you have just a few observations.

## Stemplots

A *stemplot* (also called a stem-and-leaf plot) gives a quick picture of the shape of a distribution while including the actual numerical values in the graph. Stemplots work best for small numbers of observations that are all greater than 0.

---

STEMPLOT

To make a **stemplot:**

**1.** Separate each observation into a **stem** consisting of all but the final (rightmost) digit and a **leaf,** the final digit. Stems may have as many digits as needed, but each leaf contains only a single digit.

**2.** Write the stems in a vertical column with the smallest at the top, and draw a vertical line at the right of this column.

**3.** Write each leaf in the row to the right of its stem, in increasing order out from the stem.

---

EXAMPLE

**1.7 Literacy of men and women.** The Islamic world is attracting increased attention in Europe and North America. Table 1.2 shows the percent of men and women at least 15 years old who were literate in 2002 in the major Islamic nations. We omitted countries with populations less than 3 million. Data for a few nations, such as Afghanistan and Iraq, are not available.[5]

To make a stemplot of the percents of females who are literate, use the first digits as stems and the second digits as leaves. Algeria's 60% literacy rate, for example, appears as the leaf 0 on the stem 6. Figure 1.5 shows the steps in making the plot.

| | **TABLE 1.2** | | | | | |
|---|---|---|---|---|---|---|
| | Literacy rates (percent) in Islamic nations | | | | | |
| Country | Female percent | Male percent | | Country | Female percent | Male percent |
| Algeria | 60 | 78 | | Morocco | 38 | 68 |
| Bangladesh | 31 | 50 | | Saudi Arabia | 70 | 84 |
| Egypt | 46 | 68 | | Syria | 63 | 89 |
| Iran | 71 | 85 | | Tajikistan | 99 | 100 |
| Jordan | 86 | 96 | | Tunisia | 63 | 83 |
| Kazakhstan | 99 | 100 | | Turkey | 78 | 94 |
| Lebanon | 82 | 95 | | Uzbekistan | 99 | 100 |
| Libya | 71 | 92 | | Yemen | 29 | 70 |
| Malaysia | 85 | 92 | | | | |

**FIGURE 1.5** Making a stemplot of the data in Example 1.7. (a) Write the stems. (b) Go through the data and write each leaf on the proper stem. For example, the values on the 8 stem are 86, 82, and 85 in the order of the table. (c) Arrange the leaves on each stem in order out from the stem. The 8 stem now has leaves 2 5 6.

```
2 |              2 | 9          2 | 9
3 |              3 | 1 8        3 | 1 8
4 |              4 | 6          4 | 6
5 |              5 |            5 |
6 |              6 | 0 3 3      6 | 0 3 3
7 |              7 | 1 1 0 8    7 | 0 1 1 8
8 |              8 | 6 2 5      8 | 2 5 6
9 |              9 | 9 9 9      9 | 9 9 9
   (a)               (b)            (c)
```

cluster

The overall pattern of the stemplot is irregular, as is often the case when there are only a few observations. There do appear to be two **clusters** of countries. The plot suggests that we might ask what explains the variation in literacy. For example, why do the three central Asian countries (Kazakhstan, Tajikistan, and Uzbekistan) have very high literacy rates?

## USE YOUR KNOWLEDGE

**1.5    Make a stemplot.** Here are the scores on the first exam in an introductory statistics course for 30 students in one section of the course:

| | | | | | | | | | | | | | | |
|---|---|---|---|---|---|---|---|---|---|---|---|---|---|---|
| 80 | 73 | 92 | 85 | 75 | 98 | 93 | 55 | 80 | 90 | 92 | 80 | 87 | 90 | 72 |
| 65 | 70 | 85 | 83 | 60 | 70 | 90 | 75 | 75 | 58 | 68 | 85 | 78 | 80 | 93 |

Use these data to make a stemplot. Then use the stemplot to describe the distribution of the first-exam scores for this course.

back-to-back stemplot When you wish to compare two related distributions, a **back-to-back stem-plot** with common stems is useful. The leaves on each side are ordered out from the common stem. Here is a back-to-back stemplot comparing the distributions of female and male literacy rates in the countries of Table 1.2.

| Female | | Male |
|---:|:---:|:---|
| 9 | 2 | |
| 81 | 3 | |
| 6 | 4 | |
| | 5 | 0 |
| 330 | 6 | 88 |
| 8110 | 7 | 08 |
| 652 | 8 | 3459 |
| 999 | 9 | 22456 |
| | 10 | 000 |

The values on the left are the female percents, as in Figure 1.5, but ordered out from the stem from right to left. The values on the right are the male percents. It is clear that literacy is generally higher among males than among females in these countries.

*Stemplots do not work well for large data sets, where each stem must hold a large number of leaves.* Fortunately, there are two modifications of the basic stemplot that are helpful when plotting the distribution of a moderate number of observations. You can double the number of stems in a plot by **splitting each stem** into two: one with leaves 0 to 4 and the other with leaves 5 through 9. When the observed values have many digits, it is often best to **trim** the numbers by removing the last digit or digits before making a stemplot. You must use your judgment in deciding whether to split stems and whether to trim, though statistical software will often make these choices for you. Remember that the purpose of a stemplot is to display the shape of a distribution. If a stemplot has fewer than about five stems, you should usually split the stems unless there are few observations. If there are many stems with no leaves or only one leaf, trimming will reduce the number of stems. Here is an example that makes use of both of these modifications.

splitting stems
trimming

**1.8 Stemplot for length of service calls.** Return to the 80 customer service call lengths in Table 1.1. To make a stemplot of this distribution, we first trim the call lengths to tens of seconds by dropping the last digit. For example, 56 seconds trims to 5 and 143 seconds trims to 14. (We might also round to the nearest 10 seconds, but trimming is faster than rounding if you must do it by hand.)

We can then use tens of seconds as our leaves, with the digits to the left forming stems. This gives us the single-digit leaves that a stemplot requires. For example, 56 trimmed to 5 becomes leaf 5 on the 0 stem; 143 trimmed to 14 becomes leaf 4 on the 1 stem.

Because we have 80 observations, we split the stems. Thus, 56 trimmed to 5 becomes leaf 5 on the second 0 stem, along with all leaves 5 to 9. Leaves

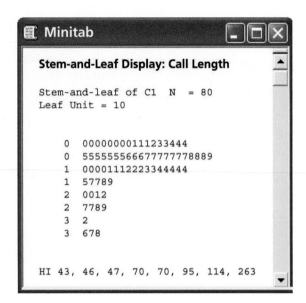

**FIGURE 1.6** Stemplot from Minitab of the 80 call lengths in Table 1.1, for Example 1.8. The software has trimmed the data by removing the last digit. It has also split stems and listed the highest observations apart from the plot.

0 to 4 go on the first 0 stem. Figure 1.6 is a stemplot of these data made by software. The software automatically did what we suggest: trimmed to tens of seconds and split stems. To save space, the software also listed the largest values as "HI" rather than create stems all the way up to 26. The stemplot shows the overall pattern of the distribution, with many short to moderate lengths and some very long calls.

## Histograms

Stemplots display the actual values of the observations. This feature makes stemplots awkward for large data sets. Moreover, the picture presented by a stemplot divides the observations into groups (stems) determined by the number system rather than by judgment. Histograms do not have these limitations. A **histogram** breaks the range of values of a variable into classes and displays only the count or percent of the observations that fall into each class. You can choose any convenient number of classes, but you should always choose classes of equal width. Histograms are slower to construct by hand than stemplots and do not display the actual values observed. For these reasons we prefer stemplots for small data sets. The construction of a histogram is best shown by example. Any statistical software package will of course make a histogram for you.

**EXAMPLE**

**1.9 Distribution of IQ scores.** You have probably heard that the distribution of scores on IQ tests is supposed to be roughly "bell-shaped." Let's look at some actual IQ scores. Table 1.3 displays the IQ scores of 60 fifth-grade students chosen at random from one school.[6]

1. Divide the range of the data into classes of equal width. The scores in Table 1.3 range from 81 to 145, so we choose as our classes

**TABLE 1.3**

IQ test scores for 60 randomly chosen fifth-grade students

| 145 | 139 | 126 | 122 | 125 | 130 | 96 | 110 | 118 | 118 |
|-----|-----|-----|-----|-----|-----|-----|-----|-----|-----|
| 101 | 142 | 134 | 124 | 112 | 109 | 134 | 113 | 81 | 113 |
| 123 | 94 | 100 | 136 | 109 | 131 | 117 | 110 | 127 | 124 |
| 106 | 124 | 115 | 133 | 116 | 102 | 127 | 117 | 109 | 137 |
| 117 | 90 | 103 | 114 | 139 | 101 | 122 | 105 | 97 | 89 |
| 102 | 108 | 110 | 128 | 114 | 112 | 114 | 102 | 82 | 101 |

$$75 \leq \text{IQ score} < 85$$

$$85 \leq \text{IQ score} < 95$$

$$\vdots$$

$$145 \leq \text{IQ score} < 155$$

Be sure to specify the classes precisely so that each individual falls into exactly one class. A student with IQ 84 would fall into the first class, but IQ 85 falls into the second.

**frequency**
**frequency table**

2. Count the number of individuals in each class. These counts are called **frequencies,** and a table of frequencies for all classes is a **frequency table.**

| Class | Count | Class | Count |
|-------|-------|-------|-------|
| 75 to 84 | 2 | 115 to 124 | 13 |
| 85 to 94 | 3 | 125 to 134 | 10 |
| 95 to 104 | 10 | 135 to 144 | 5 |
| 105 to 114 | 16 | 145 to 154 | 1 |

3. Draw the histogram. First, on the horizontal axis mark the scale for the variable whose distribution you are displaying. That's IQ score. The scale runs from 75 to 155 because that is the span of the classes we chose. The vertical axis contains the scale of counts. Each bar represents a class. The base of the bar covers the class, and the bar height is the class count. There is no horizontal space between the bars unless a class is empty, so that its bar has height zero. Figure 1.7 is our histogram. It does look roughly "bell-shaped."

Large sets of data are often reported in the form of frequency tables when it is not practical to publish the individual observations. In addition to the frequency (count) for each class, we may be interested in the fraction or percent of the observations that fall in each class. A histogram of percents looks just like a frequency histogram such as Figure 1.7. Simply relabel the vertical scale to read in percents. Use histograms of percents for comparing several distributions that have different numbers of observations.

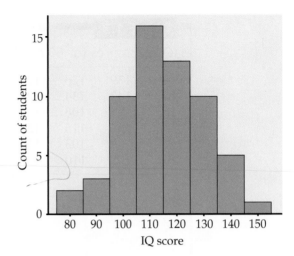

**FIGURE 1.7** Histogram of the IQ scores of 60 fifth-grade students, for Example 1.9.

## USE YOUR KNOWLEDGE

1.6    **Make a histogram.** Refer to the first-exam scores from Exercise 1.5. Use these data to make a histogram using classes 50–59, 60–69, etc. Compare the histogram with the stemplot as a way of describing this distribution. Which do you prefer for these data?

Our eyes respond to the *area* of the bars in a histogram. Because the classes are all the same width, area is determined by height and all classes are fairly represented. There is no one right choice of the classes in a histogram. Too few classes will give a "skyscraper" graph, with all values in a few classes with tall bars. Too many will produce a "pancake" graph, with most classes having one or no observations. Neither choice will give a good picture of the shape of the distribution. You must use your judgment in choosing classes to display the shape. Statistical software will choose the classes for you. The software's choice is often a good one, but you can change it if you want.

*You should be aware that the appearance of a histogram can change when you change the classes.* Figure 1.8 is a histogram of the customer service call lengths

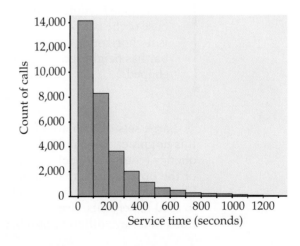

**FIGURE 1.8** The "default" histogram produced by software for the call lengths in Example 1.6. This choice of classes hides the large number of very short calls that is revealed by the histogram of the same data in Figure 1.4.

that are also displayed in Figure 1.4. It was produced by software with no special instructions from the user. The software's "default" histogram shows the overall shape of the distribution, but it hides the spike of very short calls by lumping all calls of less than 100 seconds into the first class. We produced Figure 1.4 by asking for smaller classes after Table 1.1 suggested that very short calls might be a problem. Software automates making graphs, but it can't replace thinking about your data. The histogram function in the *One-Variable Statistical Calculator* applet on the text CD and Web site allows you to change the number of classes by dragging with the mouse, so that it is easy to see how the choice of classes affects the histogram.

## USE YOUR KNOWLEDGE

**1.7** **Change the classes in the histogram.** Refer to the first-exam scores from Exercise 1.5 and the histogram you produced in Exercise 1.6. Now make a histogram for these data using classes 40–59, 60–79, and 80–100. Compare this histogram with the one that you produced in Exercise 1.6.

**1.8** **Use smaller classes.** Repeat the previous exercise using classes 55–59, 60–64, 65–69, etc.

Although histograms resemble bar graphs, their details and uses are distinct. A histogram shows the distribution of counts or percents among the values of a single variable. A bar graph compares the size of different items. The horizontal axis of a bar graph need not have any measurement scale but simply identifies the items being compared. Draw bar graphs with blank space between the bars to separate the items being compared. Draw histograms with no space, to indicate that all values of the variable are covered. *Some spreadsheet programs, which are not primarily intended for statistics, will draw histograms as if they were bar graphs, with space between the bars. Often, you can tell the software to eliminate the space to produce a proper histogram.*

## Examining distributions

Making a statistical graph is not an end in itself. The purpose of the graph is to help us understand the data. After you make a graph, always ask, "What do I see?" Once you have displayed a distribution, you can see its important features as follows.

---

EXAMINING A DISTRIBUTION

In any graph of data, look for the **overall pattern** and for striking **deviations** from that pattern.

You can describe the overall pattern of a distribution by its **shape, center,** and **spread.**

An important kind of deviation is an **outlier,** an individual value that falls outside the overall pattern.

In Section 1.2, we will learn how to describe center and spread numerically. For now, we can describe the center of a distribution by its *midpoint*, the value with roughly half the observations taking smaller values and half taking larger values. We can describe the spread of a distribution by giving the *smallest and largest values*. Stemplots and histograms display the shape of a distribution in the same way. Just imagine a stemplot turned on its side so that the larger values lie to the right. Some things to look for in describing shape are:

*modes*
*unimodal*

- Does the distribution have one or several major peaks, called **modes**? A distribution with one major peak is called **unimodal.**

*symmetric*
*skewed*

- Is it approximately symmetric or is it skewed in one direction? A distribution is **symmetric** if the values smaller and larger than its midpoint are mirror images of each other. It is **skewed to the right** if the right tail (larger values) is much longer than the left tail (smaller values).

Some variables commonly have distributions with predictable shapes. Many biological measurements on specimens from the same species and sex—lengths of bird bills, heights of young women—have symmetric distributions. Money amounts, on the other hand, usually have right-skewed distributions. There are many moderately priced houses, for example, but the few very expensive mansions give the distribution of house prices a strong right-skew.

**EXAMPLE**

**1.10 Examine the histogram.** What does the histogram of IQ scores (Figure 1.7) tell us? **Shape:** The distribution is *roughly symmetric* with a *single peak* in the center. We don't expect real data to be perfectly symmetric, so we are satisfied if the two sides of the histogram are roughly similar in shape and extent. **Center:** You can see from the histogram that the midpoint is not far from 110. Looking at the actual data shows that the midpoint is 114. **Spread:** The spread is from 81 to 145. There are no outliers or other strong deviations from the symmetric, unimodal pattern.

The distribution of call lengths in Figure 1.8, on the other hand, is strongly *skewed to the right*. The midpoint, the length of a typical call, is about 115 seconds, or just under 2 minutes. The spread is very large, from 1 second to 28,739 seconds.

The longest few calls are *outliers*. They stand apart from the long right tail of the distribution, though we can't see this from Figure 1.8, which omits the largest observations. The longest call lasted almost 8 hours—that may well be due to equipment failure rather than an actual customer call.

## USE YOUR KNOWLEDGE

**1.9   Describe the first-exam scores.** Refer to the first-exam scores from Exercise 1.5. Use your favorite graphical display to describe the shape, the center, and the spread of these data. Are there any outliers?

## Dealing with outliers

In data sets smaller than the service call data, you can spot outliers by looking for observations that stand apart (either high or low) from the overall pattern of a histogram or stemplot. *Identifying outliers is a matter for judgment. Look for points that are clearly apart from the body of the data, not just the most extreme observations in a distribution.* You should search for an explanation for any outlier. Sometimes outliers point to errors made in recording the data. In other cases, the outlying observation may be caused by equipment failure or other unusual circumstances.

**EXAMPLE**

**1.11 Semiconductor wires.** Manufacturing an electronic component requires attaching very fine wires to a semiconductor wafer. If the strength of the bond is weak, the component may fail. Here are measurements on the breaking strength (in pounds) of 23 connections:[7]

|      |      |      |      |      |      |      |      |
|------|------|------|------|------|------|------|------|
| 0    | 0    | 550  | 750  | 950  | 950  | 1150 | 1150 |
| 1150 | 1150 | 1150 | 1250 | 1250 | 1350 | 1450 | 1450 |
| 1450 | 1550 | 1550 | 1550 | 1850 | 2050 | 3150 |      |

Figure 1.9 is a histogram of these data. We expect the breaking strengths of supposedly identical connections to have a roughly symmetric overall pattern, showing chance variation among the connections. Figure 1.9 does show a symmetric pattern centered at about 1250 pounds—but it also shows three *outliers* that stand apart from this pattern, two low and one high.

The engineers were able to explain all three outliers. The two low outliers had strength 0 because the bonds between the wire and the wafer were not made. The high outlier at 3150 pounds was a measurement error. Further study of the data can simply omit the three outliers. One immediate finding is that the variation in breaking strength is too large—550 pounds to 2050 pounds when we ignore the outliers. The process of bonding wire to wafer must be improved to give more consistent results.

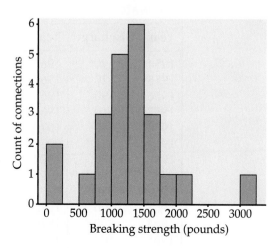

**FIGURE 1.9** Histogram of a distribution with both low and high outliers, for Example 1.11.

## Time plots

Whenever data are collected over time, it is a good idea to plot the observations in time order. *Displays of the distribution of a variable that ignore time order, such as stemplots and histograms, can be misleading when there is systematic change over time.*

---

**TIME PLOT**

A **time plot** of a variable plots each observation against the time at which it was measured. Always put time on the horizontal scale of your plot and the variable you are measuring on the vertical scale. Connecting the data points by lines helps emphasize any change over time.

---

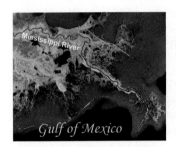

**EXAMPLE**

**1.12 Water from the Mississippi River.**    Table 1.4 lists the volume of water discharged by the Mississippi River into the Gulf of Mexico for each year from 1954 to 2001.[8] The units are cubic kilometers of water—the Mississippi is a big river. Both graphs in Figure 1.10 describe these data. The histogram in Figure 1.10(a) shows the distribution of the volume discharged. The histogram is symmetric and unimodal, with center near 550 cubic kilometers. We might think that the data show just chance year-to-year fluctuation in river level about its long-term average.

Figure 1.10(b) is a time plot of the same data. For example, the first point lies above 1954 on the "Year" scale at height 290, the volume of water discharged by the Mississippi in 1954. The time plot tells a more interesting story

---

**TABLE 1.4**

Yearly discharge of the Mississippi River (in cubic kilometers of water)

| Year | Discharge | Year | Discharge | Year | Discharge | Year | Discharge |
|------|-----------|------|-----------|------|-----------|------|-----------|
| 1954 | 290 | 1966 | 410 | 1978 | 560 | 1990 | 680 |
| 1955 | 420 | 1967 | 460 | 1979 | 800 | 1991 | 700 |
| 1956 | 390 | 1968 | 510 | 1980 | 500 | 1992 | 510 |
| 1957 | 610 | 1969 | 560 | 1981 | 420 | 1993 | 900 |
| 1958 | 550 | 1970 | 540 | 1982 | 640 | 1994 | 640 |
| 1959 | 440 | 1971 | 480 | 1983 | 770 | 1995 | 590 |
| 1960 | 470 | 1972 | 600 | 1984 | 710 | 1996 | 670 |
| 1961 | 600 | 1973 | 880 | 1985 | 680 | 1997 | 680 |
| 1962 | 550 | 1974 | 710 | 1986 | 600 | 1998 | 690 |
| 1963 | 360 | 1975 | 670 | 1987 | 450 | 1999 | 580 |
| 1964 | 390 | 1976 | 420 | 1988 | 420 | 2000 | 390 |
| 1965 | 500 | 1977 | 430 | 1989 | 630 | 2001 | 580 |

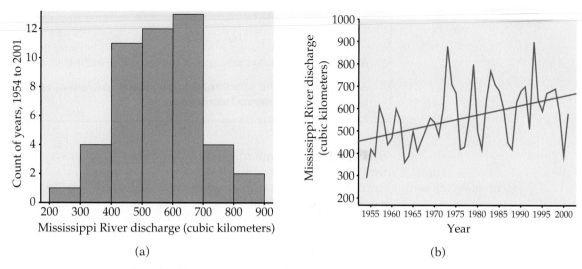

(a)                                                                    (b)

**FIGURE 1.10** (a) Histogram of the volume of water discharged by the Mississippi River over the 48 years from 1954 to 2001, for Example 1.12. Data are from Table 1.4. (b) Time plot of the volume of water discharged by the Mississippi River for the years 1954 to 2001. The line shows the trend toward increasing river flow, a trend that cannot be seen in the histogram in Figure 1.10(a).

**trend**

than the histogram. There is a great deal of year-to-year variation, but there is also a clear increasing **trend** over time. That is, there is a long-term rise in the volume of water discharged. The line on the graph is a "trend line" calculated from the data to describe this trend. The trend reflects climate change: rainfall and river flows have been increasing over most of North America.

**time series**

Many interesting data sets are **time series,** measurements of a variable taken at regular intervals over time. Government, economic, and social data are often published as time series. Some examples are the monthly unemployment rate and the quarterly gross domestic product. Weather records, the demand for electricity, and measurements on the items produced by a manufacturing process are other examples of time series. Time plots can reveal the main features of a time series.

## BEYOND THE BASICS

### Decomposing Time Series*

When you examine a time plot, again look first for overall patterns and then for striking deviations from those patterns. Here are two important types of overall patterns to look for in a time series.

---

*"Beyond the Basics" sections briefly discuss supplementary topics. Your software may make some of these topics available to you. For example, the results plotted in Figures 1.11 to 1.13 come from the Minitab statistical software.

> **TREND AND SEASONAL VARIATION**
>
> A **trend** in a time series is a persistent, long-term rise or fall.
>
> A pattern in a time series that repeats itself at known regular intervals of time is called **seasonal variation.**

seasonally adjusted

Because many economic time series show strong seasonal variation, government agencies often adjust for this variation before releasing economic data. The data are then said to be **seasonally adjusted.** Seasonal adjustment helps avoid misinterpretation. A rise in the unemployment rate from December to January, for example, does not mean that the economy is slipping. Unemployment almost always rises in January as temporary holiday help is laid off and outdoor employment in the North drops because of bad weather. The seasonally adjusted unemployment rate reports an increase only if unemployment rises more than normal from December to January.

**EXAMPLE**

**1.13 Gasoline prices.** Figure 1.11 is a time plot of the average retail price of regular gasoline each month for the years 1990 to 2003.[9] The prices are *not* seasonally adjusted. You can see the upward spike in prices due to the 1990 Iraqi invasion of Kuwait, the drop in 1998 when an economic crisis in Asia reduced demand for fuel, and rapid price increases in 2000 and 2003 due to instability in the Middle East and OPEC production limits. These deviations are so large that overall patterns are hard to see.

There is nonetheless a clear *trend* of increasing price. Much of this trend just reflects inflation, the rise in the overall price level during these years. In addition, a close look at the plot shows *seasonal variation,* a regular rise and fall that recurs each year. Americans drive more in the summer vacation season, so the price of gasoline rises each spring, then drops in the fall as demand goes down.

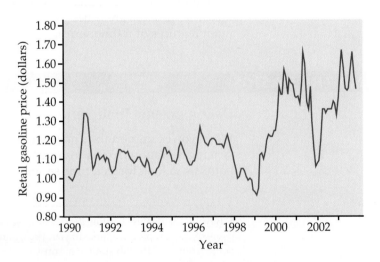

**FIGURE 1.11** Time plot of the average monthly price of regular gasoline from 1990 to 2003, for Example 1.13.

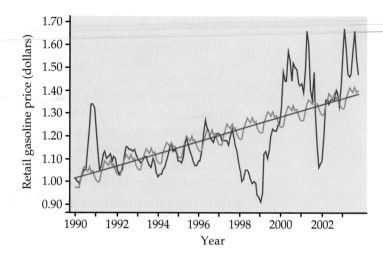

**FIGURE 1.12** Time plot of gasoline prices with a trend line and seasonal variation added. These are overall patterns extracted from the data by software.

Statistical software can help us examine a time series by "decomposing" the data into systematic patterns, such as trends and seasonal variation, and the *residuals* that remain after we remove these patterns. Figure 1.12 superimposes the trend and seasonal variation on the time plot of gasoline prices. The red line shows the increasing trend. The seasonal variation appears as the colored line that regularly rises and falls each year. This is an average of the seasonal pattern for all the years in the original data, automatically extracted by software.

The trend and seasonal variation in Figure 1.12 are overall patterns in the data. Figure 1.13 is a plot of what remains when we subtract both the trend and the seasonal variation from the original data. That is, Figure 1.13 emphasizes the deviations from the pattern. In the case of gasoline prices, the deviations are large (as much as 30 cents both up and down). It is clear that we can't use trend and seasonal variation to predict gasoline prices at all accurately.

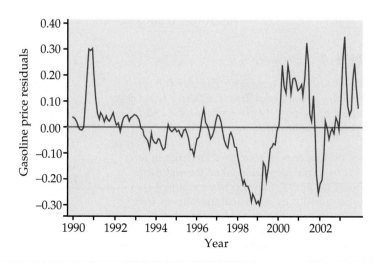

**FIGURE 1.13** The residuals that remain when we subtract both trend and seasonal variation from monthly gasoline prices.

## SECTION 1.1 Summary

A data set contains information on a collection of **individuals.** Individuals may be people, animals, or things. The data for one individual make up a **case.** For each individual, the data give values for one or more **variables.** A variable describes some characteristic of an individual, such as a person's height, gender, or salary.

Some variables are **categorical** and others are **quantitative.** A categorical variable places each individual into a category, such as male or female. A quantitative variable has numerical values that measure some characteristic of each individual, such as height in centimeters or annual salary in dollars.

**Exploratory data analysis** uses graphs and numerical summaries to describe the variables in a data set and the relations among them.

The **distribution** of a variable tells us what values it takes and how often it takes these values.

**Bar graphs** and **pie charts** display the distributions of categorical variables. These graphs use the counts or percents of the categories.

**Stemplots** and **histograms** display the distributions of quantitative variables. Stemplots separate each observation into a **stem** and a one-digit **leaf.** Histograms plot the **frequencies** (counts) or the percents of equal-width classes of values.

When examining a distribution, look for **shape, center,** and **spread** and for clear **deviations** from the overall shape.

Some distributions have simple shapes, such as **symmetric** or **skewed.** The number of **modes** (major peaks) is another aspect of overall shape. Not all distributions have a simple overall shape, especially when there are few observations.

**Outliers** are observations that lie outside the overall pattern of a distribution. Always look for outliers and try to explain them.

When observations on a variable are taken over time, make a **time plot** that graphs time horizontally and the values of the variable vertically. A time plot can reveal **trends** or other changes over time.

## SECTION 1.1 Exercises

*For Exercises 1.1 to 1.2, see page 3; for Exercise 1.3, see page 5; for Exercise 1.4, see page 7; for Exercise 1.5, see page 10; for Exercise 1.6, see page 14; for Exercises 1.7 and 1.8, see page 15; and for Exercise 1.9, see page 16.*

1.10 **Survey of students.** A survey of students in an introductory statistics class asked the following questions: (a) age; (b) do you like to dance? (yes, no); (c) can you play a musical instrument (not at all, a little, pretty well); (d) how much did you spend on food last week? (e) height; (f) do you like broccoli? (yes, no). Classify each of these variables as categorical or quantitative and give reasons for your answers.

1.11 **What questions would you ask?** Refer to the previous exercise. Make up your own survey questions with at least six questions. Include at least two categorical variables and at least two quantitative variables. Tell which variables are categorical and which are quantitative. Give reasons for your answers.

1.12 **Study habits of students.** You are planning a survey to collect information about the study habits of college students. Describe two categorical variables and two quantitative variables that you might measure for each student. Give the units of measurement for the quantitative variables.

**1.13 Physical fitness of students.** You want to measure the "physical fitness" of college students. Describe several variables you might use to measure fitness. What instrument or instruments does each measurement require?

**1.14 Choosing a college or university.** Popular magazines rank colleges and universities on their "academic quality" in serving undergraduate students. Describe five variables that you would like to see measured for each college if you were choosing where to study. Give reasons for each of your choices.

**1.15 Favorite colors.** What is your favorite color? One survey produced the following summary of responses to that question: blue, 42%; green, 14%; purple, 14%; red, 8%; black, 7%; orange, 5%; yellow, 3%; brown, 3%; gray, 2%; and white, 2%.[10] Make a bar graph of the percents and write a short summary of the major features of your graph.

**1.16 Least-favorite colors.** Refer to the previous exercise. The same study also asked people about their least-favorite color. Here are the results: orange, 30%; brown, 23%; purple, 13%; yellow, 13%; gray, 12%; green, 4%; white, 4%; red, 1%; black, 0%; and blue, 0%. Make a bar graph of these percents and write a summary of the results.

**1.17 Ages of survey respondents.** The survey about color preferences reported the age distribution of the people who responded. Here are the results:

| Age group (years) | 1–18 | 19–24 | 25–35 | 36–50 | 51–69 | 70 and over |
|---|---|---|---|---|---|---|
| Count | 10 | 97 | 70 | 36 | 14 | 5 |

(a) Add the counts and compute the percents for each age group.

(b) Make a bar graph of the percents.

(c) Describe the distribution.

(d) Explain why your bar graph is not a histogram.

**1.18 Garbage.** The formal name for garbage is "municipal solid waste." The table at the top of the next column gives a breakdown of the materials that made up American municipal solid waste.[11]

(a) Add the weights for the nine materials given, including "Other." Each entry, including the total, is separately rounded to the nearest tenth. So the sum and the total may differ slightly because of **roundoff error.**

| Material | Weight (million tons) | Percent of total |
|---|---|---|
| Food scraps | 25.9 | 11.2 |
| Glass | 12.8 | 5.5 |
| Metals | 18.0 | 7.8 |
| Paper, paperboard | 86.7 | 37.4 |
| Plastics | 24.7 | 10.7 |
| Rubber, leather, textiles | 15.8 | 6.8 |
| Wood | 12.7 | 5.5 |
| Yard trimmings | 27.7 | 11.9 |
| Other | 7.5 | 3.2 |
| Total | 231.9 | 100.0 |

(b) Make a bar graph of the percents. The graph gives a clearer picture of the main contributors to garbage if you order the bars from tallest to shortest.

(c) If you use software, also make a pie chart of the percents. Comparing the two graphs, notice that it is easier to see the small differences among "Food scraps," "Plastics," and "Yard trimmings" in the bar graph.

**1.19 Spam.** Email spam is the curse of the Internet. Here is a compilation of the most common types of spam:[12]

| Type of spam | Percent |
|---|---|
| Adult | 14.5 |
| Financial | 16.2 |
| Health | 7.3 |
| Leisure | 7.8 |
| Products | 21.0 |
| Scams | 14.2 |

Make two bar graphs of these percents, one with bars ordered as in the table (alphabetical) and the other with bars in order from tallest to shortest. Comparisons are easier if you order the bars by height. A bar graph ordered from tallest to shortest bar is sometimes called a **Pareto chart,** after the Italian economist who recommended this procedure.

**1.20 Women seeking graduate and professional degrees.** The table on the next page gives the percents of women among students seeking various graduate and professional degrees:[13]

(a) Explain clearly why we cannot use a pie chart to display these data.

(b) Make a bar graph of the data. (Comparisons are easier if you order the bars by height.)

| Degree | Percent female |
|---|---|
| Master's in business administration | 39.8 |
| Master's in education | 76.2 |
| Other master of arts | 59.6 |
| Other master of science | 53.0 |
| Doctorate in education | 70.8 |
| Other PhD degree | 54.2 |
| Medicine (MD) | 44.0 |
| Law | 50.2 |
| Theology | 20.2 |

**1.21 An aging population.** The population of the United States is aging, though less rapidly than in other developed countries. Here is a stemplot of the percents of residents aged 65 and over in the 50 states, according to the 2000 census. The stems are whole percents and the leaves are tenths of a percent.

```
 5 | 7
 6 |
 7 |
 8 | 5
 9 | 679
10 | 6
11 | 02233677
12 | 0011113445789
13 | 00012233345568
14 | 034579
15 | 36
16 |
17 | 6
```

(a) There are two outliers: Alaska has the lowest percent of older residents, and Florida has the highest. What are the percents for these two states?

(b) Ignoring Alaska and Florida, describe the shape, center, and spread of this distribution.

**1.22 Split the stems.** Make another stemplot of the percent of residents aged 65 and over in the states other than Alaska and Florida by splitting stems 8 to 15 in the plot from the previous exercise. Which plot do you prefer? Why?

**1.23 Diabetes and glucose.** People with diabetes must monitor and control their blood glucose level. The goal is to maintain "fasting plasma glucose" between about 90 and 130 milligrams per deciliter (mg/dl). Here are the fasting plasma glucose levels for 18 diabetics enrolled in a diabetes control class, five months after the end of the class:[14]

| 141 | 158 | 112 | 153 | 134 | 95 | 96 | 78 | 148 |
|---|---|---|---|---|---|---|---|---|
| 172 | 200 | 271 | 103 | 172 | 359 | 145 | 147 | 255 |

Make a stemplot of these data and describe the main features of the distribution. (You will want to trim and also split stems.) Are there outliers? How well is the group as a whole achieving the goal for controlling glucose levels?

**1.24 Compare glucose of instruction and control groups.** The study described in the previous exercise also measured the fasting plasma glucose of 16 diabetics who were given individual instruction on diabetes control. Here are the data:

| 128 | 195 | 188 | 158 | 227 | 198 | 163 | 164 |
|---|---|---|---|---|---|---|---|
| 159 | 128 | 283 | 226 | 223 | 221 | 220 | 160 |

Make a back-to-back stemplot to compare the class and individual instruction groups. How do the distribution shapes and success in achieving the glucose control goal compare?

**1.25 Vocabulary scores of seventh-grade students.** Figure 1.14 displays the scores of all 947 seventh-grade students in the public schools of Gary, Indiana, on the vocabulary part of the Iowa Test of Basic Skills.[15] Give a brief description of the overall pattern (shape, center, spread) of this distribution.

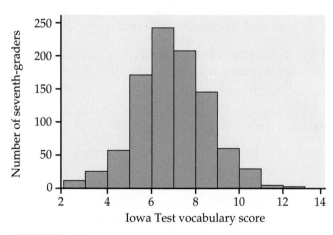

**FIGURE 1.14** Histogram of the Iowa Test of Basic Skills vocabulary scores of seventh-grade students in Gary, Indiana, for Exercise 1.25.

**1.26 Shakespeare's plays.** Figure 1.15 is a histogram of the lengths of words used in Shakespeare's plays. Because there are so many words in the plays, we use a histogram of percents. What is the overall shape of this distribution? What does this shape say about word lengths in Shakespeare? Do you expect other authors to have word length distributions of the same general shape? Why?

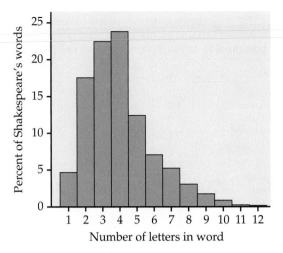

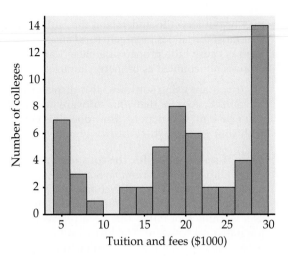

**FIGURE 1.15** Histogram of lengths of words used in Shakespeare's plays, for Exercise 1.26.

**FIGURE 1.16** Histogram of the tuition and fees charged by four-year colleges in Massachusetts, for Exercise 1.27.

**1.27 College tuition and fees.** Jeanna plans to attend college in her home state of Massachusetts. She looks up the tuition and fees for all 56 four-year colleges in Massachusetts (omitting art schools and other special colleges). Figure 1.16 is a histogram of the data. For state schools, Jeanna used the in-state tuition. What is the most important aspect of the overall pattern of this distribution? Why do you think this pattern appears?

**1.28 Tornado damage.** The states differ greatly in the kinds of severe weather that afflict them. Table 1.5

shows the average property damage caused by tornadoes per year over the period from 1950 to 1999 in each of the 50 states and Puerto Rico.[16] (To adjust for the changing buying power of the dollar over time, all damages were restated in 1999 dollars.)

(a) What are the top five states for tornado damage? The bottom five?

(b) Make a histogram of the data, by hand or using software, with classes "$0 \leq$ damage $< 10$," "$10 \leq$ damage $< 20$," and so on. Describe the shape, center, and spread of the distribution. Which states

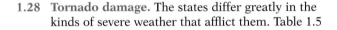

**TABLE 1.5**

Average property damage per year due to tornadoes

| State | Damage ($millions) | State | Damage ($millions) | State | Damage ($millions) |
|---|---|---|---|---|---|
| Alabama | 51.88 | Louisiana | 27.75 | Ohio | 44.36 |
| Alaska | 0.00 | Maine | 0.53 | Oklahoma | 81.94 |
| Arizona | 3.47 | Maryland | 2.33 | Oregon | 5.52 |
| Arkansas | 40.96 | Massachusetts | 4.42 | Pennsylvania | 17.11 |
| California | 3.68 | Michigan | 29.88 | Puerto Rico | 0.05 |
| Colorado | 4.62 | Minnesota | 84.84 | Rhode Island | 0.09 |
| Connecticut | 2.26 | Mississippi | 43.62 | South Carolina | 17.19 |
| Delaware | 0.27 | Missouri | 68.93 | South Dakota | 10.64 |
| Florida | 37.32 | Montana | 2.27 | Tennessee | 23.47 |
| Georgia | 51.68 | Nebraska | 30.26 | Texas | 88.60 |
| Hawaii | 0.34 | Nevada | 0.10 | Utah | 3.57 |
| Idaho | 0.26 | New Hampshire | 0.66 | Vermont | 0.24 |
| Illinois | 62.94 | New Jersey | 2.94 | Virginia | 7.42 |
| Indiana | 53.13 | New Mexico | 1.49 | Washington | 2.37 |
| Iowa | 49.51 | New York | 15.73 | West Virginia | 2.14 |
| Kansas | 49.28 | North Carolina | 14.90 | Wisconsin | 31.33 |
| Kentucky | 24.84 | North Dakota | 14.69 | Wyoming | 1.78 |

may be outliers? (To understand the outliers, note that most tornadoes in largely rural states such as Kansas cause little property damage. Damage to crops is not counted as property damage.)

(c) If you are using software, also display the "default" histogram that your software makes when you give it no instructions. How does this compare with your graph in (b)?

**1.29** ⟨APPLET⟩ **Use an applet for the tornado damage data.** The *One-Variable Statistical Calculator* applet on the text CD and Web site will make stemplots and histograms. It is intended mainly as a learning tool rather than as a replacement for statistical software. The histogram function is particularly useful because you can change the number of classes by dragging with the mouse. The tornado damage data from Table 1.5 are available in the applet. Choose this data set and go to the "Histogram" tab.

(a) Sketch the default histogram that the applet first presents. If the default graph does not have nine classes, drag it to make a histogram with nine classes and sketch the result. This should agree with your histogram in part (b) of the previous exercise.

(b) Make a histogram with one class and also a histogram with the greatest number of classes that the applet allows. Sketch the results.

(c) Drag the graph until you find the histogram that you think best pictures the data. How many classes did you choose? Sketch your final histogram.

**1.30** **Carbon dioxide from burning fuels.** Burning fuels in power plants or motor vehicles emits carbon dioxide ($CO_2$), which contributes to global warming. Table 1.6 displays $CO_2$ emissions per person from countries with population at least 20 million.[17]

(a) Why do you think we choose to measure emissions per person rather than total $CO_2$ emissions for each country?

(b) Display the data of Table 1.6 in a graph. Describe the shape, center, and spread of the distribution. Which countries are outliers?

**1.31** **California temperatures.** Table 1.7 contains data on the mean annual temperatures (degrees Fahrenheit) for the years 1951 to 2000 at two locations in California: Pasadena and Redding.[18] Make time plots of both time series and compare their main features. You can see why discussions of climate change often bring disagreement.

**1.32** **What do you miss in the histogram?** Make a histogram of the mean annual temperatures

### TABLE 1.6

Carbon dioxide emissions (metric tons per person)

| Country | CO$_2$ | Country | CO$_2$ |
|---|---|---|---|
| Algeria | 2.3 | Mexico | 3.7 |
| Argentina | 3.9 | Morocco | 1.0 |
| Australia | 17.0 | Myanmar | 0.2 |
| Bangladesh | 0.2 | Nepal | 0.1 |
| Brazil | 1.8 | Nigeria | 0.3 |
| Canada | 16.0 | Pakistan | 0.7 |
| China | 2.5 | Peru | 0.8 |
| Columbia | 1.4 | Tanzania | 0.1 |
| Congo | 0.0 | Philippines | 0.9 |
| Egypt | 1.7 | Poland | 8.0 |
| Ethiopia | 0.0 | Romania | 3.9 |
| France | 6.1 | Russia | 10.2 |
| Germany | 10.0 | Saudi Arabia | 11.0 |
| Ghana | 0.2 | South Africa | 8.1 |
| India | 0.9 | Spain | 6.8 |
| Indonesia | 1.2 | Sudan | 0.2 |
| Iran | 3.8 | Thailand | 2.5 |
| Iraq | 3.6 | Turkey | 2.8 |
| Italy | 7.3 | Ukraine | 7.6 |
| Japan | 9.1 | United Kingdom | 9.0 |
| Kenya | 0.3 | United States | 19.9 |
| Korea, North | 9.7 | Uzbekistan | 4.8 |
| Korea, South | 8.8 | Venezuela | 5.1 |
| Malaysia | 4.6 | Vietnam | 0.5 |

### TABLE 1.7

Mean annual temperatures (°F) in two California cities

| | Mean Temperature | | | Mean Temperature | |
|---|---|---|---|---|---|
| Year | Pasadena | Redding | Year | Pasadena | Redding |
| 1951 | 62.27 | 62.02 | 1976 | 64.23 | 63.51 |
| 1952 | 61.59 | 62.27 | 1977 | 64.47 | 63.89 |
| 1953 | 62.64 | 62.06 | 1978 | 64.21 | 64.05 |
| 1954 | 62.88 | 61.65 | 1979 | 63.76 | 60.38 |
| 1955 | 61.75 | 62.48 | 1980 | 65.02 | 60.04 |
| 1956 | 62.93 | 63.17 | 1981 | 65.80 | 61.95 |
| 1957 | 63.72 | 62.42 | 1982 | 63.50 | 59.14 |
| 1958 | 65.02 | 64.42 | 1983 | 64.19 | 60.66 |
| 1959 | 65.69 | 65.04 | 1984 | 66.06 | 61.72 |
| 1960 | 64.48 | 63.07 | 1985 | 64.44 | 60.50 |
| 1961 | 64.12 | 63.50 | 1986 | 65.31 | 61.76 |
| 1962 | 62.82 | 63.97 | 1987 | 64.58 | 62.94 |
| 1963 | 63.71 | 62.42 | 1988 | 65.22 | 63.70 |
| 1964 | 62.76 | 63.29 | 1989 | 64.53 | 61.50 |
| 1965 | 63.03 | 63.32 | 1990 | 64.96 | 62.22 |
| 1966 | 64.25 | 64.51 | 1991 | 65.60 | 62.73 |
| 1967 | 64.36 | 64.21 | 1992 | 66.07 | 63.59 |
| 1968 | 64.15 | 63.40 | 1993 | 65.16 | 61.55 |
| 1969 | 63.51 | 63.77 | 1994 | 64.63 | 61.63 |
| 1970 | 64.08 | 64.30 | 1995 | 65.43 | 62.62 |
| 1971 | 63.59 | 62.23 | 1996 | 65.76 | 62.93 |
| 1972 | 64.53 | 63.06 | 1997 | 66.72 | 62.48 |
| 1973 | 63.46 | 63.75 | 1998 | 64.12 | 60.23 |
| 1974 | 63.93 | 63.80 | 1999 | 64.85 | 61.88 |
| 1975 | 62.36 | 62.66 | 2000 | 66.25 | 61.58 |

at Pasadena for the years 1951 to 2000. (Data appear in Table 1.7.) Describe the distribution of temperatures. Then explain why this histogram misses very important facts about temperatures in Pasadena.

**1.33** ⚠️ CAUTION **Change the scale of the axis.** The impression that a time plot gives depends on the scales you use on the two axes. If you stretch the vertical axis and compress the time axis, change appears to be more rapid. Compressing the vertical axis and stretching the time axis make change appear slower. Make two more time plots of the data for Pasadena in Table 1.7, one that makes mean temperature appear to increase very rapidly and one that shows only a slow increase. The moral of this exercise is: *pay close attention to the scales when you look at a time plot.*

**1.34** **Fish in the Bering Sea.** "Recruitment," the addition of new members to a fish population, is an important measure of the health of ocean ecosystems. Here are data on the recruitment of rock sole in the Bering Sea between 1973 and 2000:[19]

| Year | Recruitment (millions) | Year | Recruitment (millions) |
|---|---|---|---|
| 1973 | 173 | 1987 | 4700 |
| 1974 | 234 | 1988 | 1702 |
| 1975 | 616 | 1989 | 1119 |
| 1976 | 344 | 1990 | 2407 |
| 1977 | 515 | 1991 | 1049 |
| 1978 | 576 | 1992 | 505 |
| 1979 | 727 | 1993 | 998 |
| 1980 | 1411 | 1994 | 505 |
| 1981 | 1431 | 1995 | 304 |
| 1982 | 1250 | 1996 | 425 |
| 1983 | 2246 | 1997 | 214 |
| 1984 | 1793 | 1998 | 385 |
| 1985 | 1793 | 1999 | 445 |
| 1986 | 2809 | 2000 | 676 |

(a) Make a graph to display the distribution of rock sole recruitment, then describe the pattern and any striking deviations that you see.

(b) Make a time plot of recruitment and describe its pattern. As is often the case with time series data, a time plot is needed to understand what is happening.

**1.35** **Thinness in Asia.** Asian culture does not emphasize thinness, but young Asians are often influenced by Western culture. In a study of concerns about weight among young Korean women, researchers administered the Drive for Thinness scale (a questionnaire) to 264 female college students in Seoul, South Korea.[20] Drive for Thinness measures excessive concern with weight and dieting and fear of weight gain. Roughly speaking, a score of 15 is typical of Western women with eating disorders but is unusually high (90th percentile) for other Western women. Graph the data and describe the shape, center, and spread of the distribution of Drive for Thinness scores for these Korean students. Are there any outliers?

**1.36** 🔺 CHALLENGE **Acidity of rainwater.** Changing the choice of classes can change the appearance of a histogram. Here is an example in which a small shift in the classes, with no change in the number of classes, has an important effect on the histogram. The data are the acidity levels (measured by pH) in 105 samples of rainwater. Distilled water has pH 7.00. As the water becomes more acidic, the pH goes down. The pH of rainwater is important to environmentalists because of the problem of acid rain.[21]

| | | | | | | | |
|---|---|---|---|---|---|---|---|
| 4.33 | 4.38 | 4.48 | 4.48 | 4.50 | 4.55 | 4.59 | 4.59 |
| 4.61 | 4.61 | 4.75 | 4.76 | 4.78 | 4.82 | 4.82 | 4.83 |
| 4.86 | 4.93 | 4.94 | 4.94 | 4.94 | 4.96 | 4.97 | 5.00 |
| 5.01 | 5.02 | 5.05 | 5.06 | 5.08 | 5.09 | 5.10 | 5.12 |
| 5.13 | 5.15 | 5.15 | 5.15 | 5.16 | 5.16 | 5.16 | 5.18 |
| 5.19 | 5.23 | 5.24 | 5.29 | 5.32 | 5.33 | 5.35 | 5.37 |
| 5.37 | 5.39 | 5.41 | 5.43 | 5.44 | 5.46 | 5.46 | 5.47 |
| 5.50 | 5.51 | 5.53 | 5.55 | 5.55 | 5.56 | 5.61 | 5.62 |
| 5.64 | 5.65 | 5.65 | 5.66 | 5.67 | 5.67 | 5.68 | 5.69 |
| 5.70 | 5.75 | 5.75 | 5.75 | 5.76 | 5.76 | 5.79 | 5.80 |
| 5.81 | 5.81 | 5.81 | 5.81 | 5.85 | 5.85 | 5.90 | 5.90 |
| 6.00 | 6.03 | 6.03 | 6.04 | 6.04 | 6.05 | 6.06 | 6.07 |
| 6.09 | 6.13 | 6.21 | 6.34 | 6.43 | 6.61 | 6.62 | 6.65 |
| 6.81 | | | | | | | |

(a) Make a histogram of pH with 14 classes, using class boundaries 4.2, 4.4, ..., 7.0. How many modes does your histogram show? More than one mode suggests that the data contain groups that have different distributions.

(b) Make a second histogram, also with 14 classes, using class boundaries 4.14, 4.34, ..., 6.94. The classes are those from (a) moved 0.06 to the left. How many modes does the new histogram show?

(c) Use your software's histogram function to make a histogram without specifying the number of classes or their boundaries. How does the software's default histogram compare with those in (a) and (b)?

**1.37** 🔺 CHALLENGE **Identify the histograms.** A survey of a large college class asked the following questions:

1. Are you female or male? (In the data, male = 0, female = 1.)

2. Are you right-handed or left-handed? (In the data, right = 0, left = 1.)

3. What is your height in inches?

4. How many minutes do you study on a typical weeknight?

Figure 1.17 shows histograms of the student responses, in scrambled order and without scale markings. Which histogram goes with each variable? Explain your reasoning.

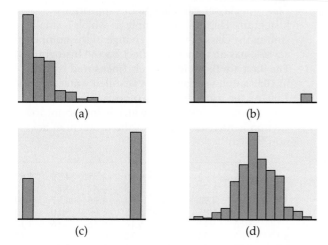

(a)

(b)

(c)

(d)

FIGURE 1.17 Match each histogram with its variable, for Exercise 1.37.

**1.38 Sketch a skewed distribution.** Sketch a histogram for a distribution that is skewed to the left. Suppose that you and your friends emptied your pockets of coins and recorded the year marked on each coin. The distribution of dates would be skewed to the left. Explain why.

**1.39 Oil wells.** How much oil the wells in a given field will ultimately produce is key information in deciding whether to drill more wells. Here are the estimated total amounts of oil recovered from 64 wells in the Devonian Richmond Dolomite area of the Michigan basin, in thousands of barrels:[22]

| | | | | | | | |
|---|---|---|---|---|---|---|---|
| 21.7 | 53.2 | 46.4 | 42.7 | 50.4 | 97.7 | 103.1 | 51.9 |
| 43.4 | 69.5 | 156.5 | 34.6 | 37.9 | 12.9 | 2.5 | 31.4 |
| 79.5 | 26.9 | 18.5 | 14.7 | 32.9 | 196.0 | 24.9 | 118.2 |
| 82.2 | 35.1 | 47.6 | 54.2 | 63.1 | 69.8 | 57.4 | 65.6 |
| 56.4 | 49.4 | 44.9 | 34.6 | 92.2 | 37.0 | 58.8 | 21.3 |
| 36.6 | 64.9 | 14.8 | 17.6 | 29.1 | 61.4 | 38.6 | 32.5 |
| 12.0 | 28.3 | 204.9 | 44.5 | 10.3 | 37.7 | 33.7 | 81.1 |
| 12.1 | 20.1 | 30.5 | 7.1 | 10.1 | 18.0 | 3.0 | 2.0 |

Graph the distribution and describe its main features.

**1.40 The density of the earth.** In 1798 the English scientist Henry Cavendish measured the density of the earth by careful work with a torsion balance. The variable recorded was the density of the earth as a multiple of the density of water. Here are Cavendish's 29 measurements:[23]

| | | | | | | | |
|---|---|---|---|---|---|---|---|
| 5.50 | 5.61 | 4.88 | 5.07 | 5.26 | 5.55 | 5.36 | 5.29 |
| 5.58 | 5.65 | 5.57 | 5.53 | 5.62 | 5.29 | 5.44 | 5.34 |
| 5.79 | 5.10 | 5.27 | 5.39 | 5.42 | 5.47 | 5.63 | 5.34 |
| 5.46 | 5.30 | 5.75 | 5.68 | 5.85 | | | |

Present these measurements graphically by either a stemplot or a histogram and explain the reason for your choice. Then briefly discuss the main features of the distribution. In particular, what is your estimate of the density of the earth based on these measurements?

**1.41 Time spent studying.** Do women study more than men? We asked the students in a large first-year college class how many minutes they studied on a typical weeknight. Here are the responses of random samples of 30 women and 30 men from the class:

| | Women | | | | | Men | | | |
|---|---|---|---|---|---|---|---|---|---|
| 180 | 120 | 180 | 360 | 240 | 90 | 120 | 30 | 90 | 200 |
| 120 | 180 | 120 | 240 | 170 | 90 | 45 | 30 | 120 | 75 |
| 150 | 120 | 180 | 180 | 150 | 150 | 120 | 60 | 240 | 300 |
| 200 | 150 | 180 | 150 | 180 | 240 | 60 | 120 | 60 | 30 |
| 120 | 60 | 120 | 180 | 180 | 30 | 230 | 120 | 95 | 150 |
| 90 | 240 | 180 | 115 | 120 | 0 | 200 | 120 | 120 | 180 |

(a) Examine the data. Why are you not surprised that most responses are multiples of 10 minutes? We eliminated one student who claimed to study 30,000 minutes per night. Are there any other responses you consider suspicious?

(b) Make a back-to-back stemplot of these data. Report the approximate midpoints of both groups. Does it appear that women study more than men (or at least claim that they do)?

**1.42 Guinea pigs.** Table 1.8 gives the survival times in days of 72 guinea pigs after they were injected with tubercle bacilli in a medical experiment.[24] Make a suitable graph and describe the shape, center, and spread of the distribution of survival times. Are there any outliers?

**1.43 Grades and self-concept.** Table 1.9 presents data on 78 seventh-grade students in a rural midwestern school.[25] The researcher was interested in the relationship between the students' "self-concept"

**TABLE 1.8**

Survival times (days) of guinea pigs in a medical experiment

| | | | | | | | | | |
|---|---|---|---|---|---|---|---|---|---|
| 43 | 45 | 53 | 56 | 56 | 57 | 58 | 66 | 67 | 73 |
| 74 | 79 | 80 | 80 | 81 | 81 | 81 | 82 | 83 | 83 |
| 84 | 88 | 89 | 91 | 91 | 92 | 92 | 97 | 99 | 99 |
| 100 | 100 | 101 | 102 | 102 | 102 | 103 | 104 | 107 | 108 |
| 109 | 113 | 114 | 118 | 121 | 123 | 126 | 128 | 137 | 138 |
| 139 | 144 | 145 | 147 | 156 | 162 | 174 | 178 | 179 | 184 |
| 191 | 198 | 211 | 214 | 243 | 249 | 329 | 380 | 403 | 511 |
| 522 | 598 | | | | | | | | |

and their academic performance. The data we give here include each student's grade point average (GPA), score on a standard IQ test, and gender, taken from school records. Gender is coded as F for female and M for male. The students are identified only by an observation number (OBS). The missing OBS numbers show that some students dropped out of the study. The final variable is each student's score on the Piers-Harris Children's Self-Concept Scale, a psychological test administered by the researcher.

(a) How many variables does this data set contain? Which are categorical variables and which are quantitative variables?

(b) Make a stemplot of the distribution of GPA, after rounding to the nearest tenth of a point.

**TABLE 1.9**

Educational data for 78 seventh-grade students

| OBS | GPA | IQ | Gender | Self-concept | OBS | GPA | IQ | Gender | Self-concept |
|---|---|---|---|---|---|---|---|---|---|
| 001 | 7.940 | 111 | M | 67 | 043 | 10.760 | 123 | M | 64 |
| 002 | 8.292 | 107 | M | 43 | 044 | 9.763 | 124 | M | 58 |
| 003 | 4.643 | 100 | M | 52 | 045 | 9.410 | 126 | M | 70 |
| 004 | 7.470 | 107 | M | 66 | 046 | 9.167 | 116 | M | 72 |
| 005 | 8.882 | 114 | F | 58 | 047 | 9.348 | 127 | M | 70 |
| 006 | 7.585 | 115 | M | 51 | 048 | 8.167 | 119 | M | 47 |
| 007 | 7.650 | 111 | M | 71 | 050 | 3.647 | 97 | M | 52 |
| 008 | 2.412 | 97 | M | 51 | 051 | 3.408 | 86 | F | 46 |
| 009 | 6.000 | 100 | F | 49 | 052 | 3.936 | 102 | M | 66 |
| 010 | 8.833 | 112 | M | 51 | 053 | 7.167 | 110 | M | 67 |
| 011 | 7.470 | 104 | F | 35 | 054 | 7.647 | 120 | M | 63 |
| 012 | 5.528 | 89 | F | 54 | 055 | 0.530 | 103 | M | 53 |
| 013 | 7.167 | 104 | M | 54 | 056 | 6.173 | 115 | M | 67 |
| 014 | 7.571 | 102 | F | 64 | 057 | 7.295 | 93 | M | 61 |
| 015 | 4.700 | 91 | F | 56 | 058 | 7.295 | 72 | F | 54 |
| 016 | 8.167 | 114 | F | 69 | 059 | 8.938 | 111 | F | 60 |
| 017 | 7.822 | 114 | F | 55 | 060 | 7.882 | 103 | F | 60 |
| 018 | 7.598 | 103 | F | 65 | 061 | 8.353 | 123 | M | 63 |
| 019 | 4.000 | 106 | M | 40 | 062 | 5.062 | 79 | M | 30 |
| 020 | 6.231 | 105 | F | 66 | 063 | 8.175 | 119 | M | 54 |
| 021 | 7.643 | 113 | M | 55 | 064 | 8.235 | 110 | M | 66 |
| 022 | 1.760 | 109 | M | 20 | 065 | 7.588 | 110 | M | 44 |
| 024 | 6.419 | 108 | F | 56 | 068 | 7.647 | 107 | M | 49 |
| 026 | 9.648 | 113 | M | 68 | 069 | 5.237 | 74 | F | 44 |
| 027 | 10.700 | 130 | F | 69 | 071 | 7.825 | 105 | M | 67 |
| 028 | 10.580 | 128 | M | 70 | 072 | 7.333 | 112 | F | 64 |
| 029 | 9.429 | 128 | M | 80 | 074 | 9.167 | 105 | M | 73 |
| 030 | 8.000 | 118 | M | 53 | 076 | 7.996 | 110 | M | 59 |
| 031 | 9.585 | 113 | M | 65 | 077 | 8.714 | 107 | F | 37 |
| 032 | 9.571 | 120 | F | 67 | 078 | 7.833 | 103 | F | 63 |
| 033 | 8.998 | 132 | F | 62 | 079 | 4.885 | 77 | M | 36 |
| 034 | 8.333 | 111 | F | 39 | 080 | 7.998 | 98 | F | 64 |
| 035 | 8.175 | 124 | M | 71 | 083 | 3.820 | 90 | M | 42 |
| 036 | 8.000 | 127 | M | 59 | 084 | 5.936 | 96 | F | 28 |
| 037 | 9.333 | 128 | F | 60 | 085 | 9.000 | 112 | F | 60 |
| 038 | 9.500 | 136 | M | 64 | 086 | 9.500 | 112 | F | 70 |
| 039 | 9.167 | 106 | M | 71 | 087 | 6.057 | 114 | M | 51 |
| 040 | 10.140 | 118 | F | 72 | 088 | 6.057 | 93 | F | 21 |
| 041 | 9.999 | 119 | F | 54 | 089 | 6.938 | 106 | M | 56 |

(c) Describe the shape, center, and spread of the GPA distribution. Identify any suspected outliers from the overall pattern.

(d) Make a back-to-back stemplot of the rounded GPAs for female and male students. Write a brief comparison of the two distributions.

**1.44  Describe the IQ scores.** Make a graph of the distribution of IQ scores for the seventh-grade students in Table 1.9. Describe the shape, center, and spread of the distribution, as well as any outliers. IQ scores are usually said to be centered at 100. Is the midpoint for these students close to 100, clearly above, or clearly below?

**1.45  Describe the self-concept scores.** Based on a suitable graph, briefly describe the distribution of self-concept scores for the students in Table 1.9. Be sure to identify any suspected outliers.

**1.46  The Boston Marathon.** Women were allowed to enter the Boston Marathon in 1972. The following table gives the times (in minutes, rounded to the nearest minute) for the winning women from 1972 to 2006.

| Year | Time | Year | Time | Year | Time | Year | Time |
|------|------|------|------|------|------|------|------|
| 1972 | 190 | 1981 | 147 | 1990 | 145 | 1999 | 143 |
| 1973 | 186 | 1982 | 150 | 1991 | 144 | 2000 | 146 |
| 1974 | 167 | 1983 | 143 | 1992 | 144 | 2001 | 144 |
| 1975 | 162 | 1984 | 149 | 1993 | 145 | 2002 | 141 |
| 1976 | 167 | 1985 | 154 | 1994 | 142 | 2003 | 145 |
| 1977 | 168 | 1986 | 145 | 1995 | 145 | 2004 | 144 |
| 1978 | 165 | 1987 | 146 | 1996 | 147 | 2005 | 145 |
| 1979 | 155 | 1988 | 145 | 1997 | 146 | 2006 | 143 |
| 1980 | 154 | 1989 | 144 | 1998 | 143 |  |  |

Make a graph that shows change over time. What overall pattern do you see? Have times stopped improving in recent years? If so, when did improvement end?

# 1.2  Describing Distributions with Numbers

Interested in a sporty car? Worried that it may use too much gas? The Environmental Protection Agency lists most such vehicles in its "two-seater" or "minicompact" categories. Table 1.10 gives the city and highway gas mileage for cars in these groups.[26] (The mileages are for the basic engine and transmission combination for each car.) We want to compare two-seaters with minicompacts and city mileage with highway mileage. We can begin with graphs, but numerical summaries make the comparisons more specific.

A brief description of a distribution should include its *shape* and numbers describing its *center* and *spread.* We describe the shape of a distribution based on inspection of a histogram or a stemplot. Now we will learn specific ways to use numbers to measure the center and spread of a distribution. We can calculate these numerical measures for any quantitative variable. But to interpret measures of center and spread, and to choose among the several measures we will learn, you must think about the shape of the distribution and the meaning of the data. The numbers, like graphs, are aids to understanding, not "the answer" in themselves.

## Measuring center: the mean

Numerical description of a distribution begins with a measure of its center or average. The two common measures of center are the *mean* and the *median.* The mean is the "average value" and the median is the "middle value." These are two different ideas for "center," and the two measures behave differently. We need precise recipes for the mean and the median.

## TABLE 1.10

Fuel economy (miles per gallon) for 2004 model vehicles

| Two-Seater Cars | | | Minicompact Cars | | |
|---|---|---|---|---|---|
| Model | City | Highway | Model | City | Highway |
| Acura NSX | 17 | 24 | Aston Martin Vanquish | 12 | 19 |
| Audi TT Roadster | 20 | 28 | Audi TT Coupe | 21 | 29 |
| BMW Z4 Roadster | 20 | 28 | BMW 325CI | 19 | 27 |
| Cadillac XLR | 17 | 25 | BMW 330CI | 19 | 28 |
| Chevrolet Corvette | 18 | 25 | BMW M3 | 16 | 23 |
| Dodge Viper | 12 | 20 | Jaguar XK8 | 18 | 26 |
| Ferrari 360 Modena | 11 | 16 | Jaguar XKR | 16 | 23 |
| Ferrari Maranello | 10 | 16 | Lexus SC 430 | 18 | 23 |
| Ford Thunderbird | 17 | 23 | Mini Cooper | 25 | 32 |
| Honda Insight | 60 | 66 | Mitsubishi Eclipse | 23 | 31 |
| Lamborghini Gallardo | 9 | 15 | Mitsubishi Spyder | 20 | 29 |
| Lamborghini Murcielago | 9 | 13 | Porsche Cabriolet | 18 | 26 |
| Lotus Esprit | 15 | 22 | Porsche Turbo 911 | 14 | 22 |
| Maserati Spyder | 12 | 17 | | | |
| Mazda Miata | 22 | 28 | | | |
| Mercedes-Benz SL500 | 16 | 23 | | | |
| Mercedes-Benz SL600 | 13 | 19 | | | |
| Nissan 350Z | 20 | 26 | | | |
| Porsche Boxster | 20 | 29 | | | |
| Porsche Carrera 911 | 15 | 23 | | | |
| Toyota MR2 | 26 | 32 | | | |

### THE MEAN $\bar{x}$

To find the **mean $\bar{x}$** of a set of observations, add their values and divide by the number of observations. If the $n$ observations are $x_1, x_2, \ldots, x_n$, their mean is

$$\bar{x} = \frac{x_1 + x_2 + \cdots + x_n}{n}$$

or, in more compact notation,

$$\bar{x} = \frac{1}{n} \sum x_i$$

The $\sum$ (capital Greek sigma) in the formula for the mean is short for "add them all up." The bar over the $x$ indicates the mean of all the $x$-values. Pronounce the mean $\bar{x}$ as "x-bar." This notation is so common that writers who are discussing data use $\bar{x}, \bar{y}$, etc. without additional explanation. The subscripts on the observations $x_i$ are just a way of keeping the $n$ observations separate. They do not necessarily indicate order or any other special facts about the data.

**EXAMPLE**

**1.14 Highway mileage for two-seaters.** The mean highway mileage for the 21 two-seaters in Table 1.10 is

$$\bar{x} = \frac{x_1 + x_2 + \cdots + x_n}{n}$$

$$= \frac{24 + 28 + 28 + \cdots + 32}{21}$$

$$= \frac{518}{21} = 24.7 \text{ miles per gallon}$$

In practice, you can key the data into your calculator and hit the $\bar{x}$ key.

## USE YOUR KNOWLEDGE

**1.47 Find the mean.** Here are the scores on the first exam in an introductory statistics course for 10 students:

80   73   92   85   75   98   93   55   80   90

Find the mean first-exam score for these students.

The data for Example 1.14 contain an outlier: the Honda Insight is a hybrid gas-electric car that doesn't belong in the same category as the 20 gasoline-powered two-seater cars. If we exclude the Insight, the mean highway mileage drops to 22.6 mpg. The single outlier adds more than 2 mpg to the mean highway mileage. This illustrates an important weakness of the mean as a measure of center: *the mean is sensitive to the influence of a few extreme observations*. These may be outliers, but a skewed distribution that has no outliers will also pull the mean toward its long tail. Because the mean cannot resist the influence of extreme observations, we say that it is not a **resistant measure** of center. A measure that is resistant does more than limit the influence of outliers. Its value does not respond strongly to changes in a few observations, no matter how large those changes may be. The mean fails this requirement because we can make the mean as large as we wish by making a large enough increase in just one observation.

resistant measure

## Measuring center: the median

We used the midpoint of a distribution as an informal measure of center in the previous section. The *median* is the formal version of the midpoint, with a specific rule for calculation.

### THE MEDIAN *M*

The **median *M*** is the midpoint of a distribution. Half the observations are smaller than the median and the other half are larger than the median. Here is a rule for finding the median:

**1.** Arrange all observations in order of size, from smallest to largest.

**2.** If the number of observations $n$ is odd, the median $M$ is the center observation in the ordered list. Find the location of the median by counting $(n + 1)/2$ observations up from the bottom of the list.

**3.** If the number of observations $n$ is even, the median $M$ is the mean of the two center observations in the ordered list. The location of the median is again $(n + 1)/2$ from the bottom of the list.

Note that the formula $(n + 1)/2$ does *not* give the median, just the location of the median in the ordered list. Medians require little arithmetic, so they are easy to find by hand for small sets of data. Arranging even a moderate number of observations in order is tedious, however, so that finding the median by hand for larger sets of data is unpleasant. Even simple calculators have an $\bar{x}$ button, but you will need computer software or a graphing calculator to automate finding the median.

**EXAMPLE**

**1.15 Find the median.** To find the median highway mileage for 2004 model two-seater cars, arrange the data in increasing order:

13 15 16 16 17 19 20 22 23 23 **23** 24 25 25 26 28 28 28 29 32 66

Be sure to list *all* observations, even if they repeat the same value. The median is the bold 23, the 11th observation in the ordered list. You can find the median by eye—there are 10 observations to the left and 10 to the right. Or you can use the recipe $(n + 1)/2 = 22/2 = 11$ to locate the median in the list.

What happens if we drop the Honda Insight? The remaining 20 cars have highway mileages

13 15 16 16 17 19 20 22 23 **23 23** 24 25 25 26 28 28 28 29 32

Because the number of observations $n = 20$ is even, there is no center observation. There is a center pair—the bold pair of 23s have 9 observations to their left and 9 to their right. The median $M$ is the mean of the center pair, which is 23. The recipe $(n + 1)/2 = 21/2 = 10.5$ for the position of the median in the list says that the median is at location "ten and one-half," that is, halfway between the 10th and 11th observations.

APPLET

You see that the median is more resistant than the mean. Removing the Honda Insight did not change the median at all. Even if we mistakenly enter the Insight's mileage as 660 rather than 66, the median remains 23. The very high value is simply one observation to the right of center. The *Mean and Median* applet on the text CD and Web site is an excellent way to compare the resistance of $M$ and $\bar{x}$. See Exercises 1.75 to 1.77 for use of this applet.

## USE YOUR KNOWLEDGE

**1.48 Find the median.** Here are the scores on the first exam in an introductory statistics course for 10 students:

80   73   92   85   75   98   93   55   80   90

Find the median first-exam score for these students.

## Mean versus median

The median and mean are the most common measures of the center of a distribution. The mean and median of a symmetric distribution are close together. If the distribution is exactly symmetric, the mean and median are exactly the same. In a skewed distribution, the mean is farther out in the long tail than is the median. The endowment for a college or university is money set aside and invested. The income from the endowment is usually used to support various programs. The distribution of the sizes of the endowments of colleges and universities is strongly skewed to the right. Most institutions have modest endowments, but a few are very wealthy. The median endowment of colleges and universities in a recent year was $70 million—but the mean endowment was over $320 million. The few wealthy institutions pulled the mean up but did not affect the median. *Don't confuse the "average" value of a variable (the mean) with its "typical" value, which we might describe by the median.*

We can now give a better answer to the question of how to deal with outliers in data. First, look at the data to identify outliers and investigate their causes. You can then correct outliers if they are wrongly recorded, delete them for good reason, or otherwise give them individual attention. The three outliers in Figure 1.9 (page 17) can all be dropped from the data once we discover why they appear. If you have no clear reason to drop outliers, you may want to use resistant methods, so that outliers have little influence over your conclusions. The choice is often a matter for judgment. The government's fuel economy guide lists the Honda Insight with the other two-seaters in Table 1.10. We might choose to report median rather than mean gas mileage for all two-seaters to avoid giving too much influence to one car model. In fact, we think that the Insight doesn't belong, so we will omit it from further analysis of these data.

## Measuring spread: the quartiles

A measure of center alone can be misleading. Two nations with the same median family income are very different if one has extremes of wealth and poverty and the other has little variation among families. A drug with the correct mean concentration of active ingredient is dangerous if some batches are much too high and others much too low. We are interested in the *spread* or *variability* of incomes and drug potencies as well as their centers. **The simplest useful numerical description of a distribution consists of both a measure of center and a measure of spread.**

We can describe the spread or variability of a distribution by giving several percentiles. The median divides the data in two; half of the observations are above the median and half are below the median. We could call the median the
quartile
50th percentile. The upper **quartile** is the median of the upper half of the data. Similarly, the lower quartile is the median of the lower half of the data. With the median, the quartiles divide the data into four equal parts; 25% of the data are in each part.

percentile
We can do a similar calculation for any percent. The **$p$th percentile** of a distribution is the value that has $p$ percent of the observations fall at or below it. To calculate a percentile, arrange the observations in increasing order and count up the required percent from the bottom of the list. Our definition of percentiles is a bit inexact because there is not always a value with exactly $p$

percent of the data at or below it. We will be content to take the nearest observation for most percentiles, but the quartiles are important enough to require an exact rule.

---

**THE QUARTILES $Q_1$ AND $Q_3$**

To calculate the quartiles:

**1.** Arrange the observations in increasing order and locate the median $M$ in the ordered list of observations.

**2.** The **first quartile $Q_1$** is the median of the observations whose position in the ordered list is to the left of the location of the overall median.

**3.** The **third quartile $Q_3$** is the median of the observations whose position in the ordered list is to the right of the location of the overall median.

---

**EXAMPLE**

**1.16 Find the median and the quartiles.** The highway mileages of the 20 gasoline-powered two-seater cars, arranged in increasing order, are

13 15 16 16 17 19 20 22 23 23 | 23 24 25 25 26 28 28 28 29 32

The median is midway between the center pair of observations. We have marked its position in the list by |. The first quartile is the median of the 10 observations to the left of the position of the median. Check that its value is $Q_1 = 18$. Similarly, the third quartile is the median of the 10 observations to the right of the |. Check that $Q_3 = 27$.

When there is an odd number of observations, the median is the unique center observation, and the rule for finding the quartiles excludes this center value. The highway mileages of the 13 minicompact cars in Table 1.10 are (in order)

19 22 23 23 23 26 **26** 27 28 29 29 31 32

The median is the bold 26. The first quartile is the median of the 6 observations falling to the left of this point in the list, $Q_1 = 23$. Similarly, $Q_3 = 29$.

We find other percentiles more informally if we are working without software. For example, we take the 90th percentile of the 13 minicompact mileages to be the 12th in the ordered list, because $0.90 \times 13 = 11.7$, which we round to 12. The 90th percentile is therefore 31 mpg.

**USE YOUR KNOWLEDGE**

**1.49 Find the quartiles.** Here are the scores on the first-exam in an introductory statistics course for 10 students:

80    73    92    85    75    98    93    55    80    90

Find the quartiles for these first-exam scores.

**EXAMPLE**

**1.17 Results from software.** Statistical software often provides several numerical measures in response to a single command. Figure 1.18 displays such output from the CrunchIt! and Minitab software for the highway mileages of two-seater cars (without the Honda Insight).

Both tell us that there are 20 observations and give the mean, median, quartiles, and smallest and largest data values. Both also give other measures, some of which we will meet soon. CrunchIt! is basic online software that offers no choice of output. Minitab allows you to choose the descriptive measures you want from a long list.

The quartiles from CrunchIt! agree with our values from Example 1.16. But Minitab's quartiles are a bit different. For example, our rule for hand calculation gives first quartile $Q_1 = 18$. Minitab's value is $Q_1 = 17.5$. *There are several rules for calculating quartiles, which often give slightly different values. The differences are always small. For describing data, just report the values that your software gives.*

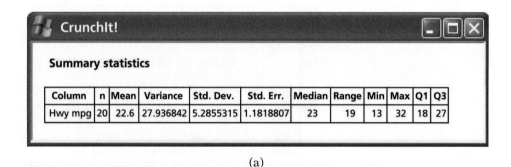

(a)

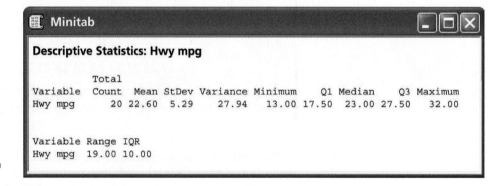

(b)

**FIGURE 1.18** Numerical descriptions of the highway gas mileage of two-seater cars from software, for Example 1.17. (a) CrunchIt! (b) Minitab.

## The five-number summary and boxplots

In Section 1.1, we used the smallest and largest observations to indicate the spread of a distribution. These single observations tell us little about the distribution as a whole, but they give information about the tails of the distribution that is missing if we know only $Q_1$, $M$, and $Q_3$. To get a quick summary of both center and spread, combine all five numbers.

---

THE FIVE-NUMBER SUMMARY

The **five-number summary** of a set of observations consists of the smallest observation, the first quartile, the median, the third quartile, and the largest observation, written in order from smallest to largest. In symbols, the five-number summary is

$$\text{Minimum} \quad Q_1 \quad M \quad Q_3 \quad \text{Maximum}$$

---

These five numbers offer a reasonably complete description of center and spread. The five-number summaries for highway gas mileages are

$$13 \ \ 18 \ \ 23 \ \ 27 \ \ 32$$

for two-seaters and

$$19 \ \ 23 \ \ 26 \ \ 29 \ \ 32$$

for minicompacts. The median describes the center of the distribution; the quartiles show the spread of the center half of the data; the minimum and maximum show the full spread of the data.

## USE YOUR KNOWLEDGE

**1.50 Find the five-number summary.** Here are the scores on the first exam in an introductory statistics course for 10 students:

$$80 \ \ 73 \ \ 92 \ \ 85 \ \ 75 \ \ 98 \ \ 93 \ \ 55 \ \ 80 \ \ 90$$

Find the five-number summary for these first-exam scores.

The five-number summary leads to another visual representation of a distribution, the *boxplot*. Figure 1.19 shows boxplots for both city and highway gas mileages for our two groups of cars.

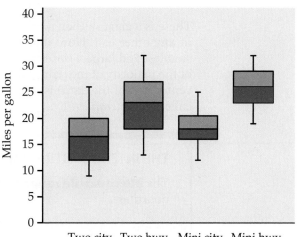

**FIGURE 1.19** Boxplots of the highway and city gas mileages for cars classified as two-seaters and as minicompacts by the Environmental Protection Agency.

---

**BOXPLOT**

A **boxplot** is a graph of the five-number summary.

- A central box spans the quartiles $Q_1$ and $Q_3$.

- A line in the box marks the median $M$.

- Lines extend from the box out to the smallest and largest observations.

---

When you look at a boxplot, first locate the median, which marks the center of the distribution. Then look at the spread. The quartiles show the spread of the middle half of the data, and the extremes (the smallest and largest observations) show the spread of the entire data set.

## USE YOUR KNOWLEDGE

**1.51  Make a boxplot.** Here are the scores on the first exam in an introductory statistics course for 10 students:

$$80 \quad 73 \quad 92 \quad 85 \quad 75 \quad 98 \quad 93 \quad 55 \quad 80 \quad 90$$

Make a boxplot for these first-exam scores.

Boxplots are particularly effective for comparing distributions as we did in Figure 1.19. We see at once that city mileages are lower than highway mileages. The minicompact cars have slightly higher median gas mileages than the two-seaters, and their mileages are markedly less variable. In particular, the low gas mileages of the Ferraris and Lamborghinis in the two-seater group pull the group minimum down.

## The 1.5 × *IQR* rule for suspected outliers

Look again at the 80 service center call lengths in Table 1.1 (page 8). Figure 1.6 (page 12) is a stemplot of their distribution. You can check that the five-number summary is

$$1 \quad 54.5 \quad 103.5 \quad 200 \quad 2631$$

There is a clear outlier, a call lasting 2631 seconds, more than twice the length of any other call. How shall we describe the spread of this distribution? The smallest and largest observations are extremes that do not describe the spread of the majority of the data. The distance between the quartiles (the range of the center half of the data) is a more resistant measure of spread. This distance is called the *interquartile range*.

---

**THE INTERQUARTILE RANGE *IQR***

The **interquartile range *IQR*** is the distance between the first and third quartiles,

$$IQR = Q_3 - Q_1$$

---

For our data on service call lengths, $IQR = 200 - 54.5 = 145.5$. The quartiles and the $IQR$ are not affected by changes in either tail of the distribution. They are therefore resistant, because changes in a few data points have no further effect once these points move outside the quartiles. However, *no single numerical measure of spread, such as IQR, is very useful for describing skewed distributions.* The two sides of a skewed distribution have different spreads, so one number can't summarize them. We can often detect skewness from the five-number summary by comparing how far the first quartile and the minimum are from the median (left tail) with how far the third quartile and the maximum are from the median (right tail). The interquartile range is mainly used as the basis for a rule of thumb for identifying suspected outliers.

CAUTION

---

**THE 1.5 × *IQR* RULE FOR OUTLIERS**

Call an observation a suspected outlier if it falls more than $1.5 \times IQR$ above the third quartile or below the first quartile.

---

EXAMPLE

**1.18 Outliers for call length data.** For the call length data in Table 1.1,

$$1.5 \times IQR = 1.5 \times 145.5 = 218.25$$

Any values below $54.5 - 218.25 = -163.75$ or above $200 + 218.25 = 418.25$ are flagged as possible outliers. There are no low outliers, but the 8 longest calls are flagged as possible high outliers. Their lengths are

$$438 \quad 465 \quad 479 \quad 700 \quad 700 \quad 951 \quad 1148 \quad 2631$$

modified boxplot

Statistical software often uses the $1.5 \times IQR$ rule. For example, the stemplot in Figure 1.6 lists these 8 observations separately. Boxplots drawn by software are often **modified boxplots** that plot suspected outliers individually. Figure 1.20 is a modified boxplot of the call length data. The lines extend out from the central box only to the smallest and largest observations that are not flagged by the $1.5 \times IQR$ rule. The 8 largest call lengths are plotted as individual points, though 2 of them are identical and so do not appear separately.

Call length (seconds)

**FIGURE 1.20** Modified boxplot of the call lengths in Table 1.1, for Example 1.18.

The distribution of call lengths is very strongly skewed. We may well decide that only the longest call is truly an outlier in the sense of deviating from the overall pattern of the distribution. The other 7 calls are just part of the long right tail. The $1.5 \times IQR$ rule does not remove the need to look at the distribution and use judgment. It is useful mainly to call our attention to unusual observations.

## USE YOUR KNOWLEDGE

**1.52  Find the *IQR*.** Here are the scores on the first exam in an introductory statistics course for 10 students:

$$80 \quad 73 \quad 92 \quad 85 \quad 75 \quad 98 \quad 93 \quad 55 \quad 80 \quad 90$$

Find the interquartile range and use the $1.5 \times IQR$ rule to check for outliers. How low would the lowest score need to be for it to be an outlier according to this rule?

The stemplot in Figure 1.6 and the modified boxplot in Figure 1.20 tell us much more about the distribution of call lengths than the five-number summary or other numerical measures. The routine methods of statistics compute numerical measures and draw conclusions based on their values. These methods are very useful, and we will study them carefully in later chapters. But they cannot be applied blindly, by feeding data to a computer program, because *statistical measures and methods based on them are generally meaningful only for distributions of sufficiently regular shape*. This principle will become clearer as we progress, but it is good to be aware at the beginning that quickly resorting to fancy calculations is the mark of a statistical amateur. Look, think, and choose your calculations selectively.

## Measuring spread: the standard deviation

The five-number summary is not the most common numerical description of a distribution. That distinction belongs to the combination of the mean to measure center and the *standard deviation* to measure spread. The standard deviation measures spread by looking at how far the observations are from their mean.

### THE STANDARD DEVIATION s

The **variance $s^2$** of a set of observations is the average of the squares of the deviations of the observations from their mean. In symbols, the variance of $n$ observations $x_1, x_2, \ldots, x_n$ is

$$s^2 = \frac{(x_1 - \overline{x})^2 + (x_2 - \overline{x})^2 + \cdots + (x_n - \overline{x})^2}{n - 1}$$

or, in more compact notation,

$$s^2 = \frac{1}{n - 1} \sum (x_i - \overline{x})^2$$

The **standard deviation** $s$ is the square root of the variance $s^2$:

$$s = \sqrt{\frac{1}{n-1} \sum (x_i - \bar{x})^2}$$

The idea behind the variance and the standard deviation as measures of spread is as follows: The deviations $x_i - \bar{x}$ display the spread of the values $x_i$ about their mean $\bar{x}$. Some of these deviations will be positive and some negative because some of the observations fall on each side of the mean. In fact, *the sum of the deviations of the observations from their mean will always be zero.* Squaring the deviations makes them all positive, so that observations far from the mean in either direction have large positive squared deviations. The variance is the average squared deviation. Therefore, $s^2$ and $s$ will be large if the observations are widely spread about their mean, and small if the observations are all close to the mean.

**EXAMPLE**

**1.19 Metabolic rate.** A person's metabolic rate is the rate at which the body consumes energy. Metabolic rate is important in studies of weight gain, dieting, and exercise. Here are the metabolic rates of 7 men who took part in a study of dieting. (The units are calories per 24 hours. These are the same calories used to describe the energy content of foods.)

$$1792 \quad 1666 \quad 1362 \quad 1614 \quad 1460 \quad 1867 \quad 1439$$

Enter these data into your calculator or software and verify that

$$\bar{x} = 1600 \text{ calories} \qquad s = 189.24 \text{ calories}$$

Figure 1.21 plots these data as dots on the calorie scale, with their mean marked by an asterisk ($*$). The arrows mark two of the deviations from the mean. If you were calculating $s$ by hand, you would find the first deviation as

$$x_1 - \bar{x} = 1792 - 1600 = 192$$

Exercise 1.70 asks you to calculate the seven deviations, square them, and find $s^2$ and $s$ directly from the deviations. Working one or two short examples by hand helps you understand how the standard deviation is obtained. In practice you will use either software or a calculator that will find $s$ from keyed-in data. The two software outputs in Figure 1.18 both give the variance and standard deviation for the highway mileage data.

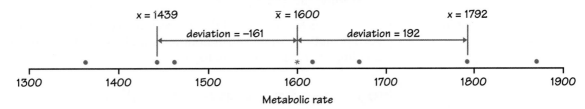

**FIGURE 1.21** Metabolic rates for seven men, with the mean ($*$) and the deviations of two observations from the mean, for Example 1.19.

## USE YOUR KNOWLEDGE

**1.53 Find the variance and the standard deviation.** Here are the scores on the first exam in an introductory statistics course for 10 students:

$$80 \quad 73 \quad 92 \quad 85 \quad 75 \quad 98 \quad 93 \quad 55 \quad 80 \quad 90$$

Find the variance and the standard deviation for these first-exam scores.

The idea of the variance is straightforward: it is the average of the squares of the deviations of the observations from their mean. The details we have just presented, however, raise some questions.

*Why do we square the deviations?*

- First, the sum of the squared deviations of any set of observations from their mean is the smallest that the sum of squared deviations from any number can possibly be. This is not true of the unsquared distances. So squared deviations point to the mean as center in a way that distances do not.

- Second, the standard deviation turns out to be the natural measure of spread for a particularly important class of symmetric unimodal distributions, the *Normal distributions*. We will meet the Normal distributions in the next section. We commented earlier that the usefulness of many statistical procedures is tied to distributions of particular shapes. This is distinctly true of the standard deviation.

*Why do we emphasize the standard deviation rather than the variance?*

- One reason why is that $s$, not $s^2$, is the natural measure of spread for Normal distributions.

- There is also a more general reason to prefer $s$ to $s^2$. Because the variance involves squaring the deviations, it does not have the same unit of measurement as the original observations. The variance of the metabolic rates, for example, is measured in squared calories. Taking the square root remedies this. The standard deviation $s$ measures spread about the mean in the original scale.

*Why do we average by dividing by $n - 1$ rather than $n$ in calculating the variance?*

- Because the sum of the deviations is always zero, the last deviation can be found once we know the other $n - 1$. So we are not averaging $n$ unrelated numbers. Only $n - 1$ of the squared deviations can vary freely, and we average by dividing the total by $n - 1$.

degrees of freedom
- The number $n - 1$ is called the **degrees of freedom** of the variance or standard deviation. Many calculators offer a choice between dividing by $n$ and dividing by $n - 1$, so be sure to use $n - 1$.

## Properties of the standard deviation

Here are the basic properties of the standard deviation $s$ as a measure of spread.

PROPERTIES OF THE STANDARD DEVIATION

- $s$ measures spread about the mean and should be used only when the mean is chosen as the measure of center.

- $s = 0$ only when there is *no spread*. This happens only when all observations have the same value. Otherwise, $s > 0$. As the observations become more spread out about their mean, $s$ gets larger.

- $s$, like the mean $\bar{x}$, is not resistant. A few outliers can make $s$ very large.

**USE YOUR KNOWLEDGE**

**1.54 A standard deviation of zero.** Construct a data set with 5 cases that has a variable with $s = 0$.

*The use of squared deviations renders $s$ even more sensitive than $\bar{x}$ to a few extreme observations.* For example, dropping the Honda Insight from our list of two-seater cars reduces the mean highway mileage from 24.7 mpg to 22.6 mpg. It cuts the standard deviation more than half, from 10.8 mpg with the Insight to 5.3 mpg without it. Distributions with outliers and strongly skewed distributions have large standard deviations. The number $s$ does not give much helpful information about such distributions.

## Choosing measures of center and spread

How do we choose between the five-number summary and $\bar{x}$ and $s$ to describe the center and spread of a distribution? Because the two sides of a strongly skewed distribution have different spreads, no single number such as $s$ describes the spread well. The five-number summary, with its two quartiles and two extremes, does a better job.

CHOOSING A SUMMARY

The five-number summary is usually better than the mean and standard deviation for describing a skewed distribution or a distribution with strong outliers. Use $\bar{x}$ and $s$ only for reasonably symmetric distributions that are free of outliers.

**EXAMPLE**

**1.20 Standard deviation as a measure of risk.** A central principle in the study of investments is that taking bigger risks is rewarded by higher returns, at least on the average over long periods of time. It is usual in finance to measure risk by the standard deviation of returns, on the grounds that investments whose returns vary a lot from year to year are less predictable and therefore more risky than those whose returns don't vary much. Compare, for example, the approximate mean and standard deviation of the annual percent

each group. Give a complete comparison of the four distributions, using both graphs and numerical summaries. How would you describe the effect of running on heart rate? Is the effect different for men and women?

*The WORKERS data set, described in the Data Appendix, contains the sex, level of education, and income of 71,076 people between the ages of 25 and 64 who were employed full-time in 2001.*

*The boxplots in Figure 1.23 compare the distributions of income for people with five levels of education. This figure is a variation on the boxplot idea: because large data sets often contain very extreme observations, the lines extend from the central box only to the 5th and 95th percentiles. Exercises 1.82 to 1.84 concern these data.*

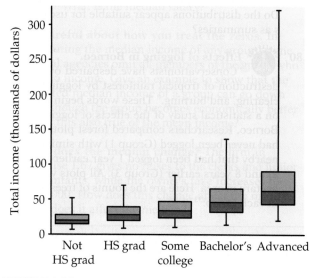

**FIGURE 1.23** Boxplots comparing the distributions of income for employed people aged 25 to 64 years with five different levels of education. The lines extend from the quartiles to the 5th and 95th percentiles.

**1.82** **Income for people with bachelor's degrees.** The data include 14,959 people whose highest level of education is a bachelor's degree.

(a) What is the position of the median in the ordered list of incomes (1 to 14,959)? From the boxplot, about what is the median income of people with a bachelor's degree?

(b) What is the position of the first and third quartiles in the ordered list of incomes for these people? About what are the numerical values of $Q_1$ and $Q_3$?

(c) You answered (a) and (b) from a boxplot that omits the lowest 5% and the highest 5% of incomes.

Explain why leaving out these values has only a very small effect on the median and quartiles.

**1.83** **Find the 5th and 95th percentiles.** About what are the positions of the 5th and 95th percentiles in the ordered list of incomes of the 14,959 people with a bachelor's degree? Incomes outside this range do not appear in the boxplot. About what are the numerical values of the 5th and 95th percentiles of income? (For comparison, the largest income among all 14,959 people was $481,720. That one person made this much tells us less about the group than does the 95th percentile.)

**1.84** **How does income change with education?** Write a brief description of how the distribution of income changes with the highest level of education reached. Be sure to discuss center, spread, and skewness. Give some specifics read from the graph to back up your statements.

**1.85** **Shakespeare's plays.** Look at the histogram of lengths of words in Shakespeare's plays, Figure 1.15 (page 25). The heights of the bars tell us what percent of words have each length. What is the median length of words used by Shakespeare? Similarly, what are the quartiles? Give the five-number summary for Shakespeare's word lengths.

**1.86** **Create a data set.** Create a set of 5 positive numbers (repeats allowed) that have median 10 and mean 7. What thought process did you use to create your numbers?

**1.87** **Create another data set.** Give an example of a small set of data for which the mean is larger than the third quartile.

**1.88** **Deviations from the mean sum to zero.** Use the definition of the mean $\bar{x}$ to show that the sum of the deviations $x_i - \bar{x}$ of the observations from their mean is always zero. This is one reason why the variance and standard deviation use squared deviations.

**1.89** **A standard deviation contest.** This is a standard deviation contest. You must choose four numbers from the whole numbers 0 to 20, with repeats allowed.

(a) Choose four numbers that have the smallest possible standard deviation.

(b) Choose four numbers that have the largest possible standard deviation.

(c) Is more than one choice possible in either (a) or (b)? Explain.

**1.90 Does your software give incorrect answers?** This exercise requires a calculator with a standard deviation button or statistical software on a computer. The observations

$$20{,}001 \quad 20{,}002 \quad 20{,}003$$

have mean $\bar{x} = 20{,}002$ and standard deviation $s = 1$. Adding a 0 in the center of each number, the next set becomes

$$200{,}001 \quad 200{,}002 \quad 200{,}003$$

The standard deviation remains $s = 1$ as more 0s are added. Use your calculator or computer to calculate the standard deviation of these numbers, adding extra 0s until you get an incorrect answer. How soon did you go wrong? This demonstrates that calculators and computers cannot handle an arbitrary number of digits correctly.

**1.91 Guinea pigs.** Table 1.8 (page 29) gives the survival times of 72 guinea pigs in a medical study. Survival times—whether of cancer patients after treatment or of car batteries in everyday use—are almost always right-skewed. Make a graph to verify that this is true of these survival times. Then give a numerical summary that is appropriate for such data. Explain in simple language, to someone who knows no statistics, what your summary tells us about the guinea pigs.

**1.92 Weight gain.** A study of diet and weight gain deliberately overfed 16 volunteers for eight weeks. The mean increase in fat was $\bar{x} = 2.39$ kilograms and the standard deviation was $s = 1.14$ kilograms. What are $\bar{x}$ and $s$ in pounds? (A kilogram is 2.2 pounds.)

**1.93 Compare three varieties of flowers.** Exercise 1.78 reports data on the lengths in millimeters of flowers of three varieties of *Heliconia*. In Exercise 1.79 you found the mean and standard deviation for each variety. Starting from the $\bar{x}$- and $s$-values in millimeters, find the means and standard deviations in inches. (A millimeter is 1/1000 of a meter. A meter is 39.37 inches.)

**1.94** ⚠ **The density of the earth.** Henry Cavendish (see Exercise 1.40, page 28) used $\bar{x}$ to summarize his 29 measurements of the density of the earth.

(a) Find $\bar{x}$ and $s$ for his data.

(b) Cavendish recorded the density of the earth as a multiple of the density of water. The density of water is almost exactly 1 gram per cubic centimeter, so his measurements have these units. In American units, the density of water is 62.43 pounds per cubic foot. This is the weight of a cube of water measuring 1 foot (that is, 30.48 cm) on each side. Express Cavendish's first result for the earth (5.50 g/cm$^3$) in pounds per cubic foot. Then find $\bar{x}$ and $s$ in pounds per cubic foot.

**1.95 Guinea pigs.** Find the **quintiles** (the 20th, 40th, 60th, and 80th percentiles) of the guinea pig survival times in Table 1.8 (page 29). For quite large sets of data, the quintiles or the **deciles** (10th, 20th, 30th, etc. percentiles) give a more detailed summary than the quartiles.

**1.96** ⚠ **Changing units from inches to centimeters.** Changing the unit of length from inches to centimeters multiplies each length by 2.54 because there are 2.54 centimeters in an inch. This change of units multiplies our usual measures of spread by 2.54. This is true of *IQR* and the standard deviation. What happens to the variance when we change units in this way?

**1.97 A different type of mean.** The **trimmed mean** is a measure of center that is more resistant than the mean but uses more of the available information than the median. To compute the 10% trimmed mean, discard the highest 10% and the lowest 10% of the observations and compute the mean of the remaining 80%. Trimming eliminates the effect of a small number of outliers. Compute the 10% trimmed mean of the guinea pig survival time data in Table 1.8 (page 29). Then compute the 20% trimmed mean. Compare the values of these measures with the median and the ordinary untrimmed mean.

**1.98** ⚠ **Changing units from centimeters to inches.** Refer to Exercise 1.56. Change the measurements from centimeters to inches by multiplying each value by 0.39. Answer the questions from the previous exercise and explain the effect of the transformation on these data.

# 1.3 Density Curves and Normal Distributions

We now have a kit of graphical and numerical tools for describing distributions. What is more, we have a clear strategy for exploring data on a single quantitative variable:

1. Always plot your data: make a graph, usually a stemplot or a histogram.

2. Look for the overall pattern and for striking deviations such as outliers.

3. Calculate an appropriate numerical summary to briefly describe center and spread.

Technology has expanded the set of graphs that we can choose for Step 1. It is possible, though painful, to make histograms by hand. Using software, clever algorithms can describe a distribution in a way that is not feasible by hand, by fitting a smooth curve to the data in addition to or instead of a histogram. The curves used are called **density curves.** Before we examine density curves in detail, here is an example of what software can do.

**EXAMPLE**

**1.23 Density curves of pH and survival times.** Figure 1.24 illustrates the use of density curves along with histograms to describe distributions.[34] Figure 1.24(a) shows the distribution of the acidity (pH) of rainwater, from Exercise 1.36 (page 27). That exercise illustrates how the choice of classes can change the shape of a histogram. The density curve and the software's default histogram agree that the distribution has a single peak and is approximately symmetric.

Figure 1.24(b) shows a strongly skewed distribution, the survival times of guinea pigs from Table 1.8 (page 29). The histogram and density curve agree on the overall shape and on the "bumps" in the long right tail. The density curve shows a higher peak near the single mode of the distribution. The histogram divides the observations near the mode into two classes, thus reducing the peak.

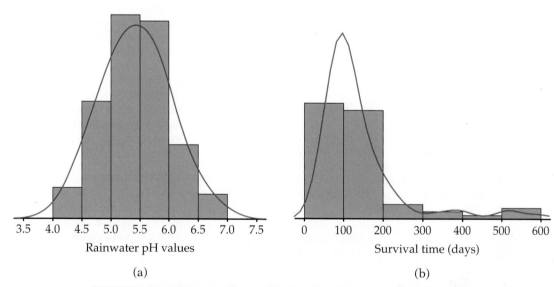

Rainwater pH values

(a)

Survival time (days)

(b)

**FIGURE 1.24** (a) The distribution of pH values measuring the acidity of 105 samples of rainwater, for Example 1.23. The roughly symmetric distribution is pictured by both a histogram and a density curve. (b) The distribution of the survival times of 72 guinea pigs in a medical experiment, for Example 1.23. The right-skewed distribution is pictured by both a histogram and a density curve.

In general, software that draws density curves describes the data in a way that is less arbitrary than choosing classes for a histogram. A smooth density curve is, however, an idealization that pictures the overall pattern of the data but ignores minor irregularities as well as any outliers. We will concentrate, not on general density curves, but on a special class, the bell-shaped Normal curves.

## Density curves

One way to think of a density curve is as a smooth approximation to the irregular bars of a histogram. Figure 1.25 shows a histogram of the scores of all 947 seventh-grade students in Gary, Indiana, on the vocabulary part of the Iowa Test of Basic Skills. Scores of many students on this national test have a very regular distribution. The histogram is symmetric, and both tails fall off quite smoothly from a single center peak. There are no large gaps or obvious outliers. The curve drawn through the tops of the histogram bars in Figure 1.25 is a good description of the overall pattern of the data.

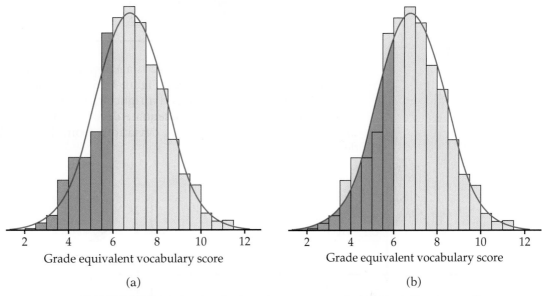

**FIGURE 1.25** (a) The distribution of Iowa Test vocabulary scores for Gary, Indiana, seventh-graders. The shaded bars in the histogram represent scores less than or equal to 6.0. The proportion of such scores in the data is 0.303. (b) The shaded area under the Normal density curve also represents scores less than or equal to 6.0. This area is 0.293, close to the true 0.303 for the actual data.

**EXAMPLE**

**1.24 Vocabulary scores.** In a histogram, the *areas* of the bars represent either counts or proportions of the observations. In Figure 1.25(a) we have shaded the bars that represent students with vocabulary scores 6.0 or lower. There are 287 such students, who make up the proportion $287/947 = 0.303$ of all Gary seventh-graders. The shaded bars in Figure 1.25(a) make up proportion 0.303 of the total area under all the bars. If we adjust the scale so that the total area of the bars is 1, the area of the shaded bars will be 0.303.

In Figure 1.25(b), we have shaded the *area under the curve* to the left of 6.0. Adjust the scale so that the total area under the curve is exactly 1.

Areas under the curve then represent proportions of the observations. That is, *area = proportion*. The curve is then a density curve. The shaded area under the density curve in Figure 1.25(b) represents the proportion of students with score 6.0 or lower. This area is 0.293, only 0.010 away from the histogram result. You can see that areas under the density curve give quite good approximations of areas given by the histogram.

---

### DENSITY CURVE

A **density curve** is a curve that

• is always on or above the horizontal axis and

• has area exactly 1 underneath it.

A density curve describes the overall pattern of a distribution. The area under the curve and above any range of values is the proportion of all observations that fall in that range.

---

The density curve in Figure 1.25 is a *Normal curve.* Density curves, like distributions, come in many shapes. Figure 1.26 shows two density curves, a symmetric Normal density curve and a right-skewed curve. A density curve of an appropriate shape is often an adequate description of the overall pattern of a distribution. Outliers, which are deviations from the overall pattern, are not described by the curve.

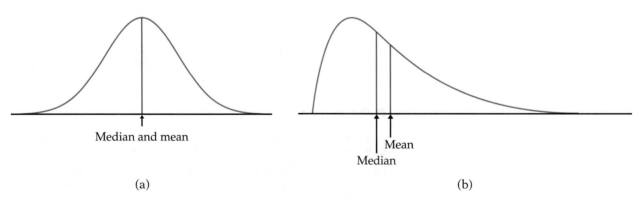

(a)                                                    (b)

**FIGURE 1.26** (a) A symmetric density curve with its mean and median marked. (b) A right-skewed density curve with its mean and median marked.

## Measuring center and spread for density curves

Our measures of center and spread apply to density curves as well as to actual sets of observations, but only some of these measures are easily seen from the curve. A **mode** of a distribution described by a density curve is a peak point of the curve, the location where the curve is highest. Because areas under a density curve represent proportions of the observations, the **median** is the point with half the total area on each side. You can roughly locate the **quartiles** by dividing the area under the curve into quarters as accurately as possible by eye. The *IQR* is then the distance between the first and third quartiles. There are mathematical ways of calculating areas under curves. These allow us to locate the median and quartiles exactly on any density curve.

What about the mean and standard deviation? The mean of a set of observations is their arithmetic average. If we think of the observations as weights strung out along a thin rod, the mean is the point at which the rod would balance. This fact is also true of density curves. The mean is the point at which the curve would balance if it were made out of solid material. Figure 1.27 illustrates this interpretation of the mean. We have marked the mean and median on the density curves in Figure 1.26. A symmetric curve, such as the Normal curve in Figure 1.26(a), balances at its center of symmetry. Half the area under a symmetric curve lies on either side of its center, so this is also the median. For a right-skewed curve, such as that shown in Figure 1.26(b), the small area in the long right tail tips the curve more than the same area near the center. The mean (the balance point) therefore lies to the right of the median. It is hard to locate the balance point by eye on a skewed curve. There are mathematical ways of calculating the mean for any density curve, so we are able to mark the mean as well as the median in Figure 1.26(b). The standard deviation can also be calculated mathematically, but it can't be located by eye on most density curves.

**FIGURE 1.27** The mean of a density curve is the point at which it would balance.

---

### MEDIAN AND MEAN OF A DENSITY CURVE

The **median** of a density curve is the equal-areas point, the point that divides the area under the curve in half.

The **mean** of a density curve is the balance point, at which the curve would balance if made of solid material.

The median and mean are the same for a symmetric density curve. They both lie at the center of the curve. The mean of a skewed curve is pulled away from the median in the direction of the long tail.

---

A density curve is an idealized description of a distribution of data. For example, the symmetric density curve in Figure 1.25 is exactly symmetric, but the histogram of vocabulary scores is only approximately symmetric. We therefore need to distinguish between the mean and standard deviation of the density curve and the numbers $\bar{x}$ and $s$ computed from the actual observations. The usual notation for the mean of an idealized distribution is $\mu$ (the Greek letter mu). We write the standard deviation of a density curve as $\sigma$ (the Greek letter sigma).

**mean $\mu$**
**standard deviation $\sigma$**

## Normal distributions

One particularly important class of density curves has already appeared in Figures 1.25 and 1.26(a). These density curves are symmetric, unimodal, and bell-shaped. They are called **Normal curves,** and they describe *Normal distributions*. All Normal distributions have the same overall shape. The exact density curve for a particular Normal distribution is specified by giving its mean $\mu$ and its standard deviation $\sigma$. The mean is located at the center of the symmetric curve and is the same as the median. Changing $\mu$ without changing $\sigma$ moves the Normal curve along the horizontal axis without changing its spread. The standard deviation $\sigma$ controls the spread of a Normal curve. Figure 1.28 shows two Normal curves with different values of $\sigma$. The curve with the larger standard deviation is more spread out.

*Normal curves*

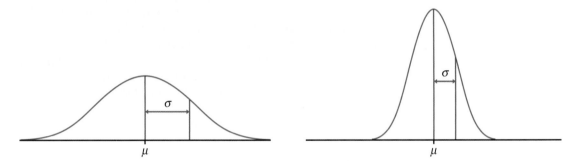

**FIGURE 1.28** Two Normal curves, showing the mean $\mu$ and standard deviation $\sigma$.

The standard deviation $\sigma$ is the natural measure of spread for Normal distributions. Not only do $\mu$ and $\sigma$ completely determine the shape of a Normal curve, but we can locate $\sigma$ by eye on the curve. Here's how. As we move out in either direction from the center $\mu$, the curve changes from falling ever more steeply

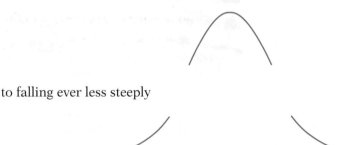

to falling ever less steeply

*The points at which this change of curvature takes place are located at distance $\sigma$ on either side of the mean $\mu$.* You can feel the change as you run your finger along a Normal curve, and so find the standard deviation. Remember that $\mu$ and $\sigma$ alone do not specify the shape of most distributions, and that the shape of density curves in general does not reveal $\sigma$. These are special properties of Normal distributions.

There are other symmetric bell-shaped density curves that are not Normal. The Normal density curves are specified by a particular equation. The height of the density curve at any point $x$ is given by

$$\frac{1}{\sigma\sqrt{2\pi}}e^{-\frac{1}{2}(\frac{x-\mu}{\sigma})^2}$$

We will not make direct use of this fact, although it is the basis of mathematical work with Normal distributions. Notice that the equation of the curve is completely determined by the mean $\mu$ and the standard deviation $\sigma$.

Why are the Normal distributions important in statistics? Here are three reasons. First, Normal distributions are good descriptions for some distributions of *real data*. Distributions that are often close to Normal include scores on tests taken by many people (such as the Iowa Test of Figure 1.25), repeated careful measurements of the same quantity, and characteristics of biological populations (such as lengths of baby pythons and yields of corn). Second, Normal distributions are good approximations to the results of many kinds of *chance outcomes,* such as tossing a coin many times. Third, and most important, we will see that many *statistical inference* procedures based on Normal distributions work well for other roughly symmetric distributions. HOWEVER . . . *even though many sets of data follow a Normal distribution, many do not.* Most income distributions, for example, are skewed to the right and so are not Normal. Non-Normal data, like non-Normal people, not only are common but are sometimes more interesting than their Normal counterparts.

## The 68–95–99.7 rule

Although there are many Normal curves, they all have common properties. Here is one of the most important.

---

### THE 68–95–99.7 RULE

In the Normal distribution with mean $\mu$ and standard deviation $\sigma$:

- Approximately **68%** of the observations fall within $\sigma$ of the mean $\mu$.

- Approximately **95%** of the observations fall within $2\sigma$ of $\mu$.

- Approximately **99.7%** of the observations fall within $3\sigma$ of $\mu$.

---

Figure 1.29 illustrates the 68–95–99.7 rule. By remembering these three numbers, you can think about Normal distributions without constantly making detailed calculations.

**EXAMPLE**

**1.25 Heights of young women.** The distribution of heights of young women aged 18 to 24 is approximately Normal with mean $\mu = 64.5$ inches and standard deviation $\sigma = 2.5$ inches. Figure 1.30 shows what the 68–95–99.7 rule says about this distribution.

Two standard deviations is 5 inches for this distribution. The 95 part of the 68–95–99.7 rule says that the middle 95% of young women are between $64.5 - 5$ and $64.5 + 5$ inches tall, that is, between 59.5 inches and 69.5 inches. This fact is exactly true for an exactly Normal distribution. It is approximately

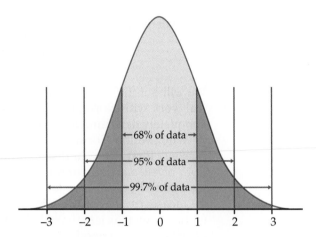

**FIGURE 1.29** The 68–95–99.7 rule for Normal distributions.

true for the heights of young women because the distribution of heights is approximately Normal.

The other 5% of young women have heights outside the range from 59.5 to 69.5 inches. Because the Normal distributions are symmetric, half of these women are on the tall side. So the tallest 2.5% of young women are taller than 69.5 inches.

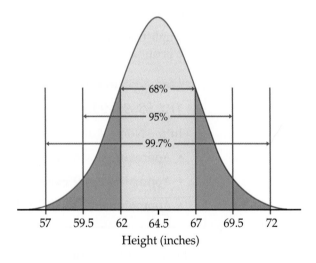

**FIGURE 1.30** The 68–95–99.7 rule applied to the heights of young women, for Example 1.25.

Because we will mention Normal distributions often, a short notation is helpful. We abbreviate the Normal distribution with mean $\mu$ and standard deviation $\sigma$ as $N(\mu, \sigma)$. For example, the distribution of young women's heights is $N(64.5, 2.5)$.

$N(\mu, \sigma)$

## USE YOUR KNOWLEDGE

**1.99    Test scores.** Many states have programs for assessing the skills of students in various grades. The Indiana Statewide Testing for Educational Progress (ISTEP) is one such program.[35] In a recent year 76,531 tenth-grade Indiana students took the English/language arts exam. The mean score was 572 and the standard deviation was 51.

Assuming that these scores are approximately Normally distributed, $N(572, 51)$, use the 68–95–99.7 rule to give a range of scores that includes 95% of these students.

**1.100** **Use the 68–95–99.7 rule.** Refer to the previous exercise. Use the 68–95–99.7 rule to give a range of scores that includes 99.7% of these students.

## Standardizing observations

As the 68–95–99.7 rule suggests, all Normal distributions share many properties. In fact, all Normal distributions are the same if we measure in units of size $\sigma$ about the mean $\mu$ as center. Changing to these units is called *standardizing*. To standardize a value, subtract the mean of the distribution and then divide by the standard deviation.

---

### STANDARDIZING AND z-SCORES

If $x$ is an observation from a distribution that has mean $\mu$ and standard deviation $\sigma$, the **standardized value** of $x$ is

$$z = \frac{x - \mu}{\sigma}$$

A standardized value is often called a **z-score**.

---

A $z$-score tells us how many standard deviations the original observation falls away from the mean, and in which direction. Observations larger than the mean are positive when standardized, and observations smaller than the mean are negative.

**EXAMPLE**

**1.26 Find some z-scores.** The heights of young women are approximately Normal with $\mu = 64.5$ inches and $\sigma = 2.5$ inches. The $z$-score for height is

$$z = \frac{\text{height} - 64.5}{2.5}$$

A woman's standardized height is the number of standard deviations by which her height differs from the mean height of all young women. A woman 68 inches tall, for example, has $z$-score

$$z = \frac{68 - 64.5}{2.5} = 1.4$$

or 1.4 standard deviations above the mean. Similarly, a woman 5 feet (60 inches) tall has $z$-score

$$z = \frac{60 - 64.5}{2.5} = -1.8$$

or 1.8 standard deviations less than the mean height.

**1.101**  **Find the $z$-score.** Consider the ISTEP scores (see Exercise 1.99), which we can assume are approximately Normal, $N(572, 51)$. Give the $z$-score for a student who received a score of 600.

**1.102**  **Find another $z$-score.** Consider the ISTEP scores, which we can assume are approximately Normal, $N(572, 51)$. Give the $z$-score for a student who received a score of 500. Explain why your answer is negative even though all of the test scores are positive.

We need a way to write variables, such as "height" in Example 1.25, that follow a theoretical distribution such as a Normal distribution. We use capital letters near the end of the alphabet for such variables. If $X$ is the height of a young woman, we can then shorten "the height of a young woman is less than 68 inches" to "$X < 68$." We will use lowercase $x$ to stand for any specific value of the variable $X$.

We often standardize observations from symmetric distributions to express them in a common scale. We might, for example, compare the heights of two children of different ages by calculating their $z$-scores. The standardized heights tell us where each child stands in the distribution for his or her age group.

Standardizing is a linear transformation that transforms the data into the standard scale of $z$-scores. We know that a linear transformation does not change the shape of a distribution, and that the mean and standard deviation change in a simple manner. In particular, *the standardized values for any distribution always have mean 0 and standard deviation 1.*

If the variable we standardize has a Normal distribution, standardizing does more than give a common scale. It makes all Normal distributions into a single distribution, and this distribution is still Normal. Standardizing a variable that has any Normal distribution produces a new variable that has the *standard Normal distribution*.

---

### THE STANDARD NORMAL DISTRIBUTION

The **standard Normal distribution** is the Normal distribution $N(0, 1)$ with mean 0 and standard deviation 1.

If a variable $X$ has any Normal distribution $N(\mu, \sigma)$ with mean $\mu$ and standard deviation $\sigma$, then the standardized variable

$$Z = \frac{X - \mu}{\sigma}$$

has the standard Normal distribution.

---

## Normal distribution calculations

Areas under a Normal curve represent proportions of observations from that Normal distribution. There is no formula for areas under a Normal curve. Calculations use either software that calculates areas or a table of areas. The table cumulative proportion — and most software calculate one kind of area: **cumulative proportions.** A cumulative proportion is the proportion of observations in a distribution that lie at or below a given value. When the distribution is given by a density curve, the cumulative proportion is the area under the curve to the left of a given value. Figure 1.31 shows the idea more clearly than words do.

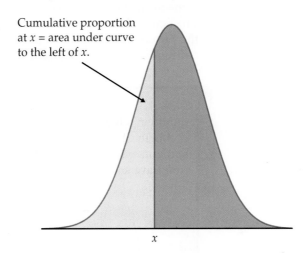

Cumulative proportion at $x$ = area under curve to the left of $x$.

**FIGURE 1.31** The *cumulative proportion* for a value $x$ is the proportion of all observations from the distribution that are less than or equal to $x$. This is the area to the left of $x$ under the Normal curve.

The key to calculating Normal proportions is to match the area you want with areas that represent cumulative proportions. Then get areas for cumulative proportions either from software or (with an extra step) from a table. The following examples show the method in pictures.

**EXAMPLE**

**1.27 The NCAA standard for SAT scores.** The National Collegiate Athletic Association (NCAA) requires Division I athletes to get a combined score of at least 820 on the SAT Mathematics and Verbal tests to compete in their first college year. (Higher scores are required for students with poor high school grades.) The scores of the 1.4 million students in the class of 2003 who took the SATs were approximately Normal with mean 1026 and standard deviation 209. What proportion of all students had SAT scores of at least 820?

Here is the calculation in pictures: the proportion of scores above 820 is the area under the curve to the right of 820. That's the total area under the curve (which is always 1) minus the cumulative proportion up to 820.

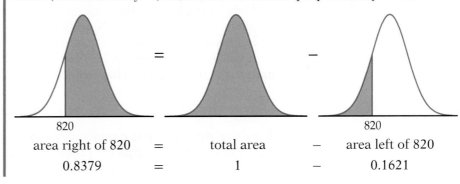

| | | |
|---|---|---|
| area right of 820 | = total area | − area left of 820 |
| 0.8379 | = 1 | − 0.1621 |

That is, the proportion of all SAT takers who would be NCAA qualifiers is 0.8379, or about 84%.

There is *no* area under a smooth curve and exactly over the point 820. Consequently, the area to the right of 820 (the proportion of scores > 820) is the same as the area at or to the right of this point (the proportion of scores ≥ 820). The actual data may contain a student who scored exactly 820 on the SAT. That the proportion of scores exactly equal to 820 is 0 for a Normal distribution is a consequence of the idealized smoothing of Normal distributions for data.

**EXAMPLE**

**1.28 NCAA partial qualifiers.**  The NCAA considers a student a "partial qualifier" eligible to practice and receive an athletic scholarship, but not to compete, if the combined SAT score is at least 720. What proportion of all students who take the SAT would be partial qualifiers? That is, what proportion have scores between 720 and 820? Here are the pictures:

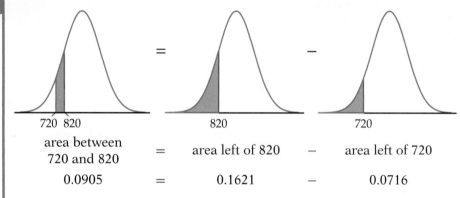

| area between<br>720 and 820 | = | area left of 820 | − | area left of 720 |
|:---:|:---:|:---:|:---:|:---:|
| 0.0905 | = | 0.1621 | − | 0.0716 |

About 9% of all students who take the SAT have scores between 720 and 820.

How do we find the numerical values of the areas in Examples 1.27 and 1.28? If you use software, just plug in mean 1026 and standard deviation 209. Then ask for the cumulative proportions for 820 and for 720. (Your software will probably refer to these as "cumulative probabilities." We will learn in Chapter 4 why the language of probability fits.) If you make a sketch of the area you want, you will never go wrong.

You can use the *Normal Curve* applet on the text CD and Web site to find Normal proportions. The applet is more flexible than most software—it will find any Normal proportion, not just cumulative proportions. The applet is an excellent way to understand Normal curves. But, because of the limitations of Web browsers, the applet is not as accurate as statistical software.

If you are not using software, you can find cumulative proportions for Normal curves from a table. That requires an extra step, as we now explain.

## Using the standard Normal table

The extra step in finding cumulative proportions from a table is that we must first standardize to express the problem in the standard scale of *z*-scores. This allows us to get by with just one table, a table of *standard Normal cumulative proportions.* Table A in the back of the book gives cumulative proportions for the standard Normal distribution. Table A also appears on the inside front cover. The pictures at the top of the table remind us that the entries are cumulative proportions, areas under the curve to the left of a value *z*.

EXAMPLE

**1.29 Find the proportion from *z*.** What proportion of observations on a standard Normal variable *Z* take values less than 1.47?

*Solution:* To find the area to the left of 1.47, locate 1.4 in the left-hand column of Table A, then locate the remaining digit 7 as .07 in the top row. The entry opposite 1.4 and under .07 is 0.9292. This is the cumulative proportion we seek. Figure 1.32 illustrates this area.

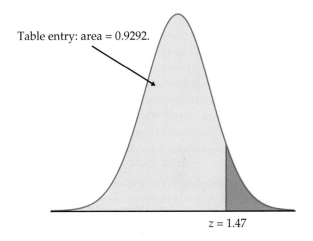

Table entry: area = 0.9292.

*z* = 1.47

**FIGURE 1.32** The area under a standard Normal curve to the left of the point *z* = 1.47 is 0.9292, for Example 1.29. Table A gives areas under the standard Normal curve.

Now that you see how Table A works, let's redo the NCAA Examples 1.27 and 1.28 using the table.

EXAMPLE

**1.30 Find the proportion from *x*.** What proportion of all students who take the SAT have scores of at least 820? The picture that leads to the answer is exactly the same as in Example 1.27. The extra step is that we first standardize in order to read cumulative proportions from Table A. If *X* is SAT score, we want the proportion of students for which $X \geq 820$.

1. *Standardize.* Subtract the mean, then divide by the standard deviation, to transform the problem about *X* into a problem about a standard Normal *Z*:

$$X \geq 820$$
$$\frac{X - 1026}{209} \geq \frac{820 - 1026}{209}$$
$$Z \geq -0.99$$

2. *Use the table.* Look at the pictures in Example 1.27. From Table A, we see that the proportion of observations less than −0.99 is 0.1611. The area to the right of −0.99 is therefore $1 - 0.1611 = 0.8389$. This is about 84%.

The area from the table in Example 1.30 (0.8389) is slightly less accurate than the area from software in Example 1.27 (0.8379) because we must round $z$ to two places when we use Table A. The difference is rarely important in practice.

**EXAMPLE**

**1.31 Proportion of partial qualifiers.** What proportion of all students who take the SAT would be partial qualifiers in the eyes of the NCAA? That is, what proportion of students have SAT scores between 720 and 820? First, sketch the areas, exactly as in Example 1.28. We again use $X$ as shorthand for an SAT score.

1. *Standardize.*

$$720 \leq X < 820$$
$$\frac{720 - 1026}{209} \leq \frac{X - 1026}{209} < \frac{820 - 1026}{209}$$
$$-1.46 \leq Z < -0.99$$

2. *Use the table.*

area between $-1.46$ and $-0.99$ = (area left of $-0.99$) − (area left of $-1.46$)
$$= 0.1611 - 0.0721 = 0.0890$$

As in Example 1.28, about 9% of students would be partial qualifiers.

Sometimes we encounter a value of $z$ more extreme than those appearing in Table A. For example, the area to the left of $z = -4$ is not given directly in the table. The $z$-values in Table A leave only area 0.0002 in each tail unaccounted for. For practical purposes, we can act as if there is zero area outside the range of Table A.

**USE YOUR KNOWLEDGE**

**1.103  Find the proportion.** Consider the ISTEP scores, which are approximately Normal, $N(572, 51)$. Find the proportion of students who have scores less than 600. Find the proportion of students who have scores greater than or equal to 600. Sketch the relationship between these two calculations using pictures of Normal curves similar to the ones given in Example 1.27.

**1.104  Find another proportion.** Consider the ISTEP scores, which are approximately Normal, $N(572, 51)$. Find the proportion of students who have scores between 600 and 650. Use pictures of Normal curves similar to the ones given in Example 1.28 to illustrate your calculations.

## Inverse Normal calculations

Examples 1.25 to 1.29 illustrate the use of Normal distributions to find the proportion of observations in a given event, such as "SAT score between 720 and

820." We may instead want to find the observed value corresponding to a given proportion.

Statistical software will do this directly. Without software, use Table A backward, finding the desired proportion in the body of the table and then reading the corresponding $z$ from the left column and top row.

**1.32 How high for the top 10%?**  Scores on the SAT Verbal test in recent years follow approximately the $N(505, 110)$ distribution. How high must a student score in order to place in the top 10% of all students taking the SAT?

Again, the key to the problem is to draw a picture. Figure 1.33 shows that we want the score $x$ with area above it 0.10. That's the same as area below $x$ equal to 0.90.

Statistical software has a function that will give you the $x$ for any cumulative proportion you specify. The function often has a name such as "inverse cumulative probability." Plug in mean 505, standard deviation 110, and cumulative proportion 0.9. The software tells you that $x = 645.97$. We see that a student must score at least 646 to place in the highest 10%.

Without software, first find the standard score $z$ with cumulative proportion 0.9, then "unstandardize" to find $x$. Here is the two-step process:

1. *Use the table.* Look in the body of Table A for the entry closest to 0.9. It is 0.8997. This is the entry corresponding to $z = 1.28$. So $z = 1.28$ is the standardized value with area 0.9 to its left.

2. *Unstandardize* to transform the solution from $z$ back to the original $x$ scale. We know that the standardized value of the unknown $x$ is $z = 1.28$. So $x$ itself satisfies

$$\frac{x - 505}{110} = 1.28$$

Solving this equation for $x$ gives

$$x = 505 + (1.28)(110) = 645.8$$

This equation should make sense: it finds the $x$ that lies 1.28 standard deviations above the mean on this particular Normal curve. That is the

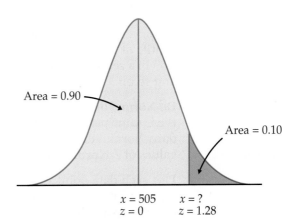

**FIGURE 1.33** Locating the point on a Normal curve with area 0.10 to its right, for Example 1.32. The result is $x = 646$, or $z = 1.28$ in the standard scale.

"unstandardized" meaning of $z = 1.28$. The general rule for unstandardizing a $z$-score is

$$x = \mu + z\sigma$$

## USE YOUR KNOWLEDGE

**1.105  What score is needed to be in the top 5%?** Consider the ISTEP scores, which are approximately Normal, $N(572, 51)$. How high a score is needed to be in the top 5% of students who take this exam?

**1.106  Find the score that 60% of students will exceed.** Consider the ISTEP scores, which are approximately Normal, $N(572, 51)$. Sixty percent of the students will score above $x$ on this exam. Find $x$.

## Normal quantile plots

The Normal distributions provide good descriptions of some distributions of real data, such as the Gary vocabulary scores. The distributions of some other common variables are usually skewed and therefore distinctly non-Normal. Examples include economic variables such as personal income and gross sales of business firms, the survival times of cancer patients after treatment, and the service lifetime of mechanical or electronic components. While experience can suggest whether or not a Normal distribution is plausible in a particular case, it is risky to assume that a distribution is Normal without actually inspecting the data.

A histogram or stemplot can reveal distinctly non-Normal features of a distribution, such as outliers (the breaking strengths in Figure 1.9, page 17), pronounced skewness (the survival times in Figure 1.24(b), page 54), or gaps and clusters (the Massachusetts college tuitions in Figure 1.16, page 25). If the stemplot or histogram appears roughly symmetric and unimodal, however, we need a more sensitive way to judge the adequacy of a Normal model. The most

**Normal quantile plot**      useful tool for assessing Normality is another graph, the **Normal quantile plot.**

Here is the basic idea of a Normal quantile plot. The graphs produced by software use more sophisticated versions of this idea. It is not practical to make Normal quantile plots by hand.

1. Arrange the observed data values from smallest to largest. Record what percentile of the data each value occupies. For example, the smallest observation in a set of 20 is at the 5% point, the second smallest is at the 10% point, and so on.

2. Do Normal distribution calculations to find the values of $z$ corresponding to these same percentiles. For example, $z = -1.645$ is the 5% point of the standard Normal distribution, and $z = -1.282$ is the 10% point. We call these

**Normal scores**      values of $Z$ **Normal scores.**

3. Plot each data point $x$ against the corresponding Normal score. If the data distribution is close to any Normal distribution, the plotted points will lie close to a straight line.

Any Normal distribution produces a straight line on the plot because standardizing turns any Normal distribution into a standard Normal distribution. Standardizing is a linear transformation that can change the slope and intercept of the line in our plot but cannot turn a line into a curved pattern.

> ## USE OF NORMAL QUANTILE PLOTS
>
> If the points on a Normal quantile plot lie close to a straight line, the plot indicates that the data are Normal. Systematic deviations from a straight line indicate a non-Normal distribution. Outliers appear as points that are far away from the overall pattern of the plot.

Figures 1.34 to 1.36 are Normal quantile plots for data we have met earlier. The data $x$ are plotted vertically against the corresponding standard Normal $z$-score plotted horizontally. The $z$-score scale extends from $-3$ to $3$ because almost all of a standard Normal curve lies between these values. These figures show how Normal quantile plots behave.

**EXAMPLE**

**1.33 Breaking strengths are Normal.** Figure 1.34 is a Normal quantile plot of the breaking strengths in Example 1.11 (page 17). Lay a transparent straightedge over the center of the plot to see that most of the points lie close to a straight line. A Normal distribution describes these points quite well. The only substantial deviations are short horizontal runs of points. Each run represents repeated observations having the same value—there are five measurements at 1150, for example. This phenomenon is called **granularity.** It is caused by the limited precision of the measurements and does not represent an important deviation from Normality.

granularity

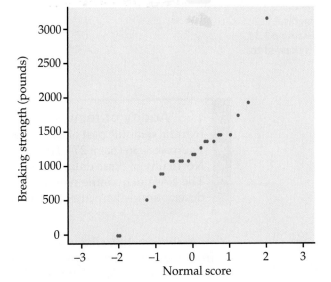

FIGURE 1.34 Normal quantile plot of the breaking strengths of wires bonded to a semiconductor wafer, for Example 1.33. This distribution has a Normal shape except for outliers in both tails.

The high outlier at 3150 pounds lies above the line formed by the center of the data—it is farther out in the high direction than we expect Normal data to be.

The two low outliers at 0 lie below the line—they are suspiciously far out in the low direction. Compare Figure 1.34 with the histogram of these data in Figure 1.9 (page 17).

**EXAMPLE**

**1.34 Survival times are not Normal.** Figure 1.35 is a Normal quantile plot of the guinea pig survival times from Table 1.8 (page 29). Figure 1.24(b) (page 54) shows that this distribution is strongly skewed to the right.

To see the right-skewness in the Normal quantile plot, draw a line through the leftmost points, which correspond to the smaller observations. The larger observations fall systematically above this line. That is, the right-of-center observations have larger values than in a Normal distribution. *In a right-skewed distribution, the largest observations fall distinctly above a line drawn through the main body of points.* Similarly, left-skewness is evident when the smallest observations fall below the line. Unlike Figure 1.34, there are no individual outliers.

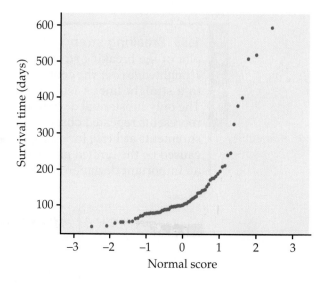

**FIGURE 1.35** Normal quantile plot of the survival times of guinea pigs in a medical experiment, for Example 1.34. This distribution is skewed to the right.

**EXAMPLE**

**1.35 Acidity of rainwater is approximately Normal.** Figure 1.36 is a Normal quantile plot of the 105 acidity (pH) measurements of rainwater from Exercise 1.36 (page 27). Histograms don't settle the question of approximate Normality of these data, because their shape depends on the choice of classes. The Normal quantile plot makes it clear that a Normal distribution is a good description—there are only minor wiggles in a generally straight-line pattern.

As Figure 1.36 illustrates, real data almost always show some departure from the theoretical Normal model. *When you examine a Normal quantile plot, look for shapes that show clear departures from Normality. Don't overreact to*

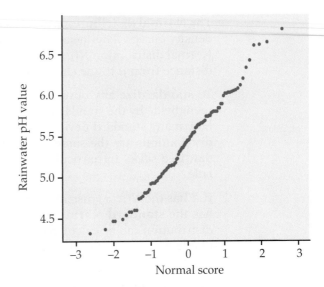

**FIGURE 1.36** Normal quantile plot of the acidity (pH) values of 105 samples of rainwater, for Example 1.35. This distribution is approximately Normal.

*minor wiggles in the plot.* When we discuss statistical methods that are based on the Normal model, we will pay attention to the sensitivity of each method to departures from Normality. Many common methods work well as long as the data are approximately Normal and outliers are not present.

## BEYOND THE BASICS

### Density Estimation

A density curve gives a compact summary of the overall shape of a distribution. Many distributions do not have the Normal shape. There are other families of density curves that are used as mathematical models for various distribution shapes. Modern software offers a more flexible option, illustrated by the two graphs in Figure 1.24 (page 54). A **density estimator** does not start with any specific shape, such as the Normal shape. It looks at the data and draws a density curve that describes the overall shape of the data. Density estimators join stemplots and histograms as useful graphical tools for exploratory data analysis.

density estimator

## SECTION 1.3  Summary

The overall pattern of a distribution can often be described compactly by a **density curve.** A density curve has total area 1 underneath it. Areas under a density curve give proportions of observations for the distribution.

The **mean** $\mu$ (balance point), the **median** (equal-areas point), and the **quartiles** can be approximately located by eye on a density curve. The **standard deviation** $\sigma$ cannot be located by eye on most density curves. The mean and median are equal for symmetric density curves, but the mean of a skewed curve is located farther toward the long tail than is the median.

The **Normal distributions** are described by bell-shaped, symmetric, unimodal density curves. The mean $\mu$ and standard deviation $\sigma$ completely specify the Normal distribution $N(\mu, \sigma)$. The mean is the center of symmetry, and $\sigma$ is the distance from $\mu$ to the change-of-curvature points on either side.

To **standardize** any observation $x$, subtract the mean of the distribution and then divide by the standard deviation. The resulting **z-score** $z = (x - \mu)/\sigma$ says how many standard deviations $x$ lies from the distribution mean. All Normal distributions are the same when measurements are transformed to the standardized scale. In particular, all Normal distributions satisfy the **68–95–99.7 rule.**

If $X$ has the $N(\mu, \sigma)$ distribution, then the standardized variable $Z = (X - \mu)/\sigma$ has the **standard Normal distribution** $N(0, 1)$. Proportions for any Normal distribution can be calculated by software or from the **standard Normal table** (Table A), which gives the **cumulative proportions** of $Z < z$ for many values of $z$.

The adequacy of a Normal model for describing a distribution of data is best assessed by a **Normal quantile plot,** which is available in most statistical software packages. A pattern on such a plot that deviates substantially from a straight line indicates that the data are not Normal.

## SECTION 1.3   Exercises

*For Exercises 1.99 and 1.100, see pages 60 and 61; for Exercises 1.101 and 1.102, see page 62; for Exercises 1.103 and 1.104, see page 66; and for Exercises 1.105 and 1.106, see page 68.*

**1.107  Sketch some density curves.** Sketch density curves that might describe distributions with the following shapes:

(a)  Symmetric, but with two peaks (that is, two strong clusters of observations).

(b)  Single peak and skewed to the right.

**1.108  A uniform distribution.** If you ask a computer to generate "random numbers" between 0 and 1, you will get observations from a **uniform distribution.** Figure 1.37 graphs the density curve for a uniform distribution. Use areas under this density curve to answer the following questions.

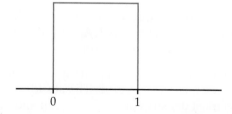

**FIGURE 1.37** The density curve of a uniform distribution, for Exercise 1.108.

(a)  Why is the total area under this curve equal to 1?

(b)  What proportion of the observations lie below 0.35?

(c)  What proportion of the observations lie between 0.35 and 0.65?

**1.109  Use a different range for the uniform distribution.** Many random number generators allow users to specify the range of the random numbers to be produced. Suppose that you specify that the outcomes are to be distributed uniformly between 0 and 4. Then the density curve of the outcomes has constant height between 0 and 4, and height 0 elsewhere.

(a)  What is the height of the density curve between 0 and 4? Draw a graph of the density curve.

(b)  Use your graph from (a) and the fact that areas under the curve are proportions of outcomes to find the proportion of outcomes that are less than 1.

(c)  Find the proportion of outcomes that lie between 0.5 and 2.5.

**1.110  Find the mean, the median, and the quartiles.** What are the mean and the median of the uniform distribution in Figure 1.37? What are the quartiles?

**1.111  Three density curves.** Figure 1.38 displays three density curves, each with three points marked on

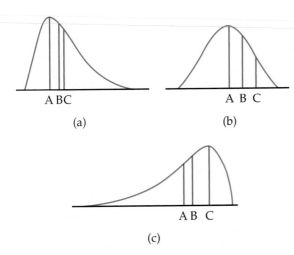

A BC

(a)

A B C

(b)

A B  C

(c)

**FIGURE 1.38** Three density curves, for Exercise 1.111.

it. At which of these points on each curve do the mean and the median fall?

**1.112  Length of pregnancies.** The length of human pregnancies from conception to birth varies according to a distribution that is approximately Normal with mean 266 days and standard deviation 16 days. Draw a density curve for this distribution on which the mean and standard deviation are correctly located.

**1.113  Use the Normal Curve applet.** The 68–95–99.7 rule for Normal distributions is a useful approximation. You can use the *Normal Curve* applet on the text CD and Web site to see how accurate the rule is. Drag one flag across the other so that the applet shows the area under the curve between the two flags.

(a) Place the flags one standard deviation on either side of the mean. What is the area between these two values? What does the 68–95–99.7 rule say this area is?

(b) Repeat for locations two and three standard deviations on either side of the mean. Again compare the 68–95–99.7 rule with the area given by the applet.

**1.114  Pregnancies and the 68–95–99.7 rule.** The length of human pregnancies from conception to birth varies according to a distribution that is approximately Normal with mean 266 days and standard deviation 16 days. Use the 68–95–99.7 rule to answer the following questions.

(a) Between what values do the lengths of the middle 95% of all pregnancies fall?

(b) How short are the shortest 2.5% of all pregnancies? How long do the longest 2.5% last?

**1.115  Horse pregnancies are longer.** Bigger animals tend to carry their young longer before birth. The length of horse pregnancies from conception to birth varies according to a roughly Normal distribution with mean 336 days and standard deviation 3 days. Use the 68–95–99.7 rule to answer the following questions.

(a) Almost all (99.7%) horse pregnancies fall in what range of lengths?

(b) What percent of horse pregnancies are longer than 339 days?

**1.116  Binge drinking survey.** One reason that Normal distributions are important is that they describe how the results of an opinion poll would vary if the poll were repeated many times. About 20% of college students say they are frequent binge drinkers. Think about taking many randomly chosen samples of 1600 students. The proportions of college students in these samples who say they are frequent binge drinkers will follow the Normal distribution with mean 0.20 and standard deviation 0.01. Use this fact and the 68–95–99.7 rule to answer these questions.

(a) In many samples, what percent of samples give results above 0.2? Above 0.22?

(b) In a large number of samples, what range contains the central 95% of proportions of students who say they are frequent binge drinkers?

**1.117  Heights of women.** The heights of women aged 20 to 29 are approximately Normal with mean 64 inches and standard deviation 2.7 inches. Men the same age have mean height 69.3 inches with standard deviation 2.8 inches. What are the *z*-scores for a woman 6 feet tall and a man 6 feet tall? What information do the *z*-scores give that the actual heights do not?

**1.118  Use the Normal Curve applet.** Use the *Normal Curve* applet for the standard Normal distribution to say how many standard deviations above and below the mean the quartiles of any Normal distribution lie.

**1.119  Acidity of rainwater.** The Normal quantile plot in Figure 1.36 (page 71) shows that the acidity (pH) measurements for rainwater samples in Exercise 1.36 are approximately Normal. How well do these scores satisfy the 68–95–99.7 rule?

To find out, calculate the mean $\bar{x}$ and standard deviation $s$ of the observations. Then calculate the percent of the 105 measurements that fall between $\bar{x} - s$ and $\bar{x} + s$ and compare your result with 68%. Do the same for the intervals covering two and three standard deviations on either side of the mean. (The 68–95–99.7 rule is exact for any theoretical Normal distribution. It will hold only approximately for actual data.)

**1.120 Find some proportions.** Using either Table A or your calculator or software, find the proportion of observations from a standard Normal distribution that satisfies each of the following statements. In each case, sketch a standard Normal curve and shade the area under the curve that is the answer to the question.

(a) $Z < 1.65$

(b) $Z > 1.65$

(c) $Z > -0.76$

(d) $-0.76 < Z < 1.65$

**1.121 Find more proportions.** Using either Table A or your calculator or software, find the proportion of observations from a standard Normal distribution for each of the following events. In each case, sketch a standard Normal curve and shade the area representing the proportion.

(a) $Z \leq -1.9$

(b) $Z \geq -1.9$

(c) $Z > 1.55$

(d) $-1.9 < Z < 1.55$

**1.122 Find some values of $z$.** Find the value $z$ of a standard Normal variable $Z$ that satisfies each of the following conditions. (If you use Table A, report the value of $z$ that comes closest to satisfying the condition.) In each case, sketch a standard Normal curve with your value of $z$ marked on the axis.

(a) 25% of the observations fall below $z$.

(b) 35% of the observations fall above $z$.

**1.123 Find more values of $z$.** The variable $Z$ has a standard Normal distribution.

(a) Find the number $z$ that has cumulative proportion 0.85.

(b) Find the number $z$ such that the event $Z > z$ has proportion 0.40.

**1.124 Find some values of $z$.** The Wechsler Adult Intelligence Scale (WAIS) is the most common "IQ test." The scale of scores is set separately for each age group and is approximately Normal with mean 100 and standard deviation 15. People with WAIS scores below 70 are considered mentally retarded when, for example, applying for Social Security disability benefits. What percent of adults are retarded by this criterion?

**1.125 High IQ scores.** The Wechsler Adult Intelligence Scale (WAIS) is the most common "IQ test." The scale of scores is set separately for each age group and is approximately Normal with mean 100 and standard deviation 15. The organization MENSA, which calls itself "the high IQ society," requires a WAIS score of 130 or higher for membership. What percent of adults would qualify for membership?

*There are two major tests of readiness for college, the ACT and the SAT. ACT scores are reported on a scale from 1 to 3. The distribution of ACT scores for more than 1 million students in a recent high school graduating class was roughly Normal with mean $\mu = 20.8$ and standard deviation $\sigma = 4.8$. SAT scores are reported on a scale from 400 to 1600. The SAT scores for 1.4 million students in the same graduating class were roughly Normal with mean $\mu = 1026$ and standard deviation $\sigma = 209$. Exercises 1.126 to 1.135 are based on this information.*

**1.126 Compare an SAT score with an ACT score.** Tonya scores 1320 on the SAT. Jermaine scores 28 on the ACT. Assuming that both tests measure the same thing, who has the higher score? Report the $z$-scores for both students.

**1.127 Make another comparison.** Jacob scores 17 on the ACT. Emily scores 680 on the SAT. Assuming that both tests measure the same thing, who has the higher score? Report the $z$-scores for both students.

**1.128 Find the ACT equivalent.** Jose scores 1380 on the SAT. Assuming that both tests measure the same thing, what score on the ACT is equivalent to Jose's SAT score?

**1.129 Find the SAT equivalent.** Maria scores 29 on the ACT. Assuming that both tests measure the same thing, what score on the SAT is equivalent to Maria's ACT score?

**1.130 Find the SAT percentile.** Reports on a student's ACT or SAT usually give the percentile as well as the actual score. The percentile is just the cumulative proportion stated as a percent: the percent of all scores that were lower than this one. Tonya scores 1320 on the SAT. What is her percentile?

**1.131  Find the ACT percentile.** Reports on a student's ACT or SAT usually give the percentile as well as the actual score. The percentile is just the cumulative proportion stated as a percent: the percent of all scores that were lower than this one. Jacob scores 17 on the ACT. What is his percentile?

**1.132  How high is the top 10%?** What SAT scores make up the top 10% of all scores?

**1.133  How low is the bottom 20%?** What SAT scores make up the bottom 20% of all scores?

**1.134  Find the ACT quartiles.** The quartiles of any distribution are the values with cumulative proportions 0.25 and 0.75. What are the quartiles of the distribution of ACT scores?

**1.135  Find the SAT quintiles.** The quintiles of any distribution are the values with cumulative proportions 0.20, 0.40, 0.60, and 0.80. What are the quintiles of the distribution of SAT scores?

**1.136  Proportion of women with high cholesterol.** Too much cholesterol in the blood increases the risk of heart disease. Young women are generally less afflicted with high cholesterol than other groups. The cholesterol levels for women aged 20 to 34 follow an approximately Normal distribution with mean 185 milligrams per deciliter (mg/dl) and standard deviation 39 mg/dl.[36]

(a) Cholesterol levels above 240 mg/dl demand medical attention. What percent of young women have levels above 240 mg/dl?

(b) Levels above 200 mg/dl are considered borderline high. What percent of young women have blood cholesterol between 200 and 240 mg/dl?

**1.137  Proportion of men with high cholesterol.** Middle-aged men are more susceptible to high cholesterol than the young women of the previous exercise. The blood cholesterol levels of men aged 55 to 64 are approximately Normal with mean 222 mg/dl and standard deviation 37 mg/dl. What percent of these men have high cholesterol (levels above 240 mg/dl)? What percent have borderline high cholesterol (between 200 and 240 mg/dl)?

**1.138  Diagnosing osteoporosis.** Osteoporosis is a condition in which the bones become brittle due to loss of minerals. To diagnose osteoporosis, an elaborate apparatus measures bone mineral density (BMD). BMD is usually reported in standardized form. The standardization is based on a population of healthy young adults. The World Health Organization (WHO) criterion for osteoporosis is a BMD 2.5 standard deviations below the mean for young adults. BMD measurements in a population of people similar in age and sex roughly follow a Normal distribution.

(a) What percent of healthy young adults have osteoporosis by the WHO criterion?

(b) Women aged 70 to 79 are of course not young adults. The mean BMD in this age is about $-2$ on the standard scale for young adults. Suppose that the standard deviation is the same as for young adults. What percent of this older population has osteoporosis?

**1.139  Length of pregnancies.** The length of human pregnancies from conception to birth varies according to a distribution that is approximately Normal with mean 266 days and standard deviation 16 days.

(a) What percent of pregnancies last less than 240 days (that's about 8 months)?

(b) What percent of pregnancies last between 240 and 270 days (roughly between 8 months and 9 months)?

(c) How long do the longest 20% of pregnancies last?

**1.140  🛆 Quartiles for Normal distributions.** The quartiles of any distribution are the values with cumulative proportions 0.25 and 0.75.

(a) What are the quartiles of the standard Normal distribution?

(b) Using your numerical values from (a), write an equation that gives the quartiles of the $N(\mu, \sigma)$ distribution in terms of $\mu$ and $\sigma$.

(c) The length of human pregnancies from conception to birth varies according to a distribution that is approximately Normal with mean 266 days and standard deviation 16 days. Apply your result from (b): what are the quartiles of the distribution of lengths of human pregnancies?

**1.141  🛆 IQR for Normal distributions.** Continue your work from the previous exercise. The interquartile range *IQR* is the distance between the first and third quartiles of a distribution.

(a) What is the value of the *IQR* for the standard Normal distribution?

1.171  **Use software to generate some data.** Most statistical software packages have routines for generating values of variables having specified distributions. Use your statistical software to generate 25 observations from the $N(20, 5)$ distribution. Compute the mean and standard deviation $\bar{x}$ and $s$ of the 25 values you obtain. How close are $\bar{x}$ and $s$ to the $\mu$ and $\sigma$ of the distribution from which the observations were drawn? Repeat 19 more times the process of generating 25 observations from the $N(20, 5)$ distribution and recording $\bar{x}$ and $s$. Make a stemplot of the 20 values of $\bar{x}$ and another stemplot of the 20 values of $s$. Make Normal quantile plots of both sets of data. Briefly describe each of these distributions. Are they symmetric or skewed? Are they roughly Normal? Where are their centers? (The distributions of measures like $\bar{x}$ and $s$ when repeated sets of observations are made from the same theoretical distribution will be very important in later chapters.)

1.172  **Distribution of income.** Each March, the Bureau of Labor Statistics collects detailed information about more than 50,000 randomly selected households. The WORKERS data set contains data on 71,076 people from the March 2002 survey. All of these people were between 25 and 64 years of age and worked throughout the year. The Data Appendix describes this data set in detail. Describe the distribution of incomes for these people. Use graphs and numbers, and briefly state your main findings. Because this is a very large randomly selected sample, your results give a good description of incomes for all working Americans aged 25 to 64.

1.173  **SAT mathematics scores and grade point averages.** The CSDATA data set described in the Data Appendix contains information on 234 computer science students. We are interested in comparing the SAT Mathematics scores and grade point averages of female students with those of male students. Make two sets of side-by-side boxplots to carry out these comparisons. Write a brief discussion of the male-female comparisons. Then make Normal quantile plots of grade point averages and SAT Math scores separately for men and women. Which students are clear outliers? Which of the four distributions are approximately Normal if we ignore outliers?

# Looking at Data— Relationships

Do large breeds of dogs have shorter lives? See Example 2.1.

## Introduction

In Chapter 1 we learned to use graphical and numerical methods to describe the distribution of a single variable. Many of the interesting examples of the use of statistics involve relationships between pairs of variables. Learning ways to describe relationships with graphical and numerical methods is the focus of this chapter.

EXAMPLE

**2.1 Large breeds of dogs have shorter lives.** Purebred dogs from breeds that are large tend to have shorter life spans than purebred dogs from breeds that are small. For example, one study found that miniature poodles lived an average of 9.3 years while Great Danes lived an average of only 4.6 years.[1] Irish wolfhounds have sometimes been referred to by the nickname "the heartbreak breed" because of their short life span relative to other breeds.[2]

We are particularly interested in situations where two variables are related in some way. To study relationships, we measure both variables on the same individuals or cases.

In dividing the states into regions, we introduced a third variable into the scatterplot. "Region" is a categorical variable that has four values, although we plotted data from only two of the four regions. The two regions are displayed by the two different plotting symbols.[4]

---

### CATEGORICAL VARIABLES IN SCATTERPLOTS

To add a categorical variable to a scatterplot, use a different plot color or symbol for each category.

---

## More examples of scatterplots

Experience in examining scatterplots is the foundation for more detailed study of relationships among quantitative variables. Here is an example with a pattern different from that in Figure 2.1.

**EXAMPLE**

**2.7 The Trans-Alaska Oil Pipeline.**   The Trans-Alaska Oil Pipeline is a tube formed from 1/2-inch-thick steel that carries oil across 800 miles of sensitive arctic and subarctic terrain. The pipe and the welds that join pipe segments were carefully examined before installation. How accurate are field measurements of the depth of small defects? Figure 2.3 compares the results of measurements on 100 defects made in the field with measurements of the same defects made in the laboratory.[5] We plot the laboratory results on the $x$ axis because they are a standard against which we compare the field results.

What is the overall pattern of this scatterplot? There is a positive linear association between the two variables. This is what we expect from two measurements of the same quantity. If field and laboratory measurements agree, the points will all fall on the $y = x$ line drawn on the plot, except for small random variations in the measurements. In fact, we see that the points for larger

**FIGURE 2.3** Depths of small defects in pipe for the Trans-Alaska Oil Pipeline, measured in the field and in the laboratory, for Example 2.7. If the two measurements were the same, the points would lie on the $y = x$ line that is drawn on the graph.

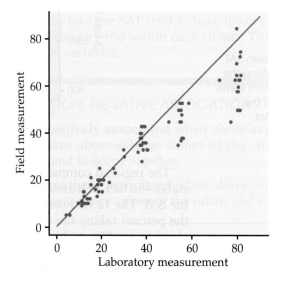

defects fall systematically below this line. That is, the field measurements are too small compared with the laboratory results as a standard. This is an important finding that can be used to adjust future field measurements.

Another part of the overall pattern is that the strength of the linear relationship decreases as the size of the defects increases. Field data show more variation (vertical spread in the plot) for large defect sizes than for small sizes. An increase in the spread in a response variable as the size of the response increases is a common pattern. It implies that predictions of the response based on the overall pattern will be less accurate for large responses.

Did you notice a fine point of graphing technique? Because both *x* and *y* measure the same thing, the graph is square and the same scales appear on both axes.

Some scatterplots appear quite different from the cloud of points in Figure 2.1 and the linear pattern in Figure 2.3. This is true, for example, in experiments in which measurements of a response variable are taken at a few selected levels of the explanatory variable. The following example illustrates the use of scatterplots in this setting.

**EXAMPLE**

**2.8 Predators and prey.** Here is one way in which nature regulates the size of animal populations: high population density attracts predators, who remove a higher proportion of the population than when the density of the prey is low. One study looked at kelp perch and their common predator, the kelp bass. The researcher set up four large circular pens on sandy ocean bottom in southern California. He chose young perch at random from a large group and placed 10, 20, 40, and 60 perch in the four pens. Then he dropped the nets protecting the pens, allowing bass to swarm in, and counted the perch left after 2 hours. Here are data on the proportions of perch eaten in four repetitions of this setup:[6]

| Perch | Proportion killed | | | |
|-------|-------|------|-----|-------|
| 10 | 0.0 | 0.1 | 0.3 | 0.3 |
| 20 | 0.2 | 0.3 | 0.3 | 0.6 |
| 40 | 0.075 | 0.3 | 0.6 | 0.725 |
| 60 | 0.517 | 0.55 | 0.7 | 0.817 |

The scatterplot in Figure 2.4 displays the results of this experiment. Because number of perch in a pen is the explanatory variable, we plot it horizontally as the *x* variable. The proportion of perch eaten by bass is the response variable *y*. Notice that there are two identical responses in the 10-perch group and also in the 20-perch group. These pairs of observations occupy the same points on the plot, so we use a different symbol for points that represent two observations. *Most software does not alert you to repeated values in your data when making scatterplots.* This can affect the impression the plot creates, especially when there are just a few points.

The vertical spread of points above each pen size shows the variation in proportions of perch eaten by bass. To see the overall pattern behind this mean or median responses at the four locations still shows the overall nature of the relationship.

Many categorical variables, like prey species or type of car, have no natural order from smallest to largest. In such situations we cannot speak of a positive or negative association with the response variable. If the mean responses in our plot increase as we go from left to right, we could make them decrease by writing the categories in the opposite order. The plot simply presents a side-by-side comparison of several distributions. The categorical variable labels the distributions. Some categorical variables do have a least-to-most order, however. We can then speak of the direction of the association between the categorical explanatory variable and the quantitative response. Look again at the boxplots of income by level of education in Figure 1.23, on page 52. Although the Census Bureau records education in categories, such as "did not graduate from high school," the categories have an order from less education to more education. The boxes in Figure 1.23 are arranged in order of increasing education. They show a positive association between education and income: people with more education tend to have higher incomes.

## SECTION 2.1  Summary

To study relationships between variables, we must measure the variables on the same group of individuals or cases.

If we think that a variable $x$ may explain or even cause changes in another variable $y$, we call $x$ an **explanatory variable** and $y$ a **response variable.**

A **scatterplot** displays the relationship between two quantitative variables. Mark values of one variable on the horizontal axis ($x$ axis) and values of the other variable on the vertical axis ($y$ axis). Plot each individual's data as a point on the graph.

Always plot the explanatory variable, if there is one, on the $x$ axis of a scatterplot. Plot the response variable on the $y$ axis.

Plot points with different colors or symbols to see the effect of a categorical variable in a scatterplot.

In examining a scatterplot, look for an overall pattern showing the **form, direction,** and **strength** of the relationship, and then for **outliers** or other deviations from this pattern.

**Form: Linear relationships,** where the points show a straight-line pattern, are an important form of relationship between two variables. Curved relationships and **clusters** are other forms to watch for.

**Direction:** If the relationship has a clear direction, we speak of either **positive association** (high values of the two variables tend to occur together) or **negative association** (high values of one variable tend to occur with low values of the other variable).

**Strength:** The **strength** of a relationship is determined by how close the points in the scatterplot lie to a simple form such as a line.

To display the relationship between a categorical explanatory variable and a quantitative response variable, make a graph that compares the distributions of the response for each category of the explanatory variable.

# SECTION 2.1 Exercises

*For Exercise 2.1, see page 84; for Exercises 2.2 and 2.3, see page 85; and for Exercise 2.4, see page 89.*

**2.5**  **Average temperatures.** Here are the average temperatures in degrees for Lafayette, Indiana, during the months of February through May:

| Month | February | March | April | May |
|---|---|---|---|---|
| Temperature (degrees F) | 30 | 41 | 51 | 62 |

(a) Explain why month should be the explanatory variable for examining this relationship.

(b) Make a scatterplot and describe the relationship.

**2.6**  **Relationship between first test and final exam.** How strong is the relationship between the score on the first test and the score on the the final exam in an elementary statistics course? Here are data for eight students from such a course:

| First-test score | 153 | 144 | 162 | 149 | 127 | 118 | 158 | 153 |
|---|---|---|---|---|---|---|---|---|
| Final-exam score | 145 | 140 | 145 | 170 | 145 | 175 | 170 | 160 |

(a) Which variable should play the role of the explanatory variable in describing this relationship?

(b) Make a scatterplot and describe the relationship.

(c) Give some possible reasons why this relationship is so weak.

**2.7**  **Relationship between second test and final exam.** Refer to the previous exercise. Here are the data for the second test and the final exam for the same students:

| Second-test score | 158 | 162 | 144 | 162 | 136 | 158 | 175 | 153 |
|---|---|---|---|---|---|---|---|---|
| Final-exam score | 145 | 140 | 145 | 170 | 145 | 175 | 170 | 160 |

(a) Explain why you should use the second-test score as the explanatory variable.

(b) Make a scatterplot and describe the relationship.

(c) Why do you think the relationship between the second-test score and the final-exam score is stronger than the relationship between the first-test score and the final-exam score?

**2.8**  **Add an outlier to the plot.** Refer to the previous exercise. Add a ninth student whose scores on the second test and final exam would lead you to classify the additional data point as an outlier. Highlight the outlier on your scatterplot and describe the performance of the student on the second exam and final exam and why that leads to the conclusion that the result is an outlier. Give a possible reason for the performance of this student.

**2.9**  **Explanatory and response variables.** In each of the following situations, is it more reasonable to simply explore the relationship between the two variables or to view one of the variables as an explanatory variable and the other as a response variable? In the latter case, which is the explanatory variable and which is the response variable?

(a) The weight of a child and the age of the child from birth to 10 years.

(b) High school English grades and high school math grades.

(c) The rental price of apartments and the number of bedrooms in the apartment.

(d) The amount of sugar added to a cup of coffee and how sweet the coffee tastes.

(e) The student evaluation scores for an instructor and the student evaluation scores for the course.

**2.10**  **Parents' income and student loans.** How well does the income of a college student's parents predict how much the student will borrow to pay for college? We have data on parents' income and college debt for a sample of 1200 recent college graduates. What are the explanatory and response variables? Are these variables categorical or quantitative? Do you expect a positive or negative association between these variables? Why?

**2.11**  **Reading ability and IQ.** A study of reading ability in schoolchildren chose 60 fifth-grade children at random from a school. The researchers had the children's scores on an IQ test and on a test of reading ability.[8] Figure 2.6 (on page 96) plots reading test score (response) against IQ score (explanatory).

(a) Explain why we should expect a positive association between IQ and reading score for children in the same grade. Does the scatterplot show a positive association?

(b) A group of four points appear to be outliers. In what way do these children's IQ and reading scores deviate from the overall pattern?

(c) Ignoring the outliers, is the association between IQ and reading scores roughly linear? Is it very strong? Explain your answers.

**2.12**  **Treasury bills and common stocks.** What is the relationship between returns from buying Treasury bills and returns from buying common stocks? The stemplots in Figure 1.22 (page 44) show the two

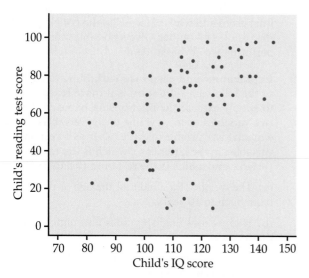

**FIGURE 2.6** IQ and reading test scores for 60 fifth-grade children, for Exercise 2.11.

individual distributions of percent returns. To see the relationship, we need a scatterplot. Figure 2.7 plots the annual returns on stocks for the years 1950 to 2003 against the returns on Treasury bills for the same years.

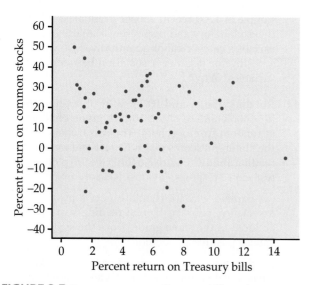

**FIGURE 2.7** Percent return on Treasury bills and common stocks for the years 1950 to 2003, for Exercise 2.12.

(a) The best year for stocks during this period was 1954. The worst year was 1974. About what were the returns on stocks in those two years?

(b) Treasury bills are a measure of the general level of interest rates. The years around 1980 saw very

high interest rates. Treasury bill returns peaked in 1981. About what was the percent return that year?

(c) Some people say that high Treasury bill returns tend to go with low returns on stocks. Does such a pattern appear clearly in Figure 2.7? Does the plot have any clear pattern?

**2.13** **Can children estimate their reading ability?** The main purpose of the study cited in Exercise 2.11 was to ask whether schoolchildren can estimate their own reading ability. The researchers had the children's scores on a test of reading ability. They asked each child to estimate his or her reading level, on a scale from 1 (low) to 5 (high). Figure 2.8 is a scatterplot of the children's estimates (response) against their reading scores (explanatory).

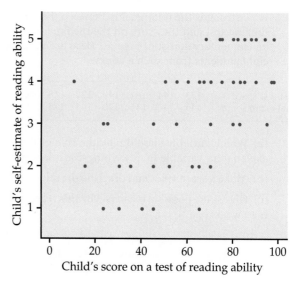

**FIGURE 2.8** Reading test scores for 60 fifth-grade children and the children's estimates of their own reading levels, for Exercise 2.13.

(a) What explains the "stair-step" pattern in the plot?

(b) Is there an overall positive association between reading score and self-estimate?

(c) There is one clear outlier. What is this child's self-estimated reading level? Does this appear to over- or underestimate the level as measured by the test?

**2.14** **Literacy of men and women.** Table 1.2 (page 10) shows the percent of men and women at least 15 years old who were literate in 2002 in the major Islamic nations for which data were available. Make a scatterplot of these data, taking male literacy as the explanatory variable. Describe the direction, form, and strength of the relationship. Are there any identical observations that plot as the same point? Are there any clear outliers?

**2.15   Small falcons in Sweden.** Often the percent of an animal species in the wild that survive to breed again is lower following a successful breeding season. This is part of nature's self-regulation, tending to keep population size stable. A study of merlins (small falcons) in northern Sweden observed the number of breeding pairs in an isolated area and the percent of males (banded for identification) who returned the next breeding season. Here are data for nine years:[9]

| Pairs   | 28 | 29 | 29 | 29 | 30 | 32 | 33 | 38 | 38 |
|---------|----|----|----|----|----|----|----|----|----|
| Percent | 82 | 83 | 70 | 61 | 69 | 58 | 43 | 50 | 47 |

(a) Why is the response variable the *percent* of males that return rather than the *number* of males that return?

(b) Make a scatterplot. To emphasize the pattern, also plot the mean response for years with 29 and 38 breeding pairs and draw lines connecting the mean responses for the six values of the explanatory variable.

(c) Describe the pattern. Do the data support the theory that a smaller percent of birds survive following a successful breeding season?

**2.16   City and highway gas mileage.** Table 1.10 (page 31) gives the city and highway gas mileages for minicompact and two-seater cars. We expect a positive association between the city and highway mileages of a group of vehicles. We have already seen that the Honda Insight is a different type of car, so omit it as you work with these data.

(a) Make a scatterplot that shows the relationship between city and highway mileage, using city mileage as the explanatory variable. Use different plotting symbols for the two types of cars.

(b) Interpret the plot. Is there a positive association? Is the form of the plot roughly linear? Is the form of the relationship similar for the two types of car? What is the most important difference between the two types?

**2.17   Social rejection and pain.** We often describe our emotional reaction to social rejection as "pain." A clever study asked whether social rejection causes activity in areas of the brain that are known to be activated by physical pain. If it does, we really do experience social and physical pain in similar ways. Subjects were first included and then deliberately

excluded from a social activity while increases in blood flow in their brains were measured. After each activity, the subjects filled out questionnaires that assessed how excluded they felt.

Below are data for 13 subjects.[10] The explanatory variable is "social distress" measured by each subject's questionnaire score after exclusion relative to the score after inclusion. (So values greater than 1 show the degree of distress caused by exclusion.) The response variable is activity in the anterior cingulate cortex, a region of the brain that is activated by physical pain.

| Subject | Social distress | Brain activity | Subject | Social distress | Brain activity |
|---------|-----------------|----------------|---------|-----------------|----------------|
| 1 | 1.26 | −0.055 | 8  | 2.18 | 0.025 |
| 2 | 1.85 | −0.040 | 9  | 2.58 | 0.027 |
| 3 | 1.10 | −0.026 | 10 | 2.75 | 0.033 |
| 4 | 2.50 | −0.017 | 11 | 2.75 | 0.064 |
| 5 | 2.17 | −0.017 | 12 | 3.33 | 0.077 |
| 6 | 2.67 | 0.017  | 13 | 3.65 | 0.124 |
| 7 | 2.01 | 0.021  |    |      |       |

Plot brain activity against social distress. Describe the direction, form, and strength of the relationship, as well as any outliers. Do the data suggest that brain activity in the "pain" region is directly related to the distress from social exclusion?

**2.18   Biological clocks.** Many plants and animals have "biological clocks" that coordinate activities with the time of day. When researchers looked at the length of the biological cycles in the plant *Arabidopsis* by measuring leaf movements, they found that the length of the cycle is not always 24 hours. The researchers suspected that the plants adapt their clocks to their north-south position. Plants don't know geography, but they do respond to light, so the researchers looked at the relationship between the plants' cycle lengths and the length of the day on June 21 at their locations. The data file has data on cycle length and day length, both in hours, for 146 plants.[11] Plot cycle length as the response variable against day length as the explanatory variable. Does there appear to be a positive association? Is it a strong association? Explain your answers.

**2.19   Business revenue and team value in the NBA.** Management theory says that the value of a business should depend on its operating income, the income produced by the business after taxes. (Operating income excludes income from sales of assets and investments, which don't reflect the actual business.) Total revenue, which ignores costs, should be less

TABLE 2.1

NBA teams as businesses

| Team | Value ($millions) | Revenue ($millions) | Income ($millions) |
|---|---|---|---|
| Los Angeles Lakers | 447 | 149 | 22.8 |
| New York Knicks | 401 | 160 | 13.5 |
| Chicago Bulls | 356 | 119 | 49.0 |
| Dallas Mavericks | 338 | 117 | −17.7 |
| Philadelphia 76ers | 328 | 109 | 2.0 |
| Boston Celtics | 290 | 97 | 25.6 |
| Detroit Pistons | 284 | 102 | 23.5 |
| San Antonio Spurs | 283 | 105 | 18.5 |
| Phoenix Suns | 282 | 109 | 21.5 |
| Indiana Pacers | 280 | 94 | 10.1 |
| Houston Rockets | 278 | 82 | 15.2 |
| Sacramento Kings | 275 | 102 | −16.8 |
| Washington Wizards | 274 | 98 | 28.5 |
| Portland Trail Blazers | 272 | 97 | −85.1 |
| Cleveland Cavaliers | 258 | 72 | 3.8 |
| Toronto Raptors | 249 | 96 | 10.6 |
| New Jersey Nets | 244 | 94 | −1.6 |
| Utah Jazz | 239 | 85 | 13.8 |
| Miami Heat | 236 | 91 | 7.9 |
| Minnesota Timberwolves | 230 | 85 | 6.9 |
| Memphis Grizzlies | 227 | 63 | −19.7 |
| Denver Nuggets | 218 | 75 | 7.9 |
| New Orleans Hornets | 216 | 80 | 21.9 |
| Los Angeles Clippers | 208 | 72 | 15.9 |
| Atlanta Hawks | 202 | 78 | −8.4 |
| Orlando Magic | 199 | 80 | 13.1 |
| Seattle Supersonics | 196 | 70 | 2.4 |
| Golden State Warriors | 188 | 70 | 7.8 |
| Milwaukee Bucks | 174 | 70 | −15.1 |

TABLE 2.2

Two measurements of foot deformities

| HAV angle | MA angle | HAV angle | MA angle | HAV angle | MA angle |
|---|---|---|---|---|---|
| 28 | 18 | 21 | 15 | 16 | 10 |
| 32 | 16 | 17 | 16 | 30 | 12 |
| 25 | 22 | 16 | 10 | 30 | 10 |
| 34 | 17 | 21 | 7 | 20 | 10 |
| 38 | 33 | 23 | 11 | 50 | 12 |
| 26 | 10 | 14 | 15 | 25 | 25 |
| 25 | 18 | 32 | 12 | 26 | 30 |
| 18 | 13 | 25 | 16 | 28 | 22 |
| 30 | 19 | 21 | 16 | 31 | 24 |
| 26 | 10 | 22 | 18 | 38 | 20 |
| 28 | 17 | 20 | 10 | 32 | 37 |
| 13 | 14 | 18 | 15 | 21 | 23 |
| 20 | 20 | 26 | 16 | | |

on 38 consecutive patients who came to a medical center for HAV surgery.[13] Using X-rays, doctors measured the angle of deformity for both MA and HAV. They speculated that there is a positive association—more serious MA is associated with more serious HAV.

(a) Make a scatterplot of the data in Table 2.2. (Which is the explanatory variable?)

(b) Describe the form, direction, and strength of the relationship between MA angle and HAV angle. Are there any clear outliers in your graph?

(c) Do you think the data confirm the doctors' speculation?

**2.21** **Body mass and metabolic rate.** Metabolic rate, the rate at which the body consumes energy, is important in studies of weight gain, dieting, and exercise. The table below gives data on the lean body mass and resting metabolic rate for 12 women and 7 men who are subjects in a study of dieting. Lean body mass, given in kilograms, is a person's weight leaving out all fat. Metabolic rate is measured in calories burned per 24 hours, the same calories

important. Table 2.1 shows the values, operating incomes, and revenues of an unusual group of businesses: the teams in the National Basketball Association (NBA).[12] Professional sports teams are generally privately owned, often by very wealthy individuals who may treat their team as a source of prestige rather than as a business.

(a) Plot team value against revenue. There are several outliers. Which teams are these, and in what way are they outliers? Is there a positive association between value and revenue? Is the pattern roughly linear?

(b) Now plot value against operating income. Are the same teams outliers? Does revenue or operating income better predict the value of an NBA team?

**2.20** **Two problems with feet.** Metatarsus adductus (call it MA) is a turning in of the front part of the foot that is common in adolescents and usually corrects itself. Hallux abducto valgus (call it HAV) is a deformation of the big toe that is not common in youth and often requires surgery. Perhaps the severity of MA can help predict the severity of HAV. Table 2.2 gives data

| Subject | Sex | Mass | Rate | Subject | Sex | Mass | Rate |
|---|---|---|---|---|---|---|---|
| 1 | M | 62.0 | 1792 | 11 | F | 40.3 | 1189 |
| 2 | M | 62.9 | 1666 | 12 | F | 33.1 | 913 |
| 3 | F | 36.1 | 995 | 13 | M | 51.9 | 1460 |
| 4 | F | 54.6 | 1425 | 14 | F | 42.4 | 1124 |
| 5 | F | 48.5 | 1396 | 15 | F | 34.5 | 1052 |
| 6 | F | 42.0 | 1418 | 16 | F | 51.1 | 1347 |
| 7 | M | 47.4 | 1362 | 17 | F | 41.2 | 1204 |
| 8 | F | 50.6 | 1502 | 18 | M | 51.9 | 1867 |
| 9 | F | 42.0 | 1256 | 19 | M | 46.9 | 1439 |
| 10 | M | 48.7 | 1614 | | | | |

**TABLE 2.3**

World record times for the 10,000-meter run

| | Men | | | Women | |
|---|---|---|---|---|---|
| Record year | Time (seconds) | Record year | Time (seconds) | Record year | Time (seconds) |
| 1912 | 1880.8 | 1962 | 1698.2 | 1967 | 2286.4 |
| 1921 | 1840.2 | 1963 | 1695.6 | 1970 | 2130.5 |
| 1924 | 1835.4 | 1965 | 1659.3 | 1975 | 2100.4 |
| 1924 | 1823.2 | 1972 | 1658.4 | 1975 | 2041.4 |
| 1924 | 1806.2 | 1973 | 1650.8 | 1977 | 1995.1 |
| 1937 | 1805.6 | 1977 | 1650.5 | 1979 | 1972.5 |
| 1938 | 1802.0 | 1978 | 1642.4 | 1981 | 1950.8 |
| 1939 | 1792.6 | 1984 | 1633.8 | 1981 | 1937.2 |
| 1944 | 1775.4 | 1989 | 1628.2 | 1982 | 1895.3 |
| 1949 | 1768.2 | 1993 | 1627.9 | 1983 | 1895.0 |
| 1949 | 1767.2 | 1993 | 1618.4 | 1983 | 1887.6 |
| 1949 | 1761.2 | 1994 | 1612.2 | 1984 | 1873.8 |
| 1950 | 1742.6 | 1995 | 1603.5 | 1985 | 1859.4 |
| 1953 | 1741.6 | 1996 | 1598.1 | 1986 | 1813.7 |
| 1954 | 1734.2 | 1997 | 1591.3 | 1993 | 1771.8 |
| 1956 | 1722.8 | 1997 | 1587.8 | | |
| 1956 | 1710.4 | 1998 | 1582.7 | | |
| 1960 | 1698.8 | 2004 | 1580.3 | | |

used to describe the energy content of foods. The researchers believe that lean body mass is an important influence on metabolic rate.

(a) Make a scatterplot of the data, using different symbols or colors for men and women.

(b) Is the association between these variables positive or negative? What is the form of the relationship? How strong is the relationship? Does the pattern of the relationship differ for women and men? How do the male subjects as a group differ from the female subjects as a group?

**2.22 Fuel consumption and speed.** How does the fuel consumption of a car change as its speed increases? Below are data for a British Ford Escort. Speed is measured in kilometers per hour, and fuel consumption is measured in liters of gasoline used per 100 kilometers traveled.[14]

| Speed (km/h) | Fuel used (liters/100 km) | Speed (km/h) | Fuel used (liter/100 km) |
|---|---|---|---|
| 10 | 21.00 | 90 | 7.57 |
| 20 | 13.00 | 100 | 8.27 |
| 30 | 10.00 | 110 | 9.03 |
| 40 | 8.00 | 120 | 9.87 |
| 50 | 7.00 | 130 | 10.79 |
| 60 | 5.90 | 140 | 11.77 |
| 70 | 6.30 | 150 | 12.83 |
| 80 | 6.95 | | |

(a) Make a scatterplot. (Which variable should go on the x axis?)

(b) Describe the form of the relationship. In what way is it not linear? Explain why the form of the relationship makes sense.

(c) It does not make sense to describe the variables as either positively associated or negatively associated. Why not?

(d) Is the relationship reasonably strong or quite weak? Explain your answer.

**2.23 World records for the 10K.** Table 2.3 shows the progress of world record times (in seconds) for the 10,000-meter run up to mid-2004.[15] Concentrate on the women's world record times. Make a scatterplot with year as the explanatory variable. Describe the pattern of improvement over time that your plot displays.

**2.24** CHALLENGE **How do icicles grow?** How fast do icicles grow? Japanese researchers measured the growth of icicles in a cold chamber under various conditions of temperature, wind, and water flow.[16] Table 2.4 contains data produced under two sets of conditions. In both cases, there was no wind and the temperature was set at $-11°C$. Water flowed over the icicle at a higher rate (29.6 milligrams per second) in Run 8905 and at a slower rate (11.9 mg/s) in Run 8903.

(b) Drag this last point down until it is opposite the group of 10 points. How small can you make the correlation? Can you make the correlation positive? *A single outlier can greatly strengthen or weaken a correlation. Always plot your data to check for outlying points.*

2.47 **What is the correlation?** Suppose that women always married men 2 years older than themselves. Draw a scatterplot of the ages of 5 married couples, with the wife's age as the explanatory variable. What is the correlation r for your data? Why?

2.48 **High correlation does not mean that the values are the same.** Investment reports often include correlations. Following a table of correlations among mutual funds, a report adds, "Two funds can have perfect correlation, yet different levels of risk. For example, Fund A and Fund B may be perfectly correlated, yet Fund A moves 20% whenever Fund B moves 10%." Write a brief explanation, for someone who knows no statistics, of how this can happen. Include a sketch to illustrate your explanation.

2.49 **Student ratings of teachers.** A college newspaper interviews a psychologist about student ratings of the teaching of faculty members. The psychologist says, "The evidence indicates that the correlation between the research productivity and teaching rating of faculty members is close to zero." The paper reports this as "Professor McDaniel said that good researchers tend to be poor teachers, and vice versa." Explain why the paper's report is wrong. Write a statement in plain language (don't use the word "correlation") to explain the psychologist's meaning.

2.50 **What's wrong?** Each of the following statements contains a blunder. Explain in each case what is wrong.

(a) "There is a high correlation between the gender of American workers and their income."

(b) "We found a high correlation (r = 1.09) between students' ratings of faculty teaching and ratings made by other faculty members."

(c) "The correlation between planting rate and yield of corn was found to be r = 0.23 bushel."

2.51 **IQ and GPA.** Table 1.9 (page 29) reports data on 78 seventh-grade students. We expect a positive association between IQ and GPA. Moreover, some people think that self-concept is related to school performance. Examine in detail the relationships between GPA and the two explanatory variables IQ and self-concept. Are the relationships roughly linear? How strong are they? Are there unusual points? What is the effect of removing these points?

2.52 **Effect of a change in units.** Consider again the correlation r between the speed of a car and its gas consumption, from the data in Exercise 2.22 (page 99).

(a) Transform the data so that speed is measured in miles per hour and fuel consumption in gallons per mile. (There are 1.609 kilometers in a mile and 3.785 liters in a gallon.) Make a scatterplot and find the correlation for both the original and the transformed data. How did the change of units affect your results?

(b) Now express fuel consumption in miles per gallon. (So each value is $1/x$ if $x$ is gallons per mile.) Again make a scatterplot and find the correlation. How did this change of units affect your results?

(*Lesson:* The effects of a linear transformation of the form $x_{\text{new}} = a + bx$ are simple. The effects of a nonlinear transformation are more complex.)

## 2.3 Least-Squares Regression

Correlation measures the direction and strength of the linear (straight-line) relationship between two quantitative variables. If a scatterplot shows a linear relationship, we would like to summarize this overall pattern by drawing a line on the scatterplot. A *regression line* summarizes the relationship between two variables, but only in a specific setting: when one of the variables helps explain or predict the other. That is, regression describes a relationship between an explanatory variable and a response variable.

REGRESSION LINE

A **regression line** is a straight line that describes how a response variable $y$ changes as an explanatory variable $x$ changes. We often use a regression line to **predict** the value of $y$ for a given value of $x$. Regression, unlike correlation, requires that we have an explanatory variable and a response variable.

EXAMPLE

**2.12 Fidgeting and fat gain.**   Does fidgeting keep you slim? Some people don't gain weight even when they overeat. Perhaps fidgeting and other "nonexercise activity" (NEA) explains why—the body might spontaneously increase nonexercise activity when fed more. Researchers deliberately overfed 16 healthy young adults for 8 weeks. They measured fat gain (in kilograms) and, as an explanatory variable, increase in energy use (in calories) from activity other than deliberate exercise—fidgeting, daily living, and the like. Here are the data:[22]

| NEA increase (cal) | −94 | −57 | −29 | 135 | 143 | 151 | 245 | 355 |
|---|---|---|---|---|---|---|---|---|
| Fat gain (kg) | 4.2 | 3.0 | 3.7 | 2.7 | 3.2 | 3.6 | 2.4 | 1.3 |

| NEA increase (cal) | 392 | 473 | 486 | 535 | 571 | 580 | 620 | 690 |
|---|---|---|---|---|---|---|---|---|
| Fat gain (kg) | 3.8 | 1.7 | 1.6 | 2.2 | 1.0 | 0.4 | 2.3 | 1.1 |

Figure 2.11 is a scatterplot of these data. The plot shows a moderately strong negative linear association with no outliers. The correlation is $r = -0.7786$. People with larger increases in nonexercise activity do indeed gain less fat. A line drawn through the points will describe the overall pattern well.

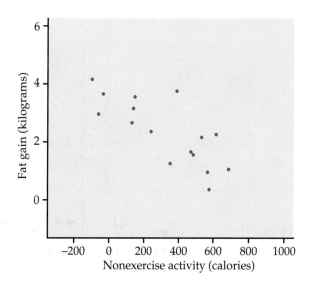

FIGURE 2.11 Fat gain after 8 weeks of overeating, plotted against the increase in nonexercise activity over the same period, for Example 2.12.

## Fitting a line to data

*fitting a line*

When a scatterplot displays a linear pattern, we can describe the overall pattern by drawing a straight line through the points. Of course, no straight line passes exactly through all of the points. **Fitting a line** to data means drawing a line that comes as close as possible to the points. The equation of a line fitted to the data gives a compact description of the dependence of the response variable $y$ on the explanatory variable $x$.

---

### STRAIGHT LINES

Suppose that $y$ is a response variable (plotted on the vertical axis) and $x$ is an explanatory variable (plotted on the horizontal axis). A straight line relating $y$ to $x$ has an equation of the form

$$y = b_0 + b_1 x$$

In this equation, $b_1$ is the **slope,** the amount by which $y$ changes when $x$ increases by one unit. The number $b_0$ is the **intercept,** the value of $y$ when $x = 0$.

---

**EXAMPLE**

**2.13 Regression line for fat gain.** Any straight line describing the nonexercise activity data has the form

$$\text{fat gain} = b_0 + (b_1 \times \text{NEA increase})$$

In Figure 2.12 we have drawn the regression line with the equation

$$\text{fat gain} = 3.505 - (0.00344 \times \text{NEA increase})$$

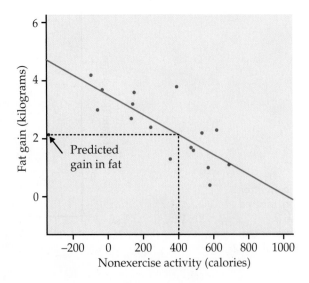

**FIGURE 2.12** A regression line fitted to the nonexercise activity data and used to predict fat gain for an NEA increase of 400 calories.

The figure shows that this line fits the data well. The slope $b_1 = -0.00344$ tells us that fat gained goes down by 0.00344 kilogram for each added calorie of NEA.

The slope $b_1$ of a line $y = b_0 + b_1x$ is the *rate of change* in the response $y$ as the explanatory variable $x$ changes. The slope of a regression line is an important numerical description of the relationship between the two variables. For Example 2.13, the intercept, $b_0 = 3.505$ kilograms, is the estimated fat gain if NEA does not change when a person overeats.

## USE YOUR KNOWLEDGE

**2.53 Plot the data with the line.** Make a sketch of the data in Example 2.12 and plot the line

$$\text{fat gain} = 4.505 - (0.00344 \times \text{NEA increase})$$

on your sketch. Explain why this line does not give a good fit to the data.

## Prediction

prediction    We can use a regression line to **predict** the response $y$ for a specific value of the explanatory variable $x$.

**EXAMPLE**

**2.14 Prediction for fat gain.** Based on the linear pattern, we want to predict the fat gain for an individual whose NEA increases by 400 calories when she overeats. To use the fitted line to predict fat gain, go "up and over" on the graph in Figure 2.12. From 400 calories on the $x$ axis, go up to the fitted line and over to the $y$ axis. The graph shows that the predicted gain in fat is a bit more than 2 kilograms.

If we have the equation of the line, it is faster and more accurate to substitute $x = 400$ in the equation. The predicted fat gain is

$$\text{fat gain} = 3.505 - (0.00344 \times 400) = 2.13 \text{ kilograms}$$

The accuracy of predictions from a regression line depends on how much scatter about the line the data show. In Figure 2.12, fat gains for similar increases in NEA show a spread of 1 or 2 kilograms. The regression line summarizes the pattern but gives only roughly accurate predictions.

## USE YOUR KNOWLEDGE

**2.54 Predict the fat gain.** Use the regression equation in Example 2.13 to predict the fat gain for a person whose NEA increases by 600 calories.

**2.15 Is this prediction reasonable?**    Can we predict the fat gain for some-one whose nonexercise activity increases by 1500 calories when she overeats? We can certainly substitute 1500 calories into the equation of the line. The prediction is

$$\text{fat gain} = 3.505 - (0.00344 \times 1500) = -1.66 \text{ kilograms}$$

That is, we predict that this individual loses fat when she overeats. This pre-diction is not trustworthy. Look again at Figure 2.12. An NEA increase of 1500 calories is far outside the range of our data. We can't say whether increases this large ever occur, or whether the relationship remains linear at such ex-treme values. Predicting fat gain when NEA increases by 1500 calories *extrap-olates* the relationship beyond what the data show.

---

### EXTRAPOLATION

**Extrapolation** is the use of a regression line for prediction far outside the range of values of the explanatory variable $x$ used to obtain the line. Such predictions are often not accurate.

---

## USE YOUR KNOWLEDGE

**2.55 Would you use the regression equation to predict?** Consider the following values for NEA increase: −400, 200, 500, 1000. For each, de-cide whether you would use the regression equation in Example 2.13 to predict fat gain or whether you would be concerned that the predic-tion would not be trustworthy because of extrapolation. Give reasons for your answers.

## Least-squares regression

Different people might draw different lines by eye on a scatterplot. This is es-pecially true when the points are widely scattered. We need a way to draw a regression line that doesn't depend on our guess as to where the line should go. No line will pass exactly through all the points, but we want one that is as close as possible. We will use the line to predict $y$ from $x$, so we want a line that is as close as possible to the points in the *vertical* direction. That's because the prediction errors we make are errors in $y$, which is the vertical direction in the scatterplot.

The line in Figure 2.12 predicts 2.13 kilograms of fat gain for an increase in nonexercise activity of 400 calories. If the actual fat gain turns out to be 2.3 kilograms, the error is

$$\text{error} = \text{observed gain} \ - \ \text{predicted gain}$$

$$= 2.3 - 2.13 = 0.17 \text{ kilograms}$$

Errors are positive if the observed response lies above the line, and negative if the response lies below the line. We want a regression line that makes these prediction errors as small as possible. Figure 2.13 illustrates the idea. For clarity, the plot shows only three of the points from Figure 2.12, along with the line, on an expanded scale. The line passes below two of the points and above one of them. The vertical distances of the data points from the line appear as vertical line segments. A "good" regression line makes these distances as small as possible. There are many ways to make "as small as possible" precise. The most common is the *least-squares* idea. The line in Figures 2.12 and 2.13 is in fact the least-squares regression line.

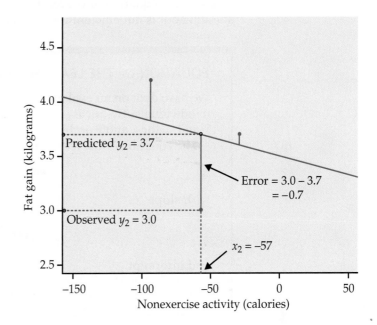

**FIGURE 2.13** The least-squares idea: make the errors in predicting *y* as small as possible by minimizing the sum of their squares.

---

### LEAST-SQUARES REGRESSION LINE

The **least-squares regression line of *y* on *x*** is the line that makes the sum of the squares of the vertical distances of the data points from the line as small as possible.

---

Here is the least-squares idea expressed as a mathematical problem. We represent *n* observations on two variables *x* and *y* as

$$(x_1, y_1), \ (x_2, y_2), \ \ldots, \ (x_n, y_n)$$

If we draw a line $y = b_0 + b_1 x$ through the scatterplot of these observations, the line predicts the value of *y* corresponding to $x_i$ as $\hat{y}_i = b_0 + b_1 x_i$. We write $\hat{y}$ (read "y-hat") in the equation of a regression line to emphasize that the line gives a *predicted* response $\hat{y}$ for any *x*. The predicted response will usually not be exactly the same as the actually *observed* response *y*. The method of least squares chooses the line that makes the sum of the squares of these errors as small as possible. To find this line, we must find the values of the intercept $b_0$

and the slope $b_1$ that minimize

$$\sum(\text{error})^2 = \sum(y_i - b_0 - b_1 x_i)^2$$

for the given observations $x_i$ and $y_i$. For the NEA data, for example, we must find the $b_0$ and $b_1$ that minimize

$$(-94 - b_0 - 4.2b_1)^2 + (-57 - b_0 - 3.0b_1)^2 + \cdots + (690 - b_0 - 1.1b_1)^2$$

These values are the intercept and slope of the least-squares line.

You will use software or a calculator with a regression function to find the equation of the least-squares regression line from data on $x$ and $y$. We will therefore give the equation of the least-squares line in a form that helps our understanding but is not efficient for calculation.

---

### EQUATION OF THE LEAST-SQUARES REGRESSION LINE

We have data on an explanatory variable $x$ and a response variable $y$ for $n$ individuals. The means and standard deviations of the sample data are $\bar{x}$ and $s_x$ for $x$ and $\bar{y}$ and $s_y$ for $y$, and the correlation between $x$ and $y$ is $r$. The equation of the least-squares regression line of $y$ on $x$ is

$$\hat{y} = b_0 + b_1 x$$

with **slope**

$$b_1 = r\frac{s_y}{s_x}$$

and **intercept**

$$b_0 = \bar{y} - b_1\bar{x}$$

---

**EXAMPLE**

**2.16 Check the calculations.** Verify from the data in Example 2.12 that the mean and standard deviation of the 16 increases in NEA are

$$\bar{x} = 324.8 \text{ calories} \quad \text{and} \quad s_x = 257.66 \text{ calories}$$

The mean and standard deviation of the 16 fat gains are

$$\bar{y} = 2.388 \text{ kg} \quad \text{and} \quad s_y = 1.1389 \text{ kg}$$

The correlation between fat gain and NEA increase is $r = -0.7786$. The least-squares regression line of fat gain $y$ on NEA increase $x$ therefore has slope

$$b_1 = r\frac{s_y}{s_x} = -0.7786\frac{1.1389}{257.66}$$

$$= -0.00344 \text{ kg per calorie}$$

and intercept

$$b_0 = \bar{y} - b_1\bar{x} = 2.388 - (-0.00344)(324.8)$$

$$= 3.505 \text{ kg}$$

The equation of the least-squares line is

$$\hat{y} = 3.505 - 0.00344x$$

*When doing calculations like this by hand, you may need to carry extra decimal places in the preliminary calculations to get accurate values of the slope and intercept.* Using software or a calculator with a regression function eliminates this worry.

## Interpreting the regression line

The slope $b_1 = -0.00344$ kilograms per calorie in Example 2.16 is the change in fat gain as NEA increases. The units "kilograms of fat gained per calorie of NEA" come from the units of $y$ (kilograms) and $x$ (calories). Although the correlation does not change when we change the units of measurement, the equation of the least-squares line does change. The slope in grams per calorie would be 1000 times as large as the slope in kilograms per calorie, because there are 1000 grams in a kilogram. The small value of the slope, $b_1 = -0.00344$, does not mean that the effect of increased NEA on fat gain is small—it just reflects the choice of kilograms as the unit for fat gain. *The slope and intercept of the least-squares line depend on the units of measurement—you can't conclude anything from their size.*

The expression $b_1 = rs_y/s_x$ for the slope says that, along the regression line, **a change of one standard deviation in $x$ corresponds to a change of $r$ standard deviations in $y$.** When the variables are perfectly correlated ($r = 1$ or $r = -1$), the change in the predicted response $\hat{y}$ is the same (in standard deviation units) as the change in $x$. Otherwise, when $-1 < r < 1$, the change in $\hat{y}$ is less than the change in $x$. As the correlation grows less strong, the prediction $\hat{y}$ moves less in response to changes in $x$.

**The least-squares regression line always passes through the point $(\overline{x}, \overline{y})$** on the graph of $y$ against $x$. Check that when you substitute $\overline{x} = 324.8$ into the equation of the regression line in Example 2.16, the result is $\hat{y} = 2.388$, equal to the mean of $y$. So the least-squares regression line of $y$ on $x$ is the line with slope $rs_y/s_x$ that passes through the point $(\overline{x}, \overline{y})$. We can describe regression entirely in terms of the basic descriptive measures $\overline{x}$, $s_x$, $\overline{y}$, $s_y$, and $r$. If both $x$ and $y$ are standardized variables, so that their means are 0 and their standard deviations are 1, then the regression line has slope $r$ and passes through the origin.

Figure 2.14 displays the basic regression output for the nonexercise activity data from two statistical software packages. Other software produces very similar output. You can find the slope and intercept of the least-squares line, calculated to more decimal places than we need, in both outputs. The software also provides information that we do not yet need, including some that we trimmed from Figure 2.14. Part of the art of using software is to ignore the extra information that is almost always present. Look for the results that you need. Once you understand a statistical method, you can read output from almost any software.

## Correlation and regression

Least-squares regression looks at the distances of the data points from the line only in the $y$ direction. So the two variables $x$ and $y$ play different roles in regression.

## SECTION 2.3 Exercises

*For Exercises 2.53 and 2.54, see page 111; for Exercise 2.55, see page 112; and for Exercise 2.56, see page 117.*

**2.57  The regression equation.** The equation of a least-squares regression line is $y = 10 + 5x$.

(a) What is the value of $y$ for $x = 5$?

(b) If $x$ increases by one unit, what is the corresponding increase in $y$?

(c) What is the intercept for this equation?

**2.58  First test and final exam.** In Exercise 2.6 you looked at the relationship between the score on the first test and the score on the final exam in an elementary statistics course. Here are data for eight students from such a course:

| First-test score | 153 | 144 | 162 | 149 | 127 | 118 | 158 | 153 |
|---|---|---|---|---|---|---|---|---|
| Final-exam score | 145 | 140 | 145 | 170 | 145 | 175 | 170 | 160 |

(a) Plot the data with the first-test scores on the $x$ axis and the final-exam scores on the $y$ axis.

(b) Find the least-squares regression line for predicting the final-exam score using the first-test score.

(c) Graph the least-squares regression line on your plot.

**2.59  Second test and final exam.** Refer to the previous exercise. Here are the data for the second test and the final exam for the same students:

| Second-test score | 158 | 162 | 144 | 162 | 136 | 158 | 175 | 153 |
|---|---|---|---|---|---|---|---|---|
| Final-exam score | 145 | 140 | 145 | 170 | 145 | 175 | 170 | 160 |

(a) Plot the data with the second-test scores on the $x$ axis and the final-exam scores on the $y$ axis.

(b) Find the least-squares regression line for predicting the final-exam score using the second-test score.

(c) Graph the least-squares regression line on your plot.

**2.60  The effect of an outlier.** Refer to the previous exercise. Add a ninth student whose scores on the second test and final exam would lead you to classify the additional data point as an outlier. Recalculate the least-squares regression line with this additional case and summarize the effect it has on the least-squares regression line.

**2.61  The effect of a different point.** Examine the data in Exercise 2.31 and add a ninth student who has low scores on the second test and the final exam, and fits the overall pattern of the other scores in the data set. Recalculate the least-squares regression line with this additional case and summarize the effect it has on the least-squares regression line.

**2.62  Revenue and value of NBA teams.** Table 2.1 (page 98) gives the values of the 29 teams in the National Basketball Association, along with their operating incomes and revenues. Plots and correlations show that revenue predicts team value much better than does operating income. The least-squares regression line for predicting value from revenue is

$$\text{value} = 21.4 + (2.59 \times \text{revenue})$$

(a) What is the slope of this line? Express in simple language what the slope says about the relationship of value to revenue.

(b) The Los Angeles Lakers are the NBA's most valuable team, valued at $447 million, with $149 million in revenue. Use the line to predict the value of the Lakers from their revenue. What is the error in this prediction?

(c) The correlation between revenue and team value is $r = 0.9265$. What does the correlation say about the success of the regression line in predicting the values of the 29 teams?

**2.63  Water discharged by the Mississippi River.** Figure 1.10(b) (page 19) is a time plot of the volume of water discharged by the Mississippi River for the years 1954 to 2001. Water volume is recorded in cubic kilometers. The trend line on the plot is the least-squares regression line. The equation of this line is

$$\text{water discharged} = -7792 + (4.2255 \times \text{year})$$

(a) How much (on the average) does the volume of water increase with each passing year?

(b) What does the equation say about the volume of water flowing out of the Mississippi in the year 1780? Why is this extrapolation clearly nonsense?

(c) What is the predicted volume discharged in 1990 (round to the nearest cubic kilometer)? What is the prediction error for 1990?

(d) Can you see evidence of the great floods of 1973 and 1993, even on the plot of annual water discharged? Explain.

**2.64  Perch and bass.** Example 2.8 (page 91) gives data from an experiment in ecology. Figure 2.4 is the scatterplot of proportion of perch eaten by bass against the number of perch in a pen before the bass were let in. There is a roughly linear pattern. The least-squares line for predicting proportion eaten from initial count of perch is

$$\text{proportion eaten} = 0.120 + (0.0086 \times \text{count})$$

(a) When 10 more perch are added to a pen, what happens to the proportion that are eaten (according to the line)? Explain your answer.

(b) If there are no perch in a pen, what proportion does the line predict will be eaten? Explain why this prediction is nonsense. What is wrong with using the regression line to predict $y$ when $x = 0$? *You see that the intercept, though it is needed to draw the line, may have no statistical interpretation if $x = 0$ is outside the range of the data.*

**2.65  Progress in math scores.** Every few years, the National Assessment of Educational Progress asks a national sample of eighth-graders to perform the same math tasks. The goal is to get an honest picture of progress in math. Here are the last few national mean scores, on a scale of 0 to 500:[25]

| Year  | 1990 | 1992 | 1996 | 2000 | 2003 | 2005 |
|-------|------|------|------|------|------|------|
| Score | 263  | 268  | 272  | 273  | 278  | 279  |

(a) Make a time plot of the mean scores, by hand. This is just a scatterplot of score against year. There is a slow linear increasing trend.

(b) Find the regression line of mean score on time step-by-step. First calculate the mean and standard deviation of each variable and their correlation (use a calculator with these functions). Then find the equation of the least-squares line from these. Draw the line on your scatterplot. What percent of the year-to-year variation in scores is explained by the linear trend?

(c) Now use software or the regression function on your calculator to verify your regression line.

**2.66  The Trans-Alaska Oil Pipeline.** Figure 2.3 (page 90) plots field measurements on the depth of 100 small defects in the Trans-Alaska Oil Pipeline against laboratory measurements of the same defects. Drawing the $y = x$ line on the graph shows that field measurements tend to be too low for larger defect depths.

(a) Find the equation of the least-squares regression line for predicting field measurement from laboratory measurement. Make a scatterplot with this line drawn on it. How does the least-squares line differ from the $y = x$ line?

(b) What is the slope of the $y = x$ line? What is the slope of the regression line? Say in simple language what these slopes mean.

**2.67  Social exclusion and pain.** Exercise 2.17 (page 97) gives data from a study that shows that social exclusion causes "real pain." That is, activity in the area of the brain that responds to physical pain goes up as distress from social exclusion goes up. Your scatterplot in Exercise 2.17 shows a moderately strong linear relationship.

(a) What is the equation of the least-squares regression line for predicting brain activity from social distress score? Make a scatterplot with this line drawn on it.

(b) On your plot, show the "up and over" lines that predict brain activity for social distress score 2.0. Use the equation of the regression line to get the predicted brain activity level. Verify that it agrees with your plot.

(c) What percent of the variation in brain activity among these subjects is explained by the straight-line relationship with social distress score?

**2.68  Problems with feet.** Your scatterplot in Exercise 2.20 (page 98) suggests that the severity of the mild foot deformity called MA can help predict the severity of the more serious deformity called HAV. Table 2.2 (page 98) gives data for 38 young patients.

(a) Find the equation of the least-squares regression line for predicting HAV angle from MA angle. Add this line to the scatterplot you made in Exercise 2.20.

(b) A new patient has MA angle 25 degrees. What do you predict this patient's HAV angle to be?

(c) Does knowing MA angle allow doctors to predict HAV angle accurately? Explain your answer from the scatterplot, then calculate a numerical measure to support your finding.

**2.69  Growth of icicles.** Table 2.4 (page 100) gives data on the growth of icicles at two rates of water flow. You examined these data in Exercise 2.24. Use least-squares regression to estimate the rate (centimeters per minute) at which icicles grow at these two flow rates. How does flow rate affect growth?

**2.70  Mutual funds.** Exercise 2.28 (page 101) gives the returns of 23 Fidelity "sector funds" for the years

2002 and 2003. These mutual funds invest in narrow segments of the stock market. They often rise faster than the overall market in up-years, such as 2003, and fall faster than the market in down-years, such as 2002. A scatterplot shows that Fidelity Gold Fund—the only fund that went up in 2002—is an outlier. In Exercise 2.38, you showed that this outlier has a strong effect on the correlation. The least-squares line, like the correlation, is not resistant.

(a) Find the equations of two least-squares lines for predicting 2003 return from 2002 return, one for all 23 funds and one omitting Fidelity Gold Fund. Make a scatterplot with both lines drawn on it. The two lines are very different.

(b) Starting with the least-squares idea, explain why adding Fidelity Gold Fund to the other 22 funds moves the line in the direction that your graph shows.

**2.71** **Stocks and Treasury bills.** The scatterplot in Figure 2.7 (page 96) suggests that returns on common stocks may be somewhat lower in years with high interest rates. Here is part of the output from software for the regression of stock returns on the Treasury bill returns for the same years:

```
Stock = 16.639318 - 0.67974913 Tbill
Sample size: 54
R (correlation coefficient) = -0.113
R-sq = 0.01275773
Estimate of error standard deviation: 17.680649
```

If you knew the return on Treasury bills for next year, do you think you could predict the return on stocks quite accurately? Use both the scatterplot in Figure 2.7 and a number from the regression output to justify your answer.

**2.72** **Icicle growth.** Find the mean and standard deviation of the times and icicle lengths for the data on Run 8903 in Table 2.4 (page 100). Find the correlation between the two variables. Use these five numbers to find the equation of the regression line for predicting length from time. Verify that your result agrees with that in Exercise 2.69. Use the same five numbers to find the equation of the regression line for predicting the time an icicle has been growing from its length. What units does the slope of each of these lines have?

**2.73** **Metabolic rate and lean body mass.** Compute the mean and the standard deviation of the metabolic rates and lean body masses in Exercise 2.21 (page 98) and the correlation between these two variables. Use these values to find the slope of the regression line of metabolic rate on lean body mass. Also find the slope of the regression line of lean body mass on metabolic rate. What are the units for each of the two slopes?

**2.74** **IQ and self-concept.** Table 1.9 (page 29) reports data on 78 seventh-grade students. We want to know how well each of IQ score and self-concept score predicts GPA using least-squares regression. We also want to know which of these explanatory variables predicts GPA better. Give numerical measures that answer these questions, and explain your answers.

**2.75** **Heights of husbands and wives.** The mean height of American women in their early twenties is about 64.5 inches and the standard deviation is about 2.5 inches. The mean height of men the same age is about 68.5 inches, with standard deviation about 2.7 inches. If the correlation between the heights of husbands and wives is about $r = 0.5$, what is the equation of the regression line of the husband's height on the wife's height in young couples? Draw a graph of this regression line. Predict the height of the husband of a woman who is 67 inches tall.

**2.76** **A property of the least-squares regression line.** Use the equation for the least-squares regression line to show that this line always passes through the point $(\bar{x}, \bar{y})$.

**2.77** **Icicle growth.** The data for Run 8903 in Table 2.4 (page 100) describe how the length $y$ in centimeters of an icicle increases over time $x$. Time is measured in minutes.

(a) What are the numerical values and units of measurement for each of $\bar{x}$, $s_x$, $\bar{y}$, $s_y$, and the correlation $r$ between $x$ and $y$?

(b) There are 2.54 centimeters in an inch. If we measure length $y$ in inches rather than in centimeters, what are the new values of $\bar{y}$, $s_y$, and the correlation $r$?

(c) If we measure length $y$ in inches rather than in centimeters, what is the new value of the slope $b_1$ of the least-squares line for predicting length from time?

**2.78** **Predict final-exam scores.** In Professor Friedman's economics course the correlation between the students' total scores before the final examination and their final-examination scores is $r = 0.55$. The pre-exam totals for all students in the course have mean 270 and standard deviation 30. The final-exam scores have mean 70 and standard deviation 9. Professor Friedman has lost Julie's final exam but knows that her total before the exam was 310. He decides to predict her final-exam score from her pre-exam total.

(a) What is the slope of the least-squares regression line of final-exam scores on pre-exam total scores in this course? What is the intercept?

(b) Use the regression line to predict Julie's final-exam score.

(c) Julie doesn't think this method accurately predicts how well she did on the final exam. Calculate $r^2$ and use the value you get to argue that her actual score could have been much higher or much lower than the predicted value.

**2.79** CHALLENGE **Class attendance and grades.** A study of class attendance and grades among first-year students at a state university showed that in general students who attended a higher percent of their classes earned higher grades. Class attendance explained 16% of the variation in grade index among the students. What is the numerical value of the correlation between percent of classes attended and grade index?

**2.80** CHALLENGE **Pesticide decay.** Fenthion is a pesticide used to control the olive fruit fly. There are government limits on the amount of pesticide residue that can be present in olive products. Because the pesticide decays over time, producers of olive oil might simply store the oil until the fenthion has decayed. The simple exponential decay model says that the concentration $C$ of pesticide remaining after time $t$ is

$$C = C_0 e^{-kt}$$

where $C_0$ is the initial concentration and $k$ is a constant that determines the rate of decay. This model is a straight line if we take the logarithm of the concentration:

$$\log C = \log C_0 - kt$$

(The logarithm here is the natural logarithm, not the common logarithm with base 10.) Here are data on the concentration (milligrams of fenthion per kilogram of oil) in specimens of Greek olive oil:[26]

| Days stored | Concentration | | | | |
|---|---|---|---|---|---|
| 28 | 0.99 | 0.99 | 0.96 | 0.95 | 0.93 |
| 84 | 0.96 | 0.94 | 0.91 | 0.91 | 0.90 |
| 183 | 0.89 | 0.87 | 0.86 | 0.85 | 0.85 |
| 273 | 0.87 | 0.86 | 0.84 | 0.83 | 0.83 |
| 365 | 0.83 | 0.82 | 0.80 | 0.80 | 0.79 |

(a) Plot the natural logarithm of concentration against days stored. Notice that there are several pairs of identical data points. Does the pattern suggest that the model of simple exponential decay describes the data reasonably well, at least over this interval of time? Explain your answer.

(b) Regress the logarithm of concentration on time. Use your result to estimate the value of the constant $k$.

**2.81** CHALLENGE **The decay product is toxic.** Unfortunately, the main product of the decay of the pesticide fenthion is fenthion sulfoxide, which is also toxic. Here are data on the total concentration of fenthion and fenthion sulfoxide in the same specimens of olive oil described in the previous exercise:

| Days stored | Concentration | | | | |
|---|---|---|---|---|---|
| 28 | 1.03 | 1.03 | 1.01 | 0.99 | 0.99 |
| 84 | 1.05 | 1.04 | 1.00 | 0.99 | 0.99 |
| 183 | 1.03 | 1.02 | 1.01 | 0.98 | 0.98 |
| 273 | 1.07 | 1.06 | 1.03 | 1.03 | 1.02 |
| 365 | 1.06 | 1.02 | 1.01 | 1.01 | 0.99 |

(a) Plot concentration against days stored. Your software may fill the available space in the plot, which in this case hides the pattern. Try a plot with vertical scale from 0.8 to 1.2. Be sure your plot takes note of the pairs of identical data points.

(b) What is the slope of the least-squares line for predicting concentration of fenthion and fenthion sulfoxide from days stored? Explain why this value agrees with the graph.

(c) What do the data say about the idea of reducing fenthion in olive oil by storing the oil before selling it?

# 2.4 Cautions about Correlation and Regression

Correlation and regression are among the most common statistical tools. They are used in more elaborate form to study relationships among many variables, a situation in which we cannot see the essentials by studying a single scatterplot.

We need a firm grasp of the use and limitations of these tools, both now and as a foundation for more advanced statistics.

## Residuals

A regression line describes the overall pattern of a linear relationship between an explanatory variable and a response variable. Deviations from the overall pattern are also important. In the regression setting, we see deviations by looking at the scatter of the data points about the regression line. The vertical distances from the points to the least-squares regression line are as small as possible in the sense that they have the smallest possible sum of squares. Because they represent "left-over" variation in the response after fitting the regression line, these distances are called *residuals*.

---

### RESIDUALS

A **residual** is the difference between an observed value of the response variable and the value predicted by the regression line. That is,

$$\text{residual} = \text{observed } y - \text{predicted } y$$
$$= y - \hat{y}$$

---

**EXAMPLE**

**2.19 Residuals for fat gain.**   Example 2.12 (page 109) describes measurements on 16 young people who volunteered to overeat for 8 weeks. Those whose nonexercise activity (NEA) spontaneously rose substantially gained less fat than others. Figure 2.20(a) is a scatterplot of these data. The pattern is linear. The least-squares line is

$$\text{fat gain} = 3.505 - (0.00344 \times \text{NEA increase})$$

One subject's NEA rose by 135 calories. That subject gained 2.7 kilograms of fat. The predicted gain for 135 calories is

$$\hat{y} = 3.505 - (0.00344 \times 135) = 3.04 \text{ kg}$$

The residual for this subject is therefore

$$\text{residual} = \text{observed } y - \text{predicted } y$$
$$= y - \hat{y}$$
$$= 2.7 - 3.04 = -0.34 \text{ kg}$$

Most regression software will calculate and store residuals for you.

### USE YOUR KNOWLEDGE

**2.82  Find the predicted value and the residual.** Another individual in the NEA data set has NEA increase equal to 143 calories and fat gain

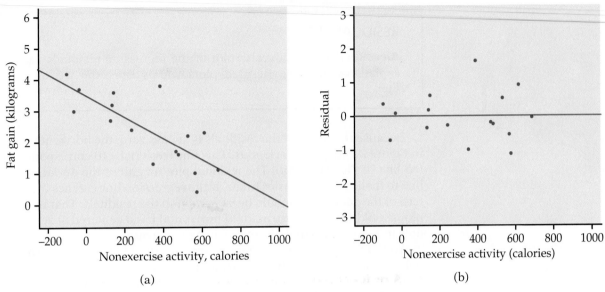

**FIGURE 2.20** (a) Scatterplot of fat gain versus increase in nonexercise activity, with the least-squares line, for Example 2.19. (b) Residual plot for the regression displayed in Figure 2.20(a). The line at $y = 0$ marks the mean of the residuals.

equal to 3.2 kg. Find the predicted value of fat gain for this individual and then calculate the residual. Explain why this residual is positive.

Because the residuals show how far the data fall from our regression line, examining the residuals helps assess how well the line describes the data. Although residuals can be calculated from any model fitted to the data, the residuals from the least-squares line have a special property: **the mean of the least-squares residuals is always zero.**

## USE YOUR KNOWLEDGE

**2.83  Find the sum of the residuals.** Here are the 16 residuals for the NEA data rounded to two decimal places:

| | | | | | | | |
|---|---|---|---|---|---|---|---|
| 0.37 | −0.70 | 0.10 | −0.34 | 0.19 | 0.61 | −0.26 | −0.98 |
| 1.64 | −0.18 | −0.23 | 0.54 | −0.54 | −1.11 | 0.93 | −0.03 |

Find the sum of these residuals. Note that the sum is not exactly zero because of roundoff error.

You can see the residuals in the scatterplot of Figure 2.20(a) by looking at the vertical deviations of the points from the line. The *residual plot* in Figure 2.20(b) makes it easier to study the residuals by plotting them against the explanatory variable, increase in NEA.

> ### RESIDUAL PLOTS
>
> A **residual plot** is a scatterplot of the regression residuals against the explanatory variable. Residual plots help us assess the fit of a regression line.

Because the mean of the residuals is always zero, the horizontal line at zero in Figure 2.20(b) helps orient us. This line (residual = 0) corresponds to the fitted line in Figure 2.20(a). The residual plot magnifies the deviations from the line to make patterns easier to see. If the regression line catches the overall pattern of the data, there should be *no pattern* in the residuals. That is, the residual plot should show an unstructured horizontal band centered at zero. The residuals in Figure 2.20(b) do have this irregular scatter.

You can see the same thing in the scatterplot of Figure 2.20(a) and the residual plot of Figure 2.20(b). It's just a bit easier in the residual plot. Deviations from an irregular horizontal pattern point out ways in which the regression line fails to catch the overall pattern. For example, if the overall pattern in the scatterplot is curved rather than straight, the residuals will magnify the curved pattern, moving up and down rather than straight across. Exercise 2.86 is an example of this. Here is a different kind of example.

**EXAMPLE**

**2.20 Patterns in the Trans-Alaska Oil Pipeline residuals.** Figure 2.3 (page 90) plots field measurements on the depth of 100 small defects in the Trans-Alaska Oil Pipeline against laboratory measurements of the same defects. The $y = x$ line on the graph shows that field measurements tend to be too low for larger defect depths. The least-squares regression line for predicting field result from lab result, unlike the $y = x$ line, goes through the center of the points. Figure 2.21 is the residual plot for these data.

Although the horizontal line at zero does go through the middle of the points, the residuals are more spread out both above and below the line as we

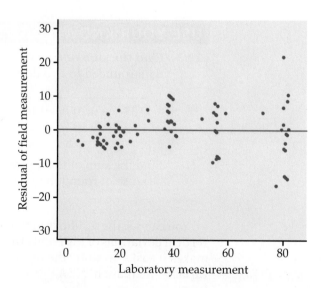

FIGURE 2.21 Residual plot for the regression of field measurements of Alaska pipeline defects on laboratory measurements of the same defects, for Example 2.20.

move to the right. The field measurements are more variable as the true defect depth measured in the lab increases. There is indeed a straight-line pattern, but the regression line doesn't catch the important fact that the variability of field measurements increases with defect depth. The scatterplot makes this clear, and the residual plot magnifies the picture.

## Outliers and influential observations

When you look at scatterplots and residual plots, look for striking individual points as well as for an overall pattern. Here is an example of data that contain some unusual cases.

**EXAMPLE**

**2.21 Diabetes and blood sugar.** People with diabetes must manage their blood sugar levels carefully. They measure their fasting plasma glucose (FPG) several times a day with a glucose meter. Another measurement, made at regular medical checkups, is called HbA. This is roughly the percent of red blood cells that have a glucose molecule attached. It measures average exposure to glucose over a period of several months. Table 2.5 gives data on both HbA and FPG for 18 diabetics five months after they had completed a diabetes education class.[27]

Because both FPG and HbA measure blood glucose, we expect a positive association. The scatterplot in Figure 2.22 shows a surprisingly weak relationship, with correlation $r = 0.4819$. The line on the plot is the least-squares regression line for predicting FPG from HbA. Its equation is

$$\hat{y} = 66.4 + 10.41x$$

It appears that one-time measurements of FPG can vary quite a bit among people with similar long-term levels, as measured by HbA.

Two unusual cases are marked in Figure 2.22. Subjects 15 and 18 are unusual in different ways. Subject 15 has dangerously high FPG and lies far from the regression line in the $y$ direction. Subject 18 is close to the line but far out

---

**TABLE 2.5**

Two measures of glucose level in diabetics

| Subject | HbA (%) | FPG (mg/ml) | Subject | HbA (%) | FPG (mg/ml) | Subject | HbA (%) | FPG (mg/ml) |
|---------|---------|-------------|---------|---------|-------------|---------|---------|-------------|
| 1 | 6.1 | 141 | 7 | 7.5 | 96 | 13 | 10.6 | 103 |
| 2 | 6.3 | 158 | 8 | 7.7 | 78 | 14 | 10.7 | 172 |
| 3 | 6.4 | 112 | 9 | 7.9 | 148 | 15 | 10.7 | 359 |
| 4 | 6.8 | 153 | 10 | 8.7 | 172 | 16 | 11.2 | 145 |
| 5 | 7.0 | 134 | 11 | 9.4 | 200 | 17 | 13.7 | 147 |
| 6 | 7.1 | 95 | 12 | 10.4 | 271 | 18 | 19.3 | 255 |

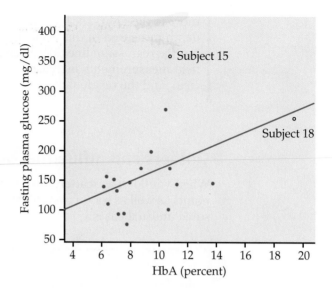

**FIGURE 2.22** Scatterplot of fasting plasma glucose against HbA (which measures long-term blood glucose), with the least-squares line, for Example 2.21.

in the *x* direction. The residual plot in Figure 2.23 confirms that Subject 15 has a large residual and that Subject 18 does not.

Points that are outliers in the *x* direction, like Subject 18, can have a strong influence on the position of the regression line. Least-squares lines make the sum of squares of the vertical distances to the points as small as possible. A point that is extreme in the *x* direction with no other points near it pulls the line toward itself.

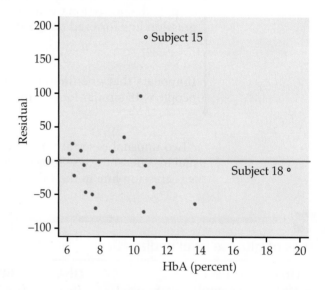

**FIGURE 2.23** Residual plot for the regression of FPG on HbA. Subject 15 is an outlier in *y*. Subject 18 is an outlier in *x* that may be influential but does not have a large residual.

## OUTLIERS AND INFLUENTIAL OBSERVATIONS IN REGRESSION

An **outlier** is an observation that lies outside the overall pattern of the other observations. Points that are outliers in the *y* direction of a scatterplot have large regression residuals, but other outliers need not have large residuals.

An observation is **influential** for a statistical calculation if removing it would markedly change the result of the calculation. Points that are outliers in the $x$ direction of a scatterplot are often influential for the least-squares regression line.

Influence is a matter of degree—how much does a calculation change when we remove an observation? It is difficult to assess influence on a regression line without actually doing the regression both with and without the suspicious observation. A point that is an outlier in $x$ is often influential. But if the point happens to lie close to the regression line calculated from the other observations, then its presence will move the line only a little and the point will not be influential. The influence of a point that is an outlier in $y$ depends on whether there are many other points with similar values of $x$ that hold the line in place. Figures 2.22 and 2.23 identify two unusual observations. How influential are they?

**EXAMPLE**

**2.22 Influential observations.** Subjects 15 and 18 both influence the correlation between FPG and HbA, in opposite directions. Subject 15 weakens the linear pattern; if we drop this point, the correlation increases from $r = 0.4819$ to $r = 0.5684$. Subject 18 extends the linear pattern; if we omit this subject, the correlation drops from $r = 0.4819$ to $r = 0.3837$.

To assess influence on the least-squares line, we recalculate the line leaving out a suspicious point. Figure 2.24 shows three least-squares lines. The solid line is the regression line of FPG on HbA based on all 18 subjects. This is the same line that appears in Figure 2.22. The dotted line is calculated from all subjects except Subject 18. You see that point 18 does pull the line down toward itself. But the influence of Subject 18 is not very large—the dotted and solid lines are close together for HbA values between 6 and 14, the range of all except Subject 18.

**FIGURE 2.24** Three regression lines for predicting FPG from HbA, for Example 2.22. The solid line uses all 18 subjects. The dotted line leaves out Subject 18. The dashed line leaves out Subject 15. "Leaving one out" calculations are the surest way to assess influence.

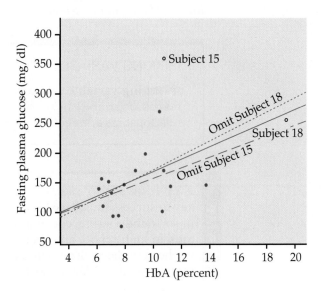

The dashed line omits Subject 15, the outlier in $y$. Comparing the solid and dashed lines, we see that Subject 15 pulls the regression line up. The influence is again not large, but it exceeds the influence of Subject 18.

The best way to see how points that are extreme in $x$ can influence the regression line is to use the *Correlation and Regression* applet on the text CD and Web site. As Exercise 2.102 demonstrates, moving one point can pull the line to almost any position on the graph.

We did not need the distinction between outliers and influential observations in Chapter 1. A single large salary that pulls up the mean salary $\bar{x}$ for a group of workers is an outlier because it lies far above the other salaries. It is also influential because the mean changes when it is removed. In the regression setting, however, not all outliers are influential. Because influential observations draw the regression line toward themselves, we may not be able to spot them by looking for large residuals.

## Beware the lurking variable

Correlation and regression are powerful tools for measuring the association between two variables and for expressing the dependence of one variable on the other. These tools must be used with an awareness of their limitations. We have seen that:

- Correlation measures *only linear association,* and fitting a straight line makes sense only when the overall pattern of the relationship is linear. Always plot your data before calculating.

- *Extrapolation* (using a fitted model far outside the range of the data that we used to fit it) often produces unreliable predictions.

- Correlation and least-squares regression are *not resistant.* Always plot your data and look for potentially influential points.

Another caution is even more important: the relationship between two variables can often be understood only by taking other variables into account. *Lurking variables* can make a correlation or regression misleading.

> **LURKING VARIABLE**
>
> A **lurking variable** is a variable that is not among the explanatory or response variables in a study and yet may influence the interpretation of relationships among those variables.

**EXAMPLE**

**2.23 High school math and success in college.** Is high school math the key to success in college? A College Board study of 15,941 high school graduates found a strong correlation between how much math minority students took in high school and their later success in college. News articles quoted the head of the College Board as saying that "math is the gatekeeper for success in college."[28] Maybe so, but we should also think about lurking variables.

Minority students from middle-class homes with educated parents no doubt take more high school math courses. They also are more likely to have a stable family, parents who emphasize education and can pay for college, and so on. These students would succeed in college even if they took fewer math courses. The family background of the students is a lurking variable that probably explains much of the relationship between math courses and college success.

**EXAMPLE**

**2.24 Imports and spending for health care.** Figure 2.25 displays a strong positive linear association. The correlation between these variables is $r = 0.9749$. Because $r^2 = 0.9504$, regression of $y$ on $x$ will explain 95% of the variation in the values of $y$.

The explanatory variable in Figure 2.25 is the dollar value of goods imported into the United States in the years between 1990 and 2001. The response variable is private spending on health in the same years. There is no economic relationship between these variables. The strong association is due entirely to the fact that both imports and health spending grew rapidly in these years. The common year for each point is a lurking variable. Any two variables that both increase over time will show a strong association. This does not mean that one variable explains or influences the other. In this example, the scatterplot and correlation are correct as exercises in following recipes, but they shed no light on any real situation.

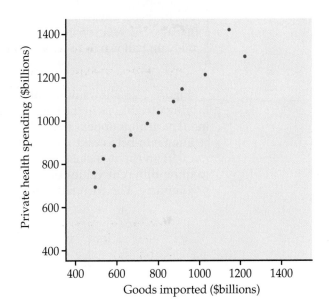

**FIGURE 2.25** The relationship between private spending on health and the value of goods imported in the same year, for Example 2.24.

Correlations such as that in Example 2.24 are sometimes called "nonsense correlations." The correlation is real. What is nonsense is the suggestion that the variables are directly related so that changing one of the variables *causes* changes in the other. The question of causation is important enough to merit separate treatment in Section 2.6. For now, just remember that an association between two variables $x$ and $y$ can reflect many types of relationship among $x$, $y$, and one or more lurking variables.

---

**ASSOCIATION DOES NOT IMPLY CAUSATION**

An association between an explanatory variable $x$ and a response variable $y$, even if it is very strong, is not by itself good evidence that changes in $x$ actually cause changes in $y$.

---

Lurking variables sometimes create a correlation between $x$ and $y$, as in Examples 2.23 and 2.24. They can also hide a true relationship between $x$ and $y$, as the following example illustrates.

**EXAMPLE**

**2.25 Overcrowding and indoor toilets.** A study of housing conditions and health in the city of Hull, England, measured a large number of variables for each of the wards into which the city is divided. Two of the variables were an index $x$ of overcrowding and an index $y$ of the lack of indoor toilets. Because $x$ and $y$ are both measures of inadequate housing, we expect a high correlation. Yet the correlation was only $r = 0.08$. How can this be? Investigation disclosed that some poor wards were dominated by public housing. These wards had high values of $x$ but low values of $y$ because public housing always includes indoor toilets. Other poor wards lacked public housing, and in these wards high values of $x$ were accompanied by high values of $y$. Because the relationship between $x$ and $y$ differed in the two types of wards, analyzing all wards together obscured the nature of the relationship.[29]

Figure 2.26 shows in simplified form how groups formed by a categorical lurking variable, as in the housing example, can make the correlation $r$ misleading. The groups appear as clusters of points in the scatterplot. There is a strong relationship between $x$ and $y$ within each of the clusters. In fact, $r = 0.85$ and $r = 0.91$ in the two clusters. However, because similar values of $x$ correspond to quite different values of $y$ in the two clusters, $x$ alone is of little value for predicting $y$. The correlation for all points displayed is therefore low: $r = 0.14$.

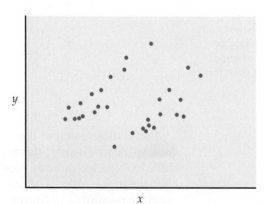

**FIGURE 2.26** This scatterplot has a low $r$ even though there is a strong correlation within each of the two clusters.

This example is another reminder to plot the data rather than simply calculate numerical measures such as the correlation.

## Beware correlations based on averaged data

Regression or correlation studies sometimes work with averages or other measures that combine information from many individuals. For example, if we plot the average height of young children against their age in months, we will see a very strong positive association with correlation near 1. But individual children of the same age vary a great deal in height. A plot of height against age for individual children will show much more scatter and lower correlation than the plot of average height against age.

*A correlation based on averages over many individuals is usually higher than the correlation between the same variables based on data for individuals.* This fact reminds us again of the importance of noting exactly what variables a statistical study involves.

## The restricted-range problem

A regression line is often used to predict the response $y$ to a given value $x$ of the explanatory variable. Successful prediction does not require a cause-and-effect relationship. If both $x$ and $y$ respond to the same underlying unmeasured variables, $x$ may help us predict $y$ even though $x$ has no direct influence on $y$. For example, the scores of SAT exams taken in high school help predict college grades. There is no cause-and-effect tie between SAT scores and college grades. Rather, both reflect a student's ability and knowledge.

How well do SAT scores, perhaps with the help of high school grades, predict college GPA? We can use the correlation $r$ and its square to get a rough answer. There is, however, a subtle difficulty.

**EXAMPLE**

**2.26 SAT scores and GPA.** Combining several studies for students graduating from college since 1980, the College Board reports these correlations between explanatory variables and the overall GPA of college students:

| SAT Math and Verbal | High school grades | SAT plus grades |
|---|---|---|
| $r = 0.36$ | $r = 0.42$ | $r = 0.52$ |

Because $0.52^2 = 0.27$, we see that SAT scores plus students' high school records explain about 27% of the variation in GPA among college students.

The subtle problem? Colleges differ greatly in the range of students they attract. Almost all students at Princeton have high SAT scores and did well in high school. At Generic State College, most students are in the middle range of SAT scores and high school performance. Both sets of students receive the full spread of grades. We suspect that if Princeton admitted weaker students they would get lower grades, and that the typical Princeton student would get very high grades at Generic State. This is the *restricted-range problem*: the data do not contain information on the full range of both explanatory and response variables. *When data suffer from restricted range, r and $r^2$ are lower than they would be if the full range could be observed.*

Thus, $r = 0.52$ understates the actual ability of SAT scores and high school grades to predict college GPA. One investigator found 21 colleges that enrolled the full range of high school graduates. Sure enough, for these colleges, $r = 0.65$.[30]

Did you notice that the correlations in Example 2.26 involve more than one explanatory variable? It is common to use several explanatory variables together to predict a response. This is called *multiple regression*. Each $r$ in the example is a *multiple correlation coefficient*, whose square is the proportion of variation in the response explained by the multiple regression. Chapter 11 introduces multiple regression.

## BEYOND THE BASICS

### Data Mining

Chapters 1 and 2 of this book are devoted to the important aspect of statistics called *exploratory data analysis* (EDA). We use graphs and numerical summaries to examine data, searching for patterns and paying attention to striking deviations from the patterns we find. In discussing regression, we advanced to using the pattern we find (in this case, a linear pattern) for prediction.

Suppose now that we have a truly enormous data base, such as all purchases recorded by the cash register scanners of a national retail chain during the past week. Surely this treasure chest of data contains patterns that might guide business decisions. If we could see clearly the types of activewear preferred in large California cities and compare the preferences of small Midwest cities—right now, not at the end of the season—we might improve profits in both parts of the country by matching stock with demand. This sounds much like EDA, and indeed it is. Exploring really large data bases in the hope of finding useful patterns is called **data mining.** Here are some distinctive features of data mining:

data mining

- When you have 100 gigabytes of data, even straightforward calculations and graphics become impossibly time-consuming. So efficient algorithms are very important.

- The structure of the data base and the process of storing the data (the fashionable term is *data warehousing*), perhaps by unifying data scattered across many departments of a large corporation, require careful consideration.

- Data mining requires automated tools that work based on only vague queries by the user. The process is too complex to do step-by-step as we have done in EDA.

All of these features point to the need for sophisticated computer science as a basis for data mining. Indeed, data mining is often thought of as a part of computer science. Yet many statistical ideas and tools—mostly tools for dealing with multidimensional data, not the sort of thing that appears in a first statistics course—are very helpful. Like many modern developments, data mining crosses the boundaries of traditional fields of study.

Do remember that the perils we associate with blind use of correlation and regression are yet more perilous in data mining, where the fog of an immense data base prevents clear vision. Extrapolation, ignoring lurking variables, and confusing association with causation are traps for the unwary data miner.

## SECTION 2.4   Summary

You can examine the fit of a regression line by plotting the **residuals,** which are the differences between the observed and predicted values of $y$. Be on the look-out for points with unusually large residuals and also for nonlinear patterns and uneven variation about the line.

Also look for **influential observations,** individual points that substantially change the regression line. Influential observations are often outliers in the $x$ direction, but they need not have large residuals.

Correlation and regression must be **interpreted with caution.** Plot the data to be sure that the relationship is roughly linear and to detect outliers and influential observations.

**Lurking variables** may explain the relationship between the explanatory and response variables. Correlation and regression can be misleading if you ignore important lurking variables.

We cannot conclude that there is a cause-and-effect relationship between two variables just because they are strongly associated. **High correlation does not imply causation.**

**A correlation based on averages** is usually higher than if we used data for individuals. A correlation based on data with a **restricted range** is often lower than would be the case if we could observe the full range of the variables.

## SECTION 2.4   Exercises

*For Exercise 2.82, see page 126; for Exercise 2.83, see page 127.*

2.84   **Price and ounces.** In Example 2.2 (page 84) and Exercise 2.3 (page 85) we examined the relationship between the price and the size of a Mocha Frappuccino©. The 12-ounce Tall drink costs $3.15, the 16-ounce Grande is $3.65, and the 24-ounce Venti is $4.15.

(a) Plot the data and describe the relationship. (Explain why you should plot size in ounces on the $x$ axis.)

(b) Find the least-squares regression line for predicting the price using size. Add the line to your plot.

(c) Draw a vertical line from the least-squares line to each data point. This gives a graphical picture of the residuals.

(d) Find the residuals and verify that they sum to zero.

(e) Plot the residuals versus size. Interpret this plot.

2.85   **Average monthly temperatures.** Here are the average monthly temperatures for Chicago, Illinois:

| Month | 1 | 2 | 3 | 4 | 5 | 6 |
|---|---|---|---|---|---|---|
| Temperature (°F) | 21.0 | 25.4 | 37.2 | 48.6 | 58.9 | 68.6 |

| Month | 7 | 8 | 9 | 10 | 11 | 12 |
|---|---|---|---|---|---|---|
| Temperature (°F) | 73.2 | 71.7 | 64.4 | 52.8 | 40.0 | 26.6 |

In this table, months are coded as integers, with January corresponding to 1 and December corresponding to 12.

(a) Plot the data with month on the $x$ axis and temperature on the $y$ axis. Describe the relationship.

| Player | 1 | 2 | 3 | 4 | 5 | 6 | 7 | 8 | 9 | 10 | 11 |
|---|---|---|---|---|---|---|---|---|---|---|---|
| Round 1 | 89 | 90 | 87 | 95 | 86 | 81 | 105 | 83 | 88 | 91 | 79 |
| Round 2 | 94 | 85 | 89 | 89 | 81 | 76 | 89 | 87 | 91 | 88 | 80 |

(a) Plot the data with the Round 1 scores on the $x$ axis and the Round 2 scores on the $y$ axis. There is a generally linear pattern except for one potentially influential observation. Circle this observation on your graph.

(b) Here are the equations of two least-squares lines. One of them is calculated from all 11 data points and the other omits the influential observation.

$$\hat{y} = 20.49 + 0.754x$$

$$\hat{y} = 50.01 + 0.410x$$

Draw both lines on your scatterplot. Which line omits the influential observation? How do you know this?

**2.96  Climate change.** Drilling down beneath a lake in Alaska yields chemical evidence of past changes in climate. Biological silicon, left by the skeletons of single-celled creatures called diatoms, measures the abundance of life in the lake. A rather complex variable based on the ratio of certain isotopes relative to ocean water gives an indirect measure of moisture, mostly from snow. As we drill down, we look farther into the past. Here are data from 2300 to 12,000 years ago:[33]

| Isotope (%) | Silicon (mg/g) | Isotope (%) | Silicon (mg/g) | Isotope (%) | Silicon (mg/g) |
|---|---|---|---|---|---|
| −19.90 | 97 | −20.71 | 154 | −21.63 | 224 |
| −19.84 | 106 | −20.80 | 265 | −21.63 | 237 |
| −19.46 | 118 | −20.86 | 267 | −21.19 | 188 |
| −20.20 | 141 | −21.28 | 296 | −19.37 | 337 |

(a) Make a scatterplot of silicon (response) against isotope (explanatory). Ignoring the outlier, describe the direction, form, and strength of the relationship. The researchers say that this and relationships among other variables they measured are evidence for cyclic changes in climate that are linked to changes in the sun's activity.

(b) The researchers single out one point: "The open circle in the plot is an outlier that was excluded in the correlation analysis." Circle this outlier on your graph. What is the correlation with and without this point? The point strongly influences the correlation.

(c) Is the outlier also strongly influential for the regression line? Calculate and draw on your graph two regression lines, and discuss what you see.

**2.97  City and highway gas mileage.** Table 1.10 (page 31) gives the city and highway gas mileages for 21 two-seater cars, including the Honda Insight gas-electric hybrid car. In Exercise 2.45 you investigated the influence of the Insight on the correlation between city and highway mileage.

(a) Make a scatterplot of highway mileage (response) against city mileage (explanatory) for all 21 cars.

(b) Use software or a graphing calculator to find the regression line for predicting highway mileage from city mileage and also the 21 residuals for this regression. Make a residual plot with a horizontal line at zero. (The "stacks" in the plot are due to the fact that mileage is measured only to the nearest mile per gallon.)

(c) Which car has the largest positive residual? The largest negative residual?

(d) The Honda Insight, an extreme outlier, does not have the largest residual in either direction. Why is this not surprising?

**2.98  Stride rate of runners.** Runners are concerned about their form when racing. One measure of form is the stride rate, the number of steps taken per second. As running speed increases, the stride rate should also increase. In a study of 21 of the best American female runners, researchers measured the stride rate for different speeds. The following table gives the speeds (in feet per second) and the mean stride rates for these runners:[34]

| Speed | 15.86 | 16.88 | 17.50 | 18.62 | 19.97 | 21.06 | 22.11 |
|---|---|---|---|---|---|---|---|
| Stride rate | 3.05 | 3.12 | 3.17 | 3.25 | 3.36 | 3.46 | 3.55 |

(a) Plot the data with speed on the $x$ axis and stride rate on the $y$ axis. Does a straight line adequately describe these data?

(b) Find the equation of the regression line of stride rate on speed. Draw this line on your plot.

(c) For each of the speeds given, obtain the predicted value of the stride rate and the residual. Verify that the residuals add to zero.

(d) Plot the residuals against speed. Describe the pattern. What does the plot indicate about the adequacy of the linear fit? Are there any potentially influential observations?

**2.99** **City and highway gas mileage.** Continue your work in Exercise 2.97. Find the regression line for predicting highway mileage from city mileage for the 20 two-seater cars other than the Honda Insight. Draw both regression lines on your scatterplot. Is the Insight very influential for the least-squares line? (Look at the position of the lines for city mileages between 10 and 30 MPG, values that cover most cars.) What explains your result?

**2.100** **Stride rate and running speed.** Exercise 2.98 gives data on the mean stride rate of a group of 21 elite female runners at various running speeds. Find the correlation between speed and stride rate. Would you expect this correlation to increase or decrease if we had data on the individual stride rates of all 21 runners at each speed? Why?

**2.101** **Use the applet.** It isn't easy to guess the position of the least-squares line by eye. Use the *Correlation and Regression* applet to compare a line you draw with the least-squares line. Click on the scatterplot to create a group of 15 to 20 points from lower left to upper right with a clear positive straight-line pattern (correlation around 0.7). Click the "Draw line" button and use the mouse to draw a line through the middle of the cloud of points from lower left to upper right. Note the "thermometer" that appears above the plot. The blue portion is the sum of the squared vertical distances from the points in the plot to the least-squares line. The green portion is the "extra" sum of squares for your line—it shows by how much your line misses the smallest possible sum of squares.

(a) You drew a line by eye through the middle of the pattern. Yet the right-hand part of the bar is probably almost entirely green. What does that tell you?

(b) Now click the "Show least-squares line" box. Is the slope of the least-squares line smaller (the new line is less steep) or larger (line is steeper) than that of your line? If you repeat this exercise several times, you will consistently get the same result. *The least-squares line minimizes the vertical distances of the points from the line. It is not the line through the "middle" of the cloud of points.* This is one reason why it is hard to draw a good regression line by eye.

**2.102** **Use the applet.** Go to the *Correlation and Regression* applet. Click on the scatterplot to create a group of 10 points in the lower-left corner of the scatterplot with a strong straight-line pattern (correlation about −0.9). In Exercise 2.46 you started here to see that correlation $r$ is not resistant. Now click the "Show least-squares line" box to display the regression line.

(a) Add one point at the upper left that is far from the other 10 points but exactly on the regression line. Why does this outlier have no effect on the line even though it changes the correlation?

(b) Now drag this last point down until it is opposite the group of 10 points. You see that one end of the least-squares line chases this single point, while the other end remains near the middle of the original group of 10. What makes the last point so influential?

**2.103** **Education and income.** There is a strong positive correlation between years of education and income for economists employed by business firms. (In particular, economists with doctorates earn more than economists with only a bachelor's degree.) There is also a strong positive correlation between years of education and income for economists employed by colleges and universities. But when all economists are considered, there is a *negative* correlation between education and income. The explanation for this is that business pays high salaries and employs mostly economists with bachelor's degrees, while colleges pay lower salaries and employ mostly economists with doctorates. Sketch a scatterplot with two groups of cases (business and academic) that illustrates how a strong positive correlation within each group and a negative overall correlation can occur together. (*Hint:* Begin by studying Figure 2.26.)

**2.104** **Dangers of not looking at a plot.** Table 2.6 presents four sets of data prepared by the statistician Frank Anscombe to illustrate the dangers of calculating without first plotting the data.[35]

(a) Without making scatterplots, find the correlation and the least-squares regression line for all four data sets. What do you notice? Use the regression line to predict $y$ for $x = 10$.

(b) Make a scatterplot for each of the data sets and add the regression line to each plot.

(c) In which of the four cases would you be willing to use the regression line to describe the dependence of $y$ on $x$? Explain your answer in each case.

**TABLE 2.6**

Four data sets for exploring correlation and regression

**Data Set A**

| $x$ | 10 | 8 | 13 | 9 | 11 | 14 | 6 | 4 | 12 | 7 | 5 |
|---|---|---|---|---|---|---|---|---|---|---|---|
| $y$ | 8.04 | 6.95 | 7.58 | 8.81 | 8.33 | 9.96 | 7.24 | 4.26 | 10.84 | 4.82 | 5.68 |

**Data Set B**

| $x$ | 10 | 8 | 13 | 9 | 11 | 14 | 6 | 4 | 12 | 7 | 5 |
|---|---|---|---|---|---|---|---|---|---|---|---|
| $y$ | 9.14 | 8.14 | 8.74 | 8.77 | 9.26 | 8.10 | 6.13 | 3.10 | 9.13 | 7.26 | 4.74 |

**Data Set C**

| $x$ | 10 | 8 | 13 | 9 | 11 | 14 | 6 | 4 | 12 | 7 | 5 |
|---|---|---|---|---|---|---|---|---|---|---|---|
| $y$ | 7.46 | 6.77 | 12.74 | 7.11 | 7.81 | 8.84 | 6.08 | 5.39 | 8.15 | 6.42 | 5.73 |

**Data Set D**

| $x$ | 8 | 8 | 8 | 8 | 8 | 8 | 8 | 8 | 8 | 8 | 19 |
|---|---|---|---|---|---|---|---|---|---|---|---|
| $y$ | 6.58 | 5.76 | 7.71 | 8.84 | 8.47 | 7.04 | 5.25 | 5.56 | 7.91 | 6.89 | 12.50 |

# 2.5 Data Analysis for Two-Way Tables

**LOOK BACK**

quantitative and
categorical variables,
page 4

When we study relationships between two variables, one of the first questions we ask is whether each variable is quantitative or categorical. For two quantitative variables, we use a scatterplot to examine the relationship, and we fit a line to the data if the relationship is approximately linear. If one of the variables is quantitative and the other is categorical, we can use the methods in Chapter 1 to describe the distribution of the quantitative variable for each value of the categorical variable. This leaves us with the situation where both variables are categorical. In this section we discuss methods for studying these relationships.

Some variables—such as gender, race, and occupation—are inherently categorical. Other categorical variables are created by grouping values of a quantitative variable into classes. Published data are often reported in grouped form to save space. To describe categorical data, we use the *counts* (frequencies) or *percents* (relative frequencies) of individuals that fall into various categories.

## The two-way table

two-way table

A key idea in studying relationships between two variables is that both variables must be measured on the same individuals or cases. When both variables are categorical, the raw data are summarized in a **two-way table** that gives counts of observations for each combination of values of the two categorical variables. Here is an example.

**EXAMPLE**

**2.27 Binge drinking by college students.** Alcohol abuse has been described by college presidents as the number one problem on campus, and it is an important cause of death in young adults. How common is it? A survey of 17,096 students in U.S. four-year colleges collected information on drinking behavior and alcohol-related problems.[36] The researchers defined "frequent binge drinking" as having five or more drinks in a row three or more times in the past two weeks. Here is the two-way table classifying students by gender and whether or not they are frequent binge drinkers:

**Two-way table for frequent binge drinking and gender**

| Frequent binge drinker | Gender | |
|---|---|---|
| | Men | Women |
| Yes | 1630 | 1684 |
| No | 5550 | 8232 |

We see that there are 1630 male students who are frequent binge drinkers and 5550 male students who are not.

## USE YOUR KNOWLEDGE

**2.105  Read the table.** How many female students are binge drinkers? How many are not?

For the binge-drinking example, we could view gender as an explanatory variable and frequent binge drinking as a response variable. This is why we put gender in the columns (like the $x$ axis in a regression) and frequent binge drinking in the rows (like the $y$ axis in a regression). We call binge drinking the **row variable** because each horizontal row in the table describes the drinking behavior. Gender is the **column variable** because each vertical column describes one gender group. Each combination of values for these two variables is called a **cell.** For example, the cell corresponding to women who are not frequent binge drinkers contains the number 8232. This table is called a $2 \times 2$ table because there are 2 rows and 2 columns.

row and
column variables

cell

To describe relationships between two categorical variables, we compute different types of percents. Our job is easier if we expand the basic two-way table by adding various totals. We illustrate the idea with our binge-drinking example.

**EXAMPLE**

**2.28  Add the margins to the table.**   We expand the table in Example 2.27 by adding the totals for each row, for each column, and the total number of all of the observations. Here is the result:

**Two-way table for frequent binge drinking and gender**

| Frequent binge drinker | Gender | | Total |
|---|---|---|---|
| | Men | Women | |
| Yes | 1,630 | 1,684 | 3,314 |
| No | 5,550 | 8,232 | 13,782 |
| Total | 7,180 | 9,916 | 17,096 |

In this study there are 7180 male students. The total number of binge drinkers is 3314 and the total number of individuals in the study is 17,096.

**2.106  Read the margins of the table.** How many women are subjects in the binge-drinking study? What is the total number of students who are not binge drinkers?

In this example, be sure that you understand how the table is obtained from the raw data. Think about a data file with one line per subject. There would be 17,096 lines or records in this data set. In the two-way table, each individual is counted once and only once. As a result, the sum of the counts in the table is the total number of individuals in the data set. *Most errors in the use of categorical-data methods come from a misunderstanding of how these tables are constructed.*

## Joint distribution

We are now ready to compute some proportions that help us understand the data in a two-way table. Suppose that we are interested in the men who are binge drinkers. The proportion of these is simply 1630 divided by 17,096, or 0.095. We would estimate that 9.5% of college students are male frequent binge drinkers. For each cell, we can compute a proportion by dividing the cell entry by the total sample size. The collection of these proportions is the **joint distribution** of the two categorical variables.

*joint distribution*

**2.29 The joint distribution.**   For the binge-drinking example, the joint distribution of binge drinking and gender is

Joint distribution of frequent binge drinking and gender

| Frequent binge drinker | Gender | |
| --- | --- | --- |
| | Men | Women |
| Yes | 0.095 | 0.099 |
| No | 0.325 | 0.482 |

Because this is a distribution, the sum of the proportions should be 1. For this example the sum is 1.001. The difference is due to roundoff error.

**2.107  Explain the computation.** Explain how the entry for the women who are not binge drinkers in Example 2.29 is computed from the table in Example 2.28.

From the joint distribution we see that the proportions of men and women frequent binge drinkers are similar in the population of college students. For the men we have 9.5%; the women are slightly higher at 9.9%. Note, however, that the proportion of women who are not frequent binge drinkers is also higher than the proportion of men. One reason for this is that there are more women in the sample than men. To understand this set of data we will need to do some additional calculations. Let's look at the distribution of gender.

## Marginal distributions

marginal distribution

When we examine the distribution of a single variable in a two-way table, we are looking at a **marginal distribution.** There are two marginal distributions, one for each categorical variable in the two-way table. They are very easy to compute.

**EXAMPLE**

**2.30 The marginal distribution of gender.** Look at the table in Example 2.28. The total numbers of men and women are given in the bottom row, labeled "Total." Our sample has 7180 men and 9916 women. To find the marginal distribution of gender we simply divide these numbers by the total sample size, 17,096. The marginal distribution of gender is

| Marginal distribution of gender | | |
| --- | --- | --- |
| | **Men** | **Women** |
| Proportion | 0.420 | 0.580 |

Note that the proportions sum to 1; there is no roundoff error.

Often we prefer to use percents rather than proportions. Here is the marginal distribution of gender described with percents:

| Marginal distribution of gender | | |
| --- | --- | --- |
| | **Men** | **Women** |
| Percent | 42.0% | 58.0% |

Which form do you prefer?

The other marginal distribution for this example is the distribution of binge drinking.

**EXAMPLE**

**2.31 The marginal distribution in percents.** Here is the marginal distribution of the frequent-binge-drinking variable (in percents):

| Marginal distribution of frequent binge drinking | | |
| --- | --- | --- |
| | **Yes** | **No** |
| Percent | 19.4% | 80.6% |

**LOOK BACK**

bar graphs and pie charts, page 6

Each marginal distribution from a two-way table is a distribution for a single categorical variable. We can use a bar graph or a pie chart to display such a distribution. For our two-way table, we will be content with numerical summaries: for example, 58% of these college students are women, and 19.4% of the students are frequent binge drinkers. When we have more rows or columns, the graphical displays are particularly useful.

## Describing relations in two-way tables

The table in Example 2.29 contains much more information than the two marginal distributions of gender alone and frequent binge drinking alone. We need to do a little more work to examine the relationship. *Relationships among categorical variables are described by calculating appropriate percents from the counts given.* What percents do you think we should use to describe the relationship between gender and frequent binge drinking?

**EXAMPLE**

**2.32 Women who are frequent binge drinkers.** What percent of the women in our sample are frequent binge drinkers? This is the count of the women who are frequent binge drinkers as a percent of the number of women in the sample:

$$\frac{1684}{9916} = 0.170 = 17.0\%$$

Recall that when we looked at the joint distribution of gender and binge drinking, we found that among all college students in the sample, 9.5% were male frequent binge drinkers and 9.9% were female frequent binge drinkers. The percents are fairly similar because the counts for these two groups, 1630 and 1684, are close. The calculations that we just performed, however, give us a different view. When we look separately at women and men, we see that the proportions of frequent binge drinkers are somewhat different, 17.0% for women versus 22.7% for men.

## Conditional distributions

In Example 2.32 we looked at the women alone and examined the distribution of the other categorical variable, frequent binge drinking. Another way to say this is that we conditioned on the value of gender being female. Similarly, we can condition on the value of gender being male. When we condition on the value of one variable and calculate the distribution of the other variable, we obtain a **conditional distribution.** Note that in Example 2.32 we calculated only the percent for frequent binge drinking. The complete conditional distribution gives the proportions or percents for all possible values of the conditioning variable.

*conditional distribution*

**EXAMPLE**

**2.33 Conditional distribution of binge drinking for women.**  For women, the conditional distribution of the binge-drinking variable in terms of percents is

| Conditional distribution of binge drinking for women | | |
| --- | --- | --- |
| | Yes | No |
| Percent | 17.0% | 83.0% |

Note that we have included the percents for both of the possible values, Yes and No, of the binge-drinking variable. These percents sum to 100%.

## USE YOUR KNOWLEDGE

**2.110  A conditional distribution.** Perform the calculations to show that the conditional distribution of binge drinking for men is

| Conditional distribution of binge drinking for men | | |
| --- | --- | --- |
| | Yes | No |
| Percent | 22.7% | 77.3% |

Comparing the conditional distributions (Example 2.33 and Exercise 2.110) reveals the nature of the association between gender and frequent binge drinking. In this set of data the men are more likely to be frequent binge drinkers than the women.

Bar graphs can help us to see relationships between two categorical variables. No single graph (such as a scatterplot) portrays the form of the relationship between categorical variables, and no single numerical measure (such as the correlation) summarizes the strength of an association. Bar graphs are flexible enough to be helpful, but you must think about what comparisons you want to display. For numerical measures, we must rely on well-chosen percents or on more advanced statistical methods.[37]

| Year | Department A | | | Department B | | |
|---|---|---|---|---|---|---|
| | Large | Small | Total | Large | Small | Total |
| First | 2 | 0 | 2 | 18 | 2 | 20 |
| Second | 9 | 1 | 10 | 40 | 10 | 50 |
| Third | 5 | 15 | 20 | 4 | 16 | 20 |
| Fourth | 4 | 16 | 20 | 2 | 14 | 16 |

**2.162 Sexual imagery in magazine ads.** In what ways do advertisers in magazines use sexual imagery to appeal to youth? One study classified each of 1509 full-page or larger ads as "not sexual" or "sexual," according to the amount and style of the clothing of the male or female model in the ad. The ads were also classified according to the target readership of the magazine.[67] Here is the two-way table of counts:

| Model clothing | Magazine readership | | | |
|---|---|---|---|---|
| | Women | Men | General interest | Total |
| Not sexual | 351 | 514 | 248 | 1113 |
| Sexual | 225 | 105 | 66 | 396 |
| Total | 576 | 619 | 314 | 1509 |

(a) Summarize the data numerically and graphically.

(b) All of the ads were taken from the March, July, and November issues of six magazines in one year. Discuss how this fact influences your interpretation of the results.

**2.163 Age of the intended readership.** The ads in the study described in the previous exercise were also classified according to the age group of the intended readership. Here is a summary of the data:

| Model clothing | Magazine readership age group | |
|---|---|---|
| | Young adult | Mature adult |
| Not sexual (percent) | 72.3% | 76.1% |
| Sexual (percent) | 27.7% | 23.9% |
| Number of ads | 1006 | 503 |

Using parts (a) and (b) of the previous exercise as a guide, analyze these data and write a report summarizing your work.

**2.164 Identity theft.** A study of identity theft looked at how well consumers protect themselves from this increasingly prevalent crime. The behaviors of 61 college students were compared with the behaviors of 59 nonstudents.[68] One of the questions was "When asked to create a password, I have used either my mother's maiden name, or my pet's name, or my birth date, or the last four digits of my social security number, or a series of consecutive numbers." For the students, 22 agreed with this statement while 30 of the nonstudents agreed.

(a) Display the data in a two-way table and analyze the data. Write a short summary of your results.

(b) The students in this study were junior and senior college students from two sections of a course in Internet marketing at a large northeastern university. The nonstudents were a group of individuals who were recruited to attend commercial focus groups on the West Coast conducted by a lifestyle marketing organization. Discuss how the method of selecting the subjects in this study relates to the conclusions that can be drawn from it.

**2.165** **Athletes and gambling.** A survey of student athletes that asked questions about gambling behavior classified students according to the National Collegiate Athletic Association (NCAA) division.[69] For male student athletes, the percents who reported wagering on collegiate sports are given here along with the numbers of respondents in each division:

| Division | I | II | III |
|---|---|---|---|
| Percent | 17.2% | 21.0% | 24.4% |
| Number | 5619 | 2957 | 4089 |

(a) Analyze the data. Give details and a short summary of your conclusion.

(b) The percents in the table above are given in the NCAA report, but the numbers of male student athletes in each division who responded to the survey question are estimated based on other information in the report. To what extent do you think this has an effect on the results?

(c) Some student athletes may be reluctant to provide this kind of information, even in a survey where there is no possibility that they can be identified. Discuss how this fact may affect your conclusions.

2.166  **Health conditions and risk behaviors.** The data set BRFSS described in the Data Appendix gives several variables related to health conditions and risk behaviors as well as demographic information for the 50 states and the District of Columbia. Pick at least three pairs of variables to analyze. Write a short report on your findings.

# Producing Data

A magazine article says that men need Pilates exercise more than women. Read the Introduction to learn more.

## Introduction

In Chapters 1 and 2 we learned some basic tools of *data analysis*. We used graphs and numbers to describe data. When we do **exploratory data analysis,** we rely heavily on plotting the data. We look for patterns that suggest interesting conclusions or questions for further study. However, *exploratory analysis alone can rarely provide convincing evidence for its conclusions, because striking patterns we find in data can arise from many sources.*

exploratory data analysis

### Anecdotal data

It is tempting to simply draw conclusions from our own experience, making no use of more broadly representative data. A magazine article about Pilates says that men need this form of exercise even more than women. The article describes the benefits that two men received from taking Pilates classes. A newspaper ad states that a particular brand of windows are "considered to be the best" and says that "now is the best time to replace your windows and doors." These types of stories, or *anecdotes*, sometimes provide quantitative data. However, this type of data does not give us a sound basis for drawing conclusions.

---

**ANECDOTAL EVIDENCE**

**Anecdotal evidence** is based on haphazardly selected individual cases, which often come to our attention because they are striking in some way. These cases need not be representative of any larger group of cases.

---

## USE YOUR KNOWLEDGE

**3.1** **Final Fu.** Your friends are big fans of "Final Fu," MTV2's martial arts competition. To what extent do you think you can generalize your preferences for this show to all students at your college?

**3.2** **Describe an example.** Find an example from some recent experience where anecdotal evidence is used to draw a conclusion that is not justified. Describe the example and explain why it cannot be used in this way.

**3.3** **Preference for Jolt Cola.** Jamie is a hard-core computer programmer. He and all his friends prefer Jolt Cola (caffeine equivalent to two cups of coffee) to either Coke or Pepsi (caffeine equivalent to less than one cup of coffee).[1] Explain why Jamie's experience is not good evidence that most young people prefer Jolt to Coke or Pepsi.

**3.4** **Automobile seat belts.** When the discussion turns to the pros and cons of wearing automobile seat belts, Herman always brings up the case of a friend who survived an accident because he was not wearing a seat belt. The friend was thrown out of the car and landed on a grassy bank, suffering only minor injuries, while the car burst into flames and was destroyed. Explain briefly why this anecdote does not provide good evidence that it is safer not to wear seat belts.

## Available data

Occasionally, data are collected for a particular purpose but can also serve as the basis for drawing sound conclusions about other research questions. We use the term **available data** for this type of data.

*available data*

---

**AVAILABLE DATA**

**Available data** are data that were produced in the past for some other purpose but that may help answer a present question.

---

The library and the Internet can be good sources of available data. Because producing new data is expensive, we all use available data whenever possible. However, the clearest answers to present questions often require that data be produced to answer those specific questions. Here are two examples:

EXAMPLE

**3.1 Causes of death.** If you visit the National Center for Health Statistics Web site, www.cdc.gov/nchs, you will learn that accidents are the most common cause of death among people aged 20 to 24, accounting for over 40% of all deaths. Homicide is next, followed by suicide. AIDS ranks seventh, behind heart disease and cancer, at 1% of all deaths. The data also show that it is dangerous to be a young man: the overall death rate for men aged 20 to 24 is three times that for women, and the death rate from homicide is more than five times higher among men.

EXAMPLE

**3.2 Math skills of children.** At the Web site of the National Center for Education Statistics, nces.ed.gov/nationsreportcard/mathematics, you will find full details about the math skills of schoolchildren in the latest National Assessment of Educational Progress (Figure 3.1). Mathematics scores have slowly but steadily increased since 1990. All racial/ethnic groups, both men and women, and students in most states are getting better in math.

Many nations have a single national statistical office, such as Statistics Canada (www.statcan.ca) or Mexico's INEGI (www.inegi.gob.mx). More than 70 different U.S. agencies collect data. You can reach most of them through the government's FedStats site (www.fedstats.gov).

### USE YOUR KNOWLEDGE

3.5   **Find some available data.** Visit the Internet and find an example of available data that is interesting to you. Explain how the data were collected and what questions the study was designed to answer.

A survey of college athletes is designed to estimate the percent who gamble. Do restaurant patrons give higher tips when their server repeats their order carefully? The validity of our conclusions from the analysis of data collected to address these issues rests on a foundation of carefully collected data. In this chapter, we will develop the skills needed to produce trustworthy data and to judge the quality of data produced by others. The techniques for producing data we will study require no formulas, but they are among the most important ideas in statistics. Statistical designs for producing data rely on either *sampling* or *experiments*.

## Sample surveys and experiments

How have the attitudes of Americans, on issues ranging from abortion to work, changed over time? **Sample surveys** are the usual tool for answering questions like these.

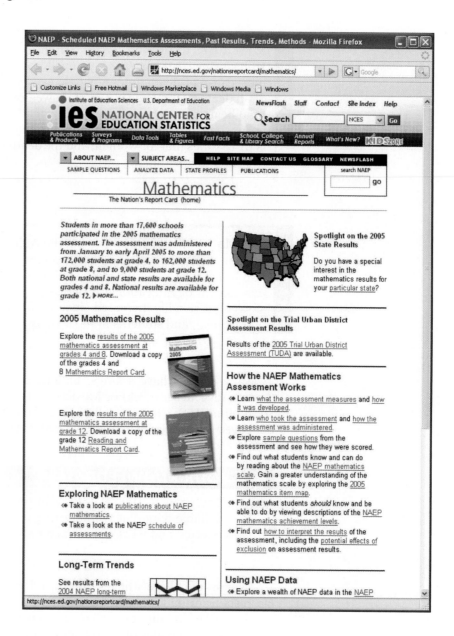

**FIGURE 3.1** The Web sites of government statistical offices are prime sources of data. Here is the home page of the National Assessment of Educational Progress.

**EXAMPLE**

**3.3 The General Social Survey.** One of the most important sample surveys is the General Social Survey (GSS) conducted by the NORC, a national organization for research and computing affiliated with the University of Chicago.[2] The GSS interviews about 3000 adult residents of the United States every second year.

*sample*
*population*

The GSS selects a **sample** of adults to represent the larger **population** of all English-speaking adults living in the United States. The idea of *sampling* is to study a part in order to gain information about the whole. Data are often produced by sampling a population of people or things. Opinion polls, for example, report the views of the entire country based on interviews with a sample of about 1000 people. Government reports on employment and unemployment are produced from a monthly sample of about 60,000 households. The quality of manufactured items is monitored by inspecting small samples each hour or each shift.

## USE YOUR KNOWLEDGE

3.6   **Find a sample survey.** Use the Internet or some printed material to find an example of a sample survey that interests you. Describe the population, how the sample was collected, and some of the conclusions.

In all of our examples, the expense of examining every item in the population makes sampling a practical necessity. Timeliness is another reason for preferring a sample to a **census,** which is an attempt to contact every individual in the entire population. We want information on current unemployment and public opinion next week, not next year. Moreover, a carefully conducted sample is often more accurate than a census. Accountants, for example, sample a firm's inventory to verify the accuracy of the records. Attempting to count every last item in the warehouse would be not only expensive but inaccurate. Bored people do not count carefully.

*census*

If conclusions based on a sample are to be valid for the entire population, a sound design for selecting the sample is required. Sampling designs are the topic of Section 3.2.

A sample survey collects information about a population by selecting and measuring a sample from the population. The goal is a picture of the population, disturbed as little as possible by the act of gathering information. Sample surveys are one kind of *observational study*.

---

### OBSERVATION VERSUS EXPERIMENT

In an **observational study** we observe individuals and measure variables of interest but do not attempt to influence the responses.

In an **experiment** we deliberately impose some treatment on individuals and we observe their responses.

---

## USE YOUR KNOWLEDGE

3.7   **Cell phones and brain cancer.** One study of cell phones and the risk of brain cancer looked at a group of 469 people who have brain cancer. The investigators matched each cancer patient with a person of the same sex, age, and race who did not have brain cancer, then asked about use of cell phones.[3] Result: "Our data suggest that use of handheld cellular telephones is not associated with risk of brain cancer." Is this an observational study or an experiment? Why? What are the explanatory and response variables?

**3.8    Violent acts on prime-time TV.** A typical hour of prime-time television shows three to five violent acts. Linking family interviews and police records shows a clear association between time spent watching TV as a child and later aggressive behavior.[4]

(a) Explain why this is an observational study rather than an experiment. What are the explanatory and response variables?

(b) Suggest several lurking variables describing a child's home life that may be related to how much TV he or she watches. Explain why this makes it difficult to conclude that more TV *causes* more aggressive behavior.

An observational study, even one based on a statistical sample, is a poor way to determine what will happen if we change something. The best way to see the effects of a change is to do an **intervention**—where we actually impose the change. When our goal is to understand cause and effect, experiments are the only source of fully convincing data.

intervention

**EXAMPLE**

**3.4  Child care and behavior.**   A study of child care enrolled 1364 infants in 1991 and planned to follow them through their sixth year in school. Twelve years later, the researchers published an article finding that "the more time children spent in child care from birth to age four-and-a-half, the more adults tended to rate them, both at age four-and-a-half and at kindergarten, as less likely to get along with others, as more assertive, as disobedient, and as aggressive."[5]

What can we conclude from this study? If parents choose to use child care, are they more likely to see these undesirable behaviors in their children?

**EXAMPLE**

**3.5  Is there a cause and effect relationship?**   Example 3.4 describes an observational study. Parents made all child care decisions and the study did not attempt to influence them. A summary of the study stated, "The study authors noted that their study was not designed to prove a cause and effect relationship. That is, the study cannot prove whether spending more time in child care causes children to have more problem behaviors."[6] Perhaps employed parents who use child care are under stress and the children react to their parents' stress. Perhaps single parents are more likely to use child care. Perhaps parents are more likely to place in child care children who already have behavior problems.

We can imagine an experiment that would remove these difficulties. From a large group of young children, choose some to be placed in child care and others to remain at home. This is an experiment because the treatment (child care or not) is imposed on the children. Of course, this particular experiment is neither practical nor ethical.

confounded

In Examples 3.4 and 3.5, we say that the effect of child care on behavior is **confounded** with (mixed up with) other characteristics of families who use child care. Observational studies that examine the effect of a single variable on an outcome can be misleading when the effects of the explanatory variable are confounded with those of other variables. Because experiments allow us to isolate the effects of specific variables, we generally prefer them. Here is an example.

EXAMPLE

**3.6 A dietary behavior experiment.** An experiment was designed to examine the effect of a 30-minute instructional session in a Food Stamp office on the dietary behavior of low-income women.[7] A group of women were randomly assigned to either the instructional session or no instruction. Two months later, data were collected on several measures of their behavior.

Experiments usually require some sort of randomization, as in this example. We begin the discussion of statistical designs for data collection in Section 3.1 with the principles underlying the design of experiments.

## USE YOUR KNOWLEDGE

3.9 **Software for teaching biology.** An educational software company wants to compare the effectiveness of its computer animation for teaching cell biology with that of a textbook presentation. The company tests the biological knowledge of each of a group of first-year college students, then randomly divides them into two groups. One group uses the animation, and the other studies the text. The company retests all the students and compares the increase in understanding of cell biology in the two groups. Is this an experiment? Why or why not? What are the explanatory and response variables?

3.10 **Find an experiment.** Use the Internet or some printed material to find an example of an experiment that interests you. Describe how the experiment was conducted and some of the conclusions.

statistical inference

Statistical techniques for producing data are the foundation for formal **statistical inference,** which answers specific questions with a known degree of confidence. In Section 3.3, we discuss some basic ideas related to inference.

ethics

Should an experiment or sample survey that could possibly provide interesting and important information always be performed? How can we safeguard the privacy of subjects in a sample survey? What constitutes the mistreatment of people or animals who are studied in an experiment? These are questions of **ethics.** In Section 3.4, we address ethical issues related to the design of studies and the analysis of data.

## 3.1 Design of Experiments

A study is an experiment when we actually do something to people, animals, or objects in order to observe the response. Here is the basic vocabulary of experiments.

---

**EXPERIMENTAL UNITS, SUBJECTS, TREATMENT**

The individuals on which the experiment is done are the **experimental units.** When the units are human beings, they are called **subjects.** A specific experimental condition applied to the units is called a **treatment.**

---

Because the purpose of an experiment is to reveal the response of one variable to changes in other variables, the distinction between explanatory and response variables is important. The explanatory variables in an experiment are often called **factors.** Many experiments study the joint effects of several factors. In such an experiment, each treatment is formed by combining a specific value (often called a **level**) of each of the factors.

factors

level of a factor

**EXAMPLE**

**3.7 Are smaller class sizes better?** Do smaller classes in elementary school really benefit students in areas such as scores on standard tests, staying in school, and going on to college? We might do an observational study that compares students who happened to be in smaller and larger classes in their early school years. Small classes are expensive, so they are more common in schools that serve richer communities. Students in small classes tend to also have other advantages: their schools have more resources, their parents are better educated, and so on. Confounding makes it impossible to isolate the effects of small classes.

The Tennessee STAR program was an experiment on the effects of class size. It has been called "one of the most important educational investigations ever carried out." The *subjects* were 6385 students who were beginning kindergarten. Each student was assigned to one of three *treatments:* regular class (22 to 25 students) with one teacher, regular class with a teacher and a full-time teacher's aide, and small class (13 to 17 students). These treatments are levels of a single *factor,* the type of class. The students stayed in the same type of class for four years, then all returned to regular classes. In later years, students from the small classes had higher scores on standard tests, were less likely to fail a grade, had better high school grades, and so on. The benefits of small classes were greatest for minority students.[8]

Example 3.7 illustrates the big advantage of experiments over observational studies. **In principle, experiments can give good evidence for causation.** In an experiment, we study the specific factors we are interested in, while controlling the effects of lurking variables. All the students in the Tennessee STAR program followed the usual curriculum at their schools. Because students were assigned to different class types within their schools, school resources and family backgrounds were not confounded with class type. The only systematic difference was the type of class. When students from the small classes did better than those in the other two types, we can be confident that class size made the difference.

EXAMPLE

**3.8 Repeated exposure to advertising.** What are the effects of repeated exposure to an advertising message? The answer may depend both on the length of the ad and on how often it is repeated. An experiment investigated this question using undergraduate students as *subjects*. All subjects viewed a 40-minute television program that included ads for a digital camera. Some subjects saw a 30-second commercial; others, a 90-second version. The same commercial was shown either 1, 3, or 5 times during the program.

This experiment has two *factors*: length of the commercial, with 2 levels, and repetitions, with 3 levels. The 6 combinations of one level of each factor form 6 *treatments*. Figure 3.2 shows the layout of the treatments. After viewing, all of the subjects answered questions about their recall of the ad, their attitude toward the camera, and their intention to purchase it. These are the *response variables*.[9]

**FIGURE 3.2** The treatments in the study of advertising, for Example 3.8. Combining the levels of the two factors forms six treatments.

|  | | Factor B Repetitions | | |
|---|---|---|---|---|
|  | | 1 time | 3 times | 5 times |
| Factor A Length | 30 seconds | 1 | 2 | 3 |
|  | 90 seconds | 4 | 5 | 6 |

Example 3.8 shows how experiments allow us to study the combined effects of several factors. The interaction of several factors can produce effects that could not be predicted from looking at the effects of each factor alone. Perhaps longer commercials increase interest in a product, and more commercials also increase interest, but if we both make a commercial longer and show it more often, viewers get annoyed and their interest in the product drops. The two-factor experiment in Example 3.8 will help us find out.

## USE YOUR KNOWLEDGE

**3.11 Food for a trip to the moon.** Storing food for long periods of time is a major challenge for those planning for human space travel beyond the moon. One problem is that exposure to radiation decreases the length of time that food can be stored. One experiment examined the effects of nine different levels of radiation on a particular type of fat, or lipid.[10] The amount of oxidation of the lipid is the measure of the extent of the damage due to the radiation. Three samples are exposed to each radiation level. Give the experimental units, the treatments, and the response variable. Describe the factor and its levels. There are many different types of lipids. To what extent do you think the results of this experiment can be generalized to other lipids?

3.12  **Learning how to draw.** A course in computer graphics technology requires students to learn multiview drawing concepts. This topic is traditionally taught using supplementary material printed on paper. The instructor of the course believes that a Web-based interactive drawing program will be more effective in increasing the drawing skills of the students.[11] The 50 students who are enrolled in the course will be randomly assigned to either the paper-based instruction or the Web-based instruction. A standardized drawing test will be given before and after the instruction. Explain why this study is an experiment and give the experimental units, the treatments, and the response variable. Describe the factor and its levels. To what extent do you think the results of this experiment can be generalized to other settings?

## Comparative experiments

Laboratory experiments in science and engineering often have a simple design with only a single treatment, which is applied to all of the experimental units. The design of such an experiment can be outlined as

$$\text{Treatment} \longrightarrow \text{Observe response}$$

For example, we may subject a beam to a load (treatment) and measure its deflection (observation). We rely on the controlled environment of the laboratory to protect us from lurking variables. When experiments are conducted in the field or with living subjects, such simple designs often yield invalid data. That is, we cannot tell whether the response was due to the treatment or to lurking variables. A medical example will show what can go wrong.

**EXAMPLE**

3.9 **Gastric freezing.**   "Gastric freezing" is a clever treatment for ulcers in the upper intestine. The patient swallows a deflated balloon with tubes attached, then a refrigerated liquid is pumped through the balloon for an hour. The idea is that cooling the stomach will reduce its production of acid and so relieve ulcers. An experiment reported in the *Journal of the American Medical Association* showed that gastric freezing did reduce acid production and relieve ulcer pain. The treatment was safe and easy and was widely used for several years. The design of the experiment was

$$\text{Gastric freezing} \longrightarrow \text{Observe pain relief}$$

placebo effect

The gastric freezing experiment was poorly designed. The patients' response may have been due to the **placebo effect.** A placebo is a dummy treatment. Many patients respond favorably to any treatment, even a placebo. This may be due to trust in the doctor and expectations of a cure or simply to the fact that medical conditions often improve without treatment. The response to a dummy treatment is the placebo effect.

A later experiment divided ulcer patients into two groups. One group was treated by gastric freezing as before. The other group received a placebo treatment in which the liquid in the balloon was at body temperature rather than freezing. The results: 34% of the 82 patients in the treatment group improved, but so did 38% of the 78 patients in the placebo group. This and other properly designed experiments showed that gastric freezing was no better than a placebo, and its use was abandoned.[12]

The first gastric freezing experiment gave misleading results because the effects of the explanatory variable were confounded with the placebo effect. We can defeat confounding by *comparing* two groups of patients, as in the second gastric freezing experiment. The placebo effect and other lurking variables now operate on both groups. The only difference between the groups is the actual effect of gastric freezing. The group of patients who received a sham treatment

**control group**  is called a **control group,** because it enables us to control the effects of outside variables on the outcome. Control is the first basic principle of statistical design of experiments. Comparison of several treatments in the same environment is the simplest form of control.

*Uncontrolled experiments in medicine and the behavioral sciences can be dominated by such influences as the details of the experimental arrangement, the selection of subjects, and the placebo effect.* The result is often *bias*.

---

**BIAS**

The design of a study is **biased** if it systematically favors certain outcomes.

---

An uncontrolled study of a new medical therapy, for example, is biased in favor of finding the treatment effective because of the placebo effect. It should not surprise you to learn that uncontrolled studies in medicine give new therapies a much higher success rate than proper comparative experiments. Well-designed experiments usually compare several treatments.

## USE YOUR KNOWLEDGE

**3.13  Does using statistical software improve exam scores?** An instructor in an elementary statistics course wants to know if using a new statistical software package will improve students' final-exam scores. He asks for volunteers and about half of the class agrees to work with the new software. He compares the final-exam scores of the students who used the new software with the scores of those who did not. Discuss possible sources of bias in this study.

## Randomization

**experiment design**  The **design of an experiment** first describes the response variable or variables, the factors (explanatory variables), and the layout of the treatments, with comparison as the leading principle. Figure 3.2 illustrates this aspect of the design of a study of response to advertising. The second aspect of design is the rule used to assign the experimental units to the treatments. Comparison of the effects of several treatments is valid only when all treatments are applied to similar groups of experimental units. If one corn variety is planted on more fertile ground, or if one cancer drug is given to more seriously ill patients, comparisons among treatments are meaningless. Systematic differences among the groups of experimental units in a comparative experiment cause bias. How can we assign experimental units to treatments in a way that is fair to all of the treatments?

Experimenters often attempt to match groups by elaborate balancing acts. Medical researchers, for example, try to match the patients in a "new drug" experimental group and a "standard drug" control group by age, sex, physical condition, smoker or not, and so on. Matching is helpful but not adequate—there are too many lurking variables that might affect the outcome. The experimenter is unable to measure some of these variables and will not think of others until after the experiment. Some important variables, such as how advanced a cancer patient's disease is, are so subjective that an experimenter might bias the study by, for example, assigning more advanced cancer cases to a promising new treatment in the unconscious hope that it will help them.

*The statistician's remedy is to rely on chance to make an assignment that does not depend on any characteristic of the experimental units and that does not rely on the judgment of the experimenter in any way.* The use of chance can be combined with matching, but the simplest design creates groups by chance alone. Here is an example.

**EXAMPLE**

**3.10 Cell phones and driving.** Does talking on a hands-free cell phone distract drivers? Undergraduate students "drove" in a high-fidelity driving simulator equipped with a hands-free cell phone. The car ahead brakes: how quickly does the subject respond? Twenty students (the control group) simply drove. Another 20 (the experimental group) talked on the cell phone while driving.

This experiment has a single factor (cell phone use) with two levels. The researchers must divide the 40 student subjects into two groups of 20. To do this in a completely unbiased fashion, put the names of the 40 students in a hat, mix them up, and draw 20. These students form the experimental group and the remaining 20 make up the control group. Figure 3.3 outlines the design of this experiment.[13]

randomization

The use of chance to divide experimental units into groups is called **randomization.** The design in Figure 3.3 combines comparison and ran-

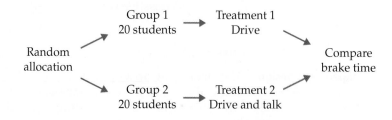

**FIGURE 3.3** Outline of a randomized comparative experiment, for Example 3.10.

domization to arrive at the simplest randomized comparative design. This "flowchart" outline presents all the essentials: randomization, the sizes of the groups and which treatment they receive, and the response variable. There are, as we will see later, statistical reasons for generally using treatment groups about equal in size.

## USE YOUR KNOWLEDGE

**3.14 Diagram the drawing experiment.** Refer to Exercise 3.12 (page 180). Draw a diagram similar to Figure 3.3 that describes the computer graphics drawing experiment.

**3.15 Diagram the food for Mars experiment.** Refer to Exercise 3.11 (page 179). Draw a diagram similar to Figure 3.3 that describes the food for space travel experiment.

## Randomized comparative experiments

The logic behind the randomized comparative design in Figure 3.3 is as follows:

- Randomization produces two groups of subjects that we expect to be similar in all respects before the treatments are applied.

- Comparative design helps ensure that influences other than the cell phone operate equally on both groups.

- Therefore, differences in average brake reaction time must be due either to talking on the cell phone or to the play of chance in the random assignment of subjects to the two groups.

That "either-or" deserves more comment. We cannot say that *any* difference in the average reaction times of the experimental and control groups is caused by talking on the cell phone. There would be some difference even if both groups were treated the same, because the natural variability among people means that some react faster than others. Chance can assign the faster-reacting students to one group or the other, so that there is a chance difference between the groups. We would not trust an experiment with just one subject in each group, for example. The results would depend too much on which group got lucky and received the subject with quicker reactions. If we assign many students to each group, however, the effects of chance will average out. There will be little difference in the average reaction times in the two groups unless talking on the cell phone causes a difference. "Use enough subjects to reduce chance variation" is the third big idea of statistical design of experiments.

### PRINCIPLES OF EXPERIMENTAL DESIGN

The basic principles of statistical design of experiments are

**1. Compare** two or more treatments. This will control the effects of lurking variables on the response.

**2. Randomize**—use impersonal chance to assign experimental units to treatments.

**3. Repeat** each treatment on many units to reduce chance variation in the results.

We hope to see a difference in the responses so large that it is unlikely to happen just because of chance variation. We can use the laws of probability, which give a mathematical description of chance behavior, to learn if the treatment effects are larger than we would expect to see if only chance were operating. If they are, we call them *statistically significant*.

---

### STATISTICAL SIGNIFICANCE

An observed effect so large that it would rarely occur by chance is called **statistically significant.**

---

You will often see the phrase "statistically significant" in reports of investigations in many fields of study. It tells you that the investigators found good evidence for the effect they were seeking. The cell phone study, for example, reported statistically significant evidence that talking on a cell phone increases the mean reaction time of drivers when the car in front of them brakes.

## How to randomize

The idea of randomization is to assign subjects to treatments by drawing names from a hat. In practice, experimenters use software to carry out randomization. Most statistical software will choose 20 out of a list of 40 at random, for example. The list might contain the names of 40 human subjects. The 20 chosen form one group, and the 20 that remain form the second group. The *Simple Random Sample* applet on the text CD and Web site makes it particularly easy to choose treatment groups at random.

You can randomize without software by using a *table of random digits*. Thinking about random digits helps you to understand randomization even if you will use software in practice. Table B at the back of the book is a table of random digits.

---

### RANDOM DIGITS

A **table of random digits** is a list of the digits 0, 1, 2, 3, 4, 5, 6, 7, 8, 9 that has the following properties:

**1.** The digit in any position in the list has the same chance of being any one of 0, 1, 2, 3, 4, 5, 6, 7, 8, 9.

**2.** The digits in different positions are independent in the sense that the value of one has no influence on the value of any other.

---

You can think of Table B as the result of asking an assistant (or a computer) to mix the digits 0 to 9 in a hat, draw one, then replace the digit drawn, mix again, draw a second digit, and so on. The assistant's mixing and drawing saves us the work of mixing and drawing when we need to randomize. Table B begins with the digits 19223950340575628713. To make the table easier to read, the digits appear in groups of five and in numbered rows. The groups and rows have no meaning—the table is just a long list of digits having the properties 1 and 2 described above.

Our goal is to use random digits for experimental randomization. We need the following facts about random digits, which are consequences of the basic properties 1 and 2:

- Any *pair* of random digits has the same chance of being any of the 100 possible pairs: 00, 01, 02, ..., 98, 99.

- Any *triple* of random digits has the same chance of being any of the 1000 possible triples: 000, 001, 002, ..., 998, 999.

- ...and so on for groups of four or more random digits.

**EXAMPLE**

**3.11 Randomize the students.**   In the cell phone experiment of Example 3.10, we must divide 40 students at random into two groups of 20 students each.

*Step 1: Label.* Give each student a numerical label, using as few digits as possible. Two digits are needed to label 40 students, so we use labels

$$01, 02, 03, \ldots, 39, 40$$

It is also correct to use labels 00 to 39 or some other choice of 40 two-digit labels.

*Step 2: Table.* Start anywhere in Table B and read two-digit groups. Suppose we begin at line 130, which is

$$69051\ 64817\ 87174\ 09517\ 84534\ 06489\ 87201\ 97245$$

The first 10 two-digit groups in this line are

$$69\ 05\ 16\ 48\ 17\ 87\ 17\ 40\ 95\ 17$$

Each of these two-digit groups is a label. The labels 00 and 41 to 99 are not used in this example, so we ignore them. The first 20 labels between 01 and 40 that we encounter in the table choose students for the experimental group. Of the first 10 labels in line 130, we ignore four because they are too high (over 40). The others are 05, 16, 17, 17, 40, and 17. The students labeled 05, 16, 17, and 40 go into the experimental group. Ignore the second and third 17s because that student is already in the group. Run your finger across line 130 (and continue to the following lines) until you have chosen 20 students. They are the students labeled

$$05, 16, 17, 40, 20, 19, 32, 04, 25, 29,$$
$$37, 39, 31, 18, 07, 13, 33, 02, 36, 23$$

You should check at least the first few of these. These students form the experimental group. The remaining 20 are the control group.

As Example 3.11 illustrates, randomization requires two steps: assign labels to the experimental units and then use Table B to select labels at random. Be sure that all labels are the same length so that all have the same chance to be chosen. Use the shortest possible labels—one digit for 9 or fewer individuals, two digits for 10 to 100 individuals, and so on. Don't try to scramble the labels as you assign them. Table B will do the required randomizing, so assign labels in any convenient manner, such as in alphabetical order for human subjects. You can read digits from Table B in any order—along a row, down a column, and so on—because the table has no order. As an easy standard practice, we recommend reading along rows.

It is easy to use statistical software or Excel to randomize. Here are the steps:

*Step 1: Label.* The first step, assigning labels to the experimental units, is similar to the procedure we described above. One difference, however, is that we are not restricted to using numerical labels. Any system where each experimental unit has a unique label identifier will work.

*Step 2: Use the computer.* Once we have the labels, we then create a data file with the labels and generate a random number for each label. In Excel, this can be done with the RAND() function. Finally, we sort the entire data set based on the random numbers. Groups are formed by selecting units in order from the sorted list.

This process is essentially the same as writing the labels on a deck of cards, shuffling the cards, and dealing them out one at a time.

**EXAMPLE**

**3.12 Using software for randomization.** Let's do a randomization similar to the one we did in Example 3.11, but this time using Excel. Here we will use 10 experimental units. We will assign 5 to the treatment group and 5 to the control group. We first create a data set with the numbers 1 to 10 in the first column. See Figure 3.4(a). Then we use RAND() to generate 10 random numbers in the second column. See Figure 3.4(b). Finally, we sort the data set based on the numbers in the second column. See Figure 3.4(c). The first 5 labels (8, 5, 9, 4, and 6) are assigned to the experimental group. The remaining 5 labels (3, 10, 7, 2, and 1) correspond to the control group.

FIGURE 3.4 Randomization of 10 experimental units using a computer, for Example 3.12. (a) Labels. (b) Random numbers. (c) Sorted list of labels.

| (a) A | (a) B | (b) A | (b) B | (c) A | (c) B |
|---|---|---|---|---|---|
| 1 | | 1 | 0.925672 | 8 | 0.077044 |
| 2 | | 2 | 0.893959 | 5 | 0.118440 |
| 3 | | 3 | 0.548247 | 9 | 0.348467 |
| 4 | | 4 | 0.349591 | 4 | 0.349591 |
| 5 | | 5 | 0.118440 | 6 | 0.390180 |
| 6 | | 6 | 0.390180 | 3 | 0.548247 |
| 7 | | 7 | 0.760262 | 10 | 0.601167 |
| 8 | | 8 | 0.077044 | 7 | 0.760262 |
| 9 | | 9 | 0.348467 | 2 | 0.893959 |
| 10 | | 10 | 0.601167 | 1 | 0.925672 |

completely randomized
design

When all experimental units are allocated at random among all treatments, as in Example 3.11, the experimental design is **completely randomized.** Completely randomized designs can compare any number of treatments. The treatments can be formed by levels of a single factor or by more than one factor.

**EXAMPLE**

**3.13 Randomization of the TV commercial experiment.** Figure 3.2 (page 179) displays six treatments formed by the two factors in an experiment on response to a TV commercial. Suppose that we have 150 students who are willing to serve as subjects. We must assign 25 students at random to each group. Figure 3.5 outlines the completely randomized design.

To carry out the random assignment, label the 150 students 001 to 150. (Three digits are needed to label 150 subjects.) Enter Table B and read three-digit groups until you have selected 25 students to receive Treatment 1 (a 30-second ad shown once). If you start at line 140, the first few labels for Treatment 1 subjects are 129, 048, and 003.

Continue in Table B to select 25 more students to receive Treatment 2 (a 30-second ad shown 3 times). Then select another 25 for Treatment 3 and so on until you have assigned 125 of the 150 students to Treatments 1 through 5. The 25 students who remain get Treatment 6. The randomization is straightforward, but very tedious to do by hand. We recommend the *Simple Random Sample* applet. Exercise 3.35 shows how to use the applet to do the randomization for this example.

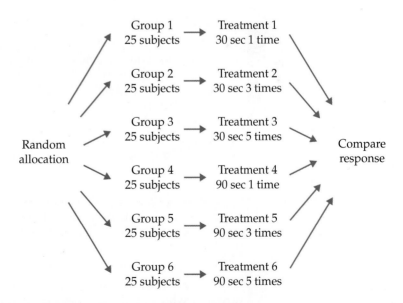

**FIGURE 3.5** Outline of a completely randomized design comparing six treatments, for Example 3.13.

---

## USE YOUR KNOWLEDGE

**3.16 Do the randomization.** Use computer software to carry out the randomization in Example 3.13.

## Cautions about experimentation

The logic of a randomized comparative experiment depends on our ability to treat all the experimental units identically in every way except for the actual treatments being compared. Good experiments therefore require careful attention to details. For example, the subjects in the second gastric freezing experiment (Example 3.9) all got the same medical attention during the study. **double-blind** Moreover, the study was **double-blind**—neither the subjects themselves nor the medical personnel who worked with them knew which treatment any subject had received. The double-blind method avoids unconscious bias by, for example, a doctor who doesn't think that "just a placebo" can benefit a patient.

*Many—perhaps most—experiments have some weaknesses in detail. The environment of an experiment can influence the outcomes in unexpected ways.* Although experiments are the gold standard for evidence of cause and effect, really convincing evidence usually requires that a number of studies in different places with different details produce similar results. Here are some brief examples of what can go wrong.

> **EXAMPLE**
>
> **3.14 Placebo for a marijuana experiment.** A study of the effects of marijuana recruited young men who used marijuana. Some were randomly assigned to smoke marijuana cigarettes, while others were given placebo cigarettes. This failed: the control group recognized that their cigarettes were phony and complained loudly. It may be quite common for blindness to fail because the subjects can tell which treatment they are receiving.[14]

> **EXAMPLE**
>
> **3.15 Knock out genes.** To study genetic influence on behavior, experimenters "knock out" a gene in one group of mice and compare their behavior with that of a control group of normal mice. The results of these experiments often don't agree as well as hoped, so investigators did exactly the same experiment with the same genetic strain of mice in Oregon, Alberta (Canada), and New York. Many results were very different.[15] It appears that small differences in the lab environments have big effects on the behavior of the mice. Remember this the next time you read that our genes control our behavior.

**lack of realism** The most serious potential weakness of experiments is **lack of realism.** The subjects or treatments or setting of an experiment may not realistically duplicate the conditions we really want to study. Here is an example.

> **EXAMPLE**
>
> **3.16 Layoffs and feeling bad.** How do layoffs at a workplace affect the workers who remain on the job? Psychologists asked student subjects to proofread text for extra course credit, then "let go" some of the workers (who were actually accomplices of the experimenters). Some subjects were told that those let go had performed poorly (Treatment 1). Others were told that not all could be kept and that it was just luck that they were kept and others let go (Treatment 2). We can't be sure that the reactions of the students are the same as those of workers who survive a layoff in which other workers

lose their jobs. Many behavioral science experiments use student subjects in a campus setting. Do the conclusions apply to the real world?

**CAUTION**

Lack of realism can limit our ability to apply the conclusions of an experiment to the settings of greatest interest. Most experimenters want to generalize their conclusions to some setting wider than that of the actual experiment. *Statistical analysis of an experiment cannot tell us how far the results will generalize to other settings.* Nonetheless, the randomized comparative experiment, because of its ability to give convincing evidence for causation, is one of the most important ideas in statistics.

## Matched pairs designs

Completely randomized designs are the simplest statistical designs for experiments. They illustrate clearly the principles of control, randomization, and repetition. However, completely randomized designs are often inferior to more elaborate statistical designs. In particular, matching the subjects in various ways can produce more precise results than simple randomization.

matched pairs design

The simplest use of matching is a **matched pairs design,** which compares just two treatments. The subjects are matched in pairs. For example, an experiment to compare two advertisements for the same product might use pairs of subjects with the same age, sex, and income. The idea is that matched subjects are more similar than unmatched subjects, so that comparing responses within a number of pairs is more efficient than comparing the responses of groups of randomly assigned subjects. Randomization remains important: which one of a matched pair sees the first ad is decided at random. One common variation of the matched pairs design imposes both treatments on the same subjects, so that each subject serves as his or her own control. Here is an example.

**EXAMPLE**

**3.17 Matched pairs for the cell phone experiment.** Example 3.10 describes an experiment on the effects of talking on a cell phone while driving. The experiment compared two treatments, driving in a simulator and driving in the simulator while talking on a hands-free cell phone. The response variable is the time the driver takes to apply the brake when the car in front brakes suddenly. In Example 3.10, 40 student subjects were assigned at random, 20 students to each treatment. This is a completely randomized design, outlined in Figure 3.3. Subjects differ in driving skill and reaction times. The completely randomized design relies on chance to create two similar groups of subjects.

In fact, the experimenters used a matched pairs design in which all subjects drove both with and without using the cell phone. They compared each individual's reaction times with and without the phone. If all subjects drove first with the phone and then without it, the effect of talking on the cell phone would be confounded with the fact that this is the first run in the simulator. The proper procedure requires that all subjects first be trained in using the simulator, that the *order* in which a subject drives with and without the phone be random, and that the two drives be on separate days to reduce the chance that the results of the second treatment will be influenced by the first treatment.

The completely randomized design uses chance to decide which 20 subjects will drive with the cell phone. The other 20 drive without it. The matched pairs design uses chance to decide which 20 subjects will drive first with and then without the cell phone. The other 20 drive first without and then with the phone.

## Block designs

The matched pairs design of Example 3.17 uses the principles of comparison of treatments, randomization, and repetition on several experimental units. However, the randomization is not complete (all subjects randomly assigned to treatment groups) but restricted to assigning the order of the treatments for each subject. *Block designs* extend the use of "similar subjects" from pairs to larger groups.

---

### BLOCK DESIGN

A **block** is a group of experimental units or subjects that are known before the experiment to be similar in some way that is expected to affect the response to the treatments. In a **block design,** the random assignment of units to treatments is carried out separately within each block.

---

Block designs can have blocks of any size. A block design combines the idea of creating equivalent treatment groups by matching with the principle of forming treatment groups at random. Blocks are another form of *control*. They control the effects of some outside variables by bringing those variables into the experiment to form the blocks. Here are some typical examples of block designs.

**EXAMPLE**

**3.18 Blocking in a cancer experiment.**   The progress of a type of cancer differs in women and men. A clinical experiment to compare three therapies for this cancer therefore treats sex as a blocking variable. Two separate randomizations are done, one assigning the female subjects to the treatments and the other assigning the male subjects. Figure 3.6 outlines the design of this experiment. Note that there is no randomization involved in making up the blocks. They are groups of subjects who differ in some way (sex in this case) that is apparent before the experiment begins.

**EXAMPLE**

**3.19 Blocking in an agriculture experiment.**   The soil type and fertility of farmland differ by location. Because of this, a test of the effect of tillage type (two types) and pesticide application (three application schedules) on soybean yields uses small fields as blocks. Each block is divided into six plots, and the six treatments are randomly assigned to plots separately within each block.

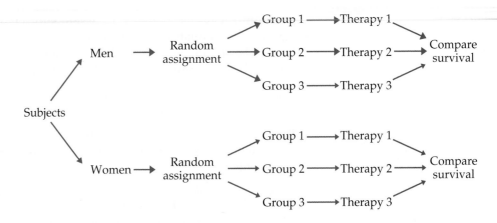

**FIGURE 3.6** Outline of a block design, for Example 3.18. The blocks consist of male and female subjects. The treatments are three therapies for cancer.

**EXAMPLE**

**3.20 Blocking in an education experiment.** The Tennessee STAR class size experiment (Example 3.7) used a block design. It was important to compare different class types in the same school because the children in a school come from the same neighborhood, follow the same curriculum, and have the same school environment outside class. In all, 79 schools across Tennessee participated in the program. That is, there were 79 blocks. New kindergarten students were randomly placed in the three types of class separately within each school.

Blocks allow us to draw separate conclusions about each block, for example, about men and women in the cancer study in Example 3.18. Blocking also allows more precise overall conclusions because the systematic differences between men and women can be removed when we study the overall effects of the three therapies. The idea of blocking is an important additional principle of statistical design of experiments. A wise experimenter will form blocks based on the most important unavoidable sources of variability among the experimental units. Randomization will then average out the effects of the remaining variation and allow an unbiased comparison of the treatments.

## SECTION 3.1 Summary

In an experiment, one or more **treatments** are imposed on the **experimental units** or **subjects.** Each treatment is a combination of **levels** of the explanatory variables, which we call **factors.**

The **design** of an experiment refers to the choice of treatments and the manner in which the experimental units or subjects are assigned to the treatments.

The basic principles of statistical design of experiments are **control, randomization,** and **repetition.**

The simplest form of control is **comparison.** Experiments should compare two or more treatments in order to prevent **confounding** the effect of a treatment with other influences, such as lurking variables.

**Randomization** uses chance to assign subjects to the treatments. Randomization creates treatment groups that are similar (except for chance variation) before the treatments are applied. Randomization and comparison together prevent **bias,** or systematic favoritism, in experiments.

You can carry out randomization by giving numerical labels to the experimental units and using a **table of random digits** to choose treatment groups.

**Repetition** of the treatments on many units reduces the role of chance variation and makes the experiment more sensitive to differences among the treatments.

Good experiments require attention to detail as well as good statistical design. Many behavioral and medical experiments are **double-blind. Lack of realism** in an experiment can prevent us from generalizing its results.

In addition to comparison, a second form of control is to restrict randomization by forming **blocks** of experimental units that are similar in some way that is important to the response. Randomization is then carried out separately within each block.

**Matched pairs** are a common form of blocking for comparing just two treatments. In some matched pairs designs, each subject receives both treatments in a random order. In others, the subjects are matched in pairs as closely as possible, and one subject in each pair receives each treatment.

## SECTION 3.1  Exercises

*For Exercises 3.1 to 3.4, see page 172; for Exercise 3.5, see page 173; for Exercises 3.6 and 3.7, see page 175; for Exercise 3.8, see page 176; for Exercises 3.9 and 3.10, see page 177; for Exercises 3.11 and 3.12, see pages 179 and 180; for Exercise 3.13, see page 181; for Exercises 3.14 and 3.15, see page 183; and for Exercise 3.16, see page 187.*

3.17  **What is needed?** Explain what is deficient in each of the following proposed experiments and explain how you would improve the experiment.

(a) Two forms of a lab exercise are to be compared. There are 10 rows in the classroom. Students who sit in the first 5 rows of the class are given the first form, and students who sit in the last 5 rows are given the second form.

(b) The effectiveness of a leadership program for high school students is evaluated by examining the change in scores on a standardized test of leadership skills.

(c) An innovative method for teaching introductory biology courses is examined by using the traditional method in the fall zoology course and the new method in the spring botany course.

3.18  **What is wrong?** Explain what is wrong with each of the following randomization procedures and describe how you would do the randomization correctly.

(a) A list of 50 subjects is entered into a computer file and then sorted by last name. The subjects are assigned to five treatments by taking the first 10 subjects for Treatment 1, the next 10 subjects for Treatment 2, and so forth.

(b) Eight subjects are to be assigned to two treatments, four to each. For each subject, a coin is tossed. If the coin comes up heads, the subject is assigned to the first treatment; if the coin comes up tails, the subject is assigned to the second treatment.

(c) An experiment will assign 80 rats to four different treatment conditions. The rats arrive from the supplier in batches of 20 and the treatment lasts two weeks. The first batch of 20 rats is randomly assigned to one of the four treatments, and data for these rats are collected. After a one-week break, another batch of 20 rats arrives and is assigned to one of the three remaining treatments. The process continues until the last batch of rats is given the treatment that has not been assigned to the three previous batches.

3.19  **Evaluate a new teaching method.** A teaching innovation is to be evaluated by randomly assigning students to either the traditional approach or the new approach. The change in a standardized test score is the response variable. Explain how this experiment should be done in a double-blind fashion.

**3.20 Can you change attitudes toward binge drinking?** A experiment designed to change attitudes about binge drinking is to be performed using college students as subjects. Discuss some variables that you might use if you were to use a block design for this experiment.

**3.21 Compost tea.** Compost tea is rich in microorganisms that help plants grow. It is made by soaking compost in water.[16] Design a comparative experiment that will provide evidence about whether or not compost tea works for a particular type of plant that interests you. Be sure to provide all details regarding your experiment, including the response variable or variables that you will measure.

**3.22** CHALLENGE **Measuring water quality in streams and lakes.** Water quality of streams and lakes is an issue of concern to the public. Although trained professionals typically are used to take reliable measurements, many volunteer groups are gathering and distributing information based on data that they collect.[17] You are part of a team to train volunteers to collect accurate water quality data. Design an experiment to evaluate the effectiveness of the training. Write a summary of your proposed design to present to your team. Be sure to include all of the details that they will need to evaluate your proposal.

*For each of the experimental situations described in Exercises 3.23 to 3.25, identify the experimental units or subjects, the factors, the treatments, and the response variables.*

**3.23 How well do pine trees grow in shade?** Ability to grow in shade may help pines in the dry forests of Arizona resist drought. How well do these pines grow in shade? Investigators planted pine seedlings in a greenhouse in either full light or light reduced to 5% of normal by shade cloth. At the end of the study, they dried the young trees and weighed them.

**3.24 Will the students do more exercise and eat better?** Most American adolescents don't eat well and don't exercise enough. Can middle schools increase physical activity among their students? Can they persuade students to eat better? Investigators designed a "physical activity intervention" to increase activity in physical education classes and during leisure periods throughout the school day.

They also designed a "nutrition intervention" that improved school lunches and offered ideas for healthy home-packed lunches. Each participating school was randomly assigned to one of the interventions, both interventions, or no intervention. The investigators observed physical activity and lunchtime consumption of fat.

**3.25 Refusals in telephone surveys.** How can we reduce the rate of refusals in telephone surveys? Most people who answer at all listen to the interviewer's introductory remarks and then decide whether to continue. One study made telephone calls to randomly selected households to ask opinions about the next election. In some calls, the interviewer gave her name, in others she identified the university she was representing, and in still others she identified both herself and the university. For each type of call, the interviewer either did or did not offer to send a copy of the final survey results to the person interviewed. Do these differences in the introduction affect whether the interview is completed?

**3.26 Does aspirin prevent strokes and heart attacks?** The Bayer Aspirin Web site claims that "Nearly five decades of research now link aspirin to the prevention of stroke and heart attacks." The most important evidence for this claim comes from the Physicians' Health Study, a large medical experiment involving 22,000 male physicians. One group of about 11,000 physicians took an aspirin every second day, while the rest took a placebo. After several years the study found that subjects in the aspirin group had significantly fewer heart attacks than subjects in the placebo group.

(a) Identify the experimental subjects, the factor and its levels, and the response variable in the Physicians' Health Study.

(b) Use a diagram to outline a completely randomized design for the Physicians' Health Study.

(c) What does it mean to say that the aspirin group had "significantly fewer heart attacks"?

**3.27 Chronic tension headaches.** Doctors identify "chronic tension-type headaches" as headaches that occur almost daily for at least six months. Can antidepressant medications or stress management training reduce the number and severity of these headaches? Are both together more effective than either alone? Investigators compared four treatments: antidepressant alone, placebo alone, antidepressant plus stress management, and placebo plus stress management. Outline the design of the experiment. The headache sufferers named below have agreed to participate in the study. Use software or Table B at line 151 to randomly assign the subjects to the treatments.

| | | | | |
|---|---|---|---|---|
| Anderson | Archberger | Bezawada | Cetin | Cheng |
| Chronopoulou | Codrington | Daggy | Daye | Engelbrecht |
| Guha | Hatfield | Hua | Kim | Kumar |
| Leaf | Li | Lipka | Lu | Martin |
| Mehta | Mi | Nolan | Olbricht | Park |
| Paul | Rau | Saygin | Shu | Tang |
| Towers | Tyner | Vassilev | Wang | Watkins |
| Xu | | | | |

**3.28 Smoking marijuana and willingness to work.** How does smoking marijuana affect willingness to work? Canadian researchers persuaded people who used marijuana to live for 98 days in a "planned environment." The subjects earned money by weaving belts. They used their earnings to pay for meals and other consumption and could keep any money left over. One group smoked two potent marijuana cigarettes every evening. The other group smoked two weak marijuana cigarettes. All subjects could buy more cigarettes but were given strong or weak cigarettes, depending on their group. Did the weak and strong groups differ in work output and earnings?[18]

(a) Outline the design of this experiment.

(b) Here are the names of the 20 subjects. Use software or Table B at line 101 to carry out the randomization your design requires.

| | | | | |
|---|---|---|---|---|
| Becker | Brifcani | Chen | Crabill | Cunningham |
| Dicklin | Fein | Gorman | Knapp | Lucas |
| McCarty | Merkulyeva | Mitchell | Ponder | Roe |
| Saeed | Seele | Truong | Wayman | Woodley |

**3.29 Eye cataracts.** Eye cataracts are responsible for over 40% of blindness worldwide. Can drinking tea regularly slow the growth of cataracts? We can't experiment on people, so we use rats as subjects. Researchers injected 21 young rats with a substance that causes cataracts. One group of the rats also received black tea extract; a second group received green tea extract; and a third got a placebo, a substance with no effect on the body. The response variable was the growth of cataracts over the next six weeks. Yes, both tea extracts did slow cataract growth.[19]

(a) Outline the design of this experiment.

(b) Use software or Table B, starting at line 120, to assign rats to treatments.

**3.30 Guilt among workers who survive a layoff.** Workers who survive a layoff of other employees at their location may suffer from "survivor guilt." A study of survivor guilt and its effects used as subjects 90 students who were offered an opportunity to earn extra course credit by doing proofreading. Each subject worked in the same cubicle as another student, who was an accomplice of the experimenters. At a break midway through the work, one of three things happened:

> Treatment 1: The accomplice was told to leave; it was explained that this was because she performed poorly.

> Treatment 2: It was explained that unforeseen circumstances meant there was only enough work for one person. By "chance," the accomplice was chosen to be laid off.

> Treatment 3: Both students continued to work after the break.

The subjects' work performance after the break was compared with performance before the break.[20]

(a) Outline the design of this completely randomized experiment.

(b) If you are using software, randomly assign the 90 students to the treatments. If not, use Table B at line 153 to choose the first four subjects for Treatment 1.

**3.31 Diagram the exercise and eating experiment.** Twenty-four public middle schools agree to participate in the experiment described in Exercise 3.24. Use a diagram to outline a completely randomized design for this experiment. Then do the randomization required to assign schools to treatments. If you use Table B, start at line 160.

**3.32 Price cuts on athletic shoes.** Stores advertise price reductions to attract customers. What type of price cut is most attractive? Market researchers prepared ads for athletic shoes announcing different levels of discounts (20%, 40%, 60%, or 80%). The student subjects who read the ads were also given "inside information" about the fraction of shoes on sale (25%, 50%, 75%, or 100%). Each subject then rated the attractiveness of the sale on a scale of 1 to 7.[21]

(a) There are two factors. Make a sketch like Figure 3.2 (page 179) that displays the treatments formed by all combinations of levels of the factors.

(b) Outline a completely randomized design using 96 student subjects. Use software or Table B at line 111 to choose the subjects for the first treatment.

**3.33  Treatment of clothing fabrics.** A maker of fabric for clothing is setting up a new line to "finish" the raw fabric. The line will use either metal rollers or natural-bristle rollers to raise the surface of the fabric; a dyeing cycle time of either 30 minutes or 40 minutes; and a temperature of either 150° or 175° Celsius. An experiment will compare all combinations of these choices. Four specimens of fabric will be subjected to each treatment and scored for quality.

(a) What are the factors and the treatments? How many individuals (fabric specimens) does the experiment require?

(b) Outline a completely randomized design for this experiment. (You need not actually do the randomization.)

**3.34  Use the simple random sample applet.** You can use the *Simple Random Sample* applet to choose a treatment group at random once you have labeled the subjects. Example 3.11 (page 185) uses Table B to choose 20 students from a group of 40 for the treatment group in a study of the effect of cell phones on driving. Use the applet to choose the 20 students for the experimental group. Which students did you choose? The remaining 20 students make up the control group.

**3.35  Use the simple random sample applet.** The *Simple Random Sample* applet allows you to randomly assign experimental units to more than two groups without difficulty. Example 3.13 (page 187) describes a randomized comparative experiment in which 150 students are randomly assigned to six groups of 25.

(a) Use the applet to randomly choose 25 out of 150 students to form the first group. Which students are in this group?

(b) The population hopper now contains the 125 students that were not chosen, in scrambled order. Click "Sample" again to choose 25 of these remaining students to make up the second group. Which students were chosen?

(c) Click "Sample" three more times to choose the third, fourth, and fifth groups. Don't take the time to write down these groups. Check that there are only 25 students remaining in the population hopper. These subjects get Treatment 6. Which students are they?

**3.36  Effectiveness of price discounts.** Experiments with more than one factor allow insight into interactions between the factors. A study

of the attractiveness of advertised price discounts had two factors: percent of all goods on sale (25%, 50%, 75%, or 100%) and whether the discount was stated precisely as 60% off or as a range, 50% to 70% off. Subjects rated the attractiveness of the sale on a scale of 1 to 7. Figure 3.7 shows the mean ratings for the eight treatments formed from the two factors.[22] Based on these results, write a careful description of how percent on sale and precise discount versus range of discounts influence the attractiveness of a sale.

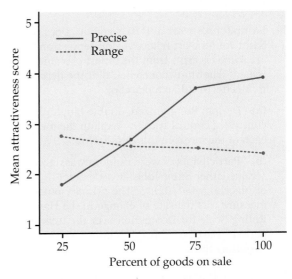

**FIGURE 3.7** Mean responses to eight treatments in an experiment with two factors, showing interaction between the factors, for Exercise 3.36.

**3.37  Health benefits of bee pollen.** "Bee pollen is effective for combating fatigue, depression, cancer, and colon disorders." So says a Web site that offers the pollen for sale. We wonder if bee pollen really does prevent colon disorders. Here are two ways to study this question. Explain why the first design will produce more trustworthy data.

1. Find 400 women who do not have colon disorders. Assign 200 to take bee pollen capsules and the other 200 to take placebo capsules that are identical in appearance. Follow both groups for 5 years.

2. Find 200 women who take bee pollen regularly. Match each with a woman of the same age, race, and occupation who does not take bee pollen. Follow both groups for 5 years.

**3.38  Treatment of pain for cancer patients.** Health care providers are giving more attention to relieving the pain of cancer patients. An article in the journal

*Cancer* surveyed a number of studies and concluded that controlled-release morphine tablets, which release the painkiller gradually over time, are more effective than giving standard morphine when the patient needs it.[23] The "methods" section of the article begins: "Only those published studies that were controlled (i.e., randomized, double blind, and comparative), repeated-dose studies with CR morphine tablets in cancer pain patients were considered for this review." Explain the terms in parentheses to someone who knows nothing about medical trials.

**3.39 Saint-John's-wort and depression.** Does the herb Saint-John's-wort relieve major depression? Here are some excerpts from the report of a study of this issue.[24] The study concluded that the herb is no more effective than a placebo.

(a) "Design: Randomized, double-blind, placebo-controlled clinical trial...." Explain the meaning of each of the terms in this description.

(b) "Participants …were randomly assigned to receive either Saint-John's-wort extract ($n = 98$) or placebo ($n = 102$).… The primary outcome measure was the rate of change in the Hamilton Rating Scale for Depression over the treatment period." Based on this information, use a diagram to outline the design of this clinical trial.

**3.40 The Monday effect on stock prices.** Puzzling but true: stocks tend to go down on Mondays. There is no convincing explanation for this fact. A recent study looked at this "Monday effect" in more detail, using data on the daily returns of stocks on several U.S. exchanges over a 30-year period. Here are some of the findings:

*To summarize, our results indicate that the well-known Monday effect is caused largely by the Mondays of the last two weeks of the month. The mean Monday return of the first three weeks of the month is, in general, not significantly different from zero and is generally significantly higher than the mean Monday return of the last two weeks. Our finding seems to make it more difficult to explain the Monday effect.*[25]

A friend thinks that "significantly" in this article has its plain English meaning, roughly "I think this is important." Explain in simple language what "significantly higher" and "not significantly different from zero" actually tell us here.

**3.41 Five-digit zip codes and delivery time of mail.** Does adding the five-digit postal zip code to an address really speed up delivery of letters? Does adding the four more digits that make up "zip + 4"

speed delivery yet more? What about mailing a letter on Monday, Thursday, or Saturday? Describe the design of an experiment on the speed of first-class mail delivery. For simplicity, suppose that all letters go from you to a friend, so that the sending and receiving locations are fixed.

**3.42** **Use the simple random sample applet.** The *Simple Random Sample* applet can demonstrate how randomization works to create similar groups for comparative experiments. Suppose that (unknown to the experimenters) the 20 even-numbered students among the 40 subjects for the cell phone study in Example 3.11 (page 185) have fast reactions, and that the odd-numbered students have slow reactions. We would like the experimental and control groups to contain similar numbers of the fast reactors. Use the applet to choose 10 samples of size 20 from the 40 students. (Be sure to click "Reset" after each sample.) Record the counts of even-numbered students in each of your 10 samples. You see that there is considerable chance variation but no systematic bias in favor of one or the other group in assigning the fast-reacting students. Larger samples from larger populations will on the average do a better job of making the two groups equivalent.

**3.43 Does oxygen help football players?** We often see players on the sidelines of a football game inhaling oxygen. Their coaches think this will speed their recovery. We might measure recovery from intense exercise as follows: Have a football player run 100 yards three times in quick succession. Then allow three minutes to rest before running 100 yards again. Time the final run. Because players vary greatly in speed, you plan a matched pairs experiment using 20 football players as subjects. Describe the design of such an experiment to investigate the effect of inhaling oxygen during the rest period. Why should each player's two trials be on different days? Use Table B at line 140 to decide which players will get oxygen on their first trial.

**3.44 Carbon dioxide in the atmosphere.** The concentration of carbon dioxide ($CO_2$) in the atmosphere is increasing rapidly due to our use of fossil fuels. Because plants use $CO_2$ to fuel photosynthesis, more $CO_2$ may cause trees and other plants to grow faster. An elaborate apparatus allows researchers to pipe extra $CO_2$ to a 30-meter circle of forest. We want to compare the growth in base area of trees in treated and untreated areas to see if extra $CO_2$ does in fact increase growth. We can afford to treat 3 circular areas.[26]

(a) Describe the design of a completely randomized experiment using 6 well-separated 30-meter circular areas in a pine forest. Sketch the forest area with the 6 circles and carry out the randomization your design calls for.

(b) Regions within the forest may differ in soil fertility. Describe a matched pairs design using three pairs of circles that will reduce the extra variation due to different fertility. Sketch the forest area with the new arrangement of circles and carry out the randomization your design calls for.

3.45 **CHALLENGE** **Calcium and the bones of young girls.** Calcium is important to the bone development of young girls. To study how the bodies of young girls process calcium, investigators used the setting of a summer camp. Calcium was given in Hawaiian Punch at either a high or a low level. The camp diet was otherwise the same for all girls. Suppose that there are 50 campers.

(a) Outline a completely randomized design for this experiment.

(b) Describe a matched pairs design in which each girl receives both levels of calcium (with a "washout period" between). What is the advantage of the matched pairs design over the completely randomized design?

(c) The same randomization can be used in different ways for both designs. Label the subjects 01 to 50. You must choose 25 of the 50. Use Table B at line 110 to choose just the first 5 of the 25. How are the 25 subjects chosen treated in the completely randomized design? How are they treated in the matched pairs design?

3.46 **CHALLENGE** **Random digits.** Table B is a table of random digits. Which of the following statements are true of a table of random digits, and which are false? Explain your answers.

(a) There are exactly four 0s in each row of 40 digits.

(b) Each pair of digits has chance 1/100 of being 00.

(c) The digits 0000 can never appear as a group, because this pattern is not random.

3.47 **Vitamin C for ultramarathon runners.** An ultramarathon, as you might guess, is a footrace longer than the 26.2 miles of a marathon. Runners commonly develop respiratory infections after an ultramarathon. Will taking 600 milligrams of vitamin C daily reduce these infections? Researchers randomly assigned ultramarathon runners to receive either vitamin C or a placebo. Separately, they also randomly assigned these treatments to a group of nonrunners the same age as the runners. All subjects were watched for 14 days after the big race to see if infections developed.[27]

(a) What is the name for this experimental design?

(b) Use a diagram to outline the design.

(c) The report of the study said:

*Sixty-eight percent of the runners in the placebo group reported the development of symptoms of upper respiratory tract infection after the race; this was significantly more than that reported by the vitamin C–supplemented group (33%).*

Explain to someone who knows no statistics why "significantly more" means there is good reason to think that vitamin C works.

# 3.2 Sampling Design

A political scientist wants to know what percent of college-age adults consider themselves conservatives. An automaker hires a market research firm to learn what percent of adults aged 18 to 35 recall seeing television advertisements for a new sport utility vehicle. Government economists inquire about average household income. In all these cases, we want to gather information about a large group of individuals. We will not, as in an experiment, impose a treatment in order to observe the response. Also, time, cost, and inconvenience forbid contacting every individual. In such cases, we gather information about only part of the group—a *sample*—in order to draw conclusions about the whole. **Sample surveys** are an important kind of observational study.

sample survey

---

**POPULATION AND SAMPLE**

The entire group of individuals that we want information about is called the **population.**

A **sample** is a part of the population that we actually examine in order to gather information.

---

Notice that "population" is defined in terms of our desire for knowledge. If we wish to draw conclusions about all U.S. college students, that group is our population even if only local students are available for questioning. The sample **sample design** is the part from which we draw conclusions about the whole. The **design** of a sample survey refers to the method used to choose the sample from the population.

**EXAMPLE**

**3.21  The Reading Recovery program.**   The Reading Recovery (RR) program has specially trained teachers work one-on-one with at-risk first-grade students to help them learn to read. A study was designed to examine the relationship between the RR teachers' beliefs about their ability to motivate students and the progress of the students whom they teach.[28] The National Data Evaluation Center (NDEC) Web site (www.ndec.us) says that there are 13,823 RR teachers. The researchers send a questionnaire to a random sample of 200 of these. The population consists of all 13,823 RR teachers, and the sample is the 200 that were randomly selected.

Unfortunately, our idealized framework of population and sample does not exactly correspond to the situations that we face in many cases. In Example 3.21, the list of teachers was prepared at a particular time in the past. It is very likely that some of the teachers on the list are no longer working as RR teachers today. New teachers have been trained in RR methods and are not on the list. In spite of these difficulties, we still view the list as the population. Also, we do not expect to get a response from every teacher in our random sample. We may have out-of-date addresses for some who are still working as RR teachers, and some teachers may choose not to respond to our survey questions.

In reporting the results of a sample survey it is important to include all details regarding the procedures used. Follow-up mailings or phone calls to those who do not initially respond can help increase the response rate. The proportion of the original sample who actually provide usable data is called **response rate** the **response rate** and should be reported for all surveys. If only 150 of the teachers who were sent questionnaires provided usable data, the response rate would be 150/200, or 75%.

**USE YOUR KNOWLEDGE**

**3.48  Job satisfaction in Mongolian universities.** A educational research team wanted to examine the relationship between faculty participation in decision making and job satisfaction in Mongolian public universities. They are planning to randomly select 300 faculty members from a list of 2500 faculty members in these universities. The Job Descriptive Index (JDI) will be used to measure job satisfaction, and the Conway Adaptation of the Alutto-Belasco Decisional Participation Scale will be used to measure decision participation. Describe the population and the sample for this study. Can you determine the response rate?

**3.49  Taxes and forestland usage.** A study was designed to assess the impact of taxes on forestland usage in part of the Upper Wabash River Watershed in Indiana.[29] A survey was sent to 772 forest owners from this region and 348 were returned. Consider the population, the sample, and the response rate for this study. Describe these based on the information given and indicate any additional information that you would need to give a complete answer.

Poor sample designs can produce misleading conclusions. Here is an example.

**EXAMPLE**

**3.22  Sampling pieces of steel.**   A mill produces large coils of thin steel for use in manufacturing home appliances. The quality engineer wants to submit a sample of 5-centimeter squares to detailed laboratory examination. She asks a technician to cut a sample of 10 such squares. Wanting to provide "good" pieces of steel, the technician carefully avoids the visible defects in the coil material when cutting the sample. The laboratory results are wonderful but the customers complain about the material they are receiving.

Online opinion polls are particularly vulnerable to bias because the sample who respond are not representative of the population at large. Here is an example that also illustrates how the results of such polls can be manipulated.

**EXAMPLE**

**3.23  The American Family Association.**   The American Family Association (AFA) is a conservative group that claims to stand for "traditional family values." It regularly posts online poll questions on its Web site—just click on a response to take part. Because the respondents are people who visit this site, the poll results always support AFA's positions. Well, almost always. In 2004, AFA's online poll asked about the heated issue of allowing same-sex marriage. Soon, email lists and social-network sites favored mostly by young liberals pointed to the AFA poll. Almost 850,000 people responded, and 60% of them favored legalization of same-sex marriage. AFA claimed that homosexual rights groups had skewed its poll.

As the AFA poll illustrates, you can't always trust poll results. People who take the trouble to respond to an open invitation are not representative of the entire adult population. That's true of regular visitors to AFA's site, of the activists who made a special effort to vote in the marriage poll, and of the people who bother to respond to write-in, call-in, or online polls in general.

In both Examples 3.22 and 3.23, the sample was selected in a manner that guaranteed that it would not be representative of the entire population. These sampling schemes display *bias,* or systematic error, in favoring some parts of the population over others. Online polls use *voluntary response samples,* a particularly common form of biased sample.

---

**VOLUNTARY RESPONSE SAMPLE**

A **voluntary response sample** consists of people who choose themselves by responding to a general appeal. Voluntary response samples are biased because people with strong opinions, especially negative opinions, are most likely to respond.

---

The remedy for bias in choosing a sample is to allow impersonal chance to do the choosing, so that there is neither favoritism by the sampler (as in Example 3.22) nor voluntary response (as in Example 3.23). Random selection of a sample eliminates bias by giving all individuals an equal chance to be chosen, just as randomization eliminates bias in assigning experimental subjects.

## Simple random samples

The simplest sampling design amounts to placing names in a hat (the population) and drawing out a handful (the sample). This is *simple random sampling.*

---

**SIMPLE RANDOM SAMPLE**

A **simple random sample (SRS)** of size $n$ consists of $n$ individuals from the population chosen in such a way that every set of $n$ individuals has an equal chance to be the sample actually selected.

---

Each treatment group in a completely randomized experimental design is an SRS drawn from the available experimental units. We select an SRS by labeling all the individuals in the population and using software or a table of random digits to select a sample of the desired size, just as in experimental randomization. Notice that an SRS not only gives each individual an equal chance to be chosen (thus avoiding bias in the choice) but gives every possible sample an equal chance to be chosen. There are other random sampling designs that give each individual, but not each sample, an equal chance. One such design, systematic random sampling, is described in Exercise 3.64.

**EXAMPLE**

**3.24 Spring break destinations.**   A campus newspaper plans a major article on spring break destinations. The authors intend to call a few randomly chosen resorts at each destination to ask about their attitudes toward groups of students as guests. Here are the resorts listed in one city. The first step is to label the members of this population as shown.

| | | | | | | | |
|----|----|----|----|----|----|----|----|
| 01 | Aloha Kai | 08 | Captiva | 15 | Palm Tree | 22 | Sea Shell |
| 02 | Anchor Down | 09 | Casa del Mar | 16 | Radisson | 23 | Silver Beach |
| 03 | Banana Bay | 10 | Coconuts | 17 | Ramada | 24 | Sunset Beach |
| 04 | Banyan Tree | 11 | Diplomat | 18 | Sandpiper | 25 | Tradewinds |
| 05 | Beach Castle | 12 | Holiday Inn | 19 | Sea Castle | 26 | Tropical Breeze |
| 06 | Best Western | 13 | Lime Tree | 20 | Sea Club | 27 | Tropical Shores |
| 07 | Cabana | 14 | Outrigger | 21 | Sea Grape | 28 | Veranda |

Now enter Table B, and read two-digit groups until you have chosen three resorts. If you enter at line 185, Banana Bay (03), Palm Tree (15), and Cabana (07) will be called.

Most statistical software will select an SRS for you, eliminating the need for Table B. The *Simple Random Sample* applet on the text CD and Web site is a convenient way to automate this task.

Excel can do the job in a way similar to what we used when we randomized experimental units to treatments in designed experiments. There are four steps:

1. Create a data set with all of the elements of the population in the first column.

2. Assign a random number to each element of the population; put these in the second column.

3. Sort the data set by the random number column.

4. The simple random sample is obtained by taking elements in the sorted list until the desired sample size is reached.

We illustrate the procedure with a simplified version of Example 3.24.

**EXAMPLE**

**3.25 Select a random sample.** Suppose that the population from Example 3.24 is only the first two rows of the display given there:

| | | | |
|----|----|----|----|
| Aloha Kai | Captiva | Palm Tree | Sea Shell |
| Anchor Down | Casa del Mar | Radisson | Silver Beach |

Note that we do not need the numerical labels to identify the individuals in the population. Suppose that we want to select a simple random sample of three resorts from this population. Figure 3.8(a) gives the spreadsheet with the population names. The random numbers generated by the RAND() function are given in the second column in Figure 3.8(b). The sorted data set is given in Figure 3.8(c). We have added a third column to the speadsheet to indicate which resorts were selected for our random sample. They are Captiva, Radisson, and Silver Beach.

(a)

(b)

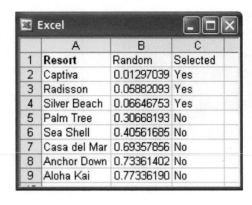
(c)

**FIGURE 3.8** Selection of a simple random sample of resorts, for Example 3.25.

## USE YOUR KNOWLEDGE

**3.50 Ringtones for cell phones.** You decide to change the ringtones for your cell phone by choosing 2 from a list of the 10 most popular ringtones.[30] Here is the list:

| | | | |
|---|---|---|---|
| Super Mario Brothers Theme | Sexy Love | Ms. New Booty | Ridin' Rims |
| I Write Sins Not Tragedies | Gasolina | My Humps | The Pink Panther |
| Down | Agarrala | | |

Select your two ringtones using a simple random sample.

**3.51 Listen to three songs.** The walk to your statistics class takes about 10 minutes, about the amount of time needed to listen to three songs on your iPod. You decide to take a simple random sample of songs from a Billboard list of Rock Songs.[31] Here is the list:

| | | | |
|---|---|---|---|
| Miss Murder | Animal I Have Become | Steady, As She Goes | Dani California |
| The Kill (Bury Me) | Original Fire | When You Were Young | MakeD—Sure |
| Vicarious | The Diary of Jane | | |

Select the three songs for your iPod using a simple random sample.

## Stratified samples

The general framework for designs that use chance to choose a sample is a *probability sample*.

### PROBABILITY SAMPLE

A **probability sample** is a sample chosen by chance. We must know what samples are possible and what chance, or probability, each possible sample has.

Some probability sampling designs (such as an SRS) give each member of the population an *equal* chance to be selected. This may not be true in more elaborate sampling designs. In every case, however, the use of chance to select the sample is the essential principle of statistical sampling.

Designs for sampling from large populations spread out over a wide area are usually more complex than an SRS. For example, it is common to sample important groups within the population separately, then combine these samples. This is the idea of a *stratified sample*.

---

**STRATIFIED RANDOM SAMPLE**

To select a **stratified random sample,** first divide the population into groups of similar individuals, called **strata.** Then choose a separate SRS in each stratum and combine these SRSs to form the full sample.

---

Choose the strata based on facts known before the sample is taken. For example, a population of election districts might be divided into urban, suburban, and rural strata. A stratified design can produce more exact information than an SRS of the same size by taking advantage of the fact that individuals in the same stratum are similar to one another. Think of the extreme case in which all individuals in each stratum are identical: just one individual from each stratum is then enough to completely describe the population. Strata for sampling are similar to blocks in experiments. We have two names because the idea of grouping similar units before randomizing arose separately in sampling and in experiments.

**EXAMPLE**

**3.26 A stratified sample of dental claims.** A dentist is suspected of defrauding insurance companies by describing some dental procedures incorrectly on claim forms and overcharging for them. An investigation begins by examining a sample of his bills for the past three years. Because there are five suspicious types of procedures, the investigators take a stratified sample. That is, they randomly select bills for each of the five types of procedures separately.

## Multistage samples

Another common means of restricting random selection is to choose the sample in stages. This is common practice for national samples of households or people. For example, data on employment and unemployment are gathered by the government's Current Population Survey, which conducts interviews in about 60,000 households each month. The cost of sending interviewers to the widely scattered households in an SRS would be too high. Moreover, the government wants data broken down by states and large cities. The Current Population Survey therefore uses a **multistage sampling design.** The final sample consists of clusters of nearby households that an interviewer can easily visit.

multistage sample

Most opinion polls and other national samples are also multistage, though interviewing in most national samples today is done by telephone rather than in person, eliminating the economic need for clustering. The Current Population Survey sampling design is roughly as follows:[32]

Stage 1.  Divide the United States into 2007 geographical areas called Primary Sampling Units, or PSUs. PSUs do not cross state lines. Select a sample of 754 PSUs. This sample includes the 428 PSUs with the largest population and a stratified sample of 326 of the others.

Stage 2.  Divide each PSU selected into smaller areas called "blocks." Stratify the blocks using ethnic and other information and take a stratified sample of the blocks in each PSU.

Stage 3.  Sort the housing units in each block into clusters of four nearby units. Interview the households in a probability sample of these clusters.

Analysis of data from sampling designs more complex than an SRS takes us beyond basic statistics. But the SRS is the building block of more elaborate designs, and analysis of other designs differs more in complexity of detail than in fundamental concepts.

## Cautions about sample surveys

Random selection eliminates bias in the choice of a sample from a list of the population. Sample surveys of large human populations, however, require much more than a good sampling design.[33] To begin, we need an accurate and complete list of the population. Because such a list is rarely available, most samples suffer from some degree of *undercoverage*. A sample survey of households, for example, will miss not only homeless people but prison inmates and students in dormitories. An opinion poll conducted by telephone will miss the 6% of American households without residential phones. The results of national sample surveys therefore have some bias if the people not covered—who most often are poor people—differ from the rest of the population.

A more serious source of bias in most sample surveys is *nonresponse*, which occurs when a selected individual cannot be contacted or refuses to cooperate. Nonresponse to sample surveys often reaches 50% or more, even with careful planning and several callbacks. Because nonresponse is higher in urban areas, most sample surveys substitute other people in the same area to avoid favoring rural areas in the final sample. If the people contacted differ from those who are rarely at home or who refuse to answer questions, some bias remains.

---

UNDERCOVERAGE AND NONRESPONSE

**Undercoverage** occurs when some groups in the population are left out of the process of choosing the sample.

**Nonresponse** occurs when an individual chosen for the sample can't be contacted or does not cooperate.

---

**EXAMPLE**

**3.27 Nonresponse in the Current Population Survey.** How bad is nonresponse? The Current Population Survey (CPS) has the lowest nonresponse rate of any poll we know: only about 4% of the households in the CPS sample refuse to take part and another 3% or 4% can't be contacted. People are more likely to respond to a government survey such as the CPS, and the CPS contacts its sample in person before doing later interviews by phone.

The General Social Survey (Figure 3.9) is the nation's most important social science research survey. The GSS also contacts its sample in person, and it is run by a university. Despite these advantages, its most recent survey had a 30% rate of nonresponse.

What about polls done by the media and by market research and opinion-polling firms? We don't know their rates of nonresponse, because they won't say. That itself is a bad sign. The Pew Research Center for People and the Press designed a careful telephone survey and published the results: out of 2879 households called, 1658 were never at home, refused, or would not finish the interview. That's a nonresponse rate of 58%.[34]

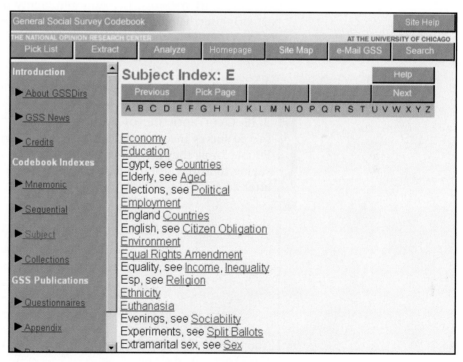

**FIGURE 3.9** Part of the subject index for the General Social Survey (GSS). The GSS has assessed attitudes on a wide variety of topics since 1972. Its continuity over time makes the GSS a valuable source for studies of changing attitudes.

Most sample surveys, and almost all opinion polls, are now carried out by telephone. This and other details of the interview method can affect the results.

**3.28 How should the data be collected?**   A Pew Research Center Poll has asked about belief in God for many years. In response to the statement "I never doubt the existence of God," subjects are asked to choose from the responses

completely agree    mostly agree    mostly disagree    completely disagree

In 1990, subjects were interviewed in person and were handed a card with the four responses on it. In 1991, the poll switched to telephone interviews. In 1990, 60% said "completely agree," in line with earlier years. In 1991, 71% completely agreed. The increase is probably explained by the effect of hearing "completely agree" read first by the interviewer.[35]

**response bias**

The behavior of the respondent or of the interviewer can cause **response bias** in sample results. Respondents may lie, especially if asked about illegal or unpopular behavior. The race or sex of the interviewer can influence responses to questions about race relations or attitudes toward feminism. Answers to questions that ask respondents to recall past events are often inaccurate because of faulty memory. For example, many people "telescope" events in the past, bringing them forward in memory to more recent time periods. "Have you visited a dentist in the last 6 months?" will often elicit a "Yes" from someone who last visited a dentist 8 months ago.[36]

**3.29 Overreporting of voter behavior.**   "One of the most frequently observed survey measurement errors is the overreporting of voting behavior."[37] People know they should vote, so those who didn't vote tend to save face by saying that they did. Here are the data from a typical sample of 663 people after an election:

|  |  | What they said: | |
|---|---|---|---|
|  |  | I voted | I didn't |
| What they did: | Voted | 358 | 13 |
|  | Didn't vote | 120 | 172 |

You can see that 478 people (72%) said that they voted, but only 371 people (56%) actually did vote.

**wording of questions**

The **wording of questions** is the most important influence on the answers given to a sample survey. Confusing or leading questions can introduce strong bias, and even minor changes in wording can change a survey's outcome. Here are some examples.

**EXAMPLE**

**3.30 The form of the question is important.**   In response to the question "Are you heterosexual, homosexual, or bisexual?" in a social science research survey, one woman answered, "It's just me and my husband, so bisexual." The issue is serious, even if the example seems silly: reporting about sexual behavior is difficult because people understand and misunderstand sexual terms in many ways.

How do Americans feel about government help for the poor? Only 13% think we are spending too much on "assistance to the poor," but 44% think we are spending too much on "welfare." How do the Scots feel about the movement to become independent from England? Well, 51% would vote for "independence for Scotland," but only 34% support "an independent Scotland separate from the United Kingdom." It seems that "assistance to the poor" and "independence" are nice, hopeful words. "Welfare" and "separate" are negative words.[38]

The statistical design of sample surveys is a science, but this science is only part of the art of sampling. Because of nonresponse, response bias, and the difficulty of posing clear and neutral questions, you should hesitate to fully trust reports about complicated issues based on surveys of large human populations. *Insist on knowing the exact questions asked, the rate of nonresponse, and the date and method of the survey before you trust a poll result.*

## SECTION 3.2   Summary

A sample survey selects a **sample** from the **population** of all individuals about which we desire information. We base conclusions about the population on data about the sample.

The **design** of a sample refers to the method used to select the sample from the population. **Probability sampling designs** use impersonal chance to select a sample.

The basic probability sample is a **simple random sample (SRS).** An SRS gives every possible sample of a given size the same chance to be chosen.

Choose an SRS by labeling the members of the population and using a **table of random digits** to select the sample. Software can automate this process.

To choose a **stratified random sample,** divide the population into **strata,** groups of individuals that are similar in some way that is important to the response. Then choose a separate SRS from each stratum and combine them to form the full sample.

**Multistage samples** select successively smaller groups within the population in stages, resulting in a sample consisting of clusters of individuals. Each stage may employ an SRS, a stratified sample, or another type of sample.

Failure to use probability sampling often results in **bias,** or systematic errors in the way the sample represents the population. **Voluntary response** samples, in which the respondents choose themselves, are particularly prone to large bias.

In human populations, even probability samples can suffer from bias due to **undercoverage** or **nonresponse,** from **response bias** due to the behavior of the interviewer or the respondent, or from misleading results due to **poorly worded questions.**

## SECTION 3.2  Exercises

*For Exercises 3.48 and 3.49, see pages 198 and 199; and for Exercises 3.50 and 3.51, see page 202.*

**3.52 What's wrong?** Explain what is wrong in each of the following scenarios.

(a) The population consists of all individuals selected in a simple random sample.

(b) In a poll of an SRS of residents in a local community, respondents are asked to indicate the level of their concern about the dangers of dihydrogen monoxide, a substance that is a major component of acid rain and in its gaseous state can cause severe burns. (*Hint:* Ask a friend who is majoring in chemistry about this substance or search the Internet for information about it.)

(c) Students in a class are asked to raise their hands if they have cheated on an exam one or more times within the past year.

**3.53 What's wrong?** Explain what is wrong with each of the following random selection procedures and explain how you would do the randomization correctly.

(a) To determine the reading level of an introductory statistics text, you evaluate all of the written material in the third chapter.

(b) You want to sample student opinions about a proposed change in procedures for changing majors. You hand out questionnaires to 100 students as they arrive for class at 7:30 A.M.

(c) A population of subjects is put in alphabetical order and a simple random sample of size 10 is taken by selecting the first 10 subjects in the list.

**3.54 Importance of students as customers.** A committee on community relations in a college town plans to survey local businesses about the importance of students as customers. From telephone book listings, the committee chooses 150 businesses at random. Of these, 73 return the questionnaire mailed by the committee. What is the population for this sample survey? What is the sample? What is the rate (percent) of nonresponse?

**3.55 Popularity of news personalities.** A Gallup Poll conducted telephone interviews with 1001 U.S. adults aged 18 and over on July 24–27, 2006. One of the questions asked whether the respondents had a favorable or an unfavorable opinion of 17 news personalities. Diane Sawyer received the highest rating, with 80% of the respondents giving her a favorable rating.[39]

(a) What is the population for this sample survey? What was the sample size?

(b) The report on the survey states that 8% of the respondents either never heard of Sawyer or had no opinion about her. When they included only those who provided an opinion, Sawyer's approval percent rose to 88% and she was still at the top of the list. Charles Gibson, on the other hand, was ranked eighth on the original list, with a 55% favorable rating. When only those providing an opinion were counted, his rank rose to second, with 87% approving. Discuss the advantages and disadvantages of the two different ways of reporting the approval percent. State which one you prefer and why.

**3.56 Identify the populations.** For each of the following sampling situations, identify the population as exactly as possible. That is, say what kind of individuals the population consists of and say exactly which individuals fall in the population. If the information given is not complete, complete the description of the population in a reasonable way.

(a) A college has changed its core curriculum and wants to obtain detailed feedback information from the students during each of the first 12 weeks of the coming semester. Each week, a random sample of 5 students will be selected to be interviewed.

(b) The American Community Survey (ACS) will replace the census "long form" starting with the 2010 census. The main part of the ACS contacts 250,000 addresses by mail each month, with follow-up by phone and in person if there is no response. Each household answers questions about their housing, economic, and social status.

(c) An opinion poll contacts 1161 adults and asks them, "Which political party do you think has better ideas for leading the country in the twenty-first century?"

**3.57 Interview residents of apartment complexes.** You are planning a report on apartment living in a college town. You decide to select 5 apartment complexes at random for in-depth interviews with residents. Select a simple random sample of 5 of the following apartment complexes. If you use Table B, start at line 137.

| | | |
|---|---|---|
| Ashley Oaks | Country View | Mayfair Village |
| Bay Pointe | Country Villa | Nobb Hill |
| Beau Jardin | Crestview | Pemberly Courts |
| Bluffs | Del-Lynn | Peppermill |
| Brandon Place | Fairington | Pheasant Run |
| Briarwood | Fairway Knolls | Richfield |
| Brownstone | Fowler | Sagamore Ridge |
| Burberry | Franklin Park | Salem Courthouse |
| Cambridge | Georgetown | Village Manor |
| Chauncey Village | Greenacres | Waterford Court |
| Country Squire | Lahr House | Williamsburg |

**3.58 Using GIS to identify mint field conditions.** A Geographic Information System (GIS) is to be used to distinguish different conditions in mint fields. Ground observations will be used to classifiy regions of each field as either healthy mint, diseased mint, or weed-infested mint. The GIS divides mint-growing areas into regions called pixels. An experimental area contains 200 pixels. For a random sample of 25 pixels, ground measurements will be made to determine the status of the mint, and these observations will be compared with information obtained by the GIS. Select the random sample. If you use Table B, start at line 112 and choose only the first 5 pixels in the sample.

**3.59** Use the simple random sample applet. After you have labeled the individuals in a population, the *Simple Random Sample* applet automates the task of choosing an SRS. Use the applet to choose the sample in the previous exercise.

**3.60** Use the simple random sample applet. There are approximately 371 active telephone area codes covering Canada, the United States, and some Caribbean areas. (More are created regularly.) You want to choose an SRS of 25 of these area codes for a study of available telephone numbers. Label the codes 001 to 371 and use the *Simple Random Sample* applet to choose your sample. (If you use Table B, start at line 120 and choose only the first 5 codes in the sample.)

**3.61 Census tracts.** The Census Bureau divides the entire country into "census tracts" that contain about 4000 people. Each tract is in turn divided into small "blocks," which in urban areas are bounded by local streets. An SRS of blocks from a census tract is often the next-to-last stage in a multistage sample. Figure 3.10 shows part of census tract 8051.12, in Cook County, Illinois, west of Chicago. The 44 blocks in this tract are divided into three "block groups." Group 1 contains 6 blocks numbered 1000 to 1005;

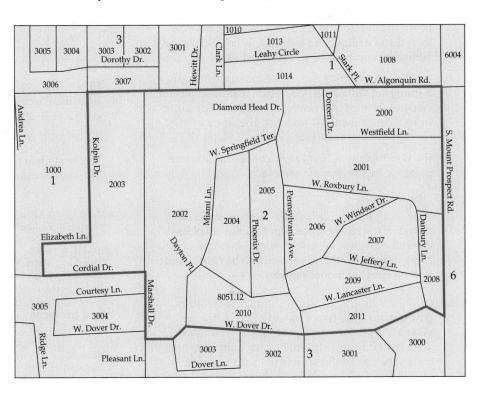

**FIGURE 3.10** Census blocks in Cook County, Illinois, for Exercises 3.61 and 3.63. The outlined area is a block group.

Group 2 (outlined in Figure 3.10) contains 12 blocks numbered 2000 to 2011; Group 3 contains 26 blocks numbered 3000 to 3025. Use Table B, beginning at line 135, to choose an SRS of 5 of the 44 blocks in this census tract. Explain carefully how you labeled the blocks.

**3.62 Repeated use of Table B.** In using Table B repeatedly to choose samples or do randomization for experiments, you should not always begin at the same place, such as line 101. Why not?

**3.63 A stratified sample.** Exercise 3.61 asks you to choose an SRS of blocks from the census tract pictured in Figure 3.10. You might instead choose a stratified sample of one block from the 6 blocks in Group 1, two from the 12 blocks in Group 2, and three from the 26 blocks in Group 3. Choose such a sample, explaining carefully how you labeled blocks and used Table B.

**3.64 Systematic random samples. Systematic random samples** are often used to choose a sample of apartments in a large building or dwelling units in a block at the last stage of a multistage sample. An example will illustrate the idea of a systematic sample. Suppose that we must choose 4 addresses out of 100. Because $100/4 = 25$, we can think of the list as four lists of 25 addresses. Choose 1 of the first 25 at random, using Table B. The sample contains this address and the addresses 25, 50, and 75 places down the list from it. If 13 is chosen, for example, then the systematic random sample consists of the addresses numbered 13, 38, 63, and 88.

(a) A study of dating among college students wanted a sample of 200 of the 9000 single male students on campus. The sample consisted of every 45th name from a list of the 9000 students. Explain why the survey chooses every 45th name.

(b) Use Table B at line 125 to choose the starting point for this systematic sample.

**3.65** CHALLENGE **Systematic random samples versus simple random samples.** The previous exercise introduces systematic random samples. Explain carefully why a systematic random sample *does* give every individual the same chance to be chosen but is *not* a simple random sample.

**3.66 Random digit telephone dialing.** An opinion poll in California uses random digit dialing to choose telephone numbers at random. Numbers are selected separately within each California area code. The size of the sample in each area code is proportional to the population living there.

(a) What is the name for this kind of sampling design?

(b) California area codes, in rough order from north to south, are

530 707 916 209 415 925 510 650 408 831 805 559 760
661 818 213 626 323 562 709 310 949 909 858 619

Another California survey does not call numbers in all area codes but starts with an SRS of 10 area codes. Choose such an SRS. If you use Table B, start at line 122.

**3.67 Stratified samples of forest areas.** Stratified samples are widely used to study large areas of forest. Based on satellite images, a forest area in the Amazon basin is divided into 14 types. Foresters studied the four most commercially valuable types: alluvial climax forests of quality levels 1, 2, and 3, and mature secondary forest. They divided the area of each type into large parcels, chose parcels of each type at random, and counted tree species in a 20- by 25-meter rectangle randomly placed within each parcel selected. Here is some detail:

| Forest type | Total parcels | Sample size |
|---|---|---|
| Climax 1 | 36 | 4 |
| Climax 2 | 72 | 7 |
| Climax 3 | 31 | 3 |
| Secondary | 42 | 4 |

Choose the stratified sample of 18 parcels. Be sure to explain how you assigned labels to parcels. If you use Table B, start at line 140.

**3.68 Select club members to go to a convention.** A club has 30 student members and 10 faculty members. The students are

| | | | | |
|---|---|---|---|---|
| Abel | Fisher | Huber | Moran | Reinmann |
| Carson | Golomb | Jimenez | Moskowitz | Santos |
| Chen | Griswold | Jones | Neyman | Shaw |
| David | Hein | Kiefer | O'Brien | Thompson |
| Deming | Hernandez | Klotz | Pearl | Utts |
| Elashoff | Holland | Liu | Potter | Vlasic |

and the faculty members are

| | | | | |
|---|---|---|---|---|
| Andrews | Fernandez | Kim | Moore | Rabinowitz |
| Besicovitch | Gupta | Lightman | Phillips | Yang |

The club can send 5 students and 3 faculty members to a convention and decides to choose those who will go by random selection. Select a stratified random sample of 5 students and 3 faculty members.

**3.69** ⚔ **Stratified samples for alcohol attitudes.** At a party there are 30 students over age 21 and 20 students under age 21. You choose at random 3 of those over 21 and separately choose at random 2 of those under 21 to interview about attitudes toward alcohol. You have given every student at the party the same chance to be interviewed: what is that chance? Why is your sample not an SRS?

**3.70** **Stratified samples for accounting audits.** Accountants use stratified samples during audits to verify a company's records of such things as accounts receivable. The stratification is based on the dollar amount of the item and often includes 100% sampling of the largest items. One company reports 5000 accounts receivable. Of these, 100 are in amounts over $50,000; 500 are in amounts between $1000 and $50,000; and the remaining 4400 are in amounts under $1000. Using these groups as strata, you decide to verify all of the largest accounts and to sample 5% of the midsize accounts and 1% of the small accounts. How would you label the two strata from which you will sample? Use Table B, starting at line 115, to select the first 5 accounts from each of these strata.

**3.71** **Nonresponse in telephone surveys.** A common form of nonresponse in telephone surveys is "ring-no-answer." That is, a call is made to an active number but no one answers. The Italian National Statistical Institute looked at nonresponse to a government survey of households in Italy during the periods January 1 to Easter and July 1 to August 31. All calls were made between 7 and 10 P.M., but 21.4% gave "ring-no-answer" in one period versus 41.5% "ring-no-answer" in the other period.[40] Which period do you think had the higher rate of no answers? Why? Explain why a high rate of nonresponse makes sample results less reliable.

**3.72** **The sampling frame.** The list of individuals from which a sample is actually selected is called the **sampling frame.** Ideally, the frame should list every individual in the population, but in practice this is often difficult. A frame that leaves out part of the population is a common source of undercoverage.

(a) Suppose that a sample of households in a community is selected at random from the telephone directory. What households are omitted from this frame? What types of people do you think are

likely to live in these households? These people will probably be underrepresented in the sample.

(b) It is usual in telephone surveys to use random digit dialing equipment that selects the last four digits of a telephone number at random after being given the area code and the exchange (the first three digits). Which of the households that you mentioned in your answer to (a) will be included in the sampling frame by random digit dialing?

**3.73** **The Excite Poll.** The Excite Poll can be found online at `poll.excite.com`. The question appears on the screen, and you simply click buttons to vote "Yes," "No," "Not sure," or "Don't care." On July 22, 2006, the question was "Do you agree or disagree with proposed legislation that would discontinue the U.S. penny coin?" In all, 631 said "Yes," another 564 said "No," and the remaining 65 indicated that they were not sure.

(a) What is the sample size for this poll?

(b) Compute the percent of responses in each of the possible response categories.

(c) Discuss the poll in terms of the population and sample framework that we have studied in this chapter.

**3.74** **Survey questions.** Comment on each of the following as a potential sample survey question. Is the question clear? Is it slanted toward a desired response?

(a) "Some cell phone users have developed brain cancer. Should all cell phones come with a warning label explaining the danger of using cell phones?"

(b) "Do you agree that a national system of health insurance should be favored because it would provide health insurance for everyone and would reduce administrative costs?"

(c) "In view of escalating environmental degradation and incipient resource depletion, would you favor economic incentives for recycling of resource-intensive consumer goods?"

**3.75** **Use of a budget surplus.** In 2000, when the federal budget showed a large surplus, the Pew Research Center asked two questions of random samples of adults. Both questions stated that Social Security would be "fixed." Here are the uses suggested for the remaining surplus:

*Should the money be used for a tax cut, or should it be used to fund new government programs?*

*Should the money be used for a tax cut, or should it be spent on programs for education, the environment, health care, crime-fighting and military defense?*

One of these questions drew 60% favoring a tax cut; the other, only 22%. Which wording pulls respondents toward a tax cut? Why?

3.76 △ **How many children are in your family?** A teacher asks her class, "How many children are there in your family, including yourself?" The mean response is about 3 children. According to the 2000 census, families that have children average 1.86 children. Why is a sample like this biased toward higher outcomes?

3.77 △ **Bad survey questions.** Write your own examples of bad sample survey questions.

(a) Write a biased question designed to get one answer rather than another.

(b) Write a question that is confusing, so that it is hard to answer.

3.78 △ **Economic attitudes of Spaniards.** Spain's Centro de Investigaciones Sociológicos carried out a sample survey on the economic attitudes of Spaniards.[41] Of the 2496 adults interviewed, 72% agreed that "Employees with higher performance must get higher pay." On the other hand, 71% agreed that "Everything a society produces should be distributed among its members as equally as possible and there should be no major differences." Use these conflicting results as an example in a short explanation of why opinion polls often fail to reveal public attitudes clearly.

# 3.3 Toward Statistical Inference

A market research firm interviews a random sample of 2500 adults. Result: 66% find shopping for clothes frustrating and time-consuming. That's the truth about the 2500 people in the sample. What is the truth about the almost 220 million American adults who make up the population? Because the sample was chosen at random, it's reasonable to think that these 2500 people represent the entire population fairly well. So the market researchers turn the *fact* that 66% of the *sample* find shopping frustrating into an *estimate* that about 66% of *all adults* feel this way. That's a basic move in statistics: use a fact about a sample to estimate the truth about the whole population. We call this **statistical inference** because we infer conclusions about the wider population from data on selected individuals. To think about inference, we must keep straight whether a number describes a sample or a population. Here is the vocabulary we use.

statistical inference

---

## PARAMETERS AND STATISTICS

A **parameter** is a number that describes the **population.** A parameter is a fixed number, but in practice we do not know its value.

A **statistic** is a number that describes a **sample.** The value of a statistic is known when we have taken a sample, but it can change from sample to sample. We often use a statistic to estimate an unknown parameter.

---

**EXAMPLE**

**3.31 Attitudes toward shopping.** Are attitudes toward shopping changing? Sample surveys show that fewer people enjoy shopping than in the past. A survey by the market research firm Yankelovich Clancy Shulman asked a nationwide random sample of 2500 adults if they agreed or disagreed that "I like buying new clothes, but shopping is often frustrating and time-consuming." Of the respondents, 1650, or 66%, said they agreed.[42] The

proportion of the sample who agree is

$$\hat{p} = \frac{1650}{2500} = 0.66 = 66\%$$

The number $\hat{p} = 0.66$ is a *statistic*. The corresponding *parameter* is the proportion (call it $p$) of all adult U.S. residents who would have said "Agree" if asked the same question. We don't know the value of the parameter $p$, so we use the statistic $\hat{p}$ to estimate it.

## USE YOUR KNOWLEDGE

**3.79  Sexual harassment of college students.** A recent survey of 2036 undergraduate college students aged 18 to 24 reports that 62% of college students say they have encountered some type of sexual harassment while at college.[43] Describe the sample and the population for this setting.

**3.80  Web polls.** If you connect to the Web site worldnetdaily.com/polls/, you will be given the opportunity to give your opinion about a different question of public interest each day. Can you apply the ideas about populations and samples that we have just discussed to this poll? Explain why or why not.

## Sampling variability

*sampling variability*

If Yankelovich took a second random sample of 2500 adults, the new sample would have different people in it. It is almost certain that there would not be exactly 1650 positive responses. That is, the value of the statistic $\hat{p}$ will vary from sample to sample. This basic fact is called **sampling variability:** the value of a statistic varies in repeated random sampling. Could it happen that one random sample finds that 66% of adults find clothes shopping frustrating and a second random sample finds that only 42% feel this way? Random samples eliminate *bias* from the act of choosing a sample, but they can still be wrong because of the *variability* that results when we choose at random. If the variation when we take repeat samples from the same population is too great, we can't trust the results of any one sample.

We are saved by the second great advantage of random samples. The first advantage is that choosing at random eliminates favoritism. That is, random sampling attacks bias. The second advantage is that if we take lots of random samples of the same size from the same population, the variation from sample to sample will follow a predictable pattern. **All of statistical inference is based on one idea: to see how trustworthy a procedure is, ask what would happen if we repeated it many times.**

To understand why sampling variability is not fatal, we ask, "What would happen if we took many samples?" Here's how to answer that question:

- Take a large number of samples from the same population.

- Calculate the sample proportion $\hat{p}$ for each sample.

- Make a histogram of the values of $\hat{p}$.

• Examine the distribution displayed in the histogram for shape, center, and spread, as well as outliers or other deviations.

In practice it is too expensive to take many samples from a large population such as all adult U.S. residents. But we can imitate many samples by using random digits. Using random digits from a table or computer software to imitate chance behavior is called **simulation.**

*simulation*

---

**EXAMPLE**

**3.32 Simulate a random sample.**   We will simulate drawing simple random samples (SRSs) of size 100 from the population of all adult U.S. residents. Suppose that in fact 60% of the population find clothes shopping time-consuming and frustrating. Then the true value of the parameter we want to estimate is $p = 0.6$. (Of course, we would not sample in practice if we already knew that $p = 0.6$. We are sampling here to understand how sampling behaves.)

We can imitate the population by a table of random digits, with each entry standing for a person. Six of the ten digits (say 0 to 5) stand for people who find shopping frustrating. The remaining four digits, 6 to 9, stand for those who do not. Because all digits in a random number table are equally likely, this assignment produces a population proportion of frustrated shoppers equal to $p = 0.6$. We then imitate an SRS of 100 people from the population by taking 100 consecutive digits from Table B. The statistic $\hat{p}$ is the proportion of 0s to 5s in the sample.

Here are the first 100 entries in Table B with digits 0 to 5 highlighted:

| | | | | | | | |
|---|---|---|---|---|---|---|---|
| 19223 | 95034 | 05756 | 28713 | 96409 | 12531 | 42544 | 82853 |
| 73676 | 47150 | 99400 | 01927 | 27754 | 42648 | 82425 | 36290 |
| 45467 | 71709 | 77558 | 00095 | | | | |

There are 64 digits between 0 and 5, so $\hat{p} = 64/100 = 0.64$. A second SRS based on the second 100 entries in Table B gives a different result, $\hat{p} = 0.55$. The two sample results are different, and neither is equal to the true population value $p = 0.6$. That's sampling variability.

---

## Sampling distributions

Simulation is a powerful tool for studying chance. Now that we see how simulation works, it is faster to abandon Table B and to use a computer programmed to generate random numbers.

---

**EXAMPLE**

**3.33 Take many random samples.**   Figure 3.11 illustrates the process of choosing many samples and finding the sample proportion $\hat{p}$ for each one. Follow the flow of the figure from the population at the left, to choosing an SRS and finding the $\hat{p}$ for this sample, to collecting together the $\hat{p}$'s from many samples. The histogram at the right of the figure shows the distribution of the values of $\hat{p}$ from 1000 separate SRSs of size 100 drawn from a population with $p = 0.6$.

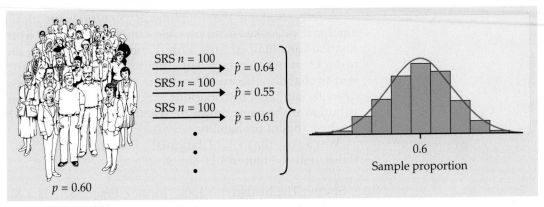

**FIGURE 3.11** The results of many SRSs have a regular pattern. Here, we draw 1000 SRSs of size 100 from the same population. The population proportion is $p = 0.60$. The histogram shows the distribution of the 1000 sample proportions.

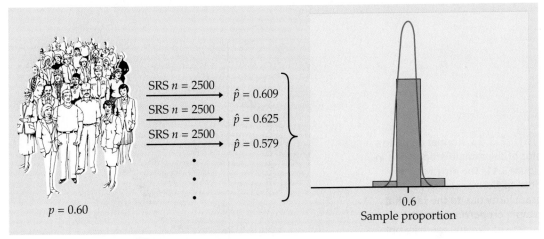

**FIGURE 3.12** The distribution of sample proportions for 1000 SRSs of size 2500 drawn from the same population as in Figure 3.11. The two histograms have the same scale. The statistic from the larger sample is less variable.

Of course, Yankelovich interviewed 2500 people, not just 100. Figure 3.12 is parallel to Figure 3.11. It shows the process of choosing 1000 SRSs, each of size 2500, from a population in which the true proportion is $p = 0.6$. The 1000 values of $\hat{p}$ from these samples form the histogram at the right of the figure. Figures 3.11 and 3.12 are drawn on the same scale. Comparing them shows what happens when we increase the size of our samples from 100 to 2500. These histograms display the *sampling distribution* of the statistic $\hat{p}$ for two sample sizes.

### SAMPLING DISTRIBUTION

The **sampling distribution** of a statistic is the distribution of values taken by the statistic in all possible samples of the same size from the same population.

Strictly speaking, the sampling distribution is the ideal pattern that would emerge if we looked at all possible samples of size 100 from our population. A distribution obtained from a fixed number of trials, like the 1000 trials in Figure 3.11, is only an approximation to the sampling distribution. We will see that probability theory, the mathematics of chance behavior, can sometimes describe sampling distributions exactly. The interpretation of a sampling distribution is the same, however, whether we obtain it by simulation or by the mathematics of probability.

We can use the tools of data analysis to describe any distribution. Let's apply those tools to Figures 3.11 and 3.12.

- **Shape:** The histograms look Normal. Figure 3.13 is a Normal quantile plot of the values of $\hat{p}$ for our samples of size 100. It confirms that the distribution in Figure 3.11 is close to Normal. The 1000 values for samples of size 2500 in Figure 3.12 are even closer to Normal. The Normal curves drawn through the histograms describe the overall shape quite well.

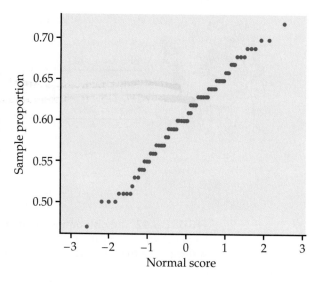

**FIGURE 3.13** Normal quantile plot of the sample proportions in Figure 3.11. The distribution is close to Normal except for some granularity due to the fact that sample proportions from a sample of size 100 can take only values that are multiples of 0.01. Because a plot of 1000 points is hard to read, this plot presents only every 10th value.

- **Center:** In both cases, the values of the sample proportion $\hat{p}$ vary from sample to sample, but the values are centered at 0.6. Recall that $p = 0.6$ is the true population parameter. Some samples have a $\hat{p}$ less than 0.6 and some greater, but there is no tendency to be always low or always high. That is, $\hat{p}$ has no **bias** as an estimator of $p$. This is true for both large and small samples. (Want the details? The mean of the 1000 values of $\hat{p}$ is 0.598 for samples of size 100 and 0.6002 for samples of size 2500. The median value of $\hat{p}$ is exactly 0.6 for samples of both sizes.)

- **Spread:** The values of $\hat{p}$ from samples of size 2500 are much less spread out than the values from samples of size 100. In fact, the standard deviations are 0.051 for Figure 3.11 and 0.0097, or about 0.01, for Figure 3.12.

Although these results describe just two sets of simulations, they reflect facts that are true whenever we use random sampling.

## USE YOUR KNOWLEDGE

**3.81 Effect of sample size on the sampling distribution.** You are planning a study and are considering taking an SRS of either 200 or 400 observations. Explain how the sampling distribution would differ for these two scenarios.

## Bias and variability

Our simulations show that a sample of size 2500 will almost always give an estimate $\hat{p}$ that is close to the truth about the population. Figure 3.12 illustrates this fact for just one value of the population proportion, but it is true for any population. Samples of size 100, on the other hand, might give an estimate of 50% or 70% when the truth is 60%.

Thinking about Figures 3.11 and 3.12 helps us restate the idea of bias when we use a statistic like $\hat{p}$ to estimate a parameter like $p$. It also reminds us that variability matters as much as bias.

---

### BIAS AND VARIABILITY

**Bias** concerns the center of the sampling distribution. A statistic used to estimate a parameter is **unbiased** if the mean of its sampling distribution is equal to the true value of the parameter being estimated.

The **variability of a statistic** is described by the spread of its sampling distribution. This spread is determined by the sampling design and the sample size $n$. Statistics from larger probability samples have smaller spreads.

---

We can think of the true value of the population parameter as the bull's-eye on a target, and of the sample statistic as an arrow fired at the bull's-eye. Bias and variability describe what happens when an archer fires many arrows at the target. *Bias* means that the aim is off, and the arrows land consistently off the bull's-eye in the same direction. The sample values do not center about the population value. Large *variability* means that repeated shots are widely scattered on the target. Repeated samples do not give similar results but differ widely among themselves. Figure 3.14 shows this target illustration of the two types of error.

Notice that small variability (repeated shots are close together) can accompany large bias (the arrows are consistently away from the bull's-eye in one direction). And small bias (the arrows center on the bull's-eye) can accompany large variability (repeated shots are widely scattered). A good sampling scheme, like a good archer, must have both small bias and small variability. Here's how we do this.

---

### MANAGING BIAS AND VARIABILITY

**To reduce bias,** use random sampling. When we start with a list of the entire population, simple random sampling produces unbiased estimates—the values of a statistic computed from an SRS neither consistently overestimate nor consistently underestimate the value of the population parameter.

**To reduce the variability** of a statistic from an SRS, use a larger sample. You can make the variability as small as you want by taking a large enough sample.

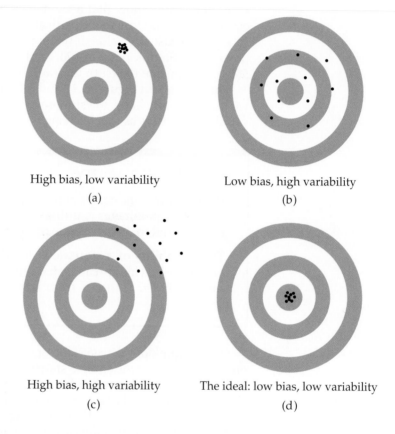

High bias, low variability
(a)

Low bias, high variability
(b)

High bias, high variability
(c)

The ideal: low bias, low variability
(d)

**FIGURE 3.14** Bias and variability in shooting arrows at a target. Bias means the archer systematically misses in the same direction. Variability means that the arrows are scattered.

In practice, Yankelovich takes only one sample. We don't know how close to the truth an estimate from this one sample is because we don't know what the truth about the population is. But *large random samples almost always give an estimate that is close to the truth.* Looking at the pattern of many samples shows that we can trust the result of one sample. The Current Population Survey's sample of 60,000 households estimates the national unemployment rate very accurately. Of course, only probability samples carry this guarantee. The American Family Association's voluntary response sample (Example 3.23, page 199) is worthless even though 850,000 people responded. Using a probability sampling design and taking care to deal with practical difficulties reduce bias in a sample. The size of the sample then determines how close to the population truth the sample result is likely to fall. Results from a sample survey usually come with a **margin of error** that sets bounds on the size of the likely error. The margin of error directly reflects the variability of the sample statistic, so it is smaller for larger samples. We will describe the details in later chapters.

margin of error

## Sampling from large populations

Yankelovich's sample of 2500 adults is only about 1 out of every 90,000 adults in the United States. Does it matter whether we sample 1-in-100 individuals in the population or 1-in-90,000?

---

### POPULATION SIZE DOESN'T MATTER

The variability of a statistic from a random sample does not depend on the size of the population, as long as the population is at least 100 times larger than the sample.

---

Why does the size of the population have little influence on the behavior of statistics from random samples? To see why this is plausible, imagine sampling harvested corn by thrusting a scoop into a lot of corn kernels. The scoop doesn't know whether it is surrounded by a bag of corn or by an entire truckload. As long as the corn is well mixed (so that the scoop selects a random sample), the variability of the result depends only on the size of the scoop.

The fact that the variability of sample results is controlled by the size of the sample has important consequences for sampling design. An SRS of size 2500 from the 220 million adult residents of the United States gives results as precise as an SRS of size 2500 from the 665,000 adult inhabitants of San Francisco. This is good news for designers of national samples but bad news for those who want accurate information about the citizens of San Francisco. If both use an SRS, both must use the same size sample to obtain equally trustworthy results.

## Why randomize?

Why randomize? The act of randomizing guarantees that the results of analyzing our data are subject to the laws of probability. The behavior of statistics is described by a sampling distribution. The form of the distribution is known, and in many cases is approximately Normal. Often the center of the distribution lies at the true parameter value, so that the notion that randomization eliminates bias is made more precise. The spread of the distribution describes the variability of the statistic and can be made as small as we wish by choosing a large enough sample. In a randomized experiment, we can reduce variability by choosing larger groups of subjects for each treatment.

These facts are at the heart of formal statistical inference. Later chapters will have much to say in more technical language about sampling distributions and the way statistical conclusions are based on them. What any user of statistics must understand is that all the technical talk has its basis in a simple question: *What would happen if the sample or the experiment were repeated many times?* The reasoning applies not only to an SRS but also to the complex sampling designs actually used by opinion polls and other national sample surveys. The same conclusions hold as well for randomized experimental designs. The details vary with the design but the basic facts are true whenever randomization is used to produce data.

Remember that proper statistical design is not the only aspect of a good sample or experiment. *The sampling distribution shows only how a statistic varies due to the operation of chance in randomization. It reveals nothing about possible bias due to undercoverage or nonresponse in a sample, or to lack of realism in an experiment.* The actual error in estimating a parameter by a statistic can be much larger than the sampling distribution suggests. What is worse, there is no way to say how large the added error is. The real world is less orderly than statistics textbooks imply.

## BEYOND THE BASICS

### Capture-Recapture Sampling

Sockeye salmon return to reproduce in the river where they were hatched four years earlier. How many salmon survived natural perils and heavy fishing to make it back this year? How many mountain sheep are there in Colorado? Are migratory songbird populations in North America decreasing or holding their own? These questions concern the size of animal populations. Biologists address them with a special kind of repeated sampling, called *capture-recapture sampling.*

**EXAMPLE**

**3.34 Estimate the number of least flycatchers.** You are interested in the number of least flycatchers migrating along a major route in the north-central United States. You set up "mist nets" that capture the birds but do not harm them. The birds caught in the net are fitted with a small aluminum leg band and released. Last year you banded and released 200 least flycatchers. This year you repeat the process. Your net catches 120 least flycatchers, 12 of which have tags from last year's catch.

The proportion of your second sample that have bands should estimate the proportion in the entire population that are banded. So if $N$ is the unknown number of least flycatchers, we should have approximately

$$\text{proportion banded in sample} = \text{proportion banded in population}$$

$$\frac{12}{120} = \frac{200}{N}$$

Solve for $N$ to estimate that the total number of flycatchers migrating while your net was up this year is approximately

$$N = 200 \times \frac{120}{12} = 2000$$

The capture-recapture idea extends the use of a sample proportion to estimate a population proportion. The idea works well if both samples are SRSs from the population and the population remains unchanged between samples. In practice, complications arise because, for example, some of the birds tagged last year died before this year's migration. Variations on capture-recapture samples are widely used in wildlife studies and are now finding other applications. One way to estimate the census undercount in a district is to consider the census as "capturing and marking" the households that respond. Census workers then visit the district, take an SRS of households, and see how many of those counted by the census show up in the sample. Capture-recapture estimates the total count of households in the district. As with estimating wildlife populations, there are many practical pitfalls. Our final word is as before: the real world is less orderly than statistics textbooks imply.

## SECTION 3.3  Summary

A number that describes a population is a **parameter.** A number that can be computed from the data is a **statistic.** The purpose of sampling or experimentation is usually **inference:** use sample statistics to make statements about unknown population parameters.

A statistic from a probability sample or randomized experiment has a **sampling distribution** that describes how the statistic varies in repeated data production. The sampling distribution answers the question "What would happen if we repeated the sample or experiment many times?" Formal statistical inference is based on the sampling distributions of statistics.

A statistic as an estimator of a parameter may suffer from **bias** or from high **variability.** Bias means that the center of the sampling distribution is not equal to the true value of the parameter. The variability of the statistic is described by the spread of its sampling distribution. Variability is usually reported by giving a **margin of error** for conclusions based on sample results.

Properly chosen statistics from randomized data production designs have no bias resulting from the way the sample is selected or the way the experimental units are assigned to treatments. We can reduce the variability of the statistic by increasing the size of the sample or the size of the experimental groups.

## SECTION 3.3  Exercises

*For Exercises 3.79 and 3.80, see page 213; and for Exercise 3.81, see page 217.*

**3.82  What's wrong?** State what is wrong in each of the following scenarios.

(a) A sampling distribution describes the distribution of some characteristic in a population.

(b) A statistic will have a large amount of bias whenever it has high variability.

(c) The variability of a statistic based on a small sample from a population will be the same as the variability of a large sample from the same population.

**3.83  Describe the population and the sample.** For each of the following situations, describe the population and the sample.

(a) A survey of 17,096 students in U.S. four-year colleges reported that 19.4% were binge drinkers.

(b) In a study of work stress, 100 restaurant workers were asked about the impact of work stress on their personal lives.

(c) A tract of forest has 584 longleaf pine trees. The diameters of 40 of these trees were measured.

**3.84  Bias and variability.** Figure 3.15 (on page 222) shows histograms of four sampling distributions of statistics intended to estimate the same parameter. Label each distribution relative to the others as high or low bias and as high or low variability.

**3.85  Opinions of Hispanics.** A New York Times News Service article on a poll concerned with the opinions of Hispanics includes this paragraph:

*The poll was conducted by telephone from July 13 to 27, with 3,092 adults nationwide, 1,074 of whom described themselves as Hispanic. It has a margin of sampling error of plus or minus three percentage points for the entire poll and plus or minus four*

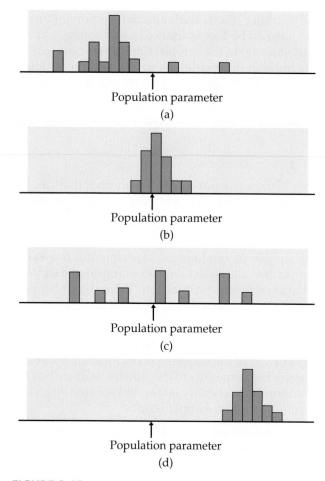

Population parameter

(a)

Population parameter

(b)

Population parameter

(c)

Population parameter

(d)

**FIGURE 3.15** Determine which of these sampling distributions displays high or low bias and high or low variability, for Exercise 3.84.

percentage points for Hispanics. Sample sizes for most Hispanic nationalities, like Cubans or Dominicans, were too small to break out the results separately.[44]

(a) Why is the "margin of sampling error" larger for Hispanics than for all 3092 respondents?

(b) Why would a very small sample size prevent a responsible news organization from breaking out results for Cubans?

**3.86 Gallup Canada polls.** Gallup Canada bases its polls of Canadian public opinion on telephone samples of about 1000 adults, the same sample size as Gallup uses in the United States. Canada's population is about one-ninth as large as that of the United States, so the percent of adults that Gallup interviews in Canada is nine times as large as in the United States. Does this mean that the margin of error for a Gallup Canada poll is smaller? Explain your answer.

**3.87 Real estate ownership.** An agency of the federal government plans to take an SRS of residents in each state to estimate the proportion of owners of real estate in each state's population. The populations of the states range from less than 500,000 people in Wyoming to about 35 million in California.

(a) Will the variability of the sample proportion vary from state to state if an SRS of size 2000 is taken in each state? Explain your answer.

(b) Will the variability of the sample proportion change from state to state if an SRS of 1/10 of 1% (0.001) of the state's population is taken in each state? Explain your answer.

**3.88 The health care system in Ontario.** The Ministry of Health in the Canadian province of Ontario wants to know whether the national health care system is achieving its goals in the province. The ministry conducted the Ontario Health Survey, which interviewed a probability sample of 61,239 adults who live in Ontario.[45]

(a) What is the population for this sample survey? What is the sample?

(b) The survey found that 76% of males and 86% of females in the sample had visited a general practitioner at least once in the past year. Do you think these estimates are close to the truth about the entire population? Why?

*The remaining exercises demonstrate the idea of a sampling distribution. Sampling distributions are the basis for statistical inference. We strongly recommend doing some of these exercises.*

**3.89** **Use the probability applet.** The *Probability* applet simulates tossing a coin, with the advantage that you can choose the true long-term proportion, or probability, of a head. Example 3.33 discusses sampling from a population in which proportion $p = 0.6$ (the parameter) find shopping frustrating. Tossing a coin with probability $p = 0.6$ of a head simulates this situation: each head is a person who finds shopping frustrating, and each tail is a person who does not. Set the "Probability of heads" in the applet to 0.6 and the number of tosses to 25. This simulates an SRS of size 25 from this population. By alternating between "Toss" and "Reset" you can take many samples quickly.

(a) Take 50 samples, recording the number of heads in each sample. Make a histogram of the 50 sample proportions (count of heads divided by 25). You are constructing the sampling distribution of this statistic.

(b) Another population contains only 20% who approve of legal gambling. Take 50 samples of size 25 from this population, record the number in each sample who approve, and make a histogram of the 50 sample proportions. How do the centers of your two histograms reflect the differing truths about the two populations?

3.90 **Use the statistical software for simulations.** Statistical software can speed simulations. We are interested in the sampling distribution of the proportion $\hat{p}$ of people who find shopping frustrating in an SRS from a population in which proportion $p$ find shopping frustrating. Here, $p$ is a parameter and $\hat{p}$ is a statistic used to estimate $p$. We will see in Chapter 5 that "binomial" is the key word to look for in the software menus. For example, in CrunchIt! go to "Simulate data" in the "Data" menu, and choose "Binomial."

(a) Set $n = 50$ and $p = 0.6$ and generate 100 binomial observations. These are the counts for 100 SRSs of size 50 when 60% of the population finds shopping frustrating. Save these counts and divide them by 50 to get values of $\hat{p}$ from 100 SRSs. Make a stemplot of the 100 values of $\hat{p}$.

(b) Repeat this process with $p = 0.3$, representing a population in which only 30% of people find shopping frustrating. Compare your two stemplots. How does changing the parameter $p$ affect the center and spread of the sampling distribution?

(c) Now generate 100 binomial observations with $n = 200$ and $p = 0.6$. This simulates 100 SRSs, each of size 200. Obtain the 100 sample proportions $\hat{p}$ and make a stemplot. Compare this with your stemplot from (a). How does changing the sample size $n$ affect the center and spread of the sampling distribution?

3.91 **Use Table B for a simulation.** We can construct a sampling distribution by hand in the case of a very small sample from a very small population. The population contains 10 students. Here are their scores on an exam:

| Student | 0 | 1 | 2 | 3 | 4 | 5 | 6 | 7 | 8 | 9 |
|---------|---|---|---|---|---|---|---|---|---|---|
| Score | 82 | 62 | 80 | 58 | 72 | 73 | 65 | 66 | 74 | 62 |

The parameter of interest is the mean score, which is 69.4. The sample is an SRS of $n = 4$ students drawn from this population. The students are labeled 0 to 9 so that a single random digit from Table B chooses one student for the sample.

(a) Use Table B to draw an SRS of size 4 from this population. Write the four scores in your sample and calculate the mean $\bar{x}$ of the sample scores. This statistic is an estimate of the population parameter.

(b) Repeat this process 9 more times. Make a histogram of the 10 values of $\bar{x}$. You are constructing the sampling distribution of $\bar{x}$. Is the center of your histogram close to 69.4? (Ten repetitions give only a crude approximation to the sampling distribution. If possible, pool your work with that of other students—using different parts of Table B—to obtain several hundred repetitions and make a histogram of the values of $\bar{x}$. This histogram is a better approximation to the sampling distribution.)

3.92 **Use the simple random sample applet.** The *Simple Random Sample* applet can illustrate the idea of a sampling distribution. Form a population labeled 1 to 100. We will choose an SRS of 10 of these numbers. That is, in this exercise, the numbers themselves are the population, not just labels for 100 individuals. The mean of the whole numbers 1 to 100 is 50.5. This is the parameter, the mean of the population.

(a) Use the applet to choose an SRS of size 10. Which 10 numbers were chosen? What is their mean? This is a statistic, the sample mean $\bar{x}$.

(b) Although the population and its mean 50.5 remain fixed, the sample mean changes as we take more samples. Take another SRS of size 10. (Use the "Reset" button to return to the original population before taking the second sample.) What are the 10 numbers in your sample? What is their mean? This is another value of $\bar{x}$.

(c) Take 8 more SRSs from this same population and record their means. You now have 10 values of the sample mean $\bar{x}$ from 10 SRSs of the same size from the same population. Make a histogram of the 10 values and mark the population mean 50.5 on the horizontal axis. Are your 10 sample values roughly centered at the population value? (If you kept going forever, your $\bar{x}$-values would form the sampling distribution of the sample mean; the population mean would indeed be the center of this distribution.)

3.93 **Analyze simple random samples.** The CSDATA data set contains the college grade point averages (GPAs) of all 224 students in a university entering class who planned to major in computer science. This is our population. Statistical software can take repeated samples to illustrate sampling variability.

(a) Using software, describe this population with a histogram and with numerical summaries. In particular, what is the mean GPA in the population? This is a parameter.

(b) Choose an SRS of 20 members from this population. Make a histogram of the GPAs in the sample and find their mean. The sample mean is a statistic. Briefly compare the distributions of GPA in the sample and in the population.

(c) Repeat the process of choosing an SRS of size 20 four more times (five in all). Record the five histograms of your sample GPAs. Does it seem reasonable to you from this small trial that an SRS will usually produce a sample that is generally representative of the population?

**3.94 Simulate the sampling distribution of the mean.** Continue the previous exercise, using software to illustrate the idea of a sampling distribution.

(a) Choose 20 more SRSs of size 20 in addition to the 5 you have already chosen. Don't make histograms of these latest samples—just record the mean GPA for each sample. Make a histogram of the 25 sample means. This histogram is a rough approximation to the sampling distribution of the mean.

(b) One sign of bias would be that the distribution of the sample means was systematically on one side of the true population mean. Mark the population mean GPA on your histogram of the 25 sample means. Is there a clear bias?

(c) Find the mean and standard deviation of your 25 sample means. We expect that the mean will be close to the true mean of the population. Is it? We also expect that the standard deviation of the sampling distribution will be smaller than the standard deviation of the population. Is it?

**3.95 Toss a coin.** Coin tossing can illustrate the idea of a sampling distribution. The population is all outcomes (heads or tails) we would get if we tossed a coin forever. The parameter $p$ is the proportion of heads in this population. We suspect that $p$ is close to 0.5. That is, we think the coin will show about one-half heads in the long run. The sample is the outcomes of 20 tosses, and the statistic $\hat{p}$ is the proportion of heads in these 20 tosses (count of heads divided by 20).

(a) Toss a coin 20 times and record the value of $\hat{p}$.

(b) Repeat this sampling process 9 more times. Make a stemplot of the 10 values of $\hat{p}$. You are constructing the sampling distribution of $\hat{p}$. Is the center of this distribution close to 0.5? (Ten repetitions give only a crude approximation to the sampling distribution. If possible, pool your work with that of other students to obtain several hundred repetitions and make a histogram of the values of $\hat{p}$.)

# 3.4 Ethics

The production and use of data, like all human endeavors, raise ethical questions. We won't discuss the telemarketer who begins a telephone sales pitch with "I'm conducting a survey." Such deception is clearly unethical. It enrages legitimate survey organizations, which find the public less willing to talk with them. Neither will we discuss those few researchers who, in the pursuit of professional advancement, publish fake data. There is no ethical question here—faking data to advance your career is just wrong. It will end your career when uncovered. But just how honest must researchers be about real, unfaked data? Here is an example that suggests the answer is "More honest than they often are."

**EXAMPLE**

**3.35 Provide all of the critical information.** Papers reporting scientific research are supposed to be short, with no extra baggage. Brevity can allow the researchers to avoid complete honesty about their data. Did they choose their subjects in a biased way? Did they report data on only some of their subjects? Did they try several statistical analyses and report only the ones that looked best? The statistician John Bailar screened more than 4000 medical papers in more than a decade as consultant to the *New England*

*Journal of Medicine.* He says, "When it came to the statistical review, it was often clear that critical information was lacking, and the gaps nearly always had the practical effect of making the authors' conclusions look stronger than they should have."[46] The situation is no doubt worse in fields that screen published work less carefully.

The most complex issues of data ethics arise when we collect data from people. The ethical difficulties are more severe for experiments that impose some treatment on people than for sample surveys that simply gather information. Trials of new medical treatments, for example, can do harm as well as good to their subjects. Here are some basic standards of data ethics that must be obeyed by any study that gathers data from human subjects, whether sample survey or experiment.

---

**BASIC DATA ETHICS**

The organization that carries out the study must have an **institutional review board** that reviews all planned studies in advance in order to protect the subjects from possible harm.

All individuals who are subjects in a study must give their **informed consent** before data are collected.

All individual data must be kept **confidential.** Only statistical summaries for groups of subjects may be made public.

---

The law requires that studies funded by the federal government obey these principles. But neither the law nor the consensus of experts is completely clear about the details of their application.

## Institutional review boards

The purpose of an institutional review board is not to decide whether a proposed study will produce valuable information or whether it is statistically sound. The board's purpose is, in the words of one university's board, "to protect the rights and welfare of human subjects (including patients) recruited to participate in research activities." The board reviews the plan of the study and can require changes. It reviews the consent form to be sure that subjects are informed about the nature of the study and about any potential risks. Once research begins, the board monitors its progress at least once a year.

The most pressing issue concerning institutional review boards is whether their workload has become so large that their effectiveness in protecting subjects drops. When the government temporarily stopped human-subject research at Duke University Medical Center in 1999 due to inadequate protection of subjects, more than 2000 studies were going on. That's a lot of review work. There are shorter review procedures for projects that involve only minimal risks to subjects, such as most sample surveys. When a board is overloaded, there is a temptation to put more proposals in the minimal-risk category to speed the work.

## USE YOUR KNOWLEDGE

*The exercises in this section on Ethics are designed to help you think about the issues that we are discussing and to formulate some opinions. In general there are no wrong or right answers but you need to give reasons for your answers.*

**3.96  Do these proposals involve minimal risk?** You are a member of your college's institutional review board. You must decide whether several research proposals qualify for lighter review because they involve only minimal risk to subjects. Federal regulations say that "minimal risk" means the risks are no greater than "those ordinarily encountered in daily life or during the performance of routine physical or psychological examinations or tests." That's vague. Which of these do you think qualifies as "minimal risk"?

(a) Draw a drop of blood by pricking a finger in order to measure blood sugar.

(b) Draw blood from the arm for a full set of blood tests.

(c) Insert a tube that remains in the arm, so that blood can be drawn regularly.

**3.97  Who should be on an institutional review board?** Government regulations require that institutional review boards consist of at least five people, including at least one scientist, one nonscientist, and one person from outside the institution. Most boards are larger, but many contain just one outsider.

(a) Why should review boards contain people who are not scientists?

(b) Do you think that one outside member is enough? How would you choose that member? (For example, would you prefer a medical doctor? A member of the clergy? An activist for patients' rights?)

## Informed consent

Both words in the phrase "informed consent" are important, and both can be controversial. Subjects must be *informed* in advance about the nature of a study and any risk of harm it may bring. In the case of a sample survey, physical harm is not possible. The subjects should be told what kinds of questions the survey will ask and about how much of their time it will take. Experimenters must tell subjects the nature and purpose of the study and outline possible risks. Subjects must then *consent* in writing.

**EXAMPLE**

**3.36  Who can give informed consent?**  Are there some subjects who can't give informed consent? It was once common, for example, to test new vaccines on prison inmates who gave their consent in return for good-behavior credit. Now we worry that prisoners are not really free to refuse, and the law forbids most medical experiments in prisons.

Very young children can't give fully informed consent, so the usual procedure is to ask their parents. A study of new ways to teach reading is about to start at a local elementary school, so the study team sends consent forms home to parents. Many parents don't return the forms. Can their children take part in the study because the parents did not say "No," or should we allow only children whose parents returned the form and said "Yes"?

What about research into new medical treatments for people with mental disorders? What about studies of new ways to help emergency room patients who may be unconscious or have suffered a stroke? In most cases, there is not time even to get the consent of the family. Does the principle of informed consent bar realistic trials of new treatments for unconscious patients?

These are questions without clear answers. Reasonable people differ strongly on all of them. There is nothing simple about informed consent.[47]

The difficulties of informed consent do not vanish even for capable subjects. Some researchers, especially in medical trials, regard consent as a barrier to getting patients to participate in research. They may not explain all possible risks; they may not point out that there are other therapies that might be better than those being studied; they may be too optimistic in talking with patients even when the consent form has all the right details. On the other hand, mentioning every possible risk leads to very long consent forms that really are barriers. "They are like rental car contracts," one lawyer said. Some subjects don't read forms that run five or six printed pages. Others are frightened by the large number of possible (but unlikely) disasters that might happen and so refuse to participate. Of course, unlikely disasters sometimes happen. When they do, lawsuits follow and the consent forms become yet longer and more detailed.

## Confidentiality

Ethical problems do not disappear once a study has been cleared by the review board, has obtained consent from its subjects, and has actually collected data about the subjects. It is important to protect the subjects' privacy by keeping all data about individuals confidential. The report of an opinion poll may say what percent of the 1500 respondents felt that legal immigration should be reduced. It may not report what *you* said about this or any other issue.

anonymity    Confidentiality is not the same as **anonymity.** Anonymity means that subjects are anonymous—their names are not known even to the director of the study. Anonymity is rare in statistical studies. Even where anonymity is possible (mainly in surveys conducted by mail), it prevents any follow-up to improve nonresponse or inform subjects of results.

Any breach of confidentiality is a serious violation of data ethics. The best practice is to separate the identity of the subjects from the rest of the data at once. Sample surveys, for example, use the identification only to check on who did or did not respond. In an era of advanced technology, however, it is no longer enough to be sure that each individual set of data protects people's privacy. The government, for example, maintains a vast amount of information about citizens in many separate data bases—census responses, tax returns, Social Security information, data from surveys such as the Current Population Survey, and so on. Many of these data bases can be searched by computers for statistical studies. A clever computer search of several data bases might be able, by combining information, to identify you and learn a great deal about you even if your name and other identification have been removed from the data available for search. A colleague from Germany once remarked that "female full professor of statistics with PhD from the United States" was enough to identify her among all the citizens of Germany. Privacy and confidentiality of data are hot issues among statisticians in the computer age.

**EXAMPLE**

**3.37 Data collected by the government.** Citizens are required to give information to the government. Think of tax returns and Social Security contributions. The government needs these data for administrative purposes—to see if we paid the right amount of tax and how large a Social Security benefit we are owed when we retire. Some people feel that individuals should be able to forbid any other use of their data, even with all identification removed. This would prevent using government records to study, say, the ages, incomes, and household sizes of Social Security recipients. Such a study could well be vital to debates on reforming Social Security.

## USE YOUR KNOWLEDGE

**3.98** **How can we obtain informed consent?** A researcher suspects that traditional religious beliefs tend to be associated with an authoritarian personality. She prepares a questionnaire that measures authoritarian tendencies and also asks many religious questions. Write a description of the purpose of this research to be read by subjects in order to obtain their informed consent. You must balance the conflicting goals of not deceiving the subjects as to what the questionnaire will tell about them and of not biasing the sample by scaring off religious people.

**3.99** **Should we allow this personal information to be collected?** In which of the circumstances below would you allow collecting personal information without the subjects' consent?

(a) A government agency takes a random sample of income tax returns to obtain information on the average income of people in different occupations. Only the incomes and occupations are recorded from the returns, not the names.

(b) A social psychologist attends public meetings of a religious group to study the behavior patterns of members.

(c) A social psychologist pretends to be converted to membership in a religious group and attends private meetings to study the behavior patterns of members.

## Clinical trials

Clinical trials are experiments that study the effectiveness of medical treatments on actual patients. Medical treatments can harm as well as heal, so clinical trials spotlight the ethical problems of experiments with human subjects. Here are the starting points for a discussion:

- Randomized comparative experiments are the only way to see the true effects of new treatments. Without them, risky treatments that are no better than placebos will become common.

- Clinical trials produce great benefits, but most of these benefits go to future patients. The trials also pose risks, and these risks are borne by the subjects of the trial. So we must balance future benefits against present risks.

- Both medical ethics and international human rights standards say that "the interests of the subject must always prevail over the interests of science and society."

The quoted words are from the 1964 Helsinki Declaration of the World Medical Association, the most respected international standard. The most outrageous examples of unethical experiments are those that ignore the interests of the subjects.

**EXAMPLE**

**3.38 The Tuskegee study.** In the 1930s, syphilis was common among black men in the rural South, a group that had almost no access to medical care. The Public Health Service Tuskegee study recruited 399 poor black sharecroppers with syphilis and 201 others without the disease in order to observe how syphilis progressed when no treatment was given. Beginning in 1943, penicillin became available to treat syphilis. The study subjects were not treated. In fact, the Public Health Service prevented any treatment until word leaked out and forced an end to the study in the 1970s.

The Tuskegee study is an extreme example of investigators following their own interests and ignoring the well-being of their subjects. A 1996 review said, "It has come to symbolize racism in medicine, ethical misconduct in human research, paternalism by physicians, and government abuse of vulnerable people." In 1997, President Clinton formally apologized to the surviving participants in a White House ceremony.[48]

Because "the interests of the subject must always prevail," medical treatments can be tested in clinical trials only when there is reason to hope that they will help the patients who are subjects in the trials. Future benefits aren't enough to justify experiments with human subjects. Of course, if there is already strong evidence that a treatment works and is safe, it is unethical *not* to give it. Here are the words of Dr. Charles Hennekens of the Harvard Medical School, who directed the large clinical trial that showed that aspirin reduces the risk of heart attacks:

> *There's a delicate balance between when to do or not do a randomized trial. On the one hand, there must be sufficient belief in the agent's potential to justify exposing half the subjects to it. On the other hand, there must be sufficient doubt about its efficacy to justify withholding it from the other half of subjects who might be assigned to placebos.*[49]

Why is it ethical to give a control group of patients a placebo? Well, we know that placebos often work. What is more, placebos have no harmful side effects. So in the state of balanced doubt described by Dr. Hennekens, the placebo group may be getting a better treatment than the drug group. If we *knew* which treatment was better, we would give it to everyone. When we don't know, it is ethical to try both and compare them. Here are some harder questions about placebos, with arguments on both sides.

**EXAMPLE**

**3.39 Is it ethical to use a placebo?**  You are testing a new drug. Is it ethical to give a placebo to a control group if an effective drug already exists?

**Yes:** The placebo gives a true baseline for the effectiveness of the new drug. There are three groups: new drug, best existing drug, and placebo. Every clinical trial is a bit different, and not even genuinely effective treatments work in every setting. The placebo control helps us see if the study is flawed so that even the best existing drug does not beat the placebo. Sometimes the placebo wins, so the doubt needed to justify its use is present. Placebo controls are ethical except for life-threatening conditions.

**No:** It isn't ethical to deliberately give patients an inferior treatment. We don't know whether the new drug is better than the existing drug, so it is ethical to give both in order to find out. If past trials showed that the existing drug is better than a placebo, it is no longer right to give patients a placebo. After all, the existing drug includes the placebo effect. A placebo group is ethical only if the existing drug is an older one that did not undergo proper clinical trials or doesn't work well or is dangerous.

## USE YOUR KNOWLEDGE

**3.100  Is this study ethical?** Researchers on aging proposed to investigate the effect of supplemental health services on the quality of life of older people. Eligible patients on the rolls of a large medical clinic were to be randomly assigned to treatment and control groups. The treatment group would be offered hearing aids, dentures, transportation, and other services not available without charge to the control group. The review board felt that providing these services to some but not other persons in the same institution raised ethical questions. Do you agree?

**3.101  Should the treatments be given to everyone?** Effective drugs for treating AIDS are very expensive, so most African nations cannot afford to give them to large numbers of people. Yet AIDS is more common in parts of Africa than anywhere else. Several clinical trials are looking at ways to prevent pregnant mothers infected with HIV from passing the infection to their unborn children, a major source of HIV infections in Africa. Some people say these trials are unethical because they do not give effective AIDS drugs to their subjects, as would be required in rich nations. Others reply that the trials are looking for treatments that can work in the real world in Africa and that they promise benefits at least to the children of their subjects. What do you think?

## Behavioral and social science experiments

When we move from medicine to the behavioral and social sciences, the direct risks to experimental subjects are less acute, but so are the possible benefits to the subjects. Consider, for example, the experiments conducted by psychologists in their study of human behavior.

**EXAMPLE**

**3.40 Personal space.** Psychologists observe that people have a "personal space" and get annoyed if others come too close to them. We don't like strangers to sit at our table in a coffee shop if other tables are available, and we see people move apart in elevators if there is room to do so. Americans tend to require more personal space than people in most other cultures. Can violations of personal space have physical, as well as emotional, effects?

Investigators set up shop in a men's public rest room. They blocked off urinals to force men walking in to use either a urinal next to an experimenter (treatment group) or a urinal separated from the experimenter (control group). Another experimenter, using a periscope from a toilet stall, measured how long the subject took to start urinating and how long he kept at it.[50]

This personal space experiment illustrates the difficulties facing those who plan and review behavioral studies.

- There is no risk of harm to the subjects, although they would certainly object to being watched through a periscope. What should we protect subjects from when physical harm is unlikely? Possible emotional harm? Undignified situations? Invasion of privacy?

- What about informed consent? The subjects in Example 3.40 did not even know they were participating in an experiment. Many behavioral experiments rely on hiding the true purpose of the study. The subjects would change their behavior if told in advance what the investigators were looking for. Subjects are asked to consent on the basis of vague information. They receive full information only after the experiment.

The "Ethical Principles" of the American Psychological Association require consent unless a study merely observes behavior in a public place. They allow deception only when it is necessary to the study, does not hide information that might influence a subject's willingness to participate, and is explained to subjects as soon as possible. The personal space study (from the 1970s) does not meet current ethical standards.

We see that the basic requirement for informed consent is understood differently in medicine and psychology. Here is an example of another setting with yet another interpretation of what is ethical. The subjects get no information and give no consent. They don't even know that an experiment may be sending them to jail for the night.

**EXAMPLE**

**3.41 Domestic violence.** How should police respond to domestic-violence calls? In the past, the usual practice was to remove the offender and order him to stay out of the household overnight. Police were reluctant to make arrests because the victims rarely pressed charges. Women's groups argued that arresting offenders would help prevent future violence even if no charges were filed. Is there evidence that arrest will reduce future offenses? That's a question that experiments have tried to answer.

A typical domestic-violence experiment compares two treatments: arrest the suspect and hold him overnight, or warn the suspect and release him. When police officers reach the scene of a domestic-violence call, they calm the participants and investigate. Weapons or death threats require an arrest. If the facts permit an arrest but do not require it, an officer radios headquarters for instructions. The person on duty opens the next envelope in a file prepared in advance by a statistician. The envelopes contain the treatments in random order. The police either arrest the suspect or warn and release him, depending on the contents of the envelope. The researchers then watch police records and visit the victim to see if the domestic violence reoccurs.

The first such experiment appeared to show that arresting domestic-violence suspects does reduce their future violent behavior. As a result of this evidence, arrest has become the common police response to domestic violence.

The domestic-violence experiments shed light on an important issue of public policy. Because there is no informed consent, the ethical rules that govern clinical trials and most social science studies would forbid these experiments. They were cleared by review boards because, in the words of one domestic-violence researcher, "These people became subjects by committing acts that allow the police to arrest them. You don't need consent to arrest someone."

## SECTION 3.4  Summary

Approval of an **institutional review board** is required for studies that involve human or animals as subjects.

Human subjects must give **informed consent** if they are to participate in experiments.

Data on human subjects must be kept **confidential.**

## SECTION 3.4  Exercises

*For Exercises 3.96 and 3.97, see page 226; for Exercises 3.98 and 3.99, see page 228; and for Exercises 3.100 and 3.101, see page 230.*

**3.102  What is wrong?** Explain what is wrong in each of the following scenarios.

(a) Clinical trials are always ethical as long as they randomly assign patients to the treatments.

(b) The job of an institutional review board is complete when they decide to allow a study to be conducted.

(c) A treatment that has no risk of physical harm to subjects is always ethical.

**3.103  Serving as an experimental subject for extra credit.** Students taking Psychology 001 are required to serve as experimental subjects. Students in Psychology 002 are not required

to serve, but they are given extra credit if they do so. Students in Psychology 003 are required either to sign up as subjects or to write a term paper. Serving as an experimental subject may be educational, but current ethical standards frown on using "dependent subjects" such as prisoners or charity medical patients. Students are certainly somewhat dependent on their teachers. Do you object to any of these course policies? If so, which ones, and why?

**3.104  Informed consent to take blood samples.** Researchers from Yale, working with medical teams in Tanzania, wanted to know how common infection with the AIDS virus is among pregnant women in that country. To do this, they planned to test blood samples drawn from pregnant women.

Yale's institutional review board insisted that the researchers get the informed consent of each woman and tell her the results of the test. This is the usual procedure in developed nations. The Tanzanian government did not want to tell the women why blood was drawn or tell them the test results. The government feared panic if many people turned out to have an incurable disease for which the country's medical system could not provide care. The study was canceled. Do you think that Yale was right to apply its usual standards for protecting subjects?

3.105 **The General Social Survey.** One of the most important nongovernment surveys in the United States is the National Opinion Research Center's General Social Survey. The GSS regularly monitors public opinion on a wide variety of political and social issues. Interviews are conducted in person in the subject's home. Are a subject's responses to GSS questions anonymous, confidential, or both? Explain your answer.

3.106 **Anonymity and confidentiality in health screening.** Texas A&M, like many universities, offers free screening for HIV, the virus that causes AIDS. The announcement says, "Persons who sign up for the HIV Screening will be assigned a number so that they do not have to give their name." They can learn the results of the test by telephone, still without giving their name. Does this practice offer *anonymity* or just *confidentiality?*

3.107 **Anonymity and confidentiality in mail surveys.** Some common practices may appear to offer anonymity while actually delivering only confidentiality. Market researchers often use mail surveys that do not ask the respondent's identity but contain hidden codes on the questionnaire that identify the respondent. A false claim of anonymity is clearly unethical. If only confidentiality is promised, is it also unethical to say nothing about the identifying code, perhaps causing respondents to believe their replies are anonymous?

3.108 **Use of stored blood.** Long ago, doctors drew a blood specimen from you as part of treating minor anemia. Unknown to you, the sample was stored. Now researchers plan to use stored samples from you and many other people to look for genetic factors that may influence anemia. It is no longer possible to ask your consent. Modern technology can read your entire genetic makeup from the blood sample.

(a) Do you think it violates the principle of informed consent to use your blood sample if your name is on it but you were not told that it might be saved and studied later?

(b) Suppose that your identity is not attached. The blood sample is known only to come from (say) "a 20-year-old white female being treated for anemia." Is it now OK to use the sample for research?

(c) Perhaps we should use biological materials such as blood samples only from patients who have agreed to allow the material to be stored for later use in research. It isn't possible to say in advance what kind of research, so this falls short of the usual standard for informed consent. Is it nonetheless acceptable, given complete confidentiality and the fact that using the sample can't physically harm the patient?

3.109 **Testing vaccines.** One of the most important goals of AIDS research is to find a vaccine that will protect against HIV. Because AIDS is so common in parts of Africa, that is the easiest place to test a vaccine. It is likely, however, that a vaccine would be so expensive that it could not (at least at first) be widely used in Africa. Is it ethical to test in Africa if the benefits go mainly to rich countries? The treatment group of subjects would get the vaccine, and the placebo group would later be given the vaccine if it proved effective. So the actual subjects would benefit—it is the future benefits that would go elsewhere. What do you think?

3.110 **Political polls.** The presidential election campaign is in full swing, and the candidates have hired polling organizations to take regular polls to find out what the voters think about the issues. What information should the pollsters be required to give out?

(a) What does the standard of informed consent require the pollsters to tell potential respondents?

(b) The standards accepted by polling organizations also require giving respondents the name and address of the organization that carries out the poll. Why do you think this is required?

(c) The polling organization usually has a professional name such as "Samples Incorporated," so respondents don't know that the poll is being paid for by a political party or candidate. Would revealing the sponsor to respondents bias the poll? Should the sponsor always be announced whenever poll results are made public?

3.111 **Should poll results be made public?** Some people think that the law should require that all political poll results be made public. Otherwise, the possessors of poll results can use the information to their own advantage. They can act on the information, release only selected parts of it, or time the release for best effect. A candidate's organization replies that they are paying for the poll in order to gain information for their own use, not to amuse the public. Do you favor requiring complete disclosure of political poll results? What about other private surveys, such as market research surveys of consumer tastes?

**3.112 The 2000 census.** The 2000 census long form asked 53 detailed questions, for example:

*Do you have COMPLETE plumbing facilities in this house, apartment, or mobile home; that is, 1) hot and cold piped water, 2) a flush toilet, and 3) a bathtub or shower?*

The form also asked your income in dollars, broken down by source, and whether any "physical, mental, or emotional condition" causes you difficulty in "learning, remembering, or concentrating." Some members of Congress objected to these questions, even though Congress had approved them.

Give brief arguments on both sides of the debate over the long form: the government has legitimate uses for such information, but the questions seem to invade people's privacy.

## CHAPTER 3   Exercises

**3.113 Select a random sample of workers.** The WORKERS data set contains information about 14,959 people aged 25 to 64 whose highest level of education is a bachelor's degree.

(a) In order to select an SRS of these people, how would you assign labels?

(b) Use Table B at line 185 to choose the first 3 members of the SRS.

**3.114 Cash bonuses for the unemployed.** Will cash bonuses speed the return to work of unemployed people? The Illinois Department of Employment Security designed an experiment to find out. The subjects were 10,065 people aged 20 to 54 who were filing claims for unemployment insurance. Some were offered $500 if they found a job within 11 weeks and held it for at least 4 months. Others could tell potential employers that the state would pay the employer $500 for hiring them. A control group got neither kind of bonus.[51]

(a) Suggest a few response variables of interest to the state and outline the design of the experiment.

(b) How will you label the subjects for random assignment? Use Table B at line 167 to choose the first 3 subjects for the first treatment.

**3.115 Name the designs.** What is the name for each of these study designs?

(a) A study to compare two methods of preserving wood started with boards of southern white pine. Each board was ripped from end to end to form two edge-matched specimens. One was assigned to Method A; the other to Method B.

(b) A survey on youth and smoking contacted by telephone 300 smokers and 300 nonsmokers, all 14 to 22 years of age.

(c) Does air pollution induce DNA mutations in mice? Starting with 40 male and 40 female mice, 20 of each sex were housed in a polluted industrial area downwind from a steel mill. The other 20 of each sex were housed at an unpolluted rural location 30 kilometers away.

**3.116 Prostate treatment study using Canada's national health records.** A large observational study used records from Canada's national health care system to compare the effectiveness of two ways to treat prostate disease. The two treatments are traditional surgery and a new method that does not require surgery. The records described many patients whose doctors had chosen one or the other method. The study found that patients treated by the new method were significantly more likely to die within 8 years.[52]

(a) Further study of the data showed that this conclusion was wrong. The extra deaths among patients who received the new treatment could be explained by lurking variables. What lurking variables might be confounded with a doctor's choice of surgical or nonsurgical treatment?

(b) You have 300 prostate patients who are willing to serve as subjects in an experiment to compare the two methods. Use a diagram to outline the design of a randomized comparative experiment.

**3.117 Price promotions and consumers' expectations.** A researcher studying the effect of price promotions on consumers' expectations makes up two different histories of the store price of a hypothetical brand of laundry detergent for the past year. Students in a marketing course view one or the other price history on a computer.

Some students see a steady price, while others see regular promotions that temporarily cut the price. Then the students are asked what price they would expect to pay for the detergent. Is this study an experiment? Why? What are the explanatory and response variables?

**3.118 What type of study?** What is the best way to answer each of the questions below: an experiment, a sample survey, or an observational study that is not a sample survey? Explain your choices.

(a) Are people generally satisfied with how things are going in the country right now?

(b) Do college students learn basic accounting better in a classroom or using an online course?

(c) How long do your teachers wait on the average after they ask their class a question?

**3.119 Choose the type of study.** Give an example of a question about college students, their behavior, or their opinions that would best be answered by

(a) a sample survey.

(b) an observational study that is not a sample survey.

(c) an experiment.

**3.120 Compare the burgers.** Do consumers prefer the taste of a cheeseburger from McDonald's or from Wendy's in a blind test in which neither burger is identified? Describe briefly the design of a matched pairs experiment to investigate this question. How will you use randomization?

**3.121 Bicycle gears.** How does the time it takes a bicycle rider to travel 100 meters depend on which gear is used and how steep the course is? It may be, for example, that higher gears are faster on the level but lower gears are faster on steep inclines. Discuss the design of a two-factor experiment to investigate this issue, using one bicycle with three gears and one rider. How will you use randomization?

**3.122 Design an experiment.** The previous two exercises illustrate the use of statistically designed experiments to answer questions that arise in everyday life. Select a question of interest to you that an experiment might answer and carefully discuss the design of an appropriate experiment.

**3.123 Design a survey.** You want to investigate the attitudes of students at your school

about the faculty's commitment to teaching. The student government will pay the costs of contacting about 500 students.

(a) Specify the exact population for your study; for example, will you include part-time students?

(b) Describe your sample design. Will you use a stratified sample?

(c) Briefly discuss the practical difficulties that you anticipate; for example, how will you contact the students in your sample?

**3.124 Compare two doses of a drug.** A drug manufacturer is studying how a new drug behaves in patients. Investigators compare two doses: 5 milligrams (mg) and 10 mg. The drug can be administered by injection, by a skin patch, or by intravenous drip. Concentration in the blood after 30 minutes (the response variable) may depend both on the dose and on the method of administration.

(a) Make a sketch that describes the treatments formed by combining dosage and method. Then use a diagram to outline a completely randomized design for this two-factor experiment.

(b) "How many subjects?" is a tough issue. We will explain the basic ideas in Chapter 6. What can you say now about the advantage of using larger groups of subjects?

**3.125 Discolored french fries.** Few people want to eat discolored french fries. Potatoes are kept refrigerated before being cut for french fries to prevent spoiling and preserve flavor. But immediate processing of cold potatoes causes discoloring due to complex chemical reactions. The potatoes must therefore be brought to room temperature before processing. Design an experiment in which tasters will rate the color and flavor of french fries prepared from several groups of potatoes. The potatoes will be fresh picked or stored for a month at room temperature or stored for a month refrigerated. They will then be sliced and cooked either immediately or after an hour at room temperature.

(a) What are the factors and their levels, the treatments, and the response variables?

(b) Describe and outline the design of this experiment.

(c) It is efficient to have each taster rate fries from all treatments. How will you use randomization in presenting fries to the tasters?

**3.126** **Would the results be different for men and women?** The drug that is the subject of the experiment in Exercise 3.124 may behave differently in men and women. How would you modify your experimental design to take this into account?

**3.127** CHALLENGE **Informed consent.** The requirement that human subjects give their informed consent to participate in an experiment can greatly reduce the number of available subjects. For example, a study of new teaching methods asks the consent of parents for their children to be randomly assigned to be taught by either a new method or the standard method. Many parents do not return the forms, so their children must continue to be taught by the standard method. Why is it not correct to consider these children as part of the control group along with children who are randomly assigned to the standard method?

**3.128** CHALLENGE **Two ways to ask sensitive questions.** Sample survey questions are usually read from a computer screen. In a Computer Aided Personal Interview (CAPI), the interviewer reads the questions and enters the responses. In a Computer Aided Self Interview (CASI), the interviewer stands aside and the respondent reads the questions and enters responses. One method almost always shows a higher percent of subjects admitting use of illegal drugs. Which method? Explain why.

**3.129** **Your institutional review board.** Your college or university has an institutional review board that screens all studies that use human subjects. Get a copy of the document that describes this board (you can probably find it online).

(a) According to this document, what are the duties of the board?

(b) How are members of the board chosen? How many members are not scientists? How many members are not employees of the college? Do these members have some special expertise, or are they simply members of the "general public"?

**3.130** **Use of data produced by the government.** Data produced by the government are often available free or at low cost to private users. For example, satellite weather data produced by the U.S. National Weather Service are available free to TV stations for their weather reports and to anyone on the Web. *Opinion 1:* Government data should be available to everyone at minimal cost. European governments, on the other hand, charge TV stations for weather data. *Opinion 2:* The satellites are expensive, and the TV stations are making a profit from their weather services, so they should share the cost. Which opinion do you support, and why?

**3.131** **Should we ask for the consent of the parents?** The Centers for Disease Control and Prevention, in a survey of teenagers, asked the subjects if they were sexually active. Those who said "Yes" were then asked,

*How old were you when you had sexual intercourse for the first time?*

Should consent of parents be required to ask minors about sex, drugs, and other such issues, or is consent of the minors themselves enough? Give reasons for your opinion.

**3.132** **A theft experiment.** Students sign up to be subjects in a psychology experiment. When they arrive, they are told that interviews are running late and are taken to a waiting room. The experimenters then stage a theft of a valuable object left in the waiting room. Some subjects are alone with the thief, and others are in pairs—these are the treatments being compared. Will the subject report the theft? The students had agreed to take part in an unspecified study, and the true nature of the experiment is explained to them afterward. Do you think this study is ethically OK?

**3.133** **A cheating experiment.** A psychologist conducts the following experiment: she measures the attitude of subjects toward cheating, then has them play a game rigged so that winning without cheating is impossible. The computer that organizes the game also records—unknown to the subjects—whether or not they cheat. Then attitude toward cheating is retested. Subjects who cheat tend to change their attitudes to find cheating more acceptable. Those who resist the temptation to cheat tend to condemn cheating more strongly on the second test of attitude. These results confirm the psychologist's theory. This experiment tempts subjects to cheat. The subjects are led to believe that they can cheat secretly when in fact they are observed. Is this experiment ethically objectionable? Explain your position.

# Probability: The Study of Randomness

Annie Duke, professional poker player, with a large stack of chips at the World Series of Poker. See Example 4.2 to learn more about probability and Texas hold 'em.

## Introduction

The reasoning of statistical inference rests on asking, "How often would this method give a correct answer if I used it very many times?" When we produce data by random sampling or randomized comparative experiments, the laws of probability answer the question "What would happen if we did this many times?" Games of chance like Texas hold 'em are exciting because the outcomes are determined by the rules of probability.

## 4.1 Randomness

Toss a coin, or choose an SRS. The result can't be predicted in advance, because the result will vary when you toss the coin or choose the sample repeatedly. But there is nonetheless a regular pattern in the results, a pattern that emerges clearly only after many repetitions. This remarkable fact is the basis for the idea of probability.

**LOOK BACK**

**sampling distributions, page 214**

EXAMPLE

**4.1 Toss a coin 5000 times.**   When you toss a coin, there are only two possible outcomes, heads or tails. Figure 4.1 shows the results of tossing a coin 5000 times twice. For each number of tosses from 1 to 5000, we have plotted the proportion of those tosses that gave a head. Trial A (solid line) begins tail, head, tail, tail. You can see that the proportion of heads for Trial A starts at 0 on the first toss, rises to 0.5 when the second toss gives a head, then falls to 0.33 and 0.25 as we get two more tails. Trial B, on the other hand, starts with five straight heads, so the proportion of heads is 1 until the sixth toss.

The proportion of tosses that produce heads is quite variable at first. Trial A starts low and Trial B starts high. As we make more and more tosses, however, the proportions of heads for both trials get close to 0.5 and stay there. If we made yet a third trial at tossing the coin a great many times, the proportion of heads would again settle down to 0.5 in the long run. We say that 0.5 is the *probability* of a head. The probability 0.5 appears as a horizontal line on the graph.

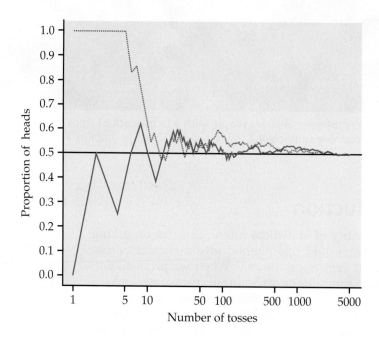

**FIGURE 4.1** The proportion of tosses of a coin that give a head varies as we make more tosses. Eventually, however, the proportion approaches 0.5, the probability of a head. This figure shows the results of two trials of 5000 tosses each.

The *Probability* applet on the text Web site animates Figure 4.1. It allows you to choose the probability of a head and simulate any number of tosses of a coin with that probability. Try it. You will see that the proportion of heads gradually settles down close to the chosen probability. Equally important, you will also see that the proportion in a small or moderate number of tosses can be far from the probability. *Probability describes only what happens in the long run. Most people expect chance outcomes to show more short-term regularity than is actually true.*

EXAMPLE

**4.2 Texas hold 'em.** In the card game Texas hold 'em, each player is dealt two cards. After a round of betting, three "community" cards, which can be used by any player, are dealt, followed by another round of betting. Then two additional community cards are dealt, with a round of betting after each. The best poker hand wins. The last community card turned is called the river. Suppose that you are dealt an ace and a king. The probability that you will get another ace or king by the river, that is, after the five community cards are dealt, is about 0.5. This means that about half of the time that you hold these cards, you will finish with a hand that has at least a pair of kings or a pair of aces.

## The language of probability

"Random" in statistics is not a synonym for "haphazard" but a description of a kind of order that emerges in the long run. We often encounter the unpredictable side of randomness in our everyday experience, but we rarely see enough repetitions of the same random phenomenon to observe the long-term regularity that probability describes. You can see that regularity emerging in Figure 4.1. In the very long run, the proportion of tosses that give a head is 0.5. This is the intuitive idea of probability. Probability 0.5 means "occurs half the time in a very large number of trials."

---

### RANDOMNESS AND PROBABILITY

We call a phenomenon **random** if individual outcomes are uncertain but there is nonetheless a regular distribution of outcomes in a large number of repetitions.

The **probability** of any outcome of a random phenomenon is the proportion of times the outcome would occur in a very long series of repetitions.

---

fair coin

Real coins have bumps and imperfections that make the probability of heads a little different from 0.5. The probability might be 0.499999 or 0.500002. We call a coin **fair** if the probability of heads is exactly 0.5. For our study of probability in this chapter, we will assume that we know the actual values of probabilities. Thus, we assume things like fair coins, even though we know that real coins are not exactly fair. We do this to learn what kinds of outcomes we are likely to see when we make such assumptions. When we study statistical inference in later chapters, we look at the situation from the opposite point of view: given that we have observed certain outcomes, what can we say about the probabilities that generated these outcomes?

### USE YOUR KNOWLEDGE

**4.1  Use Table B.** We can use the random digits in Table B in the back of the text to simulate tossing a fair coin. Start at line 109 and read the numbers from left to right. If the number is 0, 1, 2, 3, or 4, you will say that the coin toss resulted in a head; if the number is a 5, 6, 7, 8, or 9, the outcome is tails. Use the first 20 random digits on line 109 to simulate 20 tosses of a fair coin. What is the actual proportion of heads in your simulated sample? Explain why you did not get exactly 10 heads.

Probability describes what happens in very many trials, and we must actually observe many trials to pin down a probability. In the case of tossing a coin, some diligent people have in fact made thousands of tosses.

**EXAMPLE**

**4.3 Many tosses of a coin.** The French naturalist Count Buffon (1707–1788) tossed a coin 4040 times. Result: 2048 heads, or proportion 2048/4040 = 0.5069 for heads.

Around 1900, the English statistician Karl Pearson heroically tossed a coin 24,000 times. Result: 12,012 heads, a proportion of 0.5005.

While imprisoned by the Germans during World War II, the South African statistician John Kerrich tossed a coin 10,000 times. Result: 5067 heads, proportion of heads 0.5067.

## Thinking about randomness

That some things are random is an observed fact about the world. The outcome of a coin toss, the time between emissions of particles by a radioactive source, and the sexes of the next litter of lab rats are all random. So is the outcome of a random sample or a randomized experiment. Probability theory is the branch of mathematics that describes random behavior. Of course, we can never observe a probability exactly. We could always continue tossing the coin, for example. Mathematical probability is an idealization based on imagining what would happen in an indefinitely long series of trials.

The best way to understand randomness is to observe random behavior—not only the long-run regularity but the unpredictable results of short runs. You can do this with physical devices such as coins and dice, but software simulations of random behavior allow faster exploration. As you explore randomness, remember:

*independence* • You must have a long series of **independent** trials. That is, the outcome of one trial must not influence the outcome of any other. Imagine a crooked gambling house where the operator of a roulette wheel can stop it where she chooses—she can prevent the proportion of "red" from settling down to a fixed number. These trials are not independent.

• The idea of probability is empirical. Simulations start with given probabilities and imitate random behavior, but we can estimate a real-world probability only by actually observing many trials.

• Nonetheless, simulations are very useful because we need long runs of trials. In situations such as coin tossing, the proportion of an outcome often requires several hundred trials to settle down to the probability of that outcome. The kinds of physical random devices suggested in the exercises are too slow for this. Short runs give only rough estimates of a probability.

## The uses of probability

Probability theory originated in the study of games of chance. Tossing dice, dealing shuffled cards, and spinning a roulette wheel are examples of deliberate randomization. In that respect, they are similar to random sampling. Although games of chance are ancient, they were not studied by mathematicians until the sixteenth and seventeenth centuries. It is only a mild simplification to say that probability as a branch of mathematics arose when seventeenth-century French gamblers asked the mathematicians Blaise Pascal and Pierre de Fermat for help. Gambling is still with us, in casinos and state lotteries. We will make use of games of chance as simple examples that illustrate the principles of probability.

Careful measurements in astronomy and surveying led to further advances in probability in the eighteenth and nineteenth centuries because the results of repeated measurements are random and can be described by distributions much like those arising from random sampling. Similar distributions appear in data on human life span (mortality tables) and in data on lengths or weights in a population of skulls, leaves, or cockroaches.[1] Now, we employ the mathematics of probability to describe the flow of traffic through a highway system, the Internet, or a computer processor; the genetic makeup of individuals or populations; the energy states of subatomic particles; the spread of epidemics or rumors; and the rate of return on risky investments. Although we are interested in probability because of its usefulness in statistics, the mathematics of chance is important in many fields of study.

## SECTION 4.1 Summary

A **random phenomenon** has outcomes that we cannot predict but that nonetheless have a regular distribution in very many repetitions.

The **probability** of an event is the proportion of times the event occurs in many repeated trials of a random phenomenon.

## SECTION 4.1 Exercises

*For Exercise 4.1, see page 239.*

**4.2 Is music playing on the radio?** Turn on your favorite music radio station 10 times at least 10 minutes apart. Each time record whether or not music is playing. Calculate the number of times music is playing divided by 10. This number is an estimate of the probability that music is playing when you turn on this station. It is also an estimate of the proportion of time that music is playing on this station.

**4.3 Wait 5 seconds between each observation.** Refer to the previous exercise. Explain why you would not want to wait only 5 seconds between each time you turn the radio station on.

**4.4 Winning at craps.** The game of craps starts with a "come-out" roll where the shooter rolls a pair of dice. If the total is 7 or 11, the shooter wins immediately (there are ways that the shooter can win on later rolls if other numbers are rolled on the come-out roll). Roll a pair of dice 25 times and estimate the probability that the shooter wins immediately on the come-out roll. For a pair of perfectly made dice, the probability is 0.2222.

**4.5 The color of candy.** It is reasonable to think that packages of M&M's Milk Chocolate Candies are filled at the factory with candies chosen at random

## USE YOUR KNOWLEDGE

**4.12  Phone-related accidents on Monday or Friday.** Find the probability that a phone-related accident occurred on a Monday or a Friday.

**4.13  Not on Wednesday.** Find the probability that a phone-related accident occurred on a day other than a Wednesday.

## Assigning probabilities: finite number of outcomes

The individual outcomes of a random phenomenon are always disjoint. So the addition rule provides a way to assign probabilities to events with more than one outcome: start with probabilities for individual outcomes and add to get probabilities for events. This idea works well when there are only a finite (fixed and limited) number of outcomes.

---

**PROBABILITIES IN A FINITE SAMPLE SPACE**

Assign a probability to each individual outcome. These probabilities must be numbers between 0 and 1 and must have sum 1.

The probability of any event is the sum of the probabilities of the outcomes making up the event.

---

**EXAMPLE**

Benford's law

**4.12 Benford's law.** Faked numbers in tax returns, payment records, invoices, expense account claims, and many other settings often display patterns that aren't present in legitimate records. Some patterns, like too many round numbers, are obvious and easily avoided by a clever crook. Others are more subtle. It is a striking fact that the first digits of numbers in legitimate records often follow a distribution known as **Benford's law.** Here it is (note that a first digit can't be 0):[3]

| First digit | 1 | 2 | 3 | 4 | 5 | 6 | 7 | 8 | 9 |
|---|---|---|---|---|---|---|---|---|---|
| Probability | 0.301 | 0.176 | 0.125 | 0.097 | 0.079 | 0.067 | 0.058 | 0.051 | 0.046 |

Benford's law usually applies to the first digits of the sizes of similar quantities, such as invoices, expense account claims, and county populations. Investigators can detect fraud by comparing the first digits in records such as invoices paid by a business with these probabilities.

**EXAMPLE**

**4.13 Find some probabilities for Benford's law.** Consider the events

$$A = \{\text{first digit is 1}\}$$

$$B = \{\text{first digit is 6 or greater}\}$$

From the table of probabilities,

$$P(A) = P(1) = 0.301$$
$$P(B) = P(6) + P(7) + P(8) + P(9)$$
$$= 0.067 + 0.058 + 0.051 + 0.046 = 0.222$$

Note that $P(B)$ is not the same as the probability that a first digit is strictly greater than 6. The probability $P(6)$ that a first digit is 6 is included in "6 or greater" but not in "greater than 6."

## USE YOUR KNOWLEDGE

**4.14 Benford's law.** Using the probabilities for Benford's law, find the probability that a first digit is anything other than 1.

**4.15 Use the addition rule.** Use the addition rule with the probabilities for the events $A$ and $B$ from Example 4.13 to find the probability that a first digit is either 1 or 6 or greater.

Be careful to apply the addition rule only to disjoint events.

**EXAMPLE**

**4.14 Apply the addition rule to Benford's law.** Check that the probability of the event $C$ that a first digit is odd is

$$P(C) = P(1) + P(3) + P(5) + P(7) + P(9) = 0.609$$

The probability

$$P(B \text{ or } C) = P(1) + P(3) + P(5) + P(6) + P(7) + P(8) + P(9) = 0.727$$

is *not* the sum of $P(B)$ and $P(C)$, because events $B$ and $C$ are not disjoint. Outcomes 7 and 9 are common to both events.

## Assigning probabilities: equally likely outcomes

Assigning correct probabilities to individual outcomes often requires long observation of the random phenomenon. In some circumstances, however, we are willing to assume that individual outcomes are equally likely because of some balance in the phenomenon. Ordinary coins have a physical balance that should make heads and tails equally likely, for example, and the table of random digits comes from a deliberate randomization.

**EXAMPLE**

**4.15 First digits that are equally likely.** You might think that first digits are distributed "at random" among the digits 1 to 9 in business records. The 9 possible outcomes would then be equally likely. The sample space for a single digit is

$$S = \{1, 2, 3, 4, 5, 6, 7, 8, 9\}$$

Because the total probability must be 1, the probability of each of the 9 outcomes must be 1/9. That is, the assignment of probabilities to outcomes is

| First digit | 1 | 2 | 3 | 4 | 5 | 6 | 7 | 8 | 9 |
|---|---|---|---|---|---|---|---|---|---|
| Probability | 1/9 | 1/9 | 1/9 | 1/9 | 1/9 | 1/9 | 1/9 | 1/9 | 1/9 |

The probability of the event $B$ that a randomly chosen first digit is 6 or greater is

$$P(B) = P(6) + P(7) + P(8) + P(9)$$

$$= \frac{1}{9} + \frac{1}{9} + \frac{1}{9} + \frac{1}{9} = \frac{4}{9} = 0.444$$

Compare this with the Benford's law probability in Example 4.13. A crook who fakes data by using "random" digits will end up with too many first digits 6 or greater and too few 1s and 2s.

In Example 4.15 all outcomes have the same probability. Because there are 9 equally likely outcomes, each must have probability 1/9. Because exactly 4 of the 9 equally likely outcomes are 6 or greater, the probability of this event is 4/9. In the special situation where all outcomes are equally likely, we have a simple rule for assigning probabilities to events.

---

### EQUALLY LIKELY OUTCOMES

If a random phenomenon has $k$ possible outcomes, all equally likely, then each individual outcome has probability $1/k$. The probability of any event $A$ is

$$P(A) = \frac{\text{count of outcomes in } A}{\text{count of outcomes in } S}$$

$$= \frac{\text{count of outcomes in } A}{k}$$

---

Most random phenomena do not have equally likely outcomes, so the general rule for finite sample spaces is more important than the special rule for equally likely outcomes.

## USE YOUR KNOWLEDGE

**4.16  Possible outcomes for rolling a die.** A die has six sides with 1 to 6 "spots" on the sides. Give the probability distribution for the six possible outcomes that can result when a perfect die is rolled.

## Independence and the multiplication rule

Rule 3, the addition rule for disjoint events, describes the probability that *one or the other* of two events *A* and *B* will occur in the special situation when *A* and *B* cannot occur together because they are disjoint. Our final rule describes the probability that *both* events *A* and *B* occur, again only in a special situation. More general rules appear in Section 4.5, but in our study of statistics we will need only the rules that apply to special situations.

Suppose that you toss a balanced coin twice. You are counting heads, so two events of interest are

$$A = \{\text{first toss is a head}\}$$

$$B = \{\text{second toss is a head}\}$$

The events *A* and *B* are not disjoint. They occur together whenever both tosses give heads. We want to compute the probability of the event {*A* and *B*} that *both* tosses are heads. The Venn diagram in Figure 4.4 illustrates the event {*A* and *B*} as the overlapping area that is common to both *A* and *B*.

The coin tossing of Buffon, Pearson, and Kerrich described in Example 4.3 makes us willing to assign probability 1/2 to a head when we toss a coin. So

$$P(A) = 0.5$$

$$P(B) = 0.5$$

What is $P(A \text{ and } B)$? Our common sense says that it is 1/4. The first coin will give a head half the time and then the second will give a head on half of those trials, so both coins will give heads on $1/2 \times 1/2 = 1/4$ of all trials in the long run. This reasoning assumes that the second coin still has probability 1/2 of a head after the first has given a head. This is true—we can verify it by tossing two coins many times and observing the proportion of heads on the second toss after the first toss has produced a head. We say that the events "head on the first toss" and "head on the second toss" are *independent*. Here is our final probability rule.

---

**THE MULTIPLICATION RULE FOR INDEPENDENT EVENTS**

**Rule 5.** Two events *A* and *B* are **independent** if knowing that one occurs does not change the probability that the other occurs. If *A* and *B* are independent,

$$P(A \text{ and } B) = P(A)P(B)$$

This is the **multiplication rule for independent events.**

---

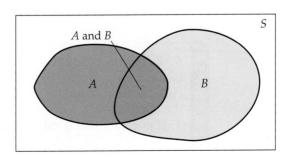

**FIGURE 4.4** Venn diagram showing the event {*A* and *B*}. This event consists of outcomes common to *A* and *B*.

Our definition of independence is rather informal. We will make this informal idea precise in Section 4.5. In practice, though, we rarely need a precise definition of independence, because independence is usually *assumed* as part of a probability model when we want to describe random phenomena that seem to be physically unrelated to each other. Here is an example of independence.

**EXAMPLE**

**4.16 Coins do not have memory.**   Because a coin has no memory and most coin tossers cannot influence the fall of the coin, it is safe to assume that successive coin tosses are independent. For a balanced coin this means that after we see the outcome of the first toss, we still assign probability 1/2 to heads on the second toss.

**USE YOUR KNOWLEDGE**

**4.17    Two tails in two tosses.** What is the probability of obtaining two tails on two tosses of a fair coin?

Here is an example of a situation where there are dependent events.

**EXAMPLE**

**4.17 Dependent events in cards.**   The colors of successive cards dealt from the same deck are not independent. A standard 52-card deck contains 26 red and 26 black cards. For the first card dealt from a shuffled deck, the probability of a red card is $26/52 = 0.50$ because the 52 possible cards are equally likely. Once we see that the first card is red, we know that there are only 25 reds among the remaining 51 cards. The probability that the second card is red is therefore only $25/51 = 0.49$. Knowing the outcome of the first deal changes the probabilities for the second.

**USE YOUR KNOWLEDGE**

**4.18    The probability of a second ace.** A deck of 52 cards contains 4 aces, so the probability that a card drawn from this deck is an ace is 4/52. If we know that the first card drawn is an ace, what is the probability that the second card drawn is also an ace? Using the idea of independence, explain why this probability is not 4/52.

Here is another example of a situation where events are dependent.

**EXAMPLE**

**4.18 Taking a test twice.**   If you take an IQ test or other mental test twice in succession, the two test scores are not independent. The learning that occurs on the first attempt influences your second attempt. If you learn a lot, then your second test score might be a lot higher than your first test score. This phenomenon is called a carry-over effect.

When independence is part of a probability model, the multiplication rule applies. Here is an example.

**EXAMPLE**

**4.19 Mendel's peas.**   Gregor Mendel used garden peas in some of the experiments that revealed that inheritance operates randomly. The seed color of Mendel's peas can be either green or yellow. Two parent plants are "crossed" (one pollinates the other) to produce seeds. Each parent plant carries two genes for seed color, and each of these genes has probability 1/2 of being passed to a seed. The two genes that the seed receives, one from each parent, determine its color. The parents contribute their genes independently of each other.

Suppose that both parents carry the $G$ and the $Y$ genes. The seed will be green if both parents contribute a $G$ gene; otherwise it will be yellow. If $M$ is the event that the male contributes a $G$ gene and $F$ is the event that the female contributes a $G$ gene, then the probability of a green seed is

$$P(M \text{ and } F) = P(M)P(F)$$
$$= (0.5)(0.5) = 0.25$$

In the long run, 1/4 of all seeds produced by crossing these plants will be green.

*The multiplication rule applies only to independent events; you cannot use it if events are not independent.* Here is a distressing example of misuse of the multiplication rule.

**EXAMPLE**

**4.20 Sudden infant death syndrome.**   Sudden infant death syndrome (SIDS) causes babies to die suddenly (often in their cribs) with no explanation. Deaths from SIDS have been greatly reduced by placing babies on their backs, but as yet no cause is known.

When more than one SIDS death occurs in a family, the parents are sometimes accused. One "expert witness" popular with prosecutors in England told juries that there is only a 1 in 73 million chance that two children in the same family could have died naturally. Here's his calculation: the rate of SIDS in a nonsmoking middle-class family is 1 in 8500. So the probability of two deaths is

$$\frac{1}{8500} \times \frac{1}{8500} = \frac{1}{72,250,000}$$

Several women were convicted of murder on this basis, without any direct evidence that they had harmed their children.

As the Royal Statistical Society said, this reasoning is nonsense. It assumes that SIDS deaths in the same family are independent events. The cause of SIDS is unknown: "There may well be unknown genetic or environmental factors that predispose families to SIDS, so that a second case within the family becomes much more likely."[4] The British government decided to review the cases of 258 parents convicted of murdering their babies.

probability that our portfolio rises in price in any one year is 0.65. (This probability is approximately correct for a portfolio containing equal dollar amounts of all common stocks listed on the New York Stock Exchange.)

(a) What is the probability that our portfolio goes up for 3 consecutive years?

(b) If you know that the portfolio has risen in price 2 years in a row, what probability do you assign to the event that it will go down next year?

(c) What is the probability that the portfolio's value moves in the same direction in both of the next 2 years?

4.40 ⊕ **Axioms of probability.** Show that any assignment of probabilities to events that obeys Rules 2 and 3 on page 246 automatically obeys the complement rule (Rule 4). This implies that a mathematical treatment of probability can start from just Rules 1, 2, and 3. These rules are sometimes called *axioms* of probability.

4.41 ⊕ **Independence of complements.** Show that if events $A$ and $B$ obey the multiplication rule, $P(A \text{ and } B) = P(A)P(B)$, then $A$ and the complement $B^c$ of $B$ also obey the multiplication rule, $P(A \text{ and } B^c) = P(A)P(B^c)$. That is, if events $A$ and $B$ are independent, then $A$ and $B^c$ are also independent. (*Hint:* Start by drawing a Venn diagram and noticing that the events "$A$ and $B$" and "$A$ and $B^c$" are disjoint.)

*Mendelian inheritance. Some traits of plants and animals depend on inheritance of a single gene. This is called Mendelian inheritance, after Gregor Mendel (1822–1884). Exercises 4.42 to 4.45 are based on the following information about Mendelian inheritance of blood type.*

*Each of us has an ABO blood type, which describes whether two characteristics called A and B are present. Every human being has two blood type alleles (gene forms), one inherited from our mother and one from our father. Each of these alleles can be A, B, or O. Which two we inherit*

*determines our blood type. The following table shows what our blood type is for each combination of two alleles:*

| Alleles inherited | Blood type |
| --- | --- |
| A and A | A |
| A and B | AB |
| A and O | A |
| B and B | B |
| B and O | B |
| O and O | O |

*We inherit each of a parent's two alleles with probability 0.5. We inherit independently from our mother and father.*

4.42 **Blood types of children.** Hannah and Jacob both have alleles A and B.

(a) What blood types can their children have?

(b) What is the probability that their next child has each of these blood types?

4.43 **Parents with alleles B and O.** Nancy and David both have alleles B and O.

(a) What blood types can their children have?

(b) What is the probability that their next child has each of these blood types?

4.44 **Two children.** Jennifer has alleles A and O. José has alleles A and B. They have two children. What is the probability that both children have blood type A? What is the probability that both children have the same blood type?

4.45 **Three children.** Jasmine has alleles A and O. Joshua has alleles B and O.

(a) What is the probability that a child of these parents has blood type O?

(b) If Jasmine and Joshua have three children, what is the probability that all three have blood type O? What is the probability that the first child has blood type O and the next two do not?

# 4.3 Random Variables

Sample spaces need not consist of numbers. When we toss a coin four times, we can record the outcome as a string of heads and tails, such as HTTH. In statistics, however, we are most often interested in numerical outcomes such as the count of heads in the four tosses. It is convenient to use a shorthand notation: Let $X$ be the number of heads. If our outcome is HTTH, then $X = 2$. If the next outcome is TTTH, the value of $X$ changes to $X = 1$. The possible values of $X$ are 0, 1, 2, 3, and 4. Tossing a coin four times will give $X$ one of these possible values. Tossing four more times will give $X$ another and probably different value. We call $X$ a *random variable* because its values vary when the coin tossing is repeated.

---

**RANDOM VARIABLE**

A **random variable** is a variable whose value is a numerical outcome of a random phenomenon.

---

We usually denote random variables by capital letters near the end of the alphabet, such as $X$ or $Y$. Of course, the random variables of greatest interest to us are outcomes such as the mean $\bar{x}$ of a random sample, for which we will keep the familiar notation.[11] As we progress from general rules of probability toward statistical inference, we will concentrate on random variables. When a random variable $X$ describes a random phenomenon, the sample space $S$ just lists the possible values of the random variable. We usually do not mention $S$ separately. There remains the second part of any probability model, the assignment of probabilities to events. There are two main ways of assigning probabilities to the values of a random variable. The two types of probability models that result will dominate our application of probability to statistical inference.

## Discrete random variables

We have learned several rules of probability but only one method of assigning probabilities: state the probabilities of the individual outcomes and assign probabilities to events by summing over the outcomes. The outcome probabilities must be between 0 and 1 and have sum 1. When the outcomes are numerical, they are values of a random variable. We will now attach a name to random variables having probability assigned in this way.[12]

---

**DISCRETE RANDOM VARIABLE**

A **discrete random variable** $X$ has a finite number of possible values. The **probability distribution** of $X$ lists the values and their probabilities:

| Value of $X$ | $x_1$ | $x_2$ | $x_3$ | $\cdots$ | $x_k$ |
|---|---|---|---|---|---|
| Probability | $p_1$ | $p_2$ | $p_3$ | $\cdots$ | $p_k$ |

The probabilities $p_i$ must satisfy two requirements:

**1.** Every probability $p_i$ is a number between 0 and 1.

**2.** $p_1 + p_2 + \cdots + p_k = 1$.

Find the probability of any event by adding the probabilities $p_i$ of the particular values $x_i$ that make up the event.

---

**EXAMPLE**

**4.22 Grade distributions.** North Carolina State University posts the grade distributions for its courses online.[13] Students in one section of English 210 in the spring 2006 semester received 31% A's, 40% B's, 20% C's, 4% D's, and 5% F's. Choose an English 210 student at random. To "choose at random" means to give every student the same chance to be chosen. The student's grade on a four-point scale (with A = 4) is a random variable $X$.

The value of $X$ changes when we repeatedly choose students at random, but it is always one of 0, 1, 2, 3, or 4. Here is the distribution of $X$:

| Value of $X$ | 0 | 1 | 2 | 3 | 4 |
|---|---|---|---|---|---|
| Probability | 0.05 | 0.04 | 0.20 | 0.40 | 0.31 |

The probability that the student got a B or better is the sum of the probabilities of an A and a B. In the language of random variables,

$$P(X \geq 3) = P(X = 3) + P(X = 4)$$
$$= 0.40 + 0.31 = 0.71$$

## USE YOUR KNOWLEDGE

**4.46 Will the course satisfy the requirement?** Refer to Example 4.22. Suppose that a grade of D or F in English 210 will not count as satisfying a requirement for a major in linguistics. What is the probability that a randomly selected student will not satisfy this requirement?

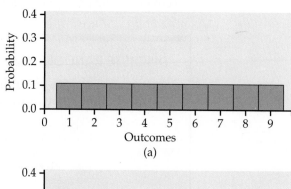

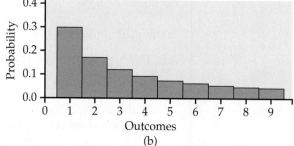

**FIGURE 4.5** Probability histograms for (a) random digits 1 to 9 and (b) Benford's law. The height of each bar shows the probability assigned to a single outcome.

We can use histograms to show probability distributions as well as distri-

**probability histogram** butions of data. Figure 4.5 displays **probability histograms** that compare the probability model for random digits for business records (Example 4.15) with the model given by Benford's law (Example 4.12). The height of each bar shows the probability of the outcome at its base. Because the heights are probabilities, they add to 1. As usual, all the bars in a histogram have the same width. So the areas also display the assignment of probability to outcomes. Think of these histograms as idealized pictures of the results of very many trials. The histograms make it easy to quickly compare the two distributions.

**EXAMPLE**

**4.23 Number of heads in four tosses of a coin.** What is the probability distribution of the discrete random variable $X$ that counts the number of heads in four tosses of a coin? We can derive this distribution if we make two reasonable assumptions:

- The coin is balanced, so it is fair and each toss is equally likely to give H or T.

- The coin has no memory, so tosses are independent.

The outcome of four tosses is a sequence of heads and tails such as HTTH. There are 16 possible outcomes in all. Figure 4.6 lists these outcomes along with the value of $X$ for each outcome. The multiplication rule for independent events tells us that, for example,

$$P(\text{HTTH}) = \frac{1}{2} \times \frac{1}{2} \times \frac{1}{2} \times \frac{1}{2} = \frac{1}{16}$$

Each of the 16 possible outcomes similarly has probability 1/16. That is, these outcomes are equally likely.

The number of heads $X$ has possible values 0, 1, 2, 3, and 4. These values are *not* equally likely. As Figure 4.6 shows, there is only one way that $X = 0$ can occur: namely, when the outcome is TTTT. So

$$P(X = 0) = \frac{1}{16} = 0.0625$$

The event $\{X = 2\}$ can occur in six different ways, so that

$$P(X = 2) = \frac{\text{count of ways } X = 2 \text{ can occur}}{16}$$

$$= \frac{6}{16} = 0.375$$

We can find the probability of each value of $X$ from Figure 4.6 in the same way. Here is the result:

| Value of $X$ | 0 | 1 | 2 | 3 | 4 |
|---|---|---|---|---|---|
| Probability | 0.0625 | 0.25 | 0.375 | 0.25 | 0.0625 |

**FIGURE 4.6** Possible outcomes in four tosses of a coin, for Example 4.23. The outcomes are arranged by the values of the random variable $X$, the number of heads.

|  |  | HTTH |  |  |
|  |  | HTHT |  |  |
|  | HTTT | THTH | HHHT |  |
|  | THTT | HHTT | HHTH |  |
|  | TTHT | THHT | HTHH |  |
| TTTT | TTTH | TTHH | THHH | HHHH |
| $X = 0$ | $X = 1$ | $X = 2$ | $X = 3$ | $X = 4$ |

Figure 4.7 is a probability histogram for the distribution in Example 4.23. The probability distribution is exactly symmetric. The probabilities (bar heights) are idealizations of the proportions after very many tosses of four coins. The actual distribution of proportions observed would be nearly symmetric but is unlikely to be exactly symmetric.

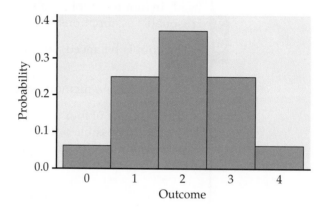

**FIGURE 4.7** Probability histogram for the number of heads in four tosses of a coin.

**EXAMPLE**

**4.24 Probability of at least two heads.** Any event involving the number of heads observed can be expressed in terms of $X$, and its probability can be found from the distribution of $X$. For example, the probability of tossing at least two heads is

$$P(X \geq 2) = 0.375 + 0.25 + 0.0625 = 0.6875$$

The probability of at least one head is most simply found by use of the complement rule:

$$P(X \geq 1) = 1 - P(X = 0)$$
$$= 1 - 0.0625 = 0.9375$$

Recall that tossing a coin $n$ times is similar to choosing an SRS of size $n$ from a large population and asking a yes-or-no question. We will extend the results of Example 4.23 when we return to sampling distributions in the next chapter.

## USE YOUR KNOWLEDGE

**4.47 Two tosses of a fair coin.** Find the probability distribution for the number of heads that appear in two tosses of a fair coin.

## Continuous random variables

When we use the table of random digits to select a digit between 0 and 9, the result is a discrete random variable. The probability model assigns probability 1/10 to each of the 10 possible outcomes. Suppose that we want to choose a number at random between 0 and 1, allowing *any* number between 0 and 1 as the outcome. Software random number generators will do this. You can visualize such a random number by thinking of a spinner (Figure 4.8) that turns freely on its axis and slowly comes to a stop. The pointer can come to rest anywhere on a circle that is marked from 0 to 1. The sample space is now an entire interval of numbers:

$$S = \{\text{all numbers } x \text{ such that } 0 \le x \le 1\}$$

How can we assign probabilities to events such as $\{0.3 \le x \le 0.7\}$? As in the case of selecting a random digit, we would like all possible outcomes to be equally likely. But we cannot assign probabilities to each individual value of $x$ and then sum, because there are infinitely many possible values. Instead, we use a new way of assigning probabilities directly to events—as *areas under a density curve*. Any density curve has area exactly 1 underneath it, corresponding to total probability 1.

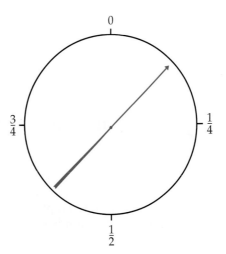

**FIGURE 4.8** A spinner that generates a random number between 0 and 1.

EXAMPLE

uniform distribution

**4.25 Uniform random numbers.** The random number generator will spread its output uniformly across the entire interval from 0 to 1 as we allow it to generate a long sequence of numbers. The results of many trials are represented by the density curve of a **uniform distribution.** This density curve appears in red in Figure 4.9. It has height 1 over the interval from 0 to 1, and height 0 everywhere else. The area under the density curve is 1: the area of a square with base 1 and height 1. The probability of any event is the area under the density curve and above the event in question.

As Figure 4.9(a) illustrates, the probability that the random number generator produces a number $X$ between 0.3 and 0.7 is

$$P(0.3 \le X \le 0.7) = 0.4$$

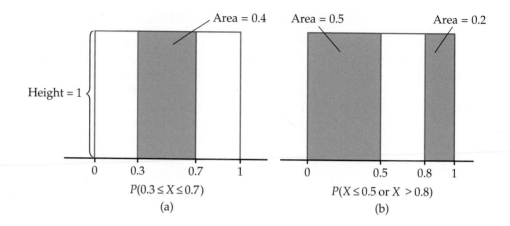

**FIGURE 4.9** Assigning probabilities for generating a random number between 0 and 1, for Example 4.25. The probability of any interval of numbers is the area above the interval and under the density curve.

because the area under the density curve and above the interval from 0.3 to 0.7 is 0.4. The height of the density curve is 1, and the area of a rectangle is the product of height and length, so the probability of any interval of outcomes is just the length of the interval.

Similarly,

$$P(X \leq 0.5) = 0.5$$

$$P(X > 0.8) = 0.2$$

$$P(X \leq 0.5 \text{ or } X > 0.8) = 0.7$$

Notice that the last event consists of two nonoverlapping intervals, so the total area above the event is found by adding two areas, as illustrated by Figure 4.9(b). This assignment of probabilities obeys all of our rules for probability.

Probability as area under a density curve is a second important way of assigning probabilities to events. Figure 4.10 illustrates this idea in general form. We call $X$ in Example 4.25 a *continuous random variable* because its values are not isolated numbers but an entire interval of numbers.

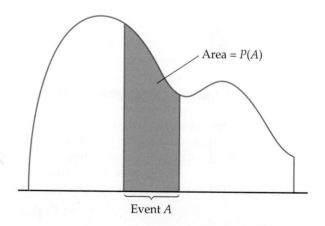

**FIGURE 4.10** The probability distribution of a continuous random variable assigns probabilities as areas under a density curve. The total area under any density curve is 1.

## USE YOUR KNOWLEDGE

**4.48 Find the probability.** For the uniform distribution described in Example 4.25, find the probability that $X$ is between 0.1 and 0.4.

---

### CONTINUOUS RANDOM VARIABLE

A **continuous random variable** $X$ takes all values in an interval of numbers. The **probability distribution** of $X$ is described by a density curve. The probability of any event is the area under the density curve and above the values of $X$ that make up the event.

---

The probability model for a continuous random variable assigns probabilities to intervals of outcomes rather than to individual outcomes. In fact, **all continuous probability distributions assign probability 0 to every individual outcome.** Only intervals of values have positive probability. To see that this is true, consider a specific outcome such as $P(X = 0.8)$ in the context of Example 4.25. The probability of any interval is the same as its length. The point 0.8 has no length, so its probability is 0.

Although this fact may seem odd, it makes intuitive, as well as mathematical, sense. The random number generator produces a number between 0.79 and 0.81 with probability 0.02. An outcome between 0.799 and 0.801 has probability 0.002. A result between 0.799999 and 0.800001 has probability 0.000002. You see that as we approach 0.8, the probability gets closer to 0. To be consistent, the probability of outcome *exactly* equal to 0.8 must be 0. Because there is no probability exactly at $X = 0.8$, the two events $\{X > 0.8\}$ and $\{X \geq 0.8\}$ have the same probability. We can ignore the distinction between $>$ and $\geq$ when finding probabilities for continuous (but not discrete) random variables.

## Normal distributions as probability distributions

The density curves that are most familiar to us are the Normal curves. Because any density curve describes an assignment of probabilities, *Normal distributions are probability distributions*. Recall that $N(\mu, \sigma)$ is our shorthand for the Normal distribution having mean $\mu$ and standard deviation $\sigma$. In the language of random variables, if $X$ has the $N(\mu, \sigma)$ distribution, then the standardized variable

$$Z = \frac{X - \mu}{\sigma}$$

is a standard Normal random variable having the distribution $N(0, 1)$.

---

**EXAMPLE**

**4.26 Cheating.** Students are reluctant to report cheating by other students. A sample survey puts this question to an SRS of 400 undergraduates: "You witness two students cheating on a quiz. Do you go to the professor?" Suppose that, if we could ask all undergraduates, 12% would answer "Yes."[14]

The proportion $p = 0.12$ is a *parameter* that describes the population of all undergraduates. The proportion $\hat{p}$ of the sample who answer "Yes" is a

know only that the mean number of dimples is $\mu_X = 0.7$. The number of paint sags is a second random variable $Y$ having mean $\mu_Y = 1.4$. (As usual, the subscripts keep straight which variable we are talking about.) The total number of both dimples and sags is another random variable, the sum $X + Y$. Its mean $\mu_{X+Y}$ is the average number of dimples and sags together. It is just the sum of the individual means $\mu_X$ and $\mu_Y$. That's an important rule for how means of random variables behave.

Here's another rule. The crickets living in a field have mean length 1.2 inches. What is the mean in centimeters? There are 2.54 centimeters in an inch, so the length of a cricket in centimeters is 2.54 times its length in inches. If we multiply every observation by 2.54, we also multiply their average by 2.54. The mean in centimeters must be 2.54 × 1.2, or about 3.05 centimeters. More formally, the length in inches of a cricket chosen at random from the field is a random variable $X$ with mean $\mu_X$. The length in centimeters is 2.54$X$, and this new random variable has mean 2.54$\mu_X$.

The point of these examples is that means behave like averages. Here are the rules we need.

---

**RULES FOR MEANS**

**Rule 1.** If $X$ is a random variable and $a$ and $b$ are fixed numbers, then

$$\mu_{a+bX} = a + b\mu_X$$

**Rule 2.** If $X$ and $Y$ are random variables, then

$$\mu_{X+Y} = \mu_X + \mu_Y$$

---

**EXAMPLE**

**4.32 Sales of cars, trucks, and SUVs.** Linda is a sales associate at a large auto dealership. At her commission rate of 25% of gross profit on each vehicle she sells, Linda expects to earn $350 for each car sold and $400 for each truck or SUV sold. Linda motivates herself by using probability estimates of her sales. For a sunny Saturday in April, she estimates her car sales as follows:

| Cars sold | 0 | 1 | 2 | 3 |
|---|---|---|---|---|
| Probability | 0.3 | 0.4 | 0.2 | 0.1 |

Linda's estimate of her truck or SUV sales is

| Vehicles sold | 0 | 1 | 2 |
|---|---|---|---|
| Probability | 0.4 | 0.5 | 0.1 |

Take $X$ to be the number of cars Linda sells and $Y$ the number of trucks or SUVs. The means of these random variables are

$$\mu_X = (0)(0.3) + (1)(0.4) + (2)(0.2) + (3)(0.1)$$

$$= 1.1 \ \text{cars}$$

$$\mu_Y = (0)(0.4) + (1)(0.5) + (2)(0.1)$$

$$= 0.7 \ \text{trucks or SUVs}$$

Linda's earnings, at \$350 per car and \$400 per truck or SUV, are

$$Z = 350X + 400Y$$

Combining Rules 1 and 2, her mean earnings are

$$\mu_Z = 350\mu_X + 400\mu_Y$$

$$= (350)(1.1) + (400)(0.7) = \$665$$

This is Linda's best estimate of her earnings for the day. It's a bit unusual for individuals to use probability estimates, but they are a common tool for business planners.

personal probability    The probabilities in Example 4.32 are **personal probabilities** that describe Linda's informed opinion about her sales in the coming weekend. Although personal probabilities need not be based on observing many repetitions of a random phenomenon, they must obey the rules of probability if they are to make sense. Personal probability extends the usefulness of probability models to one-time events, but remember that they are subject to the follies of human opinion. Overoptimism is common: 40% of college students think that they will eventually reach the top 1% in income.

## USE YOUR KNOWLEDGE

**4.69  Find $\mu_Y$.** The random variable $X$ has mean $\mu_X = 10$. If $Y = 15 + 8X$, what is $\mu_Y$?

**4.70  Find $\mu_W$.** The random variable $U$ has mean $\mu_U = 20$ and the random variable $V$ has mean $\mu_V = 20$. If $W = 0.5U + 0.5V$, find $\mu_W$.

## The variance of a random variable

The mean is a measure of the center of a distribution. A basic numerical description requires in addition a measure of the spread or variability of the distribution. The variance and the standard deviation are the measures of spread that accompany the choice of the mean to measure center. Just as for the mean, we need a distinct symbol to distinguish the variance of a random variable from the variance $s^2$ of a data set. We write the variance of a random variable $X$ as $\sigma_X^2$. Once again the subscript reminds us which variable we have in mind. The definition of the variance $\sigma_X^2$ of a random variable is similar to the definition of the sample variance $s^2$ given in Chapter 1. That is, the variance is an average value of the squared deviation $(X - \mu_X)^2$ of the variable $X$ from its mean $\mu_X$. As for the mean, the average we use is a weighted average in which each outcome is weighted by its probability in order to take account of outcomes that are not equally likely. Calculating this weighted average is straightforward for discrete random variables but requires advanced mathematics in the continuous case. Here is the definition.

---

**VARIANCE OF A DISCRETE RANDOM VARIABLE**

Suppose that $X$ is a discrete random variable whose distribution is

| Value of $X$ | $x_1$ | $x_2$ | $x_3$ | $\cdots$ | $x_k$ |
|---|---|---|---|---|---|
| Probability | $p_1$ | $p_2$ | $p_3$ | $\cdots$ | $p_k$ |

and that $\mu_X$ is the mean of $X$. The **variance** of $X$ is

$$\sigma_X^2 = (x_1 - \mu_X)^2 p_1 + (x_2 - \mu_X)^2 p_2 + \cdots + (x_k - \mu_X)^2 p_k$$
$$= \sum (x_i - \mu_X)^2 p_i$$

The **standard deviation** $\sigma_X$ of $X$ is the square root of the variance.

---

**EXAMPLE**

**4.33 Find the mean and the variance.** In Example 4.32 we saw that the number $X$ of cars that Linda hopes to sell has distribution

| Cars sold | 0 | 1 | 2 | 3 |
|---|---|---|---|---|
| Probability | 0.3 | 0.4 | 0.2 | 0.1 |

We can find the mean and variance of $X$ by arranging the calculation in the form of a table. Both $\mu_X$ and $\sigma_X^2$ are sums of columns in this table.

| $x_i$ | $p_i$ | $x_i p_i$ | $(x_i - \mu_X)^2 p_i$ |
|---|---|---|---|
| 0 | 0.3 | 0.0 | $(0 - 1.1)^2(0.3) = 0.363$ |
| 1 | 0.4 | 0.4 | $(1 - 1.1)^2(0.4) = 0.004$ |
| 2 | 0.2 | 0.4 | $(2 - 1.1)^2(0.2) = 0.162$ |
| 3 | 0.1 | 0.3 | $(3 - 1.1)^2(0.1) = 0.361$ |
| | | $\mu_X = 1.1$ | $\sigma_X^2 = 0.890$ |

We see that $\sigma_X^2 = 0.89$. The standard deviation of $X$ is $\sigma_X = \sqrt{0.89} = 0.943$. The standard deviation is a measure of the variability of the number of cars Linda sells. As in the case of distributions for data, the standard deviation of a probability distribution is easiest to understand for Normal distributions.

**USE YOUR KNOWLEDGE**

**4.71 Find the variance and the standard deviation.** The random variable $X$ has the following probability distribution:

| Value of $X$ | 0 | 2 |
|---|---|---|
| Probability | 0.5 | 0.5 |

Find the variance $\sigma_X^2$ and the standard deviation $\sigma_X$ for this random variable.

## Rules for variances and standard deviations

What are the facts for variances that parallel Rules 1 and 2 for means? *The mean of a sum of random variables is always the sum of their means, but this addition rule is true for variances only in special situations.* To understand why, take $X$ to be the percent of a family's after-tax income that is spent and $Y$ the percent that is saved. When $X$ increases, $Y$ decreases by the same amount. Though $X$ and $Y$ may vary widely from year to year, their sum $X + Y$ is always 100% and does not vary at all. It is the association between the variables $X$ and $Y$ that prevents their variances from adding. If random variables are independent, this kind of association between their values is ruled out and their variances do add. Two **independence** random variables $X$ and $Y$ are **independent** if knowing that any event involving $X$ alone did or did not occur tells us nothing about the occurrence of any event involving $Y$ alone. Probability models often assume independence when the random variables describe outcomes that appear unrelated to each other. You should ask in each instance whether the assumption of independence seems reasonable.

When random variables are not independent, the variance of their sum de- **correlation** pends on the **correlation** between them as well as on their individual variances. In Chapter 2, we met the correlation $r$ between two observed variables measured on the same individuals. We defined (page 102) the correlation $r$ as an average of the products of the standardized $x$ and $y$ observations. The correlation between two random variables is defined in the same way, once again using a weighted average with probabilities as weights. We won't give the details—it is enough to know that the correlation between two random variables has the same basic properties as the correlation $r$ calculated from data. We use $\rho$, the Greek letter rho, for the correlation between two random variables. The correlation $\rho$ is a number between $-1$ and 1 that measures the direction and strength of the linear relationship between two variables. **The correlation between two independent random variables is zero.**

Returning to family finances, if $X$ is the percent of a family's after-tax income that is spent and $Y$ the percent that is saved, then $Y = 100 - X$. This is a perfect linear relationship with a negative slope, so the correlation between $X$ and $Y$ is $\rho = -1$. With the correlation at hand, we can state the rules for manipulating variances.

---

### RULES FOR VARIANCES AND STANDARD DEVIATIONS

**Rule 1.** If $X$ is a random variable and $a$ and $b$ are fixed numbers, then

$$\sigma^2_{a+bX} = b^2 \sigma^2_X$$

**Rule 2.** If $X$ and $Y$ are independent random variables, then

$$\sigma^2_{X+Y} = \sigma^2_X + \sigma^2_Y$$

$$\sigma^2_{X-Y} = \sigma^2_X + \sigma^2_Y$$

This is the **addition rule for variances of independent random variables.**

**Rule 3.** If $X$ and $Y$ have correlation $\rho$, then

$$\sigma^2_{X+Y} = \sigma^2_X + \sigma^2_Y + 2\rho\sigma_X\sigma_Y$$

$$\sigma^2_{X-Y} = \sigma^2_X + \sigma^2_Y - 2\rho\sigma_X\sigma_Y$$

This is the **general addition rule for variances of random variables.**

To find the standard deviation, take the square root of the variance.

---

*Because a variance is the average of squared deviations from the mean, multiplying X by a constant b multiplies $\sigma^2_X$ by the square of the constant.* Adding a constant $a$ to a random variable changes its mean but does not change its variability. The variance of $X + a$ is therefore the same as the variance of $X$. Because the square of $-1$ is 1, the addition rule says that the variance of a difference of independent random variables is the *sum* of the variances. For independent random variables, the difference $X - Y$ is more variable than either $X$ or $Y$ alone because variations in both $X$ and $Y$ contribute to variation in their difference.

As with data, we prefer the standard deviation to the variance as a measure of the variability of a random variable. *Rule 2 for variances implies that standard deviations of independent random variables do not add. To combine standard deviations, use the rules for variances.* For example, the standard deviations of $2X$ and $-2X$ are both equal to $2\sigma_X$ because this is the square root of the variance $4\sigma^2_X$.

---

**EXAMPLE**

**4.34 Payoff in the Tri-State Pick 3 lottery.**   The payoff $X$ of a $1 ticket in the Tri-State Pick 3 game is $500 with probability 1/1000 and 0 the rest of the time. Here is the combined calculation of mean and variance:

| $x_i$ | $p_i$ | $x_i p_i$ | $(x_i - \mu_X)^2 p_i$ | |
|---|---|---|---|---|
| 0 | 0.999 | 0 | $(0 - 0.5)^2(0.999) =$ | 0.24975 |
| 500 | 0.001 | 0.5 | $(500 - 0.5)^2(0.001) =$ | 249.50025 |
| | | $\mu_X = 0.5$ | $\sigma^2_X =$ | 249.75 |

The mean payoff is 50 cents. The standard deviation is

$$\sigma_X = \sqrt{249.75} = \$15.80.$$

It is usual for games of chance to have large standard deviations because large variability makes gambling exciting.

If you buy a Pick 3 ticket, your winnings are $W = X - 1$ because the dollar you paid for the ticket must be subtracted from the payoff. Let's find the mean and variance for this random variable.

**EXAMPLE**

**4.35 Winnings in the Tri-State Pick 3 lottery.** By the rules for means, the mean amount you win is

$$\mu_W = \mu_X - 1 = -\$0.50$$

That is, you lose an average of 50 cents on a ticket. The rules for variances remind us that the variance and standard deviation of the winnings $W = X - 1$ are the same as those of $X$. Subtracting a fixed number changes the mean but not the variance.

Suppose now that you buy a \$1 ticket on each of two different days. The payoffs $X$ and $Y$ on the two tickets are independent because separate drawings are held each day. Your total payoff is $X + Y$. Let's find the mean and standard deviation for this payoff.

**EXAMPLE**

**4.36 Two tickets.** The mean for the payoff for the two tickets is

$$\mu_{X+Y} = \mu_X + \mu_Y = \$0.50 + \$0.50 = \$1.00$$

Because $X$ and $Y$ are independent, the variance of $X + Y$ is

$$\sigma^2_{X+Y} = \sigma^2_X + \sigma^2_Y = 249.75 + 249.75 = 499.5$$

The standard deviation of the total payoff is

$$\sigma_{X+Y} = \sqrt{499.5} = \$22.35$$

This is not the same as the sum of the individual standard deviations, which is \$15.80 + \$15.80 = \$31.60. Variances of independent random variables add; standard deviations do not.

When we add random variables that are correlated, we need to use the correlation for the calculation of the variance, but not for the calculation of the mean. Here is an example.

**EXAMPLE**

**4.37 The SAT Math score and the SAT Verbal score are dependent.**
Scores on the Mathematics part of the SAT college entrance exam in a recent year had mean 519 and standard deviation 115. Scores on the Verbal part of the SAT had mean 507 and standard deviation 111. What are the mean and standard deviation of total SAT score?[20]

Think of choosing one student's scores at random. Expressed in the language of random variables,

$$\text{SAT Math score } X \qquad \mu_X = 519 \qquad \sigma_X = 115$$
$$\text{SAT Verbal score } Y \qquad \mu_Y = 507 \qquad \sigma_Y = 111$$

The total score is $X + Y$. The mean is easy:

$$\mu_{X+Y} = \mu_X + \mu_Y = 519 + 507 = 1026$$

The variance and standard deviation of the total *cannot be computed* from the information given. SAT Verbal and Math scores are not independent, because students who score high on one exam tend to score high on the other also. Therefore, Rule 2 does not apply. We need to know $\rho$, the correlation between $X$ and $Y$, to apply Rule 3.

The correlation between SAT Math and Verbal scores was $\rho = 0.71$. By Rule 3,

$$\sigma^2_{X+Y} = \sigma^2_X + \sigma^2_Y + 2\rho\sigma_X\sigma_Y$$
$$= (115)^2 + (111)^2 + (2)(0.71)(115)(111)$$
$$= 43{,}672$$

The variance of the sum $X + Y$ is greater than the sum of the variances $\sigma^2_X + \sigma^2_Y$ because of the positive correlation between SAT Math scores and SAT Verbal scores. That is, $X$ and $Y$ tend to move up together and down together, which increases the variability of their sum. Find the standard deviation from the variance,

$$\sigma_{X+Y} = \sqrt{43{,}672} = 209$$

Total SAT scores had mean 1026 and standard deviation 209.

There are situations where we need to combine several of our rules to find means and standard deviations. Here is an example.

**EXAMPLE**

**4.38 Investing in Treasury bills and an index fund.** Zadie has invested 20% of her funds in Treasury bills and 80% in an "index fund" that represents all U.S. common stocks. The rate of return of an investment over a time period is the percent change in the price during the time period, plus any income received. If $X$ is the annual return on T-bills and $Y$ the annual return on stocks, the portfolio rate of return is

$$R = 0.2X + 0.8Y$$

The returns $X$ and $Y$ are random variables because they vary from year to year. Based on annual returns between 1950 and 2003, we have

$$X = \text{annual return on T-bills} \qquad \mu_X = 5.0\% \quad \sigma_X = 2.9\%$$

$$Y = \text{annual return on stocks} \qquad \mu_Y = 13.2\% \quad \sigma_Y = 17.6\%$$

$$\text{Correlation between } X \text{ and } Y \qquad \rho = -0.11$$

Stocks had higher returns than T-bills on the average, but the standard deviations show that returns on stocks varied much more from year to year. That is, the risk of investing in stocks is greater than the risk for T-bills because their returns are less predictable.

For the return $R$ on Zadie's portfolio of 20% T-bills and 80% stocks,

$$R = 0.2X + 0.8Y$$

$$\mu_R = 0.2\mu_X + 0.8\mu_Y$$

$$= (0.2 \times 5.0) + (0.8 \times 13.2) = 11.56\%$$

To find the variance of the portfolio return, combine Rules 1 and 3. Use the fact that, for example, the variance of $0.2X$ is $(0.2)^2$ times the variance of $X$. Also use the fact that changing scales does not change the correlation, so that the correlation between $0.2X$ and $0.8Y$ is the same as the correlation between $X$ and $Y$.

$$\sigma_R^2 = \sigma_{0.2X}^2 + \sigma_{0.8Y}^2 + 2\rho\sigma_{0.2X}\sigma_{0.8Y}$$

$$= (0.2)^2\sigma_X^2 + (0.8)^2\sigma_Y^2 + 2\rho(0.2 \times \sigma_X)(0.8 \times \sigma_Y)$$

$$= (0.2)^2(2.9)^2 + (0.8)^2(17.6)^2 + (2)(-0.11)(0.2 \times 2.9)(0.8 \times 17.6)$$

$$= 196.786$$

$$\sigma_R = \sqrt{196.786} = 14.03\%$$

The portfolio has a smaller mean return than an all-stock portfolio, but it is also less risky. That's why Zadie put some funds into Treasury bills.

## SECTION 4.4   Summary

The probability distribution of a random variable $X$, like a distribution of data, has a **mean $\mu_X$** and a **standard deviation $\sigma_X$**.

The **law of large numbers** says that the average of the values of $X$ observed in many trials must approach $\mu$.

The **mean $\mu$** is the balance point of the probability histogram or density curve. If $X$ is discrete with possible values $x_i$ having probabilities $p_i$, the mean is the average of the values of $X$, each weighted by its probability:

$$\mu_X = x_1p_1 + x_2p_2 + \cdots + x_kp_k$$

The **variance $\sigma_X^2$** is the average squared deviation of the values of the variable from their mean. For a discrete random variable,

$$\sigma_X^2 = (x_1 - \mu)^2p_1 + (x_2 - \mu)^2p_2 + \cdots + (x_k - \mu)^2p_k$$

The **standard deviation** $\sigma_X$ is the square root of the variance. The standard deviation measures the variability of the distribution about the mean. It is easiest to interpret for Normal distributions.

The mean and variance of a continuous random variable can be computed from the density curve, but to do so requires more advanced mathematics.

The means and variances of random variables obey the following rules. If $a$ and $b$ are fixed numbers, then

$$\mu_{a+bX} = a + b\mu_X$$
$$\sigma^2_{a+bX} = b^2\sigma^2_X$$

If $X$ and $Y$ are any two random variables having correlation $\rho$, then

$$\mu_{X+Y} = \mu_X + \mu_Y$$
$$\sigma^2_{X+Y} = \sigma^2_X + \sigma^2_Y + 2\rho\sigma_X\sigma_Y$$
$$\sigma^2_{X-Y} = \sigma^2_X + \sigma^2_Y - 2\rho\sigma_X\sigma_Y$$

If $X$ and $Y$ are **independent,** then $\rho = 0$. In this case,

$$\sigma^2_{X+Y} = \sigma^2_X + \sigma^2_Y$$
$$\sigma^2_{X-Y} = \sigma^2_X + \sigma^2_Y$$

To find the standard deviation, take the square root of the variance.

## SECTION 4.4 Exercises

*For Exercise 4.67, see page 271; for Exercise 4.68, see page 275; for Exercises 4.69 and 4.70, see page 279; and for Exercise 4.71, see page 281.*

**4.72 Mean of the distribution for the number of aces.** In Exercise 4.50 you examined the probability distribution for the number of aces when you are dealt two cards in the game of Texas hold 'em. Let $X$ represent the number of aces in a randomly selected deal of two cards in this game. Here is the probability distribution for the random variable $X$:

| Value of $X$ | 0 | 1 | 2 |
|---|---|---|---|
| Probability | 0.559 | 0.382 | 0.059 |

Find $\mu_X$, the mean of the probability distribution of $X$.

**4.73 Mean of the grade distribution.** Example 4.22 gives the distribution of grades (A = 4, B = 3, and so

on) in English 210 at North Carolina State University as

| Value of $X$ | 0 | 1 | 2 | 3 | 4 |
|---|---|---|---|---|---|
| Probability | 0.05 | 0.04 | 0.20 | 0.40 | 0.31 |

Find the average (that is, the mean) grade in this course.

**4.74 Mean of the distributions of errors.** Typographical and spelling errors can be either "nonword errors" or "word errors." A nonword error is not a real word, as when "the" is typed as "teh." A word error is a real word, but not the right word, as when "lose" is typed as "loose." When undergraduates are asked to write a 250-word essay (without spell-checking), the number of nonword errors has the following distribution:

| Errors | 0 | 1 | 2 | 3 | 4 |
|---|---|---|---|---|---|
| Probability | 0.1 | 0.3 | 0.3 | 0.2 | 0.1 |

The number of word errors has this distribution:

| Errors | 0 | 1 | 2 | 3 |
|---|---|---|---|---|
| Probability | 0.4 | 0.3 | 0.2 | 0.1 |

What are the mean numbers of nonword errors and word errors in an essay?

**4.75** **Means of the numbers of rooms in housing units.** How do rented housing units differ from units occupied by their owners? Exercise 4.53 (page 268) gives the distributions of the number of rooms for owner-occupied units and renter-occupied units in San Jose, California. Find the mean number of rooms for both types of housing unit. How do the means reflect the differences between the distributions that you found in Exercise 4.53?

**4.76** **Find the mean of the sum.** Figure 4.12 (page 269) displays the density curve of the sum $Y = X_1 + X_2$ of two independent random numbers, each uniformly distributed between 0 and 1.

(a) The mean of a continuous random variable is the balance point of its density curve. Use this fact to find the mean of $Y$ from Figure 4.12.

(b) Use the same fact to find the means of $X_1$ and $X_2$. (They have the density curve pictured in Figure 4.9, page 264.) Verify that the mean of $Y$ is the sum of the mean of $X_1$ and the mean of $X_2$.

**4.77** **Standard deviations of numbers of rooms in housing units.** Which of the two distributions of room counts appears more spread out in the probability histograms you made in Exercise 4.53 (page 268)? Why? Find the standard deviation for both distributions. The standard deviation provides a numerical measure of spread.

**4.78** **The effect of correlation.** Find the mean and standard deviation of the total number of errors (nonword errors plus word errors) in an essay if the error counts have the distributions given in Exercise 4.74 and

(a) the counts of nonword and word errors are independent.

(b) students who make many nonword errors also tend to make many word errors, so that the correlation between the two error counts is 0.4.

**4.79** **Means and variances of sums.** The rules for means and variances allow you to find the mean and variance of a sum of random variables without first finding the distribution of the sum, which is usually much harder to do.

(a) A single toss of a balanced coin has either 0 or 1 head, each with probability 1/2. What are the mean and standard deviation of the number of heads?

(b) Toss a coin four times. Use the rules for means and variances to find the mean and standard deviation of the total number of heads.

(c) Example 4.23 (page 261) finds the distribution of the number of heads in four tosses. Find the mean and standard deviation from this distribution. Your results in (b) and (c) should agree.

**4.80** CHALLENGE **Toss a 4-sided die twice.** Role-playing games like Dungeons & Dragons use many different types of dice. Suppose that a four-sided die has faces marked 1, 2, 3, 4. The intelligence of a character is determined by rolling this die twice and adding 1 to the sum of the spots. The faces are equally likely and the two rolls are independent. What is the average (mean) intelligence for such characters? How spread out are their intelligences, as measured by the standard deviation of the distribution?

**4.81** **A mechanical assembly.** A mechanical assembly (Figure 4.15) consists of a rod with a bearing on each end. The three parts are manufactured independently, and all vary a bit from part to part. The length of the rod has mean 12 centimeters (cm) and standard deviation 0.004 millimeters (mm). The length of a bearing has mean 2 cm and standard deviation 0.001 mm. What are the mean and standard deviation of the total length of the assembly?

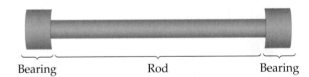

Bearing                 Rod                 Bearing

**FIGURE 4.15** Sketch of a mechanical assembly, for Exercise 4.81.

**4.82** **Sums of Normal random variables.** Continue your work in the previous exercise. Dimensions of mechanical parts are often roughly Normal. According to the 68–95–99.7 rule, 95% of rods have lengths within $\pm d_1$ of 12 cm and 95% of bearings have lengths within $\pm d_2$ of 2 cm.

(a) What are the values of $d_1$ and $d_2$? These are often called the "natural tolerances" of the parts.

(b) Statistical theory says that any sum of independent Normal random variables has a

Note that this agrees with the result obtained from the "Liberal Arts" row of Table 4.1:

$$P(\text{A grade} \mid \text{liberal arts course}) = \frac{2142}{6300} = 0.34$$

### USE YOUR KNOWLEDGE

**4.99 Find the conditional probability.** Refer to Table 4.1. What is the conditional probability that a grade is a B, given that it comes from Engineering and Physical Sciences? Find the answer by dividing two numbers from Table 4.1 and using the multiplication rule according to the method in Example 4.45.

## General multiplication rules

The definition of conditional probability reminds us that in principle all probabilities, including conditional probabilities, can be found from the assignment of probabilities to events that describe random phenomena. More often, however, conditional probabilities are part of the information given to us in a probability model, and the multiplication rule is used to compute $P(A \text{ and } B)$. This rule extends to more than two events.

The union of a collection of events is the event that *any* of them occur. Here is the corresponding term for the event that *all* of them occur.

---

INTERSECTION

The **intersection** of any collection of events is the event that *all* of the events occur.

---

To extend the multiplication rule to the probability that all of several events occur, the key is to condition each event on the occurrence of *all* of the preceding events. For example, the intersection of three events $A$, $B$, and $C$ has probability

$$P(A \text{ and } B \text{ and } C) = P(A)P(B \mid A)P(C \mid A \text{ and } B)$$

**4.46 High school athletes and professional careers.** Only 5% of male high school basketball, baseball, and football players go on to play at the college level. Of these, only 1.7% enter major league professional sports. About 40% of the athletes who compete in college and then reach the pros have a career of more than 3 years.[25] Define these events:

$$A = \{\text{competes in college}\}$$

$$B = \{\text{competes professionally}\}$$

$$C = \{\text{pro career longer than 3 years}\}$$

What is the probability that a high school athlete competes in college and then goes on to have a pro career of more than 3 years? We know that

$$P(A) = 0.05$$

$$P(B \mid A) = 0.017$$

$$P(C \mid A \text{ and } B) = 0.4$$

The probability we want is therefore

$$P(A \text{ and } B \text{ and } C) = P(A)P(B \mid A)P(C \mid A \text{ and } B)$$

$$= 0.05 \times 0.017 \times 0.4 = 0.00034$$

Only about 3 of every 10,000 high school athletes can expect to compete in college and have a professional career of more than 3 years. High school students would be wise to concentrate on studies rather than on unrealistic hopes of fortune from pro sports.

## Tree diagrams

Probability problems often require us to combine several of the basic rules into a more elaborate calculation. Here is an example that illustrates how to solve problems that have several stages.

**EXAMPLE**

**4.47 Online chat rooms.** Online chat rooms are dominated by the young. Teens are the biggest users. If we look only at adult Internet users (aged 18 and over), 47% of the 18 to 29 age group chat, as do 21% of the 30 to 49 age group and just 7% of those 50 and over. To learn what percent of all Internet users participate in chat, we also need the age breakdown of users. Here it is: 29% of adult Internet users are 18 to 29 years old (event $A_1$), another 47% are 30 to 49 (event $A_2$), and the remaining 24% are 50 and over (event $A_3$).[26]

What is the probability that a randomly chosen user of the Internet participates in chat rooms (event $C$)? To find out, use the **tree diagram** in Figure 4.20 to organize your thinking. Each segment in the tree is one stage of the problem. Each complete branch shows a path through the two stages. The probability written on each segment is the conditional probability of an Internet user following that segment, given that he or she has reached the node from which it branches.

Starting at the left, an Internet user falls into one of the three age groups. The probabilities of these groups

$$P(A_1) = 0.29 \qquad P(A_2) = 0.47 \qquad P(A_3) = 0.24$$

mark the leftmost branches in the tree. Conditional on being 18 to 29 years old, the probability of participating in chat is $P(C \mid A_1) = 0.47$. So the conditional probability of *not* participating is

$$P(C^c \mid A_1) = 1 - 0.47 = 0.53$$

tree diagram

# Sampling Distributions

The heights of young women are approximately Normal. See Example 5.1.

**5.1 Sampling Distributions for Counts and Proportions**

**5.2 The Sampling Distribution of a Sample Mean**

## Introduction

Statistical inference draws conclusions about a population or process on the basis of data. The data are summarized by *statistics* such as means, proportions, and the slopes of least-squares regression lines. When the data are produced by random sampling or randomized experimentation, a statistic is a random variable that obeys the laws of probability theory. *Sampling distributions* of statistics provide the link between probability and data. A sampling distribution shows how a statistic would vary in repeated data production. That is, a sampling distribution is a probability distribution that answers the question "What would happen if we did this many times?" The sampling distribution tells us about the results we are likely to see if, for example, we survey a sample of 2000 college students. In Section 3.3 we simulated a large number of random samples to illustrate the idea of a sampling distribution.

**LOOK BACK**
**sampling distribution,**
**page 215**

The accuracy of the Normal approximations improves as the sample size $n$ increases. They are most accurate for any fixed $n$ when $p$ is close to $1/2$, and least accurate when $p$ is near 0 or 1. You can compare binomial distributions with their Normal approximations by using the *Normal Approximation to Binomial* applet. This applet allows you to change $n$ or $p$ while watching the effect on the binomial probability histogram and the Normal curve that approximates it.

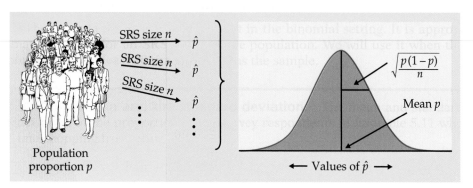

**FIGURE 5.4** The sampling distribution of a sample proportion $\hat{p}$ is approximately Normal with mean $p$ and standard deviation $\sqrt{p(1-p)/n}$.

Figure 5.4 summarizes the distribution of a sample proportion in a form that emphasizes the big idea of a sampling distribution. Sampling distributions answer the question "What would happen if we took many samples from the same population?"

- Keep taking random samples of size $n$ from a population that contains proportion $p$ of successes.

- Find the sample proportion $\hat{p}$ for each sample.

- Collect all the $\hat{p}$'s and display their distribution.

That's the sampling distribution of $\hat{p}$.

**EXAMPLE**

**5.13 Compare the Normal approximation with the exact calculation.**
Let's compare the Normal approximation for the calculation of Example 5.11 with the exact calculation from software. We want to calculate $P(\hat{p} \geq 0.58)$ when the sample size is $n = 2500$ and the population proportion is $p = 0.6$. Example 5.12 shows that

$$\mu_{\hat{p}} = p = 0.6$$

$$\sigma_{\hat{p}} = \sqrt{\frac{p(1-p)}{n}} = 0.0098$$

Act as if $\hat{p}$ were Normal with mean 0.6 and standard deviation 0.0098. The approximate probability, as illustrated in Figure 5.5, is

$$P(\hat{p} \geq 0.58) = P\left(\frac{\hat{p} - 0.6}{0.0098} \geq \frac{0.58 - 0.6}{0.0098}\right)$$

$$\doteq P(Z \geq -2.04) = 0.9793$$

That is, about 98% of all samples have a sample proportion that is at least 0.58. Because the sample was large, this Normal approximation is quite accurate. It misses the software value 0.9802 by only 0.0009.

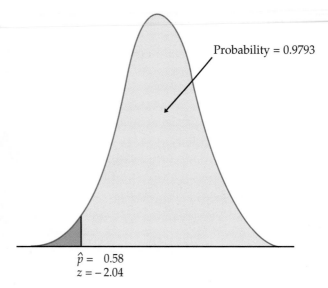

Probability = 0.9793

$\hat{p} = 0.58$
$z = -2.04$

**FIGURE 5.5** The Normal probability calculation for Example 5.13.

---

EXAMPLE

**5.14 Using the Normal appoximation.** The audit described in Example 5.6 examined an SRS of 150 sales records for compliance with sales tax laws. In fact, 8% of all the company's sales records have an incorrect sales tax classification. The count $X$ of bad records in the sample has approximately the $B(150, 0.08)$ distribution.

According to the Normal approximation to the binomial distributions, the count $X$ is approximately Normal with mean and standard deviation

$$\mu_X = np = (150)(0.08) = 12$$

$$\sigma_X = \sqrt{np(1-p)} = \sqrt{(150)(0.08)(0.92)}$$
$$= 3.3226$$

The Normal approximation for the probability of no more than 10 misclassified records is the area to the left of $X = 10$ under the Normal curve. Using Table A,

$$P(X \le 10) = P\left(\frac{X-12}{3.3226} \le \frac{10-12}{3.3226}\right)$$
$$\doteq P(Z \le -0.60) = 0.2743$$

Software tells us that the actual binomial probability that no more than 10 of the records in the sample are misclassified is $P(X \le 10) = 0.3384$. The Normal approximation is only roughly accurate. Because $np = 12$, this combination of $n$ and $p$ is close to the border of the values for which we are willing to use the approximation.

The distribution of the count of bad records in a sample of 15 is distinctly non-Normal, as Figure 5.2 showed. When we increase the sample size to 150, however, the shape of the binomial distribution becomes roughly Normal.

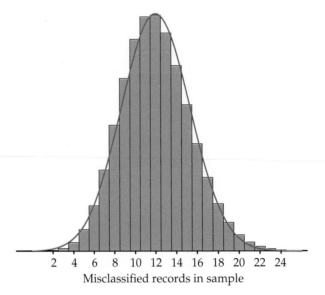

**FIGURE 5.6** Probability histogram and Normal approximation for the binomial distribution with $n = 150$ and $p = 0.08$, for Example 5.14.

Misclassified records in sample

Figure 5.6 displays the probability histogram of the binomial distribution with the density curve of the approximating Normal distribution superimposed. Both distributions have the same mean and standard deviation, and both the area under the histogram and the area under the curve are 1. The Normal curve fits the histogram reasonably well. Look closely: the histogram is slightly skewed to the right, a property that the symmetric Normal curve can't match.

## USE YOUR KNOWLEDGE

**5.7** **Use the Normal approximation.** Suppose we toss a fair coin 100 times. Use the Normal approximation to find the probability that the sample proportion is

(a) between 0.4 and 0.6.        (b) between 0.45 and 0.55.

## The continuity correction*

Figure 5.7 illustrates an idea that greatly improves the accuracy of the Normal approximation to binomial probabilities. The binomial probability $P(X \leq 10)$ is the area of the histogram bars for values 0 to 10. The bar for $X = 10$ actually extends from 9.5 to 10.5. Because the discrete binomial distribution puts probability only on whole numbers, the probabilities $P(X \leq 10)$ and $P(X \leq 10.5)$ are the same. The Normal distribution spreads probability continuously, so these two Normal probabilities are different. The Normal approximation is more accurate if we consider $X = 10$ to extend from 9.5 to 10.5, matching the bar in the probability histogram.

The event $\{X \leq 10\}$ includes the outcome $X = 10$. Figure 5.7 shades the area under the Normal curve that matches all the histogram bars for outcomes 0 to

---

*This material can be omitted if desired.

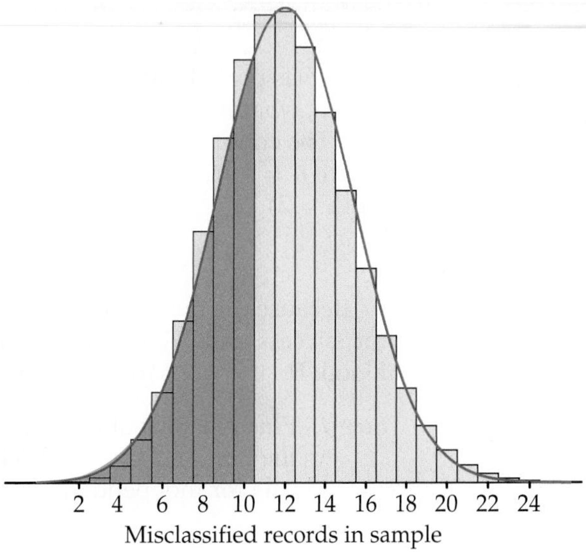
2 4 6 8 10 12 14 16 18 20 22 24
Misclassified records in sample

**FIGURE 5.7** Area under the Normal approximation curve for the probability in Example 5.14.

10, bounded on the right not by 10, but by 10.5. So $P(X \leq 10)$ is calculated as $P(X \leq 10.5)$. On the other hand, $P(X < 10)$ excludes the outcome $X = 10$, so we exclude the entire interval from 9.5 to 10.5 and calculate $P(X \leq 9.5)$ from the Normal table. Here is the result of the Normal calculation in Example 5.14 improved in this way:

$$\begin{aligned} P(X \leq 10) &= P(X \leq 10.5) \\ &= P\left(\frac{X - 12}{3.3226} \leq \frac{10.5 - 12}{3.3226}\right) \\ &\doteq P(Z \leq -0.45) = 0.3264 \end{aligned}$$

The improved approximation misses the binomial probability by only 0.012. Acting as though a whole number occupies the interval from 0.5 below to 0.5 above the number is called the **continuity correction** to the Normal approximation. If you need accurate values for binomial probabilities, try to use software to do exact calculations. If no software is available, use the continuity correction unless $n$ is very large. Because most statistical purposes do not require extremely accurate probability calculations, we do not emphasize use of the continuity correction.

continuity correction

## Binomial formula*

We can find a formula for the probability that a binomial random variable takes any value by adding probabilities for the different ways of getting exactly that many successes in $n$ observations. Here is the example we will use to show the idea.

*The formula for binomial probabilities is useful in many settings, but we will not need it in our study of statistical inference. This section can therefore be omitted if desired.

**EXAMPLE**

**5.15 Blood types of children.** Each child born to a particular set of parents has probability 0.25 of having blood type O. If these parents have 5 children, what is the probability that exactly 2 of them have type O blood?

The count of children with type O blood is a binomial random variable $X$ with $n = 5$ tries and probability $p = 0.25$ of a success on each try. We want $P(X = 2)$.

Because the method doesn't depend on the specific example, we will use "S" for success and "F" for failure. In Example 5.15, "S" would stand for type O blood. Do the work in two steps.

*Step 1:* Find the probability that a specific 2 of the 5 tries give successes, say the first and the third. This is the outcome SFSFF. The multiplication rule for independent events tells us that

$$P(\text{SFSFF}) = P(\text{S})P(\text{F})P(\text{S})P(\text{F})P(\text{F})$$
$$= (0.25)(0.75)(0.25)(0.75)(0.75)$$
$$= (0.25)^2(0.75)^3$$

*Step 2:* Observe that the probability of *any one* arrangement of 2 S's and 3 F's has this same probability. That's true because we multiply together 0.25 twice and 0.75 three times whenever we have 2 S's and 3 F's. The probability that $X = 2$ is the probability of getting 2 S's and 3 F's in any arrangement whatsoever. Here are all the possible arrangements:

SSFFF   SFSFF   SFFSF   SFFFS   FSSFF
FSFSF   FSFFS   FFSSF   FFSFS   FFFSS

There are 10 of them, all with the same probability. The overall probability of 2 successes is therefore

$$P(X = 2) = 10(0.25)^2(0.75)^3 = 0.2637$$

The pattern of this calculation works for any binomial probability. To use it, we need to be able to count the number of arrangements of $k$ successes in $n$ observations without actually listing them. We use the following fact to do the counting.

---

**BINOMIAL COEFFICIENT**

The number of ways of arranging $k$ successes among $n$ observations is given by the **binomial coefficient**

$$\binom{n}{k} = \frac{n!}{k!\,(n-k)!}$$

for $k = 0, 1, 2, \ldots, n$.

---

factorial   The formula for binomial coefficients uses the **factorial** notation. The factorial $n!$ for any positive whole number $n$ is

$$n! = n \times (n-1) \times (n-2) \times \cdots \times 3 \times 2 \times 1$$

Also, $0! = 1$. Notice that the larger of the two factorials in the denominator of a binomial coefficient will cancel much of the $n!$ in the numerator. For example, the binomial coefficient we need for Example 5.15 is

$$\binom{5}{2} = \frac{5!}{2!\,3!}$$

$$= \frac{(5)(4)(3)(2)(1)}{(2)(1) \times (3)(2)(1)}$$

$$= \frac{(5)(4)}{(2)(1)} = \frac{20}{2} = 10$$

This agrees with our previous calculation.

*The notation* $\binom{n}{k}$ *is not related to the fraction* $\frac{n}{k}$. A helpful way to remember its meaning is to read it as "binomial coefficient $n$ choose $k$." Binomial coefficients have many uses in mathematics, but we are interested in them only as an aid to finding binomial probabilities. The binomial coefficient $\binom{n}{k}$ counts the number of ways in which $k$ successes can be distributed among $n$ observations. The binomial probability $P(X = k)$ is this count multiplied by the probability of any specific arrangement of the $k$ successes. Here is the formula we seek.

---

### BINOMIAL PROBABILITY

If $X$ has the binomial distribution $B(n, p)$ with $n$ observations and probability $p$ of success on each observation, the possible values of $X$ are $0, 1, 2, \ldots, n$. If $k$ is any one of these values, the **binomial probability** is

$$P(X = k) = \binom{n}{k} p^k (1-p)^{n-k}$$

---

Here is an example of the use of the binomial probability formula.

**EXAMPLE**

**5.16 Using the binomial probability formula.** The number $X$ of misclassified sales records in the auditor's sample in Example 5.8 has the $B(15, 0.08)$ distribution. The probability of finding no more than 1 misclassified record is

$$P(X \leq 1) = P(X = 0) + P(X = 1)$$

$$= \binom{15}{0}(0.08)^0(0.92)^{15} + \binom{15}{1}(0.08)^1(0.92)^{14}$$

$$= \frac{15!}{0!\,15!}(1)(0.2863) + \frac{15!}{1!\,14!}(0.08)(0.3112)$$

$$= (1)(1)(0.2863) + (15)(0.08)(0.3112)$$

$$= 0.2863 + 0.3734 = 0.6597$$

The calculation used the facts that $0! = 1$ and that $a^0 = 1$ for any number $a \neq 0$. The result agrees with that obtained from Table C in Example 5.8.

## USE YOUR KNOWLEDGE

**5.8** **A bent coin.** A coin is slightly bent, and as a result the probability of a head is 0.52. Suppose that you toss the coin four times.

(a) Use the binomial formula to find the probability of 3 or more heads.

(b) Compare your answer with the one that you would obtain if the coin were fair.

## SECTION 5.1 Summary

A **count** $X$ of successes has the **binomial distribution** $B(n, p)$ in the **binomial setting:** there are $n$ trials, all independent, each resulting in a success or a failure, and each having the same probability $p$ of a success.

**Binomial probabilities** are most easily found by software. There is an exact formula that is practical for calculations when $n$ is small. Table C contains binomial probabilities for some values of $n$ and $p$. For large $n$, you can use the Normal approximation.

The binomial distribution $B(n, p)$ is a good approximation to the **sampling distribution of the count of successes** in an SRS of size $n$ from a large population containing proportion $p$ of successes. We will use this approximation when the population is at least 20 times larger than the sample.

The mean and standard deviation of a **binomial count** $X$ and a **sample proportion** of successes $\hat{p} = X/n$ are

$$\mu_X = np \qquad\qquad \mu_{\hat{p}} = p$$

$$\sigma_X = \sqrt{np(1-p)} \qquad \sigma_{\hat{p}} = \sqrt{\frac{p(1-p)}{n}}$$

The sample proportion $\hat{p}$ is therefore an unbiased estimator of the population proportion $p$.

The **Normal approximation** to the binomial distribution says that if $X$ is a count having the $B(n, p)$ distribution, then when $n$ is large,

$$X \text{ is approximately } N(np, \sqrt{np(1-p)})$$

$$\hat{p} \text{ is approximately } N\left(p, \sqrt{\frac{p(1-p)}{n}}\right)$$

We will use these approximations when $np \geq 10$ and $n(1-p) \geq 10$. The **continuity correction** improves the accuracy of the Normal approximations.

The exact **binomial probability formula** is

$$P(X = k) = \binom{n}{k} p^k (1-p)^{n-k}$$

where the possible values of $X$ are $k = 0, 1, \ldots, n$. The binomial probability formula uses the **binomial coefficient**

$$\binom{n}{k} = \frac{n!}{k!\,(n-k)!}$$

Here the **factorial** $n!$ is

$$n! = n \times (n-1) \times (n-2) \times \cdots \times 3 \times 2 \times 1$$

for positive whole numbers $n$ and $0! = 1$. The binomial coefficient counts the number of ways of distributing $k$ successes among $n$ trials.

## SECTION 5.1   Exercises

*For Exercises 5.1 and 5.2, see page 314; for Exercises 5.3 and 5.4, see page 315; for Exercise 5.5, see page 319; for Exercise 5.6, see page 322; for Exercise 5.7, see page 326; and for Exercise 5.8, see page 330.*

Most binomial probability calculations required in these exercises can be done by using Table C or the Normal approximation. Your instructor may request that you use the binomial probability formula or software. In exercises requiring the Normal approximation, you should use the continuity correction if you studied that topic.

5.9   **What is wrong?** Explain what is wrong in each of the following scenarios.

(a) If you toss a fair coin three times and a head appears each time, then the next toss is more likely to be a tail than a head.

(b) If you toss a fair coin three times and a head appears each time, then the next toss is more likely to be a head than a tail.

(c) $\hat{p}$ is one of the parameters for a binomial distribution.

5.10   **What is wrong?** Explain what is wrong in each of the following scenarios.

(a) In the binomial setting $X$ is a proportion.

(b) The variance for a binomial count is $\sqrt{p(1-p)/n}$.

(c) The Normal approximation to the binomial distribution is always accurate when $n$ is greater than 1000.

5.11   **Should you use the binomial distribution?** In each situation below, is it reasonable to use a

binomial distribution for the random variable $X$? Give reasons for your answer in each case. If a binomial distribution applies, give the values of $n$ and $p$.

(a) A poll of 200 college students asks whether or not you are usually irritable in the morning. $X$ is the number who reply that they are usually irritable in the morning.

(b) You toss a fair coin until a head appears. $X$ is the count of the number of tosses that you make.

(c) Most calls made at random by sample surveys don't succeed in talking with a live person. Of calls to New York City, only 1/12 succeed. A survey calls 500 randomly selected numbers in New York City. $X$ is the number that reach a live person.

5.12   **Should you use the binomial distribution?** In each situation below, is it reasonable to use a binomial distribution for the random variable $X$? Give reasons for your answer in each case.

(a) A random sample of students in a fitness study. $X$ is the mean systolic blood pressure of the sample.

(b) A manufacturer of running shoes picks a random sample of the production of shoes each day for a detailed inspection. Today's sample of 20 pairs of shoes includes 1 pair with a defect.

(c) A nutrition study chooses an SRS of college students. They are asked whether or not they usually eat at least five servings of fruits or vegetables per day. $X$ is the number who say that they do.

5.13   **Typographic errors.** Typographic errors in a text are either nonword errors (as when "the" is typed as "teh") or word errors that result in a real but incorrect word. Spell-checking software will catch nonword errors but not word errors. Human proofreaders catch 70% of word errors. You ask a fellow student to proofread an essay in which you have deliberately made 10 word errors.

(a) If the student matches the usual 70% rate, what is the distribution of the number of errors caught? What is the distribution of the number of errors missed?

(b) Missing 4 or more out of 10 errors seems a poor performance. What is the probability that a proofreader who catches 70% of word errors misses 4 or more out of 10?

5.14 **Visits to Web sites.** What kinds of Web sites do males aged 18 to 34 visit most often? Pornographic sites take first place, but about 50% of male Internet users in this age group visit an auction site such as eBay at least once a month.[2] Interview a random sample of 15 male Internet users aged 18 to 34.

(a) What is the distribution of the number who have visited an online auction site in the past month?

(b) What is the probability that at least 8 of the 15 have visited an auction site in the past month?

5.15 **Typographic errors.** Return to the proofreading setting of Exercise 5.13.

(a) What is the mean number of errors caught? What is the mean number of errors missed? You see that these two means must add to 10, the total number of errors.

(b) What is the standard deviation $\sigma$ of the number of errors caught?

(c) Suppose that a proofreader catches 90% of word errors, so that $p = 0.9$. What is $\sigma$ in this case? What is $\sigma$ if $p = 0.99$? What happens to the standard deviation of a binomial distribution as the probability of a success gets close to 1?

5.16 **Visits to Web sites.** Suppose that 50% of male Internet users aged 18 to 34 have visited an auction site at least once in the past month.

(a) If you interview 15 at random, what is the mean of the count $X$ who have visited an auction site? What is the mean of the proportion $\hat{p}$ in your sample who have visited an auction site?

(b) Repeat the calculations in (a) for samples of size 150 and 1500. What happens to the mean count of successes as the sample size increases? What happens to the mean proportion of successes?

5.17 **Typographic errors.** In the proofreading setting of Exercise 5.13, what is the smallest number of misses $m$ with $P(X \geq m)$ no larger than 0.05? You might consider $m$ or more misses as evidence that a proofreader actually catches fewer than 70% of word errors.

5.18 **Attitudes toward drinking and behavior studies.** Some of the methods in this section are approximations rather than exact probability results. We have given rules of thumb for safe use of these approximations.

(a) You are interested in attitudes toward drinking among the 75 members of a fraternity. You choose 30 members at random to interview. One question is "Have you had five or more drinks at one time during the last week?" Suppose that in fact 30% of the 75 members would say "Yes." Explain why you *cannot* safely use the $B(30, 0.3)$ distribution for the count $X$ in your sample who say "Yes."

(b) The National AIDS Behavioral Surveys found that 0.2% (that's 0.002 as a decimal fraction) of adult heterosexuals had both received a blood transfusion and had a sexual partner from a group at high risk of AIDS. Suppose that this national proportion holds for your region. Explain why you *cannot* safely use the Normal approximation for the sample proportion who fall in this group when you interview an SRS of 1000 adults.

5.19 **Random digits.** Each entry in a table of random digits like Table B has probability 0.1 of being a 0, and digits are independent of each other.

(a) What is the probability that a group of five digits from the table will contain at least one 5?

(b) What is the mean number of 5s in lines 40 digits long?

5.20 Use the *Probability* applet. The *Probability* applet simulates tosses of a coin. You can choose the number of tosses $n$ and the probability $p$ of a head. You can therefore use the applet to simulate binomial random variables.

The count of misclassified sales records in Example 5.8 (page 317) has the binomial distribution with $n = 15$ and $p = 0.08$. Set these values for the number of tosses and probability of heads in the applet. Table C shows that the probability of getting a sample with exactly 0 misclassified records is 0.2863. This is the long-run proportion of samples with exactly 1 bad record. Click "Toss" and "Reset" repeatedly to simulate 25 samples. Record the number of bad records (the count of heads) in each of the 25 samples. What proportion of the 25 samples had exactly 0 bad records? Remember that probability tells us only what happens in the long run.

**5.21 Inheritance of blood types.** Children inherit their blood type from their parents, with probabilities that reflect the parents' genetic makeup. Children of Juan and Maria each have probability 1/4 of having blood type A and inherit independently of each other. Juan and Maria plan to have 4 children; let $X$ be the number who have blood type A.

(a) What are $n$ and $p$ in the binomial distribution of $X$?

(b) Find the probability of each possible value of $X$, and draw a probability histogram for this distribution.

(c) Find the mean number of children with type A blood, and mark the location of the mean on your probability histogram.

**5.22 The ideal number of children.** "What do you think is the ideal number of children for a family to have?" A Gallup Poll asked this question of 1016 randomly chosen adults. Almost half (49%) thought two children was ideal.[3] Suppose that $p = 0.49$ is exactly true for the population of all adults. Gallup announced a margin of error of $\pm 3$ percentage points for this poll. What is the probability that the sample proportion $\hat{p}$ for an SRS of size $n = 1016$ falls between 0.46 and 0.52? You see that it is likely, but not certain, that polls like this give results that are correct within their margin of error. We will say more about margins of error in Chapter 6.

**5.23 Visiting a casino and betting on college sports.** A Gallup Poll finds that 30% of adults visited a casino in the past 12 months, and that 6% bet on college sports.[4] These results come from a random sample of 1011 adults. For an SRS of size $n = 1011$:

(a) What is the probability that the sample proportion $\hat{p}$ is between 0.28 and 0.32 if the population proportion is $p = 0.30$?

(b) What is the probability that the sample proportion $\hat{p}$ is between 0.04 and 0.08 if the population proportion is $p = 0.06$?

(c) How does the probability that $\hat{p}$ falls within $\pm 0.02$ of the true $p$ change as $p$ gets closer to 0?

**5.24 How do the results depend on the sample size?** Return to the Gallup Poll setting of Exercise 5.22. We are supposing that the proportion of all adults who think that two children is ideal is $p = 0.49$. What is the probability that a sample proportion $\hat{p}$ falls between 0.46 and 0.52 (that is, within $\pm 3$

percentage points of the true $p$) if the sample is an SRS of size $n = 300$? Of size $n = 5000$? Combine these results with your work in Exercise 5.22 to make a general statement about the effect of larger samples in a sample survey.

**5.25** CHALLENGE **A college alcohol study.** The Harvard College Alcohol Study finds that 67% of college students support efforts to "crack down on underage drinking." The study took a sample of almost 15,000 students, so the population proportion who support a crackdown is very close to $p = 0.67$.[5] The administration of your college surveys an SRS of 200 students and finds that 140 support a crackdown on underage drinking.

(a) What is the sample proportion who support a crackdown on underage drinking?

(b) If in fact the proportion of all students on your campus who support a crackdown is the same as the national 67%, what is the probability that the proportion in an SRS of 200 students is as large or larger than the result of the administration's sample?

(c) A writer in the student paper says that support for a crackdown is higher on your campus than nationally. Write a short letter to the editor explaining why the survey does not support this conclusion.

**5.26** CHALLENGE **How large a sample is needed?** The changing probabilities you found in Exercises 5.22 and 5.24 are due to the fact that the standard deviation of the sample proportion $\hat{p}$ gets smaller as the sample size $n$ increases. If the population proportion is $p = 0.49$, how large a sample is needed to reduce the standard deviation of $\hat{p}$ to $\sigma_{\hat{p}} = 0.004$? (The 68–95–99.7 rule then says that about 95% of all samples will have $\hat{p}$ within 0.01 of the true $p$.)

**5.27 A test for ESP.** In a test for ESP (extrasensory perception), the experimenter looks at cards that are hidden from the subject. Each card contains either a star, a circle, a wave, or a square. As the experimenter looks at each of 20 cards in turn, the subject names the shape on the card.

(a) If a subject simply guesses the shape on each card, what is the probability of a successful guess on a single card? Because the cards are independent, the count of successes in 20 cards has a binomial distribution.

(b) What is the probability that a subject correctly guesses at least 10 of the 20 shapes?

(c) In many repetitions of this experiment with a subject who is guessing, how many cards will the subject guess correctly on the average? What is the standard deviation of the number of correct guesses?

(d) A standard ESP deck actually contains 25 cards. There are five different shapes, each of which appears on 5 cards. The subject knows that the deck has this makeup. Is a binomial model still appropriate for the count of correct guesses in one pass through this deck? If so, what are $n$ and $p$? If not, why not?

**5.28 Admitting students to college.** A selective college would like to have an entering class of 950 students. Because not all students who are offered admission accept, the college admits more than 950 students. Past experience shows that about 75% of the students admitted will accept. The college decides to admit 1200 students. Assuming that students make their decisions independently, the number who accept has the $B(1200, 0.75)$ distribution. If this number is less than 950, the college will admit students from its waiting list.

(a) What are the mean and the standard deviation of the number $X$ of students who accept?

(b) The college does not want more than 950 students. What is the probability that more than 950 will accept?

(c) If the college decides to increase the number of admission offers to 1300, what is the probability that more than 950 will accept?

**5.29** CHALLENGE **Is the ESP result better than guessing?** When the ESP study of Exercise 5.27 discovers a subject whose performance appears to be better than guessing, the study continues at greater length. The experimenter looks at many cards bearing one of five shapes (star, square, circle, wave, and cross) in an order determined by random numbers. The subject cannot see the experimenter as he looks at each card in turn, in order to avoid any possible nonverbal clues. The answers of a subject who does not have ESP should be independent observations, each with probability 1/5 of success. We record 900 attempts.

(a) What are the mean and the standard deviation of the count of successes?

(b) What are the mean and standard deviation of the proportion of successes among the 900 attempts?

(c) What is the probability that a subject without ESP will be successful in at least 24% of 900 attempts?

(d) The researcher considers evidence of ESP to be a proportion of successes so large that there is only probability 0.01 that a subject could do this well or better by guessing. What proportion of successes must a subject have to meet this standard? (Example 1.32 shows how to do an inverse calculation for the Normal distribution that is similar to the type required here.)

**5.30** CHALLENGE **Scuba-diving trips.** The mailing list of an agency that markets scuba-diving trips to the Florida Keys contains 60% males and 40% females. The agency calls 30 people chosen at random from its list.

(a) What is the probability that 20 of the 30 are men? (Use the binomial probability formula.)

(b) What is the probability that the first woman is reached on the fourth call? (That is, the first 4 calls give MMMF.)

**5.31 Checking for problems with a sample survey.** One way of checking the effect of undercoverage, nonresponse, and other sources of error in a sample survey is to compare the sample with known demographic facts about the population. The 2000 census found that 23,772,494 of the 209,128,094 adults (aged 18 and over) in the United States called themselves "Black or African American."

(a) What is the population proportion $p$ of blacks among American adults?

(b) An opinion poll chooses 1200 adults at random. What is the mean number of blacks in such samples? (Explain the reasoning behind your calculation.)

(c) Use a Normal approximation to find the probability that such a sample will contain 100 or fewer blacks. Be sure to check that you can safely use the approximation.

**5.32** CHALLENGE **Show that these facts are true.** Use the definition of binomial coefficients to show that each of the following facts is true. Then restate each fact in words in terms of the number of ways that $k$ successes can be distributed among $n$ observations.

(a) $\binom{n}{n} = 1$ for any whole number $n \geq 1$.

(b) $\binom{n}{n-1} = n$ for any whole number $n \geq 1$.

(c) $\binom{n}{k} = \binom{n}{n-k}$ for any $n$ and $k$ with $k \leq n$.

**5.33 Multiple-choice tests.** Here is a simple probability model for multiple-choice tests. Suppose that each student has probability $p$ of correctly answering a question chosen at random from a universe of possible questions. (A strong student has a higher $p$ than a weak student.) The correctness of an answer to a question is independent of the correctness of answers to other questions. Jodi is a good student for whom $p = 0.85$.

(a) Use the Normal approximation to find the probability that Jodi scores 80% or lower on a 100-question test.

(b) If the test contains 250 questions, what is the probability that Jodi will score 80% or lower?

(c) How many questions must the test contain in order to reduce the standard deviation of Jodi's proportion of correct answers to half its value for a 100-item test?

(d) Laura is a weaker student for whom $p = 0.75$. Does the answer you gave in (c) for the standard deviation of Jodi's score apply to Laura's standard deviation also?

**5.34 Tossing a die.** You are tossing a balanced die that has probability 1/6 of coming up 1 on each toss.

Tosses are independent. We are interested in how long we must wait to get the first 1.

(a) The probability of a 1 on the first toss is 1/6. What is the probability that the first toss is not a 1 and the second toss is a 1?

(b) What is the probability that the first two tosses are not 1s and the third toss is a 1? This is the probability that the first 1 occurs on the third toss.

(c) Now you see the pattern. What is the probability that the first 1 occurs on the fourth toss? On the fifth toss?

**5.35 CHALLENGE The geometric distribution.** Generalize your work in Exercise 5.34. You have independent trials, each resulting in a success or a failure. The probability of a success is $p$ on each trial. The binomial distribution describes the count of successes in a fixed number of trials. Now the number of trials is not fixed; instead, continue until you get a success. The random variable $Y$ is the number of the trial on which the first success occurs. What are the possible values of $Y$? What is the probability $P(Y = k)$ for any of these values? (*Comment:* The distribution of the number of trials to the first success is called a **geometric distribution.**)

# 5.2 The Sampling Distribution of a Sample Mean

Counts and proportions are discrete random variables that describe categorical data. The statistics most often used to describe quantitative data, on the other hand, are continuous random variables. The sample mean, percentiles, and standard deviation are examples of statistics based on quantitative data. Statistical theory describes the sampling distributions of these statistics. In this section we will concentrate on the sample mean. Because sample means are just averages of observations, they are among the most common statistics.

**EXAMPLE**

**5.17 Sample means are approximately Normal.** Figure 5.8 illustrates two striking facts about the sampling distribution of a sample mean. Figure 5.8(a) displays the distribution of customer service call lengths for a bank service center for a month. There are more than 30,000 calls in this population.[6] (We omitted a few extreme outliers, calls that lasted more than 20 minutes.) The distribution is extremely skewed to the right. The population mean is $\mu = 173.95$ seconds.

Table 1.1 (page 8) contains the lengths of a sample of 80 calls from this population. The mean of these 80 calls is $\bar{x} = 196.6$ seconds. If we take more samples of size 80, we will get different values of $\bar{x}$. To find the sampling distribution of $\bar{x}$, take many random samples of size 80 and calculate $\bar{x}$

of $m$ observations from a control group. Suppose that the response to the treatment has the $N(\mu_X, \sigma_X)$ distribution and that the response of control subjects has the $N(\mu_Y, \sigma_Y)$ distribution. Inference about the difference $\mu_Y - \mu_X$ between the population means is based on the difference $\bar{y} - \bar{x}$ between the sample means in the two groups.

(a) Under the assumptions given, what is the distribution of $\bar{y}$? Of $\bar{x}$?

(b) What is the distribution of $\bar{y} - \bar{x}$?

5.62 **Investments in two funds.** Linda invests her money in a portfolio that consists of 70% Fidelity 500 Index Fund and 30% Fidelity Diversified International Fund. Suppose that in the long run the annual real return $X$ on the 500 Index Fund has mean 9% and standard deviation 19%, the annual real return $Y$ on the Diversified International Fund has mean 11% and standard deviation 17%, and the correlation between $X$ and $Y$ is 0.6.

(a) The return on Linda's portfolio is $R = 0.7X + 0.3Y$. What are the mean and standard deviation of $R$?

(b) The distribution of returns is typically roughly symmetric but with more extreme high and low observations than a Normal distribution. The average return over a number of years, however, is close to Normal. If Linda holds her portfolio for 20

years, what is the approximate probability that her average return is less than 5%?

(c) The calculation you just made is not overly helpful, because Linda isn't really concerned about the mean return $\bar{R}$. To see why, suppose that her portfolio returns 12% this year and 6% next year. The mean return for the two years is 9%. If Linda starts with $1000, how much does she have at the end of the first year? At the end of the second year? How does this amount compare with what she would have if both years had the mean return, 9%? Over 20 years, there may be a large difference between the ordinary mean $\bar{R}$ and the *geometric mean*, which reflects the fact that returns in successive years multiply rather than add.

5.63 **Concrete blocks and mortar.** You are building a wall from precast concrete blocks. Standard "8 inch" blocks are $7\frac{5}{8}$ inches high to allow for a $\frac{3}{8}$ inch layer of mortar under each row of blocks. In practice, the height of a block-plus-mortar row varies according to a Normal distribution with mean 8 inches and standard deviation 0.1 inch. Heights of successive rows are independent. Your wall has four rows of blocks. What is the distribution of the height of the wall? What is the probability that the height differs from the design height of 32 inches by more than half an inch?

## CHAPTER 5  Exercises

5.64 **The effect of sample size on the standard deviation.** Assume that the standard deviation in a very large population is 100.

(a) Calculate the standard deviation for the sample mean for samples of size 1, 4, 25, 100, 250, 500, 1000, and 5000.

(b) Graph your results with the sample size on the $x$ axis and the standard deviation on the $y$ axis.

(c) Summarize the relationship between the sample size and the standard deviation that you showed in your graph.

5.65 **Auto accidents.** The probability that a randomly chosen driver will be involved in an accident in the next year is about 0.2. This is based on the proportion of millions of drivers who have accidents. "Accident" includes things like crumpling a fender in your own driveway, not just highway accidents. Carlos, David, Jermaine, Ramon, and Scott are college students who live together in an off-campus apartment. Last year, 3 of the 5 had

accidents. What is the probability that 3 or more of 5 randomly chosen drivers have an accident in the same year? Why does your calculation not apply to drivers like the 5 students?

5.66 **SAT scores.** Example 4.37 (page 284) notes that the total SAT scores of high school seniors in a recent year had mean $\mu = 1026$ and standard deviation $\sigma = 209$. The distribution of SAT scores is roughly Normal.

(a) Julie scored 1110. If scores have a Normal distribution, what percentile of the distribution is this?

(b) Now consider the mean $\bar{x}$ of the scores of 80 randomly chosen students. If $\bar{x} = 1110$, what percentile of the sampling distribution of $\bar{x}$ is this?

(c) Which of your calculations, (a) or (b), is less accurate because SAT scores do not have an exactly Normal distribution? Explain your answer.

**5.67 Carpooling.** Although cities encourage carpooling to reduce traffic congestion, most vehicles carry only one person. For example, 70% of vehicles on the roads in the Minneapolis–St. Paul metropolitan area are occupied by just the driver.

(a) If you choose 12 vehicles at random, what is the probability that more than half (that is, 7 or more) carry just one person?

(b) If you choose 80 vehicles at random, what is the probability that more than half (that is, 41 or more) carry just one person?

**5.68 Common last names.** The Census Bureau says that the 10 most common names in the United States are (in order) Smith, Johnson, Williams, Jones, Brown, Davis, Miller, Wilson, Moore, and Taylor. These names account for 5.6% of all U.S. residents. Out of curiosity, you look at the authors of the textbooks for your current courses. There are 12 authors in all. Would you be surprised if none of the names of these authors were among the 10 most common? Give a probability to support your answer and explain the reasoning behind your calculation.

**5.69 Benford's law.** It is a striking fact that the first digits of numbers in legitimate records often follow a distribution known as Benford's law. Here it is:

| First digit | 1 | 2 | 3 | 4 | 5 | 6 | 7 | 8 | 9 |
|---|---|---|---|---|---|---|---|---|---|
| Proportion | 0.301 | 0.176 | 0.125 | 0.097 | 0.079 | 0.067 | 0.058 | 0.051 | 0.046 |

Fake records usually have fewer first digits 1, 2, and 3. What is the approximate probability, if Benford's law holds, that among 1000 randomly chosen invoices there are 560 or fewer in amounts with first digit 1, 2, or 3?

**5.70 Genetics of peas.** According to genetic theory, the blossom color in the second generation of a certain cross of sweet peas should be red or white in a 3:1 ratio. That is, each plant has probability 3/4 of having red blossoms, and the blossom colors of separate plants are independent.

(a) What is the probability that exactly 9 out of 12 of these plants have red blossoms?

(b) What is the mean number of red-blossomed plants when 120 plants of this type are grown from seeds?

(c) What is the probability of obtaining at least 80 red-blossomed plants when 120 plants are grown from seeds?

**5.71 The weight of a dozen eggs.** The weight of the eggs produced by a certain breed of hen is Normally distributed with mean 65 grams (g) and standard deviation 5 g. If cartons of such eggs can be considered to be SRSs of size 12 from the population of all eggs, what is the probability that the weight of a carton falls between 755 and 830 g?

**5.72 Losses of British aircraft in World War II.** Serving in a bomber crew in World War II was dangerous. The British estimated that the probability of an aircraft loss due to enemy action was 1/20 for each mission. A tour of duty for British airmen in Bomber Command was 30 missions. What is the probability that an airman would complete a tour of duty without being on an aircraft lost from enemy action?

**5.73 A survey of college women.** A sample survey interviews an SRS of 280 college women. Suppose (as is roughly true) that 70% of all college women have been on a diet within the last 12 months. What is the probability that 75% or more of the women in the sample have been on a diet?

**5.74 Plastic caps for motor oil containers.** A machine fastens plastic screw-on caps onto containers of motor oil. If the machine applies more torque than the cap can withstand, the cap will break. Both the torque applied and the strength of the caps vary. The capping-machine torque has the Normal distribution with mean 7.0 inch-pounds and standard deviation 0.9 inch-pounds. The cap strength (the torque that would break the cap) has the Normal distribution with mean 10.1 inch-pounds and standard deviation 1.2 inch-pounds.

(a) Explain why it is reasonable to assume that the cap strength and the torque applied by the machine are independent.

(b) What is the probability that a cap will break while being fastened by the capping machine?

**5.75 Colors of cashmere sweaters.** The unique colors of the cashmere sweaters your firm makes result from heating undyed yarn in a kettle with a dye liquor. The pH (acidity) of the liquor is critical for regulating dye uptake and hence the final color. There are 5 kettles, all of which receive dye liquor from a common source. Past data show that pH varies according to a Normal distribution with $\mu = 4.25$ and $\sigma = 0.135$. You use statistical process control to check the stability of the process. Twice each day, the pH of the liquor in each kettle is measured, each time giving a sample of size 5. The mean pH $\bar{x}$ is compared with "control limits" given by the 99.7 part of the 68–95–99.7 rule for Normal distributions, namely, $\mu_{\bar{x}} \pm 3\sigma_{\bar{x}}$. What are the numerical values of these control limits for $\bar{x}$?

**5.76** **CHALLENGE** **Learning a foreign language.** Does delaying oral practice hinder learning a foreign language? Researchers randomly assigned 25 beginning students of Russian to begin speaking practice immediately and another 25 to delay speaking for 4 weeks. At the end of the semester both groups took a standard test of comprehension of spoken Russian. Suppose that in the population of all beginning students, the test scores for early speaking vary according to the $N(32, 6)$ distribution and scores for delayed speaking have the $N(29, 5)$ distribution.

(a) What is the sampling distribution of the mean score $\bar{x}$ in the early-speaking group in many repetitions of the experiment? What is the sampling distribution of the mean score $\bar{y}$ in the delayed-speaking group?

(b) If the experiment were repeated many times, what would be the sampling distribution of the difference $\bar{y} - \bar{x}$ between the mean scores in the two groups?

(c) What is the probability that the experiment will find (misleadingly) that the mean score for delayed speaking is at least as large as that for early speaking?

**5.77** **CHALLENGE** **Summer employment of college students.** Suppose (as is roughly true) that 88% of college men and 82% of college women were employed last summer. A sample survey interviews SRSs of 400 college men and 400 college women. The two samples are of course independent.

(a) What is the approximate distribution of the proportion $\hat{p}_F$ of women who worked last summer? What is the approximate distribution of the proportion $\hat{p}_M$ of men who worked?

(b) The survey wants to compare men and women. What is the approximate distribution of the difference in the proportions who worked, $\hat{p}_M - \hat{p}_F$? Explain the reasoning behind your answer.

(c) What is the probability that in the sample a higher proportion of women than men worked last summer?

**5.78** **Income of working couples.** A study of working couples measures the income $X$ of the husband and the income $Y$ of the wife in a large number of couples in which both partners are employed. Suppose that you knew the means $\mu_X$ and $\mu_Y$ and the variances $\sigma_X^2$ and $\sigma_Y^2$ of both variables in the population.

(a) Is it reasonable to take the mean of the total income $X + Y$ to be $\mu_X + \mu_Y$? Explain your answer.

(b) Is it reasonable to take the variance of the total income to be $\sigma_X^2 + \sigma_Y^2$? Explain your answer.

**5.79** **CHALLENGE** **A random walk.** A particle moves along the line in a random walk. That is, the particle starts at the origin (position 0) and moves either right or left in independent steps of length 1. If the particle moves to the right with probability 0.6, its movement at the $i$th step is a random variable $X_i$ with distribution

$$P(X_i = 1) = 0.6$$
$$P(X_i = -1) = 0.4$$

The position of the particle after $k$ steps is the sum of these random movements,

$$Y = X_1 + X_2 + \cdots + X_k$$

Use the central limit theorem to find the approximate probability that the position of the particle after 500 steps is at least 200 to the right.

# Introduction to Inference

Undergraduate student loan debt has been increasing steadily during the past decade. Is the debt becoming too much of a burden upon graduation? Example 6.4 discusses the average debt of undergraduate borrowers.

## Introduction

Statistical inference draws conclusions about a population or process based on sample data. It also provides a statement, expressed in terms of probability, of how much confidence we can place in our conclusions. Although there are many specific recipes for inference, there are only a few general types of statistical inference. This chapter introduces the two most common types: *confidence intervals* and *tests of significance*.

Because the underlying reasoning for these types of inference remains the same across different settings, this chapter considers a single simple setting: inference about the mean of a Normal population whose standard deviation is known. Later chapters will present the recipes for inference in other situations.

a 95% confidence interval, we know that the probability that the interval we compute will cover the parameter is 0.95. That's the meaning of 95% confidence. If we use several such intervals, however, our confidence that *all* of them give correct results is less than 95%. Suppose we take independent samples each month for five months and report a 95% confidence interval for each set of data.

(a) What is the probability that all five intervals cover the true means? This probability (expressed as a percent) is our overall confidence level for the five simultaneous statements.

(b) What is the probability that at least four of the five intervals cover the true means?

6.34 **Telemarketing wages.** An advertisement in the student newspaper asks you to consider working for a telemarketing company. The ad states, "Earn between $500 and $1000 per week." Do you think that the ad is describing a confidence interval? Explain your answer.

6.35 **Like your job?** A Gallup Poll asked working adults about their job satisfaction. One question was "All in all, which best describes how you feel about your job?" The possible answers were "love job," "like job," "dislike job," and "hate job." Fifty-nine percent of the sample responded that they liked their job. Material provided with the results of the poll noted:

*Results are based on telephone interviews with 1,001 national adults, aged 18 and older, conducted Aug. 8–11, 2005. For results based on the total sample of national adults, one can say with 95% confidence that the maximum margin of sampling error is ±3 percentage points.*[11]

The Gallup Poll uses a complex multistage sample design, but the sample percent has approximately a Normal sampling distribution.

(a) The announced poll result was 59% ± 3%. Can we be certain that the true population percent falls in this interval?

(b) Explain to someone who knows no statistics what the announced result 59% ± 3% means.

(c) This confidence interval has the same form we have met earlier:

$$\text{estimate} \pm z^* \sigma_{\text{estimate}}$$

What is the standard deviation $\sigma_{\text{estimate}}$ of the estimated percent?

(d) Does the announced margin of error include errors due to practical problems such as undercoverage and nonresponse?

# 6.2 Tests of Significance

The confidence interval is appropriate when our goal is to estimate population parameters. The second common type of inference is directed at a quite different goal: to assess the evidence provided by the data in favor of some claim about the population parameters.

## The reasoning of significance tests

A significance test is a formal procedure for comparing observed data with a hypothesis whose truth we want to assess. The hypothesis is a statement about the population parameters. The results of a test are expressed in terms of a probability that measures how well the data and the hypothesis agree. We use the following examples to illustrate these concepts.

**EXAMPLE**

**6.8 Debt levels of private and public college borrowers.** One purpose of the National Student Loan Survey described in Example 6.4 (page 361) is to compare the debt of different subgroups of students. For example, the 525 borrowers who last attended a private four-year college had a mean debt of $21,200, while those who last attended a public four-year college had a mean debt of $17,100. The difference of $4100 is fairly large, but we know that these numbers are estimates of the true means. If we took different samples, we would get different estimates. Can we conclude from these data that the average debt of borrowers who attended a private college is different than the average debt of borrowers who attended a public college?

One way to answer this question is to compute the probability of obtaining a difference as large or larger than the observed $4100 assuming that, in fact, there is no difference in the true means. This probability is 0.17. Because this probability is not particularly small, we conclude that observing a difference of $4100 is not very surprising when the true means are equal. The data do not provide evidence for us to conclude that the mean debts for private four-year borrowers and public four-year borrowers are different.

Here is an example with a different conclusion.

**EXAMPLE**

**6.9 Change in average debt levels between 1997 and 2002.**   Another purpose of the National Student Loan Survey is to look for changes over time. For example, in 1997, the survey found that the mean debt for undergraduate study was $11,400. How does this compare with the value of $18,900 in the 2002 study? The difference is $7500. As we learned in the previous example, an observed difference in means is not necessarily sufficient for us to conclude that the true means are different. Do the data provide evidence that there is an increase in borrowing? Again, we answer this question with a probability calculated under the assumption that there is *no difference in the true means*. The probability is 0.00004 of observing an increase in mean debt that is $7500 or more when there really is no difference. Because this probability is so small, we have sufficient evidence in the data to conclude that there has been a change in borrowing between 1997 and 2002.

What are the key steps in these examples?

- We started each with a question about the difference between two mean debts. In Example 6.8, we compare private four-year borrowers with public four-year borrowers. In Example 6.9, we compare borrowers in 2002 with borrowers in 1997. In both cases, we ask whether or not the data are compatible with no difference, that is, a difference of $0.

- Next we compared the data, $4100 in the first case and $7500 in the second, with the value that comes from the question, $0.

- The results of the comparisons are probabilities, 0.17 in the first case and 0.00004 in the second.

The 0.17 probability is not particularly small, so we have no evidence to question the possibility that the true difference is zero. In the second case, however, the probability is quite small. Something that happens with probability 0.00004 occurs only about 4 times out of 100,000. In this case we have two possible explanations:

---

**TWO-SIDED SIGNIFICANCE TESTS AND CONFIDENCE INTERVALS**

A level $\alpha$ two-sided significance test rejects a hypothesis $H_0: \mu = \mu_0$ exactly when the value $\mu_0$ falls outside a level $1 - \alpha$ confidence interval for $\mu$.

---

## USE YOUR KNOWLEDGE

**6.45 Two-sided significance tests and confidence intervals.** The $P$-value for a two-sided test of the null hypothesis $H_0: \mu = 30$ is 0.08.

(a) Does the 95% confidence interval include the value 30? Explain.

(b) Does the 90% confidence interval include the value 30? Explain.

**6.46 More on two-sided tests and confidence intervals.** A 95% confidence interval for a population mean is $(57, 65)$.

(a) Can you reject the null hypothesis that $\mu = 68$ at the 5% significance level? Explain.

(b) Can you reject the null hypothesis that $\mu = 62$ at the 5% significance level? Explain.

## $P$-values versus fixed $\alpha$

The observed result in Example 6.17 was $z = -4.99$. The conclusion that this result is significant at the 1% level does not tell the whole story. The observed $z$ is far beyond the $z$ corresponding to 1%, and the evidence against $H_0$ is far stronger than 1% significance suggests. The $P$-value

$$2P(Z \geq 4.99) = 0.0000006$$

gives a better sense of how strong the evidence is. *The P-value is the smallest level $\alpha$ at which the data are significant.* Knowing the $P$-value allows us to assess significance at any level.

---

**EXAMPLE**

**6.19 Test of the mean SATM score: significance.** In Example 6.16, we tested the hypotheses

$$H_0: \mu = 450$$

$$H_a: \mu > 450$$

concerning the mean SAT Mathematics score $\mu$ of California high school seniors. The test had the $P$-value $P = 0.0069$. This result is significant at the $\alpha = 0.01$ level because $0.0069 \leq 0.01$. It is not significant at the $\alpha = 0.005$ level, because the $P$-value is larger than 0.005. See Figure 6.14.

---

A $P$-value is more informative than a reject-or-not finding at a fixed significance level. But assessing significance at a fixed level $\alpha$ is easier, because no

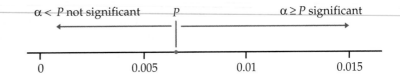

$\alpha < P$ not significant —————— $P$ —————— $\alpha \geq P$ significant

0    0.005    0.01    0.015

**FIGURE 6.14** An outcome with $P$-value $P$ is significant at all levels $\alpha$ at or above $P$ and is not significant at smaller levels $\alpha$.

probability calculation is required. You need only look up a number in a table. A value $z^*$ with a specified area to its right under the standard Normal curve is called a **critical value** of the standard Normal distribution. Because the practice of statistics almost always employs computer software that calculates $P$-values automatically, the use of tables of critical values is becoming outdated. We include the usual tables of critical values (such as Table D) at the end of the book for learning purposes and to rescue students without good computing facilities. The tables can be used directly to carry out fixed $\alpha$ tests. They also allow us to approximate $P$-values quickly without a probability calculation. The following example illustrates the use of Table D to find an approximate $P$-value.

**critical value**

**EXAMPLE**

**6.20 Debt levels of private and public college borrowers: assessing significance.** In Example 6.11 we found the test statistic $z = 1.37$ for testing the null hypothesis that there was no difference in the mean debt between borrowers who attended a private college and those who attended a public college. The alternative was two-sided. Under the null hypothesis, $z$ has a standard Normal distribution, and from the last row in Table D we can see that there is a 95% chance that $z$ is between $\pm1.96$. Therefore, we reject $H_0$ in favor of $H_a$ whenever $z$ is outside this range. Since our calculated value is 1.37, we are within the range and we do not reject the null hypothesis at the 5% level of significance.

## USE YOUR KNOWLEDGE

**6.47** **$P$-value and the significance level.** The $P$-value for a significance test is 0.026.

  (a) Do you reject the null hypothesis at level $\alpha = 0.05$?

  (b) Do you reject the null hypothesis at level $\alpha = 0.01$?

  (c) Explain your answers.

**6.48** **More on the $P$-value and the significance level.** The $P$-value for a significance test is 0.074.

  (a) Do you reject the null hypothesis at level $\alpha = 0.05$?

  (b) Do you reject the null hypothesis at level $\alpha = 0.01$?

  (c) Explain your answers.

**6.49** **One-sided and two-sided $P$-values.** The $P$-value for a two-sided significance test is 0.06.

  (a) State the $P$-values for the one-sided tests.

later (1940s). Because decision theory in its pure form leaves you with two error probabilities and no simple rule on how to balance them, it has been used less often than either tests of significance or tests of hypotheses. Decision ideas have been applied in testing problems mainly by way of the Neyman-Pearson hypothesis-testing theory. That theory asks you first to choose $\alpha$, and the influence of Fisher often has led users of hypothesis testing comfortably back to $\alpha = 0.05$ or $\alpha = 0.01$. Fisher, who was exceedingly argumentative, violently attacked the Neyman-Pearson decision-oriented ideas, and the argument still continues.

## SECTION 6.4  Summary

The **power** of a significance test measures its ability to detect an alternative hypothesis. The power against a specific alternative is calculated as the probability that the test will reject $H_0$ when that alternative is true. This calculation requires knowledge of the sampling distribution of the test statistic under the alternative hypothesis. Increasing the size of the sample increases the power when the significance level remains fixed.

An alternative to significance testing regards $H_0$ and $H_a$ as two statements of equal status that we must decide between. This **decision theory** point of view regards statistical inference in general as giving rules for making decisions in the presence of uncertainty.

In the case of testing $H_0$ versus $H_a$, decision analysis chooses a decision rule on the basis of the probabilities of two types of error. A **Type I error** occurs if $H_0$ is rejected when it is in fact true. A **Type II error** occurs if $H_0$ is accepted when in fact $H_a$ is true.

In a fixed level $\alpha$ significance test, the significance level $\alpha$ is the probability of a Type I error, and the power against a specific alternative is 1 minus the probability of a Type II error for that alternative.

## SECTION 6.4  Exercises

**6.106  Make a recommendation.** Your manager has asked you to review a research proposal that includes a section on sample size justification. A careful reading of this section indicates that the power is 20% for detecting an effect that you would consider important. Write a short report for your manager explaining what this means and make a recommendation on whether or not this study should be run.

**6.107  Explain power and sample size.** Two studies are identical in all respects except for the sample sizes. Consider the power versus a particular sample size. Will the study with the larger sample size have more power or less power than the one with the smaller sample size? Explain your answer in terms that could be understood by someone with very little knowledge of statistics.

**6.108  Power for a different alternative.** The power for a two-sided test of the null hypothesis $\mu_0 = 0$ versus the alternative $\mu = 5$ is 0.82. What is the power versus the alternative $\mu = -5$? Explain your answer.

**6.109  More on the power for a different alternative.** A one-sided test of the null hypothesis $\mu = 50$ versus the alternative $\mu = 60$ has power equal to 0.5. Will the power for the alternative $\mu = 70$ be higher or lower than 0.5? Draw a picture and use this to explain your answer.

**6.110** APPLET  **Power of the random north-south distribution of trees test.** In Exercise 6.66 (page 392) you performed a two-sided significance test of the null hypothesis that the average north-south location of the longleaf pine trees sampled in the Wade Tract was $\mu = 100$. There were 584 trees in the sample and the standard deviation was assumed to be 58. The sample mean in that analysis was $\bar{x} = 99.74$. Use the *Power* applet to compute the power for the alternative $\mu = 99$ using a two-sided test at the 5% level of significance.

**6.111** **Power of the random east-west distribution of trees test.** Refer to the previous exercise. Note that in the east-west direction, the average location was 113.8. Use the *Power* applet to find the power for the alternative $\mu = 110$.

**6.112** **Mail-order catalog sales.** You want to see if a redesign of the cover of a mail-order catalog will increase sales. A very large number of customers will receive the original catalog, and a random sample of customers will receive the one with the new cover. For planning purposes, you are willing to assume that the sales from the new catalog will be approximately Normal with $\sigma = 50$ dollars and that the mean for the original catalog will be $\mu = 25$ dollars. You decide to use a sample size of $n = 900$. You wish to test

$$H_0: \mu = 25$$
$$H_a: \mu > 25$$

You decide to reject $H_0$ if $\bar{x} \geq 26$.

(a) Find the probability of a Type I error, that is, the probability that your test rejects $H_0$ when in fact $\mu = 25$ dollars.

(b) Find the probability of a Type II error when $\mu = 28$ dollars. This is the probability that your test accepts $H_0$ when in fact $\mu = 28$.

(c) Find the probability of a Type II error when $\mu = 30$.

(d) The distribution of sales is not Normal, because many customers buy nothing. Why is it nonetheless reasonable in this circumstance to assume that the mean will be approximately Normal?

**6.113** **Power of the mean SAT score test.** Example 6.16 (page 385) gives a test of a hypothesis about the SAT scores of California high school students based on an SRS of 500 students. The hypotheses are

$$H_0: \mu = 450$$
$$H_a: \mu > 450$$

Assume that the population standard deviation is $\sigma = 100$. The test rejects $H_0$ at the 1% level of significance when $z \geq 2.326$, where

$$z = \frac{\bar{x} - 450}{100/\sqrt{500}}$$

Is this test sufficiently sensitive to usually detect an increase of 10 points in the population mean SAT score? Answer this question by calculating the power of the test against the alternative $\mu = 460$.

**6.114** **Choose the appropriate distribution.** You must decide which of two discrete distributions a random variable $X$ has. We will call the distributions $p_0$ and $p_1$. Here are the probabilities they assign to the values $x$ of $X$:

| $x$ | 0 | 1 | 2 | 3 | 4 | 5 | 6 |
|-----|-----|-----|-----|-----|-----|-----|-----|
| $p_0$ | 0.1 | 0.1 | 0.1 | 0.2 | 0.1 | 0.1 | 0.3 |
| $p_1$ | 0.3 | 0.1 | 0.1 | 0.2 | 0.1 | 0.1 | 0.1 |

You have a single observation on $X$ and wish to test

$$H_0: p_0 \text{ is correct}$$
$$H_a: p_1 \text{ is correct}$$

One possible decision procedure is to reject $H_0$ only if $X = 0$ or $X = 1$.

(a) Find the probability of a Type I error, that is, the probability that you reject $H_0$ when $p_0$ is the correct distribution.

(b) Find the probability of a Type II error.

**6.115** **A Web-based business.** You are in charge of marketing for a Web site that offers automated medical diagnoses. The program will scan the results of routine medical tests (pulse rate, blood pressure, urinalysis, etc.) and either clear the patient or refer the case to a doctor. You are marketing the program for use as part of a preventive-medicine system to screen many thousands of persons who do not have specific medical complaints. The program makes a decision about each patient.

(a) What are the two hypotheses and the two types of error that the program can make? Describe the two types of error in terms of "false-positive" and "false-negative" test results.

(b) The program can be adjusted to decrease one error probability at the cost of an increase in the other error probability. Which error probability would you choose to make smaller, and why? (This is a matter of judgment. There is no single correct answer.)

## CHAPTER 6 Exercises

**6.116** **Full-time employment and age.** A study of late adolescents and early adults reported average months of full-time employment for individuals aged 18 to 26.[23] Here are the means:

| Age | 18 | 19 | 20 | 21 | 22 | 23 | 24 | 25 | 26 |
|---|---|---|---|---|---|---|---|---|---|
| Months employed | 2.9 | 4.2 | 5.0 | 5.3 | 6.4 | 7.4 | 8.5 | 8.9 | 9.3 |

Assume that the standard deviation for each of these means is 4.5 months and that each sample size is 750.

(a) Calculate the 95% confidence interval for each mean.

(b) Plot the means versus age. Draw a vertical line through the first mean extending up to the upper confidence limit and down to the lower limit. At the ends of the line, draw a short dash. Do the same for each of the other means.

(c) Write a summary of what the data show. Note that in circumstances such as this, it is common practice not to make any adjustments for the fact that several confidence intervals are being reported. Be sure to include comments about this in your summary.

**6.117** **Workers' perceptions about safety.** The Safety Climate Index (SCI) measures workers' perceptions about the safety of their work environment. A study of safe work practices of industrial workers reported mean SCI scores for workers classified by workplace size.[24] Here are the means:

| Workplace size | Fewer than 50 workers | 50 to 200 workers | More than 200 workers |
|---|---|---|---|
| Mean SCI | 67.23 | 70.37 | 74.83 |

Assume that the standard deviation is 19 and the sample sizes are all 180. (We will discuss ways to compare three means such as these in Chapter 12.)

(a) Calculate the 95% confidence interval for each mean.

(b) Plot the means versus workplace size. Draw a vertical line through the first mean extending up to the upper confidence limit and down to the lower limit. At the ends of the line, draw a short dash. Do the same for each of the other means.

(c) One way to adjust for the fact that we are reporting three confidence intervals is a procedure that uses a larger value of $z^*$ in the calculation of the margin of error. For this problem one recommendation would be to use $z^* = 2.40$. Repeat parts (a) and (b) making this adjustment.

(d) Summarize your results. Be sure to include comments on the effects of the adjustment on your results.

**6.118** **Coverage percent of 95% confidence interval.** For this exercise you will use the *Confidence Interval* applet. Set the confidence level at 95% and click the "Sample" button 10 times to simulate 10 confidence intervals. Record the percent hit. Simulate another 10 intervals by clicking another 10 times (do not click the "Reset" button). Record the percent hit for your 20 intervals. Repeat the process of simulating 10 additional intervals and recording the results until you have a total of 200 intervals. Plot your results and write a summary of what you have found.

**6.119** **Coverage percent of 90% confidence interval.** Refer to the previous exercise. Do the simulations and report the results for 90% confidence.

**6.120** **Effect of sample size on significance.** You are testing the null hypothesis that $\mu = 0$ versus the alternative $\mu > 0$ using $\alpha = 0.05$. Assume $\sigma = 14$. Suppose $\bar{x} = 4$ and $n = 10$. Calculate the test statistic and its $P$-value. Repeat assuming the same value of $\bar{x}$ but with $n = 20$. Do the same for sample sizes of 30, 40, and 50. Plot the values of the test statistic versus the sample size. Do the same for the $P$-values. Summarize what this demonstration shows about the effect of the sample size on significance testing.

**6.121** **Blood phosphorus level in dialysis patients.** Patients with chronic kidney failure may be treated by dialysis, using a machine that removes toxic wastes from the blood, a function normally performed by the kidneys. Kidney failure and dialysis can cause other changes, such as retention of phosphorus, that must be corrected by changes in diet. A study of the nutrition of dialysis patients measured the level of phosphorus in the blood of several patients on six occasions. Here are the data for one patient (in milligrams of phosphorus per deciliter of blood):[25]

$$5.4 \quad 5.2 \quad 4.5 \quad 4.9 \quad 5.7 \quad 6.3$$

The measurements are separated in time and can be considered an **SRS** of the patient's blood phosphorus level. Assume that this level varies Normally with $\sigma = 0.9$ mg/dl.

(a) Give a 95% confidence interval for the mean blood phosphorus level.

(b) The normal range of phosphorus in the blood is considered to be 2.6 to 4.8 mg/dl. Is there strong evidence that this patient has a mean phosphorus level that exceeds 4.8?

**6.122 Cellulose content in alfalfa hay.** An agronomist examines the cellulose content of a variety of alfalfa hay. Suppose that the cellulose content in the population has standard deviation $\sigma = 8$ milligrams per gram (mg/g). A sample of 15 cuttings has mean cellulose content $\bar{x} = 145$ mg/g.

(a) Give a 90% confidence interval for the mean cellulose content in the population.

(b) A previous study claimed that the mean cellulose content was $\mu = 140$ mg/g, but the agronomist believes that the mean is higher than that figure. State $H_0$ and $H_a$ and carry out a significance test to see if the new data support this belief.

(c) The statistical procedures used in (a) and (b) are valid when several assumptions are met. What are these assumptions?

**6.123 Odor threshold of future wine experts.** Many food products contain small quantities of substances that would give an undesirable taste or smell if they are present in large amounts. An example is the "off-odors" caused by sulfur compounds in wine. Oenologists (wine experts) have determined the odor threshold, the lowest concentration of a compound that the human nose can detect. For example, the odor threshold for dimethyl sulfide (DMS) is given in the oenology literature as 25 micrograms per liter of wine ($\mu$g/l). Untrained noses may be less sensitive, however. Here are the DMS odor thresholds for 10 beginning students of oenology:

$$31 \quad 31 \quad 43 \quad 36 \quad 23 \quad 34 \quad 32 \quad 30 \quad 20 \quad 24$$

Assume (this is not realistic) that the standard deviation of the odor threshold for untrained noses is known to be $\sigma = 7$ $\mu$g/l.

(a) Make a stemplot to verify that the distribution is roughly symmetric with no outliers. (A Normal quantile plot confirms that there are no systematic departures from Normality.)

(b) Give a 95% confidence interval for the mean DMS odor threshold among all beginning oenology students.

(c) Are you convinced that the mean odor threshold for beginning students is higher than the published threshold, 25 $\mu$g/l? Carry out a significance test to justify your answer.

**6.124** **Where do you buy?** Consumers can purchase nonprescription medications at food stores, mass merchandise stores such as Kmart and Wal-Mart, or pharmacies. About 45% of consumers make such purchases at pharmacies. What accounts for the popularity of pharmacies, which often charge higher prices?

A study examined consumers' perceptions of overall performance of the three types of stores, using a long questionnaire that asked about such things as "neat and attractive store," "knowledgeable staff," and "assistance in choosing among various types of nonprescription medication." A performance score was based on 27 such questions. The subjects were 201 people chosen at random from the Indianapolis telephone directory. Here are the means and standard deviations of the performance scores for the sample:[26]

| Store type | $\bar{x}$ | $s$ |
|---|---|---|
| Food stores | 18.67 | 24.95 |
| Mass merchandisers | 32.38 | 33.37 |
| Pharmacies | 48.60 | 35.62 |

We do not know the population standard deviations, but a sample standard deviation $s$ from so large a sample is usually close to $\sigma$. Use $s$ in place of the unknown $\sigma$ in this exercise.

(a) What population do you think the authors of the study want to draw conclusions about? What population are you certain they can draw conclusions about?

(b) Give 95% confidence intervals for the mean performance for each type of store.

(c) Based on these confidence intervals, are you convinced that consumers think that pharmacies offer higher performance than the other types of stores? (In Chapter 12, we will study a statistical method for comparing means of several groups.)

**6.125  CEO pay.** A study of the pay of corporate chief executive officers (CEOs) examined the increase in cash compensation of the CEOs of 104 companies, adjusted for inflation, in a recent year. The mean increase in real compensation was $\bar{x} = 6.9\%$, and the standard deviation of the increases was $s = 55\%$. Is this good evidence that the mean real compensation $\mu$ of all CEOs increased that year? The hypotheses are

$$H_0: \mu = 0 \quad \text{(no increase)}$$

$$H_a: \mu > 0 \quad \text{(an increase)}$$

Because the sample size is large, the sample $s$ is close to the population $\sigma$, so take $\sigma = 55\%$.

(a) Sketch the Normal curve for the sampling distribution of $\bar{x}$ when $H_0$ is true. Shade the area that represents the $P$-value for the observed outcome $\bar{x} = 6.9\%$.

(b) Calculate the $P$-value.

(c) Is the result significant at the $\alpha = 0.05$ level? Do you think the study gives strong evidence that the mean compensation of all CEOs went up?

**6.126  Meaning of "statistically significant."** When asked to explain the meaning of "statistically significant at the $\alpha = 0.01$ level," a student says, "This means there is only probability 0.01 that the null hypothesis is true." Is this an essentially correct explanation of statistical significance? Explain your answer.

**6.127  More on the meaning of "statistically significant."** Another student, when asked why statistical significance appears so often in research reports, says, "Because saying that results are significant tells us that they cannot easily be explained by chance variation alone." Do you think that this statement is essentially correct? Explain your answer.

**6.128  Roulette.** A roulette wheel has 18 red slots among its 38 slots. You observe many spins and record the number of times that red occurs. Now you want to use these data to test whether the probability of a red has the value that is correct for a fair roulette wheel. State the hypotheses $H_0$ and $H_a$ that you will test. (We will describe the test for this situation in Chapter 8.)

**6.129  Simulation study of the confidence interval.** Use a computer to generate $n = 12$ observations from a Normal distribution with mean 25 and standard deviation 4: $N(25, 4)$.

Find the 95% confidence interval for $\mu$. Repeat this process 100 times and then count the number of times that the confidence interval includes the value $\mu = 25$. Explain your results.

**6.130  Simulation study of a test of significance.** Use a computer to generate $n = 12$ observations from a Normal distribution with mean 25 and standard deviation 4: $N(25, 4)$. Test the null hypothesis that $\mu = 25$ using a two-sided significance test. Repeat this process 100 times and then count the number of times that you reject $H_0$. Explain your results.

**6.131  Another simulation study of a test of significance.** Use the same procedure for generating data as in the previous exercise. Now test the null hypothesis that $\mu = 23$. Explain your results.

**6.132  Older customer concerns in restaurants.** Persons aged 55 and over represented 21.3% of the U.S. population in the year 2000. This group is expected to increase to 30.5% by 2025. In terms of actual numbers of people, the increase is from 58.6 million to 101.4 million. Restauranteurs have found this market to be important and would like to make their businesses attractive to older customers. One study used a questionnaire to collect data from people aged 50 and over.[27] For one part of the analysis, individuals were classified into two age groups: 50 to 64 and 65 to 79. There were 267 people in the first group and 263 in the second. One set of items concerned ambiance, menu design, and service. A series of statements were rated on a 1 to 5 scale with 1 representing "strongly disagree" and 5 representing "strongly agree." In some cases the wording has been shortened in the table below. Here are the means:

| Statement | 50–64 | 65–79 |
|---|---|---|
| Ambiance: | | |
| Most restaurants are too dark | 2.75 | 2.93 |
| Most restaurants are too noisy | 3.33 | 3.43 |
| Background music is often too loud | 3.27 | 3.55 |
| Restaurants are too smoky | 3.17 | 3.12 |
| Tables are too small | 3.00 | 3.19 |
| Tables are too close together | 3.79 | 3.81 |
| | | |
| Menu design: | | |
| Print size is not large enough | 3.68 | 3.77 |
| Glare makes menus difficult to read | 2.81 | 3.01 |
| Colors of menus make them difficult to read | 2.53 | 2.72 |

| Statement | 50–64 | 65–79 |
|---|---|---|
| Service: | | |
| It is difficult to hear the service staff | 2.65 | 3.00 |
| I would rather be served than serve myself | 4.23 | 4.14 |
| I would rather pay the server than a cashier | 3.88 | 3.48 |
| Service is too slow | 3.13 | 3.10 |

First examine the means of the people who are 50 to 64. Order the statements according to the means and describe the results. Then do the same for the older group. For each statement compute the $z$ statistic and the associated $P$-value for the comparison between the two groups. For these calculations you can assume that the standard deviation of the difference is 0.08, so $z$ is simply the difference in the means divided by 0.08. Note that you are performing 13 significance tests in this exercise. Keep this in mind when you interpret your results. Write a report summarizing your work.

6.133 **Find published studies with confidence intervals.** Search the Internet or some journals that report research in your field and find two reports that provide an estimate with a margin of error or a confidence interval. For each report:

(a) Describe the method used to collect the data.

(b) Describe the variable being studied.

(c) Give the estimate and the confidence interval.

(d) Describe any practical difficulties that may have led to errors in addition to the sampling errors quantified by the margin of error.

# Inference for Distributions

Some people feel that a full moon causes strange and aggressive behavior in people. Is there any scientific evidence to support this? Example 7.7 describes one such study.

## Introduction

We began our study of data analysis in Chapter 1 by learning graphical and numerical tools for describing the distribution of a single variable and for comparing several distributions. Our study of the practice of statistical inference begins in the same way, with inference about a single distribution and comparison of two distributions. Comparing more than two distributions requires more elaborate methods, which are presented in Chapters 12 and 13.

Two important aspects of any distribution are its center and spread. If the distribution is Normal, we describe its center by the mean $\mu$ and its spread by the standard deviation $\sigma$. In this chapter, we will meet confidence intervals and significance tests for inference about a population mean $\mu$ and for comparing the means or spreads of two populations. The previous chapter emphasized the reasoning of tests and confidence intervals; now we emphasize statistical practice, so we no longer assume that population standard deviations are known. The $t$ procedures for inference about means are among the most common statistical methods. Inference about the spreads, as we will see, poses some difficult practical problems.

The methods in this chapter will allow us to address questions like:

- Does cellular phone use, specifically the number of hours listening to music tracks, differ between cell phone users in the United States and the United Kingdom?

- Do male and female college students differ in terms of "social insight," the ability to appraise other people?

- Does the daily number of disruptive behaviors in dementia patients change when there is a full moon?

# 7.1 Inference for the Mean of a Population

**LOOK BACK**

**sampling distribution of $\bar{x}$, page 339**

Both confidence intervals and tests of significance for the mean $\mu$ of a Normal population are based on the sample mean $\bar{x}$, which estimates the unknown $\mu$. The sampling distribution of $\bar{x}$ depends on $\sigma$. This fact causes no difficulty when $\sigma$ is known. When $\sigma$ is unknown, however, we must estimate $\sigma$ even though we are primarily interested in $\mu$. The sample standard deviation $s$ is used to estimate the population standard deviation $\sigma$.

## The $t$ distributions

Suppose that we have a simple random sample (SRS) of size $n$ from a Normally distributed population with mean $\mu$ and standard deviation $\sigma$. The sample mean $\bar{x}$ is then Normally distributed with mean $\mu$ and standard deviation $\sigma/\sqrt{n}$. When $\sigma$ is not known, we estimate it with the sample standard deviation $s$, and then we estimate the standard deviation of $\bar{x}$ by $s/\sqrt{n}$. This quantity is called the *standard error* of the sample mean $\bar{x}$ and we denote it by $\mathrm{SE}_{\bar{x}}$.

---

### STANDARD ERROR

When the standard deviation of a statistic is estimated from the data, the result is called the **standard error** of the statistic. The standard error of the sample mean is

$$\mathrm{SE}_{\bar{x}} = \frac{s}{\sqrt{n}}$$

---

The term "standard error" is sometimes used for the actual standard deviation of a statistic. The estimated value is then called the "estimated standard error." In this book we will use the term "standard error" only when the standard deviation of a statistic is estimated from the data. The term has this meaning in the output of many statistical computer packages and in research reports that apply statistical methods.

The standardized sample mean, or one-sample $z$ statistic,

$$z = \frac{\bar{x} - \mu}{\sigma/\sqrt{n}}$$

is the basis of the $z$ procedures for inference about $\mu$ when $\sigma$ is known. This statistic has the standard Normal distribution $N(0, 1)$. When we substitute the standard error $s/\sqrt{n}$ for the standard deviation $\sigma/\sqrt{n}$ of $\bar{x}$, the statistic does *not* have a Normal distribution. It has a distribution that is new to us, called a *t distribution*.

---

### THE *t* DISTRIBUTIONS

Suppose that an SRS of size $n$ is drawn from an $N(\mu, \sigma)$ population. Then the **one-sample *t* statistic**

$$t = \frac{\bar{x} - \mu}{s/\sqrt{n}}$$

has the ***t* distribution** with $n - 1$ **degrees of freedom.**

---

A particular $t$ distribution is specified by giving the *degrees of freedom*. We use $t(k)$ to stand for the $t$ distribution with $k$ degrees of freedom. The degrees of freedom for this $t$ statistic come from the sample standard deviation $s$ in the denominator of $t$. We showed earlier that $s$ has $n - 1$ degrees of freedom. Thus, there is a different $t$ distribution for each sample size. There are also other $t$ statistics with different degrees of freedom, some of which we will meet later in this chapter.

**LOOK BACK**
**degrees of freedom, page 42**

The $t$ distributions were discovered in 1908 by William S. Gosset. Gosset was a statistician employed by the Guinness brewing company, which prohibited its employees from publishing their discoveries that were brewing related. In this case, the company let him publish under the pen name "Student" using an example that did not involve brewing. The $t$ distribution is often called "Student's $t$" in his honor.

The density curves of the $t(k)$ distributions are similar in shape to the standard Normal curve. That is, they are symmetric about 0 and are bell-shaped. Figure 7.1 compares the density curves of the standard Normal distribution and the $t$ distributions with 5 and 10 degrees of freedom. The similarity in shape is apparent, as is the fact that the $t$ distributions have more probability in the tails and less in the center. This greater spread is due to the extra variability caused by substituting the random variable $s$ for the fixed parameter $\sigma$. Figure 7.1 also shows that as the degrees of freedom $k$ increase, the $t(k)$ density curve gets closer to the $N(0, 1)$ curve. This reflects the fact that $s$ will likely be closer to $\sigma$ as the sample size increases.

Table D in the back of the book gives critical values $t^*$ for the $t$ distributions. For convenience, we have labeled the table entries both by the value of $p$ needed for significance tests and by the confidence level $C$ (in percent) required for confidence intervals. The standard Normal critical values are in the bottom row of entries and labeled $z^*$. As in the case of the Normal table (Table A), computer software often makes Table D unnecessary.

### USE YOUR KNOWLEDGE

**7.1** **Apartment rents.** You randomly choose 15 unfurnished one-bedroom apartments from a large number of advertisements in your local

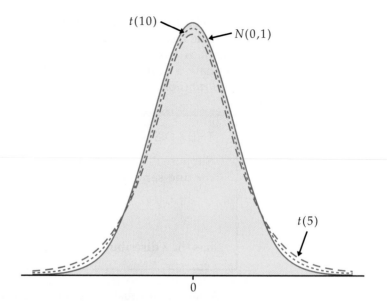

**FIGURE 7.1** Density curves for the standard Normal, $t(10)$, and $t(5)$ distributions. All are symmetric with center 0. The $t$ distributions have more probability in the tails than the standard Normal distribution.

newspaper. You calculate that their mean monthly rent is $570 and their standard deviation is $105.

(a) What is the standard error of the mean?

(b) What are the degrees of freedom for a one-sample $t$ statistic?

7.2 **Finding critical $t^*$ values.** What critical value $t^*$ from Table D should be used to construct

(a) a 95% confidence interval when $n = 12$?

(b) a 99% confidence interval when $n = 24$?

(c) a 90% confidence interval when $n = 200$?

## The one-sample $t$ confidence interval

**LOOK BACK**
**z confidence interval, page 361**

With the $t$ distributions to help us, we can now analyze a sample from a Normal population with unknown $\sigma$. The one-sample $t$ confidence interval is similar in both reasoning and computational detail to the $z$ confidence interval of Chapter 6. There, the margin of error for the population mean was $z^*\sigma/\sqrt{n}$. Here, we replace $\sigma$ by its estimate $s$ and $z^*$ by $t^*$. This means that the margin of error for the population mean when we use the data to estimate $\sigma$ is $t^*s/\sqrt{n}$.

---

**THE ONE-SAMPLE $t$ CONFIDENCE INTERVAL**

Suppose that an SRS of size $n$ is drawn from a population having unknown mean $\mu$. A level $C$ **confidence interval** for $\mu$ is

$$\bar{x} \pm t^* \frac{s}{\sqrt{n}}$$

where $t^*$ is the value for the $t(n-1)$ density curve with area $C$ between $-t^*$ and $t^*$. The quantity

$$t^* \frac{s}{\sqrt{n}}$$

is the **margin of error.** This interval is exact when the population distribution is Normal and is approximately correct for large $n$ in other cases.

**EXAMPLE**

**7.1 Listening to music on cell phones.** Founded in 1998, Telephia provides a wide variety of information on cellular phone use. In 2006, Telephia reported that, on average, United Kingdom (U.K.) subscribers with third-generation technology (3G) phones spent an average of 8.3 hours per month listening to full-track music on their cell phones.[1] Suppose we want to determine a 95% confidence interval for the U.S. average and draw the following random sample of size 8 from the U.S. population of 3G subscribers:

$$5 \ 6 \ 0 \ 4 \ 11 \ 9 \ 2 \ 3$$

The sample mean is $\bar{x} = 5$ and the standard deviation is $s = 3.63$ with degrees of freedom $n - 1 = 7$. The standard error is

$$SE_{\bar{x}} = s/\sqrt{n} = 3.63/\sqrt{8} = 1.28$$

From Table D we find $t^* = 2.365$. The 95% confidence interval is

$$\bar{x} \pm t^* \frac{s}{\sqrt{n}} = 5.0 \pm 2.365 \frac{3.63}{\sqrt{8}}$$

$$= 5.0 \pm (2.365)(1.28)$$

$$= 5.0 \pm 3.0$$

$$= (2.0, 8.0)$$

We are 95% confident that the U.S. population's average time spent listening to full-track music on a cell phone is between 2.0 and 8.0 hours per month. Since this interval does not contain 8.3 hours, these data suggest that, on average, a U.S. subscriber listens to less full-track music.

In this example we have given the actual interval (2.0, 8.0) as our answer. Sometimes we prefer to report the mean and margin of error: the mean time is 5.0 hours per month with a margin of error of 3.0 hours.

The use of the $t$ confidence interval in Example 7.1 rests on assumptions that appear reasonable here. First, we assume our random sample is an SRS from the U.S. population of cell phone users. Second, we assume the distribution of listening times is Normal. With only 8 observations, this assumption cannot be effectively checked. In fact, because the listening time cannot be negative, we might expect this distribution to be skewed to the right. With these data, however, there are no extreme outliers to suggest a severe departure from Normality.

**USE YOUR KNOWLEDGE**

**7.3    More on apartment rents.** Recall Exercise 7.1 (page 419). Construct a 95% confidence interval for the mean monthly rent of all advertised one-bedroom apartments.

**7.4    90% versus 95% confidence interval.** If you were to use 90% confidence, rather than 95% confidence, would the margin of error be larger or smaller? Explain your answer.

## The one-sample *t* test

**LOOK BACK**

*z* **significance test, page 383**

Significance tests using the standard error are also very similar to the *z* test that we studied in the last chapter.

---

### THE ONE-SAMPLE *t* TEST

Suppose that an **SRS** of size $n$ is drawn from a population having unknown mean $\mu$. To test the hypothesis $H_0: \mu = \mu_0$ based on an **SRS** of size $n$, compute the one-sample *t* statistic

$$t = \frac{\bar{x} - \mu_0}{s/\sqrt{n}}$$

In terms of a random variable $T$ having the $t(n-1)$ distribution, the *P*-value for a test of $H_0$ against

$H_a: \mu > \mu_0$ is $P(T \geq t)$

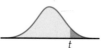

$H_a: \mu < \mu_0$ is $P(T \leq t)$

$H_a: \mu \neq \mu_0$ is $2P(T \geq |t|)$

These *P*-values are exact if the population distribution is Normal and are approximately correct for large $n$ in other cases.

---

**EXAMPLE**

**7.2  Significance test for cell phone use.**  Suppose that, for the U.S. data in Example 7.1, we want to test whether the U.S. average is different from the reported U.K. average. Specifically, we want to test

$$H_0: \mu = 8.3$$
$$H_a: \mu \neq 8.3$$

Recall that $n = 8$, $\bar{x} = 5.0$, and $s = 3.63$. The $t$ test statistic is

$$t = \frac{\bar{x} - \mu_0}{s/\sqrt{n}} = \frac{5.0 - 8.3}{3.63/\sqrt{8}}$$
$$= -2.57$$

This means that the sample mean $\bar{x} = 5.0$ is slightly over 2.5 standard deviations away from the null hypothesized value $\mu = 8.3$. Because the degrees of freedom are $n - 1 = 7$, this $t$ statistic has the $t(7)$ distribution. Figure 7.2 shows that the $P$-value is $2P(T \geq 2.57)$, where $T$ has the $t(7)$ distribution. From Table D we see that $P(T \geq 2.517) = 0.02$ and $P(T \geq 2.998) = 0.01$. Therefore, we conclude that the $P$-value is between $2 \times 0.01 = 0.02$ and $2 \times 0.02 = 0.04$. Software gives the exact value as $P = 0.037$. These data are incompatible with a mean of 8.3 hours per month at the $\alpha = 0.05$ level.

**df = 7**

| $p$ | 0.02 | 0.01 |
|-----|------|------|
| $t^*$ | 2.517 | 2.998 |

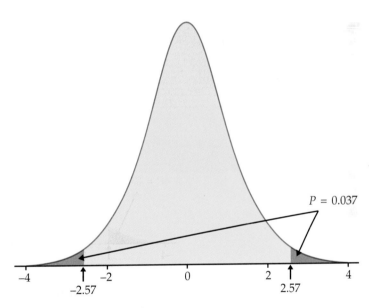

FIGURE 7.2 The $P$-value for Example 7.2.

In this example we tested the null hypothesis $\mu = 8.3$ hours per month against the two-sided alternative $\mu \neq 8.3$ hours per month because we had no prior suspicion that the average in the United States would be larger or smaller. If we had suspected that the U.S. average would be smaller, we would have used a one-sided test. *It is wrong, however, to examine the data first and then decide to do a one-sided test in the direction indicated by the data.* If in doubt, use a two-sided test. In the present circumstance, however, we could use our results from Example 7.2 to justify a one-sided test for *another* sample from the same population.

The assumption that these 39 monthly returns represent an SRS from the population of monthly returns is certainly questionable. If the monthly S&P 500 returns were available, an alternative analysis would be to compare the average difference between the monthly returns for this account and for the S&P 500. This method of analysis is discussed next.

## USE YOUR KNOWLEDGE

**7.5 Significance test using the *t* distribution.** A test of a null hypothesis versus a two-sided alternative gives $t = 2.35$.

    (a) The sample size is 15. Is the test result significant at the 5% level? Explain how you obtained your answer.

    (b) The sample size is 6. Is the test result significant at the 5% level? Explain how you obtained your answer.

    (c) Sketch the two *t* distributions to illustrate your answers.

**7.6 Significance test for apartment rents.** Recall Exercise 7.1 (page 419). Does this SRS give good reason to believe that the mean rent of all advertised one-bedroom apartments is greater than $550? State the hypotheses, find the *t* statistic and its *P*-value, and state your conclusion.

**7.7 Using software.** In Example 7.1 (page 421) we calculated the 95% confidence interval for the U.S. average of hours per month spent listening to full-track music on a cell phone. Use software to compute this interval and verify that you obtain the same interval.

## Matched pairs *t* procedures

The cell phone problem of Example 7.1 concerns only a single population. We know that comparative studies are usually preferred to single-sample investigations because of the protection they offer against confounding. For that reason, inference about a parameter of a single distribution is less common than comparative inference. One common comparative design, however, makes use of single-sample procedures. In a matched pairs study, subjects are matched in pairs and the outcomes are compared within each matched pair. The experimenter can toss a coin to assign two treatments to the two subjects in each pair. Matched pairs are also common when randomization is not possible. One situation calling for matched pairs is when observations are taken on the same subjects, under different conditions.

**LOOK BACK**

matched pairs design, page 189

**EXAMPLE**

**7.7 Does a full moon affect behavior?** Many people believe that the moon influences the actions of some individuals. A study of dementia patients in nursing homes recorded various types of disruptive behaviors every day for 12 weeks. Days were classified as moon days if they were in a three-day period centered at the day of the full moon. For each patient the average number of disruptive behaviors was computed for moon days and for all other days. The data for the 15 subjects whose behaviors were classified as aggressive are presented in Table 7.2.[3] The patients in this study are not a

**TABLE 7.2**

Aggressive behaviors of dementia patients

| Patient | Moon days | Other days | Difference | Patient | Moon days | Other days | Difference |
|---------|-----------|------------|------------|---------|-----------|------------|------------|
| 1 | 3.33 | 0.27 | 3.06 | 9 | 6.00 | 1.59 | 4.41 |
| 2 | 3.67 | 0.59 | 3.08 | 10 | 4.33 | 0.60 | 3.73 |
| 3 | 2.67 | 0.32 | 2.35 | 11 | 3.33 | 0.65 | 2.68 |
| 4 | 3.33 | 0.19 | 3.14 | 12 | 0.67 | 0.69 | −0.02 |
| 5 | 3.33 | 1.26 | 2.07 | 13 | 1.33 | 1.26 | 0.07 |
| 6 | 3.67 | 0.11 | 3.56 | 14 | 0.33 | 0.23 | 0.10 |
| 7 | 4.67 | 0.30 | 4.37 | 15 | 2.00 | 0.38 | 1.62 |
| 8 | 2.67 | 0.40 | 2.27 | | | | |

```
4 | 4 4
3 | 1 1 1 6 7
2 | 1 3 4 7
1 |
0 | 0 1 1
```

**FIGURE 7.8** Stemplot of differences in aggressive behaviors for Examples 7.7 and 7.8.

random sample of dementia patients. However, we examine their data in the hope that what we find is not unique to this particular group of individuals and applies to other patients who have similar characteristics.

To analyze these paired data, we first subtract the disruptive behaviors for moon days from the disruptive behaviors for other days. These 15 differences form a single sample. They appear in the "Difference" columns in Table 7.2. The first patient, for example, averaged 3.33 aggressive behaviors on moon days but only 0.27 aggressive behaviors on other days. The difference $3.33 - 0.27 = 3.06$ is what we will use in our analysis.

Next, we examine the distribution of these differences. Figure 7.8 gives a stemplot of the differences. This plot indicates that there are three patients with very small differences but there are no indications of extreme outliers or strong skewness. We will proceed with our analysis using the Normality-based methods of this section.

To assess whether there is a difference in aggressive behaviors on moon days versus other days, we test

$$H_0 : \mu = 0$$
$$H_a : \mu \neq 0$$

Here $\mu$ is the mean difference in aggressive behaviors, moon versus other days, for patients of this type. The null hypothesis says that aggressive behaviors occur at the same frequency for both types of days, and $H_a$ says that the behaviors on moon days are not the same as on other days.

The 15 differences have

$$\bar{x} = 2.433 \quad \text{and} \quad s = 1.460$$

The one-sample $t$ statistic is therefore

$$t = \frac{\bar{x} - 0}{s/\sqrt{n}} = \frac{2.433}{1.460/\sqrt{15}}$$
$$= 6.45$$

**df = 14**

| $p$ | 0.001 | 0.0005 |
|-----|-------|--------|
| $t^*$ | 3.787 | 4.140 |

The *P*-value is found from the $t(14)$ distribution (remember that the degrees of freedom are 1 less than the sample size). Table D shows that 6.45 lies beyond the upper 0.0005 critical value of the $t(14)$ distribution. Since we are using a two-sided alternative, we know that the *P*-value is less than two times this value, or 0.0010. Software gives a value that is much smaller, $P = 0.000015$. In practice, there is little difference between these two *P*-values; the data provide clear evidence in favor of the alternative hypothesis. A difference this large is very unlikely to occur by chance if there is, in fact, no effect of the moon on aggressive behaviors. In scholarly publications, the details of routine statistical procedures are omitted; our test would be reported in the form: "There was more aggressive behavior on moon days than on other days ($t = 6.45$, df $= 14$, $P < 0.001$)."

Note that we could have justified a one-sided alternative in this example. Based on previous research, we expect more aggressive behaviors on moon days, and the alternative $H_a: \mu > 0$ is reasonable in this setting. The choice of the alternative here, however, has no effect on the conclusion: from Table D we determine that *P* is less than 0.0005; from software it is 0.000008. These are very small values and we would still report $P < 0.001$. *In most circumstances we cannot be absolutely certain about the direction and the safest strategy is to use the two-sided alternative.*

The results of the significance test allow us to conclude that dementia patients exhibit more aggressive behaviors in the days around a full moon. What are the implications of the study for the administrators who run the facilities where these patients live? For example, should they increase staff on these days? To make these kinds of decisions, an estimate of the magnitude of the problem, with a margin of error, would be helpful.

**EXAMPLE**

**7.8 95% confidence interval for the full-moon study.** A 95% confidence interval for the mean difference in aggressive behaviors per day requires the critical value $t^* = 2.145$ from Table D. The margin of error is

$$t^* \frac{s}{\sqrt{n}} = 2.145 \frac{1.460}{\sqrt{15}}$$

$$= 0.81$$

and the confidence interval is

$$\bar{x} \pm t^* \frac{s}{\sqrt{n}} = 2.43 \pm 0.81$$

$$= (1.62, 3.24)$$

The estimated average difference is 2.43 aggressive behaviors per day, with margin of error 0.81 for 95% confidence. The increase needs to be interpreted in terms of the baseline values. The average number of aggressive behaviors per day on other days is 0.59; on moon days it is 3.02. This is approximately a 400% increase. If aggressive behaviors require a substantial amount of attention by staff, then administrators should be aware of the increased level of these activities during the full-moon period. Additional staff may be needed.

The following are key points to remember concerning matched pairs:

1. A matched pairs analysis is called for when subjects are matched in pairs or there are two measurements or observations on each individual and we want to examine the difference.

2. For each pair or individual, use the difference between the two measurements as the data for your analysis.

3. Use the one-sample confidence interval and significance-testing procedures that we learned in this section.

Use of the *t* procedures in Examples 7.7 and 7.8 faces several issues. First, no randomization is possible in a study like this. Our inference procedures assume that there is a process that generates these aggressive behaviors and that the process produces them at possibly different rates during the days near the full moon. Second, many of the patients in these nursing homes did not exhibit any disruptive behaviors. These were not included in our analysis. So our inference is restricted to patients who do exhibit disruptive behaviors.

A final difficulty is that the data show departures from Normality. In a matched pairs analysis, when the *t* procedures are applied to the differences, we are assuming that the differences are Normally distributed. Figure 7.8 gives a stemplot of the differences. There are 3 patients with very small differences in aggressive behaviors while the other 12 have a large increase. We have a dilemma here similar to that in Example 7.1. *The data may not be Normal, and our sample size is very small.* We can try an alternative procedure that does not require the Normality assumption—but there is a price to pay. The alternative procedures have less power to detect differences. Despite these caveats, for Example 7.7 the *P*-value is so small that we are very confident that we have found an effect of the moon phase on behavior.

## USE YOUR KNOWLEDGE

7.8 **Comparison of two energy drinks.** Consider the following study to compare two popular energy drinks. Each drink was rated on a 0 to 100 scale, with 100 being the highest rating.

| Subject | 1 | 2 | 3 | 4 | 5 |
|---------|-----|-----|-----|-----|-----|
| Drink A | 43 | 79 | 66 | 88 | 78 |
| Drink B | 45 | 78 | 61 | 77 | 70 |

Is there a difference in preference? State appropriate hypotheses and carry out a matched pairs *t* test for these data.

7.9 **95% confidence interval for the difference in energy drinks.** For the companies producing these drinks, the real question is how much difference there is between the two preferences. Use the data above to give a 95% confidence interval for the difference in preference between Drink A and Drink B.

## Robustness of the *t* procedures

The results of one-sample *t* procedures are exactly correct only when the population is Normal. Real populations are never exactly Normal. The usefulness of the *t* procedures in practice therefore depends on how strongly they are affected by non-Normality. Procedures that are not strongly affected are called *robust*.

---

**ROBUST PROCEDURES**

A statistical inference procedure is called **robust** if the required probability calculations are insensitive to violations of the assumptions made.

---

**LOOK BACK**

**resistant measure, page 32**

The assumption that the population is Normal rules out outliers, so the presence of outliers shows that this assumption is not valid. The *t* procedures are not robust against outliers, because $\bar{x}$ and *s* are not resistant to outliers.

In Example 7.7, there are three patients with fairly low values of the difference. Whether or not these are outliers is a matter of judgment. If we rerun the analysis without these three patients, the *t* statistic would increase to 11.89 and the *P*-value would be much lower. Careful inspection of the records may reveal some characteristic of these patients which distinguishes them from the others in the study. Without such information, it is difficult to justify excluding them from the analysis. *In general, we should be very cautious about discarding suspected outliers, particularly when they make up a substantial proportion of the data, as they do in this example.*

Fortunately, the *t* procedures are quite robust against non-Normality of the population except in the case of outliers or strong skewness. Larger samples improve the accuracy of *P*-values and critical values from the *t* distributions when the population is not Normal. This is true for two reasons:

1. The sampling distribution of the sample mean $\bar{x}$ from a large sample is close to Normal (that's the central limit theorem). Normality of the individual observations is of little concern when the sample is large.

2. As the sample size *n* grows, the sample standard deviation *s* will be an accurate estimate of $\sigma$ whether or not the population has a Normal distribution. This fact is closely related to the law of large numbers.

**LOOK BACK**

**central limit theorem, page 339**

**law of large numbers, page 274**

Constructing a Normal quantile plot, stemplot, or boxplot to check for skewness and outliers is an important preliminary to the use of *t* procedures for small samples. For most purposes, the one-sample *t* procedures can be safely used when $n \geq 15$ unless an outlier or clearly marked skewness is present. *Except in the case of small samples, the assumption that the data are an SRS from the population of interest is more crucial than the assumption that the population distribution is Normal.* Here are practical guidelines for inference on a single mean:[4]

- *Sample size less than 15:* Use *t* procedures if the data are close to Normal. If the data are clearly non-Normal or if outliers are present, do not use *t*.

- *Sample size at least 15:* The *t* procedures can be used except in the presence of outliers or strong skewness.

- *Large samples:* The $t$ procedures can be used even for clearly skewed distributions when the sample is large, roughly $n \geq 40$.

Consider, for example, some of the data we studied in Chapter 1. The breaking-strength data in Figure 1.34 (page 69) contain three outliers in a sample of size 23, which makes the use of $t$ procedures risky. The guinea pig survival times in Figure 1.35 (page 70) are strongly skewed to the right with no outliers. Since there are 72 observations, we could use the $t$ procedures here. On the other hand, many would prefer to use a transformation to make these data more nearly Normal. (See the material on inference for non-Normal populations on page 435 and in Chapter 16.) Figure 1.36 (page 71) gives the Normal quantile plot for 105 acidity measurements of rainwater. These data appear to be Normal and we would apply the $t$ procedures in this case.

## USE YOUR KNOWLEDGE

**7.10 Significance test for $CO_2$ emissions?** Consider the $CO_2$ emissions data presented in Figure 1.40 (page 76). Would you feel comfortable applying the $t$ procedures in this case? Explain your answer.

**7.11 Significance test for mounting holes data?** Consider data on the distance between mounting holes presented in Figure 1.41 (page 76). Would you feel comfortable applying the $t$ procedures in this case? Explain your answer.

## The power of the $t$ test*

The power of a statistical test measures its ability to detect deviations from the null hypothesis. In practice, we carry out the test in the hope of showing that the null hypothesis is false, so high power is important. The power of the one-sample $t$ test for a specific alternative value of the population mean $\mu$ is the probability that the test will reject the null hypothesis when the alternative value of the mean is true. To calculate the power, we assume a fixed level of significance, often $\alpha = 0.05$.

Calculation of the exact power of the $t$ test takes into account the estimation of $\sigma$ by $s$ and is a bit complex. But an approximate calculation that acts as if $\sigma$ were known is almost always adequate for planning a study. This calculation is very much like that for the $z$ test:

**LOOK BACK**

power of the $z$ test, page 402

1. Decide on a standard deviation, significance level, whether the test is one-sided or two-sided, and an alternative value of $\mu$ to detect.

2. Write the event that the test rejects $H_0$ in terms of $\bar{x}$.

3. Find the probability of this event when the population mean has this alternative value.

Consider Example 7.7, where we examined the effect of the moon on the aggressive behavior of dementia patients in nursing homes. Suppose that we wanted to perform a similar study in a different setting. How many patients should we include in our new study? To answer this question, we do a power calculation.

---

*This section can be omitted without loss of continuity.

In Example 7.7, we found $\bar{x} = 2.433$ and $s = 1.460$. Let's use $s = 1.5$ for our calculations. *It is always better to use a value of the standard deviation that is a little larger than what we expect than one that is smaller.* This may give a sample size that is a little larger than we need. We want to avoid a situation where we fail to find the effect that we are looking for because we did not have enough data. Let's use $\mu = 1.0$ as the alternative value to detect. We are very confident that the effect was larger than this in our previous study, and this amount of an increase in aggressive behavior would still be important to those who work in these facilities. Finally, based on the previous study, we can justify using a one-sided alternative; we expect the moon days to be associated with an increase in aggressive behavior.

**EXAMPLE**

**7.9 Computing the power of a *t* test.**   Let's compute the power of the *t* test for

$$H_0: \mu = 0$$

$$H_a: \mu > 0$$

when the alternative $\mu = 1.0$. We will use a 5% level of significance. The *t* test with *n* observations rejects $H_0$ at the 5% significance level if the *t* statistic

$$t = \frac{\bar{x} - 0}{s/\sqrt{n}}$$

exceeds the upper 5% point of $t(n - 1)$. Taking $n = 20$ and $s = 1.5$, the upper 5% point of $t(19)$ is 1.729. The event that the test rejects $H_0$ is therefore

$$t = \frac{\bar{x}}{1.5/\sqrt{20}} \geq 1.729$$

$$\bar{x} \geq 1.729\frac{1.5}{\sqrt{20}}$$

$$\bar{x} \geq 0.580$$

The power is the probability that $\bar{x} \geq 0.580$ when $\mu = 1.0$. Taking $\sigma = 1.5$, this probability is found by standardizing $\bar{x}$:

$$P(\bar{x} \geq 0.580 \text{ when } \mu = 1.0) = P\left(\frac{\bar{x} - 1.0}{1.5/\sqrt{20}} \geq \frac{0.580 - 1.0}{1.5/\sqrt{20}}\right)$$

$$= P(Z \geq -1.25)$$

$$= 1 - 0.1056 = 0.89$$

The power is 89% that we will detect an increase of 1.0 aggressive behaviors per day during moon days. This is sufficient power for most situations. For many studies, 80% is considered the standard value for desirable power. We could repeat the calculations for some smaller values of *n* to determine the smallest value that would meet the 80% criterion.

Power calculations are used in planning studies to ensure that we have a reasonable chance of detecting effects of interest. They give us some guidance in selecting a sample size. In making these calculations, we need assumptions about the standard deviation and the alternative of interest. In our example we assumed that the standard deviation would be 1.5, but in practice we are hoping that the value will be somewhere around this value. Similarly, we have used a somewhat arbitrary alternative of 1.0. This is a guess based on the results of the previous study. *Beware of putting too much trust in fine details of the results of these calculations.* They serve as a guide, not a mandate.

### USE YOUR KNOWLEDGE

**7.12 Power and the alternative mean $\mu$.** If you were to repeat the power calculation in Example 7.9 for a value of $\mu$ that is smaller than 1, would you expect the power to be higher or lower than 89%? Why?

**7.13 More on power and the alternative mean $\mu$.** Verify your answer to the previous question by doing the calculation for the alternative $\mu = 0.75$.

## Inference for non-Normal populations*

We have not discussed how to do inference about the mean of a clearly non-Normal distribution based on a small sample. If you face this problem, you should consult an expert. Three general strategies are available:

1. In some cases a distribution other than a Normal distribution will describe the data well. There are many non-Normal models for data, and inference procedures for these models are available.

2. Because skewness is the chief barrier to the use of $t$ procedures on data without outliers, you can attempt to transform skewed data so that the distribution is symmetric and as close to Normal as possible. Confidence levels and $P$-values from the $t$ procedures applied to the transformed data will be quite accurate for even moderate sample sizes.

distribution-free procedures

nonparametric procedures

3. Use a **distribution-free** inference procedure. Such procedures do not assume that the population distribution has any specific form, such as Normal. Distribution-free procedures are often called **nonparametric procedures.** Chapter 15 discusses several of these procedures.

Each of these strategies can be effective, but each quickly carries us beyond the basic practice of statistics. We emphasize procedures based on Normal distributions because they are the most common in practice, because their robustness makes them widely useful, and (most important) because we are first of all concerned with understanding the principles of inference. We will therefore not discuss procedures for non-Normal continuous distributions. We will be content with illustrating by example the use of a transformation and of a simple distribution-free procedure.

**Transforming data**   When the distribution of a variable is skewed, it often happens that a simple transformation results in a variable whose distribution is symmetric and even close to Normal. The most common transformation is the

log transformation

**logarithm,** or **log.** The logarithm tends to pull in the right tail of a distribution.

---

*This section can be omitted without loss of continuity.

For example, the data 2, 3, 4, 20 show an outlier in the right tail. Their logarithms 0.30, 0.48, 0.60, 1.30 are much less skewed. Taking logarithms is a possible remedy for right-skewness. Instead of analyzing values of the original variable $X$, we first compute their logarithms and analyze the values of $\log X$. Here is an example of this approach.

**EXAMPLE**

**7.10 Length of audio files on an iPod.** Table 7.3 presents data on the length (in seconds) of audio files found on an iPod. There were a total of 10,003 audio files and 50 files were randomly selected using the "shuffle songs" command.[5] We would like to give a confidence interval for the average audio file length $\mu$ for this iPod.

A Normal quantile plot of the audio data from Table 7.3 (Figure 7.9) shows that the distribution is skewed to the right. Because there are no extreme outliers, the sample mean of the 50 observations will nonetheless have an approximately Normal sampling distribution. The $t$ procedures could be used for approximate inference. For more exact inference, we will seek to transform the data so that the distribution is more nearly Normal. Figure 7.10 is a Normal quantile plot of the logarithms of the time measurements. The transformed data are very close to Normal, so $t$ procedures will give quite exact results.

### TABLE 7.3

Length (in seconds) of audio files sampled from an iPod

| | | | | | | |
|---|---|---|---|---|---|---|
| 240 | 316 | 259 | 46 | 871 | 411 | 1366 |
| 233 | 520 | 239 | 259 | 535 | 213 | 492 |
| 315 | 696 | 181 | 357 | 130 | 373 | 245 |
| 305 | 188 | 398 | 140 | 252 | 331 | 47 |
| 309 | 245 | 69 | 293 | 160 | 245 | 184 |
| 326 | 612 | 474 | 171 | 498 | 484 | 271 |
| 207 | 169 | 171 | 180 | 269 | 297 | 266 |
| 1847 | | | | | | |

The application of the $t$ procedures to the transformed data is straightforward. Call the original length values from Table 7.3 the variable $X$. The transformed data are values of $X_{\text{new}} = \log X$. In most software packages, it is an easy task to transform data in this way and then analyze the new variable.

**EXAMPLE**

**7.11 Software output of audio length data.** Analysis of the logs of the length values in Minitab produces the following output:

```
N     MEAN     STDEV     SE MEAN     95.0 PERCENT C.I.
50   5.6315   0.6840    0.0967      ( 5.4371, 5.8259)
```

For comparison, the 95% $t$ confidence interval for the original mean $\mu$ is found from the original data as follows:

| N | MEAN | STDEV | SE MEAN | 95.0 PERCENT C.I. |
|---|------|-------|---------|-------------------|
| 50 | 354.1 | 307.9 | 43.6 | (266.6, 441.6) |

The advantage of analyzing transformed data is that use of procedures based on the Normal distributions is better justified and the results are more exact.

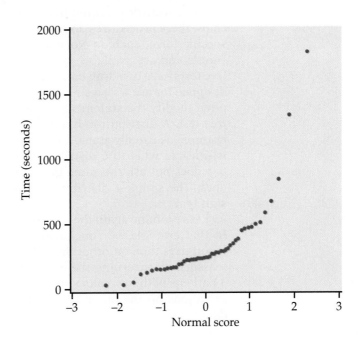

**FIGURE 7.9** Normal quantile plot of audio file length, for Example 7.10. The distribution is skewed to the right.

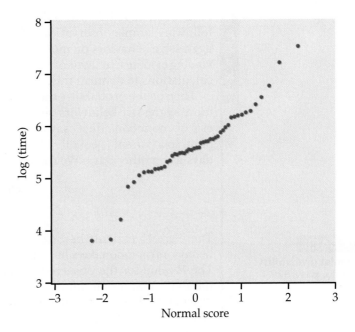

**FIGURE 7.10** Normal quantile plot of the logarithms of the audio file lengths, for Example 7.10. This distribution is close to Normal.

The disadvantage is that a confidence interval for the mean $\mu$ in the original scale (in our example, seconds) cannot be easily recovered from the confidence interval for the mean of the logs. One approach based on the lognormal distribution[6] results in an interval of (290.33, 428.30), which is much narrower than the $t$ interval.

**The sign test**    Perhaps the most straightforward way to cope with non-Normal data is to use a *distribution-free*, or *nonparametric*, procedure. As the name indicates, these procedures do not require the population distribution to have any specific form, such as Normal. Distribution-free significance tests are quite simple and are available in most statistical software packages. Distribution-free tests have two drawbacks. First, they are generally less powerful than tests designed for use with a specific distribution, such as the $t$ test. Second, we must often modify the statement of the hypotheses in order to use a distribution-free test. A distribution-free test concerning the center of a distribution, for example, is usually stated in terms of the median rather than the mean. This is sensible when the distribution may be skewed. But the distribution-free test does not ask the same question (Has the mean changed?) that the $t$ test does. The simplest distribution-free test, and one of the most useful, is the **sign test.**

sign test

Let's examine again the aggressive-behavior data of Example 7.7 (page 428). In that example we concluded that there was more aggressive behavior on moon days than on other days. The stemplot given in Figure 7.8 was not very reassuring concerning the assumption that the data are Normal. There were 3 patients with low values that seemed to be somewhat different from the observations on the other 12 patients. How does the sign test deal with these data?

---

**EXAMPLE**

**7.12 Sign test for the full-moon effect.**    The sign test is based on the following simple observation: of the 15 patients in our sample, 14 had more aggressive behaviors on moon days than on other days. This sounds like convincing evidence in favor of a moon effect on behavior, but we need to do some calculations to confirm this.

Let $p$ be the probability that a randomly chosen dementia patient will have more aggressive behaviors on moon days than on other days. The null hypothesis of "no moon effect" says that the moon days are no different from other days, so a patient is equally likely to have more aggressive behaviors on moon days as on other days. We therefore want to test

$$H_0 : p = 1/2$$
$$H_a : p > 1/2$$

**LOOK BACK**
binomial probability
formula, page 329

There are 15 patients in the study, so the number who have more aggressive behaviors on moon days has the binomial distribution $B(15, 1/2)$ if $H_0$ is true. The $P$-value for the observed count 14 is therefore $P(X \geq 14)$, where $X$ has the $B(15, 1/2)$ distribution. You can compute this probability with software or from the binomial probability formula:

$$P(X \geq 14) = P(X = 14) + P(X = 15)$$

$$= \binom{15}{14}\left(\frac{1}{2}\right)^{14}\left(\frac{1}{2}\right)^{1} + \binom{15}{15}\left(\frac{1}{2}\right)^{15}\left(\frac{1}{2}\right)^{0}$$

$$= (15)\left(\frac{1}{2}\right)^{15} + \left(\frac{1}{2}\right)^{15}$$

$$= 0.000488$$

Using Table C we would approximate this value as 0.0005. As in Example 7.7, there is very strong evidence in favor of an increase in aggressive behavior on moon days.

There are several varieties of sign test, all based on counts and the binomial distribution. The sign test for matched pairs (Example 7.12) is the most useful. The null hypothesis of "no effect" is then always $H_0: p = 1/2$. The alternative can be one-sided in either direction or two-sided, depending on the type of change we are looking for. The test gets its name from the fact that we look only at the signs of the differences, not their actual values.

---

**THE SIGN TEST FOR MATCHED PAIRS**

Ignore pairs with difference 0; the number of trials $n$ is the count of the remaining pairs. The test statistic is the count $X$ of pairs with a positive difference. $P$-values for $X$ are based on the binomial $B(n, 1/2)$ distribution.

---

The matched pairs $t$ test in Example 7.7 tested the hypothesis that the mean of the distribution of differences (moon days minus other days) is 0. The sign test in Example 7.12 is in fact testing the hypothesis that the *median* of the differences is 0. If $p$ is the probability that a difference is positive, then $p = 1/2$ when the median is 0. This is true because the median of the distribution is the point with probability 1/2 lying to its right. As Figure 7.11 illustrates, $p > 1/2$ when the median is greater than 0, again because the probability to the right of the median is always 1/2. The sign test of $H_0: p = 1/2$ against $H_a: p > 1/2$ is a test of

**FIGURE 7.11** Why the sign test tests the median difference: when the median is greater than 0, the probability $p$ of a positive difference is greater than 1/2, and vice versa.

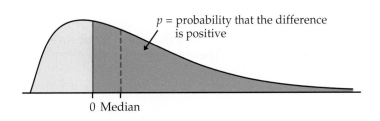

$p$ = probability that the difference is positive

0 Median

$$H_0: \text{population median} = 0$$

$$H_a: \text{population median} > 0$$

The sign test in Example 7.12 makes no use of the actual differences—it just counts how many patients had more aggressive behaviors on moon days than on other days. Because the sign test uses so little of the available information, it is much less powerful than the $t$ test when the population is close to Normal. *It is better to use a test that is powerful when we believe our assumptions are approximately satisfied than a less powerful test with fewer assumptions.* There are other distribution-free tests that are more powerful than the sign test.[7]

## USE YOUR KNOWLEDGE

**7.14  Sign test for energy drink comparison.** Exercise 7.8 (page 431) gives data on the appeal of two popular energy drinks. Is there evidence that the medians are different? State the hypotheses, carry out the sign test, and report your conclusion.

## SECTION 7.1  Summary

Significance tests and confidence intervals for the mean $\mu$ of a Normal population are based on the sample mean $\bar{x}$ of an SRS. Because of the central limit theorem, the resulting procedures are approximately correct for other population distributions when the sample is large.

The standardized sample mean, or **one-sample $z$ statistic,**

$$z = \frac{\bar{x} - \mu}{\sigma/\sqrt{n}}$$

has the $N(0, 1)$ distribution. If the standard deviation $\sigma/\sqrt{n}$ of $\bar{x}$ is replaced by the **standard error** $s/\sqrt{n}$, the **one-sample $t$ statistic**

$$t = \frac{\bar{x} - \mu}{s/\sqrt{n}}$$

has the ***t* distribution** with $n - 1$ degrees of freedom.

There is a $t$ distribution for every positive **degrees of freedom $k$.** All are symmetric distributions similar in shape to Normal distributions. The $t(k)$ distribution approaches the $N(0, 1)$ distribution as $k$ increases.

A level $C$ **confidence interval for the mean** $\mu$ of a Normal population is

$$\bar{x} \pm t^* \frac{s}{\sqrt{n}}$$

where $t^*$ is the value for the $t(n - 1)$ density curve with area $C$ between $-t^*$ and $t^*$. The quantity

$$t^* \frac{s}{\sqrt{n}}$$

is the **margin of error.**

Significance tests for $H_0: \mu = \mu_0$ are based on the $t$ statistic. $P$-values or fixed significance levels are computed from the $t(n-1)$ distribution.

These one-sample procedures are used to analyze **matched pairs** data by first taking the differences within the matched pairs to produce a single sample.

The $t$ procedures are relatively **robust** against non-Normal populations. The $t$ procedures are useful for non-Normal data when $15 \le n < 40$ unless the data show outliers or strong skewness. When $n \ge 40$, the $t$ procedures can be used even for clearly skewed distributions.

The **power** of the $t$ test is calculated like that of the $z$ test, using an approximate value for both $\sigma$ and $s$.

Small samples from skewed populations can sometimes be analyzed by first applying a **transformation** (such as the logarithm) to obtain an approximately Normally distributed variable. The $t$ procedures then apply to the transformed data.

The **sign test** is a **distribution-free test** because it uses probability calculations that are correct for a wide range of population distributions.

The sign test for "no treatment effect" in matched pairs counts the number of positive differences. The $P$-value is computed from the $B(n, 1/2)$ distribution, where $n$ is the number of non-0 differences. The sign test is less powerful than the $t$ test in cases where use of the $t$ test is justified.

# SECTION 7.1  Exercises

*For Exercises 7.1 and 7.2, see pages 419 and 420; for Exercises 7.3 and 7.4, see page 422; for Exercises 7.5 to 7.7, see page 428; for Exercises 7.8 and 7.9, see page 431; for Exercises 7.10 and 7.11, see page 433; for Exercises 7.12 and 7.13, see page 435; and for Exercise 7.14, see page 440.*

**7.15  Finding the critical value $t^*$.** What critical value $t^*$ from Table D should be used to calculate the margin of error for a confidence interval for the mean of the population in each of the following situations?

(a) A 95% confidence interval based on $n = 15$ observations.

(b) A 95% confidence interval from an SRS of 25 observations.

(c) A 90% confidence interval from a sample of size 25.

(d) These cases illustrate how the size of the margin of error depends upon the confidence level and the sample size. Summarize these relationships.

**7.16  Distribution of the $t$ statistic.** Assume a sample size of $n = 20$. Draw a picture of the distribution of the $t$ statistic under the null hypothesis. Use Table D and your picture to illustrate the values of the test statistic that would lead to rejection of the null hypothesis at the 5% level for a two-sided alternative.

**7.17  More on the distribution of the $t$ statistic.** Repeat the previous exercise for the two situations where the alternative is one-sided.

**7.18  One-sided versus two-sided $P$-values.** Computer software reports $\bar{x} = 15.3$ and $P = 0.04$ for a $t$ test of $H_0: \mu = 0$ versus $H_a: \mu \neq 0$. Based on prior knowledge, you can justify testing the alternative $H_a: \mu > 0$. What is the $P$-value for your significance test?

**7.19  More on one-sided versus two-sided $P$-values.** Suppose that $\bar{x} = -15.3$ in the setting of the previous exercise. Would this change your $P$-value? Use a sketch of the distribution of the test statistic under the null hypothesis to illustrate and explain your answer.

**7.20  A one-sample $t$ test.** The one-sample $t$ statistic for testing

$$H_0: \mu = 10$$
$$H_a: \mu > 10$$

from a sample of $n = 20$ observations has the value $t = 2.10$.

(a) What are the degrees of freedom for this statistic?

(b) Give the two critical values $t^*$ from Table D that bracket $t$.

(c) Between what two values does the $P$-value of the test fall?

(d) Is the value $t = 2.10$ significant at the 5% level? Is it significant at the 1% level?

(e) If you have software available, find the exact $P$-value.

**7.21 Another one-sample $t$ test.** The one-sample $t$ statistic for testing

$$H_0: \mu = 60$$

$$H_a: \mu \neq 60$$

from a sample of $n = 24$ observations has the value $t = 2.40$.

(a) What are the degrees of freedom for $t$?

(b) Locate the two critical values $t^*$ from Table D that bracket $t$.

(c) Between what two values does the $P$-value of the test fall?

(d) Is the value $t = 2.40$ statistically significant at the 5% level? At the 1% level?

(e) If you have software available, find the exact $P$-value.

**7.22 A final one-sample $t$ test.** The one-sample $t$ statistic for testing

$$H_0: \mu = 20$$

$$H_a: \mu < 20$$

based on $n = 115$ observations has the value $t = -1.55$.

(a) What are the degrees of freedom for this statistic?

(b) Between what two values does the $P$-value of the test fall?

(c) If you have software available, find the exact $P$-value.

**7.23 Two-sided to one-sided $P$-value.** Most software gives $P$-values for two-sided alternatives. Explain why you cannot always divide these $P$-values by 2 to obtain $P$-values for one-sided alternatives.

**7.24** CHALLENGE **Fuel efficiency $t$ test.** Computers in some vehicles calculate various quantities related to performance. One of these is the fuel efficiency,

or gas mileage, usually expressed as miles per gallon (mpg). For one vehicle equipped in this way, the mpg were recorded each time the gas tank was filled, and the computer was then reset.[8] Here are the mpg values for a random sample of 20 of these records:

| | | | | | | | | | |
|---|---|---|---|---|---|---|---|---|---|
| 41.5 | 50.7 | 36.6 | 37.3 | 34.2 | 45.0 | 48.0 | 43.2 | 47.7 | 42.2 |
| 43.2 | 44.6 | 48.4 | 46.4 | 46.8 | 39.2 | 37.3 | 43.5 | 44.3 | 43.3 |

(a) Describe the distribution using graphical methods. Is it appropriate to analyze these data using methods based on Normal distributions? Explain why or why not.

(b) Find the mean, standard deviation, standard error, and margin of error for 95% confidence.

(c) Report the 95% confidence interval for $\mu$, the mean mpg for this vehicle based on these data.

**7.25 Random distribution of trees $t$ test.** A study of 584 longleaf pine trees in the Wade Tract in Thomas County, Georgia, is described in Example 6.1 (page 354). For each tree in the tract, the researchers measured the diameter at breast height (DBH). This is the diameter of the tree at 4.5 feet and the units are centimeters (cm). Only trees with DBH greater than 1.5 cm were sampled. Here are the diameters of a random sample of 40 of these trees:

| | | | | | | | | | |
|---|---|---|---|---|---|---|---|---|---|
| 10.5 | 13.3 | 26.0 | 18.3 | 52.2 | 9.2 | 26.1 | 17.6 | 40.5 | 31.8 |
| 47.2 | 11.4 | 2.7 | 69.3 | 44.4 | 16.9 | 35.7 | 5.4 | 44.2 | 2.2 |
| 4.3 | 7.8 | 38.1 | 2.2 | 11.4 | 51.5 | 4.9 | 39.7 | 32.6 | 51.8 |
| 43.6 | 2.3 | 44.6 | 31.5 | 40.3 | 22.3 | 43.3 | 37.5 | 29.1 | 27.9 |

(a) Use a histogram or stemplot and a boxplot to examine the distribution of DBHs. Include a Normal quantile plot if you have the necessary software. Write a careful description of the distribution.

(b) Is it appropriate to use the methods of this section to find a 95% confidence interval for the mean DBH of all trees in the Wade Tract? Explain why or why not.

(c) Report the mean with the margin of error and the confidence interval. Write a short summary describing the meaning of the confidence interval.

(d) Do you think these results would apply to other similar trees in the same area? Give reasons for your answer.

**7.26 C-reactive protein in children.** C-reactive protein (CRP) is a substance that can be measured in the blood. Values increase substantially within 6 hours of an infection and reach a peak within 24 to 48 hours after. In adults, chronically high values have been linked to an increased risk of cardiovascular disease. In a study of apparently healthy children aged 6 to 60 months in Papua New Guinea, CRP was measured in 90 children.[9] The units are milligrams per liter (mg/l). Here are the data from a random sample of 40 of these children.

| | | | | | | | | | |
|---|---|---|---|---|---|---|---|---|---|
| 0.00 | 3.90 | 5.64 | 8.22 | 0.00 | 5.62 | 3.92 | 6.81 | 30.61 | 0.00 |
| 73.20 | 0.00 | 46.70 | 0.00 | 0.00 | 26.41 | 22.82 | 0.00 | 0.00 | 3.49 |
| 0.00 | 0.00 | 4.81 | 9.57 | 5.36 | 0.00 | 5.66 | 0.00 | 59.76 | 12.38 |
| 15.74 | 0.00 | 0.00 | 0.00 | 0.00 | 9.37 | 20.78 | 7.10 | 7.89 | 5.53 |

(a) Look carefully at the data above. Do you think that there are outliers or is this a skewed distribution? Now use a histogram or stemplot to examine the distribution. Write a short summary describing the distribution.

(b) Do you think that the mean is a good characterization of the center of this distribution? Explain why or why not.

(c) Find a 95% confidence interval for the mean CRP. Discuss the appropriateness of using this methodology for these data.

7.27 **More on C-reactive protein in children.** Refer to the previous exercise. With strongly skewed distributions such as this, we frequently reduce the skewness by taking a log transformation. We have a bit of a problem here, however, because some of the data are recorded as 0.00 and the logarithm of zero is not defined. For this variable, the value 0.00 is recorded whenever the amount of CRP in the blood is below the level that the measuring instrument is capable of detecting. The usual procedure in this circumstance is to add a small number to each observation before taking the logs. Transform these data by adding 1 to each observation and then taking the logarithm. Use the questions in the previous exercise as a guide to your analysis, and prepare a summary contrasting this analysis with the one that you performed in the previous exercise.

7.28 **Serum retinol in children.** In the Papua New Guinea study that provided the data for the previous two exercises, the researchers also measured serum retinol. A low value of this variable can be an indicator of vitamin A deficiency. Following are the data on the same sample of 40 children from this study. The units are micromoles per liter ($\mu$mol/l).

| | | | | | | | | | |
|---|---|---|---|---|---|---|---|---|---|
| 1.15 | 1.36 | 0.38 | 0.34 | 0.35 | 0.37 | 1.17 | 0.97 | 0.97 | 0.67 |
| 0.31 | 0.99 | 0.52 | 0.70 | 0.88 | 0.36 | 0.24 | 1.00 | 1.13 | 0.31 |
| 1.44 | 0.35 | 0.34 | 1.90 | 1.19 | 0.94 | 0.34 | 0.35 | 0.33 | 0.69 |
| 0.69 | 1.04 | 0.83 | 1.11 | 1.02 | 0.56 | 0.82 | 1.20 | 0.87 | 0.41 |

Analyze these data. Use the questions in the previous two exercises as a guide.

7.29 **Do you feel lucky?** Children in a psychology study were asked to solve some puzzles and were then given feedback on their performance. Then they were asked to rate how luck played a role in determining their scores.[10] This variable was recorded on a 1 to 10 scale with 1 corresponding to very lucky and 10 corresponding to very unlucky. Here are the scores for 60 children:

| | | | | | | | | | | | | | | |
|---|---|---|---|---|---|---|---|---|---|---|---|---|---|---|
| 1 | 10 | 1 | 10 | 1 | 1 | 10 | 5 | 1 | 1 | 8 | 1 | 10 | 2 | 1 |
| 9 | 5 | 2 | 1 | 8 | 10 | 5 | 9 | 10 | 10 | 9 | 6 | 10 | 1 | 5 |
| 1 | 9 | 2 | 1 | 7 | 10 | 9 | 5 | 10 | 10 | 10 | 1 | 8 | 1 | 6 |
| 10 | 1 | 6 | 10 | 10 | 8 | 10 | 3 | 10 | 8 | 1 | 8 | 10 | 4 | 2 |

(a) Use graphical methods to display the distribution. Describe any unusual characteristics. Do you think that these would lead you to hesitate before using the Normality-based methods of this section?

(b) Give a 95% confidence interval for the mean luck score.

(c) The children in this study were volunteers whose parents agreed to have them participate in the study. To what extent do you think your results would apply to all similar children in this community?

7.30 **Perceived organizational skills.** In a study of children with attention deficit hyperactivity disorder (ADHD), parents were asked to rate their child on a variety of items related to how well their child performs different tasks.[11] One item was "Has difficulty organizing work," rated on a five-point scale of 0 to 4 with 0 corresponding to "not at all" and 4 corresponding to "very much." The mean rating for 282 boys with ADHD was reported as 2.22 with a standard deviation of 1.03.

(a) Do you think that these data are Normally distributed? Explain why or why not.

(b) Is it appropriate to use the methods of this section to compute a 99% confidence interval? Explain why or why not.

(c) Find the 99% margin of error and the corresponding confidence interval. Write a sentence explaining the interval and the meaning of the 99% confidence level.

(d) The boys in this study were all evaluated at the Western Psychiatric Institute and Clinic at the University of Pittsburgh. To what extent do you think the results could be generalized to boys with ADHD in other locations?

**7.31 Confidence level and interval width.** Refer to the previous exercise. Compute the 90% and the 95% confidence intervals. Display the three intervals graphically and write a short explanation of the effect of the confidence level on the width of the interval using your display as an example.

**7.32** ⬧ **Food intake and weight gain.** If we increase our food intake, we generally gain weight. Nutrition scientists can calculate the amount of weight gain that would be associated with a given increase in calories. In one study, 16 nonobese adults, aged 25 to 36 years, were fed 1000 calories per day in excess of the calories needed to maintain a stable body weight. The subjects maintained this diet for 8 weeks, so they consumed a total of 56,000 extra calories.[12] According to theory, 3500 extra calories will translate into a weight gain of 1 pound. Therefore, we expect each of these subjects to gain 56,000/3500 = 16 pounds (lb). Here are the weights before and after the 8-week period expressed in kilograms (kg):

| Subject | 1 | 2 | 3 | 4 | 5 | 6 | 7 | 8 |
|---|---|---|---|---|---|---|---|---|
| Weight before | 55.7 | 54.9 | 59.6 | 62.3 | 74.2 | 75.6 | 70.7 | 53.3 |
| Weight after | 61.7 | 58.8 | 66.0 | 66.2 | 79.0 | 82.3 | 74.3 | 59.3 |

| Subject | 9 | 10 | 11 | 12 | 13 | 14 | 15 | 16 |
|---|---|---|---|---|---|---|---|---|
| Weight before | 73.3 | 63.4 | 68.1 | 73.7 | 91.7 | 55.9 | 61.7 | 57.8 |
| Weight after | 79.1 | 66.0 | 73.4 | 76.9 | 93.1 | 63.0 | 68.2 | 60.3 |

(a) For each subject, subtract the weight before from the weight after to determine the weight change.

(b) Find the mean and the standard deviation for the weight change.

(c) Calculate the standard error and the margin of error for 95% confidence. Report the 95% confidence interval in a sentence that explains the meaning of the 95%.

(d) Convert the mean weight gain in kilograms to mean weight gain in pounds. Because there are 2.2 kg per pound, multiply the value in kilograms by

2.2 to obtain pounds. Do the same for the standard deviation and the confidence interval.

(e) Test the null hypothesis that the mean weight gain is 16 lb. Be sure to specify the null and alternative hypotheses, the test statistic with degrees of freedom, and the P-value. What do you conclude?

(f) Write a short paragraph explaining your results.

**7.33 Food intake and NEAT.** Nonexercise activity thermogenesis (NEAT) provides a partial explanation for the results you found in the previous analysis. NEAT is energy burned by fidgeting, maintenance of posture, spontaneous muscle contraction, and other activities of daily living. In the study of the previous exercise, the 16 subjects increased their NEAT by 328 calories per day, on average, in response to the additional food intake. The standard deviation was 256.

(a) Test the null hypothesis that there was no change in NEAT versus the two-sided alternative. Summarize the results of the test and give your conclusion.

(b) Find a 95% confidence interval for the change in NEAT. Discuss the additional information provided by the confidence interval that is not evident from the results of the significance test.

**7.34 Potential insurance fraud?** Insurance adjusters are concerned about the high estimates they are receiving from Jocko's Garage. To see if the estimates are unreasonably high, each of 10 damaged cars was taken to Jocko's and to another garage and the estimates recorded. Here are the results:

| Car | 1 | 2 | 3 | 4 | 5 |
|---|---|---|---|---|---|
| Jocko's | 500 | 1550 | 1250 | 1300 | 750 |
| Other | 400 | 1500 | 1300 | 1300 | 800 |

| Car | 6 | 7 | 8 | 9 | 10 |
|---|---|---|---|---|---|
| Jocko's | 1000 | 1250 | 1300 | 800 | 2500 |
| Other | 800 | 1000 | 1100 | 650 | 2200 |

Test the null hypothesis that there is no difference between the two garages. Be sure to specify the null and alternative hypotheses, the test statistic with degrees of freedom, and the P-value. What do you conclude?

**7.35 Fuel efficiency comparison *t* test.** Refer to Exercise 7.24. In addition to the computer calculating mpg, the driver also recorded the mpg by dividing the miles driven by the amount of gallons at fill-up. The driver wants to determine if these calculations are different.

| Fill-up | 1 | 2 | 3 | 4 | 5 | 6 | 7 | 8 | 9 | 10 |
|---|---|---|---|---|---|---|---|---|---|---|
| Computer | 41.5 | 50.7 | 36.6 | 37.3 | 34.2 | 45.0 | 48.0 | 43.2 | 47.7 | 42.2 |
| Driver | 36.5 | 44.2 | 37.2 | 35.6 | 30.5 | 40.5 | 40.0 | 41.0 | 42.8 | 39.2 |

| Fill-up | 11 | 12 | 13 | 14 | 15 | 16 | 17 | 18 | 19 | 20 |
|---|---|---|---|---|---|---|---|---|---|---|
| Computer | 43.2 | 44.6 | 48.4 | 46.4 | 46.8 | 39.2 | 37.3 | 43.5 | 44.3 | 43.3 |
| Driver | 38.8 | 44.5 | 45.4 | 45.3 | 45.7 | 34.2 | 35.2 | 39.8 | 44.9 | 47.5 |

(a) State the appropriate $H_0$ and $H_a$.

(b) Carry out the test. Give the $P$-value, and then interpret the result.

7.36 **Level of phosphate in the blood.** The level of various substances in the blood of kidney dialysis patients is of concern because kidney failure and dialysis can lead to nutritional problems. A researcher performed blood tests on several dialysis patients on 6 consecutive clinic visits. One variable measured was the level of phosphate in the blood. Phosphate levels for an individual tend to vary Normally over time. The data on one patient, in milligrams of phosphate per deciliter (mg/dl) of blood, are given below:[13]

$$5.6 \quad 5.1 \quad 4.6 \quad 4.8 \quad 5.7 \quad 6.4$$

(a) Calculate the sample mean $\bar{x}$ and its standard error.

(b) Use the $t$ procedures to give a 90% confidence interval for this patient's mean phosphate level.

7.37 **More on the level of phosphate in the blood.** The normal range of values for blood phosphate levels is 2.6 to 4.8 mg/dl. The sample mean for the patient in the previous exercise falls above this range. Is this good evidence that the patient's mean level in fact falls above 4.8? State $H_0$ and $H_a$ and use the data in the previous exercise to carry out a $t$ test. Between which levels from Table D does the $P$-value lie? Are you convinced that the patient's phosphate level is higher than normal?

7.38 **A customer satisfaction survey.** Many organizations are doing surveys to determine the satisfaction of their customers. Attitudes toward various aspects of campus life were the subject of one such study conducted at Purdue University. Each item was rated on a 1 to 5 scale, with 5 being the highest rating. The average response of 1406 first-year students to "Feeling welcomed at Purdue" was 3.9 with a standard deviation of 0.98. Assuming that the respondents are an SRS, give a 90% confidence interval for the mean of all first-year students.

7.39 **Comparing operators of a DXA machine.** Dual-energy X-ray absorptiometry (DXA) is a technique for measuring bone health. One of the most common measures is total body bone mineral content (TBBMC). A highly skilled operator is required to take the measurements. Recently, a new DXA machine was purchased by a research lab and two operators were trained to take the measurements. TBBMC for eight subjects was measured by both operators.[14] The units are grams (g). A comparison of the means for the two operators provides a check on the training they received and allows us to determine if one of the operators is producing measurements that are consistently higher than the other. Here are the data:

| | Subject | | | | | | | |
|---|---|---|---|---|---|---|---|---|
| Operator | 1 | 2 | 3 | 4 | 5 | 6 | 7 | 8 |
| 1 | 1.328 | 1.342 | 1.075 | 1.228 | 0.939 | 1.004 | 1.178 | 1.286 |
| 2 | 1.323 | 1.322 | 1.073 | 1.233 | 0.934 | 1.019 | 1.184 | 1.304 |

(a) Take the difference between the TBBMC recorded for Operator 1 and the TBBMC for Operator 2. Describe the distribution of these differences.

(b) Use a significance test to examine the null hypothesis that the two operators have the same mean. Be sure to give the test statistic with its degrees of freedom, the $P$-value, and your conclusion.

(c) The sample here is rather small, so we may not have much power to detect differences of interest. Use a 95% confidence interval to provide a range of differences that are compatible with these data.

(d) The eight subjects used for this comparison were not a random sample. In fact, they were friends of the researchers whose ages and weights were similar to the types of people who would be measured with this DXA. Comment on the appropriateness of this procedure for selecting a sample, and discuss any consequences regarding the interpretation of the significance testing and confidence interval results.

7.40 **Another comparison of DXA machine operators.** Refer to the previous exercise. TBBMC measures the total amount of mineral in the bones. Another important variable is total body bone mineral density (TBBMD). This variable is calculated by dividing TBBMC by the area corresponding to bone in the DXA scan. The units are grams per squared centimeter ($g/cm^2$). Here are the TBBMD values for the same subjects:

| Operator | Subject | | | | | | | |
|---|---|---|---|---|---|---|---|---|
|  | 1 | 2 | 3 | 4 | 5 | 6 | 7 | 8 |
| 1 | 4042 | 3703 | 2626 | 2673 | 1724 | 2136 | 2808 | 3322 |
| 2 | 4041 | 3697 | 2613 | 2628 | 1755 | 2140 | 2836 | 3287 |

Analyze these data using the questions in the previous exercise as a guide.

7.41  **Assessment of a foreign-language institute.** The National Endowment for the Humanities sponsors summer institutes to improve the skills of high school teachers of foreign languages. One such institute hosted 20 French teachers for 4 weeks. At the beginning of the period, the teachers were given the Modern Language Association's listening test of understanding of spoken French. After 4 weeks of immersion in French in and out of class, the listening test was given again. (The actual French spoken in the two tests was different, so that simply taking the first test should not improve the score on the second test.) The maximum possible score on the test is 36.[15] Here are the data:

| Teacher | Pretest | Posttest | Gain | Teacher | Pretest | Posttest | Gain |
|---|---|---|---|---|---|---|---|
| 1 | 32 | 34 | 2 | 11 | 30 | 36 | 6 |
| 2 | 31 | 31 | 0 | 12 | 20 | 26 | 6 |
| 3 | 29 | 35 | 6 | 13 | 24 | 27 | 3 |
| 4 | 10 | 16 | 6 | 14 | 24 | 24 | 0 |
| 5 | 30 | 33 | 3 | 15 | 31 | 32 | 1 |
| 6 | 33 | 36 | 3 | 16 | 30 | 31 | 1 |
| 7 | 22 | 24 | 2 | 17 | 15 | 15 | 0 |
| 8 | 25 | 28 | 3 | 18 | 32 | 34 | 2 |
| 9 | 32 | 26 | −6 | 19 | 23 | 26 | 3 |
| 10 | 20 | 26 | 6 | 20 | 23 | 26 | 3 |

To analyze these data, we first subtract the pretest score from the posttest score to obtain the improvement for each teacher. These 20 differences form a single sample. They appear in the "Gain" columns. The first teacher, for example, improved from 32 to 34, so the gain is $34 - 32 = 2$.

(a) State appropriate null and alternative hypotheses for examining the question of whether or not the course improves French spoken-language skills.

(b) Describe the gain data. Use numerical and graphical summaries.

(c) Perform the significance test. Give the test statistic, the degrees of freedom, and the *P*-value. Summarize your conclusion.

(d) Give a 95% confidence interval for the mean improvement.

7.42  **Length of calls to customer service center.** Refer to the lengths of calls to a customer service center in Table 1.1 (page 8). Give graphical and numerical summaries for these data. Compute a 95% confidence interval for the mean call length. Comment on the validity of your interval.

7.43  **IQ test scores.** Refer to the IQ test scores for fifth-grade students in Table 1.3 (page 13). Give numerical and graphical summaries of the data and compute a 95% confidence interval. Comment on the validity of the interval.

7.44  **Property damage due to tornadoes.** Table 1.5 (page 25) gives the average property damage per year due to tornadoes for each of the 50 states and Puerto Rico. It does not make sense to use the *t* procedures (or any other statistical procedures) to give a 95% confidence interval for the mean property damage per year due to tornadoes in the United States. Explain why not.

*The following exercises concern the optional material in the sections on the power of the t test and on non-Normal populations.*

7.45  **Sign test for potential insurance fraud.** The differences in the repair estimates in Exercise 7.34 can also be analyzed using a sign test. Set up the appropriate null and alternative hypotheses, carry out the test, and summarize the results. How do these results compare with those that you obtained in Exercise 7.34?

7.46  **Sign test for the comparison of operators.** The differences in the TBBMC measures in Exercise 7.39 can also be analyzed using a sign test. Set up the appropriate null and alternative hypotheses, carry out the test, and summarize the results. How do these results compare with those that you obtained in Exercise 7.39?

7.47  **Another sign test for the comparison of operators.** TBBMD values for the same subjects that you studied in the previous exercise are given in Exercise 7.40. Answer the questions given in the previous exercise for TBBMD.

**7.48 Sign test for assessment of foreign-language institute.** Use the sign test to assess whether the summer institute of Exercise 7.41 improves French listening skills. State the hypotheses, give the *P*-value using the binomial table (Table C), and report your conclusion.

**7.49 Sign test for fuel efficiency comparison.** Use the sign test to assess whether the computer calculates a higher mpg than the driver in Exercise 7.35. State the hypotheses, give the *P*-value using the binomial table (Table C), and report your conclusion.

**7.50 Insulation study.** A manufacturer of electric motors tests insulation at a high temperature (250°C) and records the number of hours until the insulation fails.[16] The data for 5 specimens are

$$446 \quad 326 \quad 372 \quad 377 \quad 310$$

The small sample size makes judgment from the data difficult, but engineering experience suggests that the logarithm of the failure time will have a Normal distribution. Take the logarithms of the 5 observations, and use *t* procedures to give a 90% confidence interval for the mean of the log failure time for insulation of this type.

**7.51 Power of the comparison of DXA machine operators.** Suppose that the bone researchers in Exercise 7.39 wanted to be able to detect an alternative mean difference of 0.002. Find the power for this alternative for a sample size of 15. Use the standard deviation that you found in Exercise 7.39 for these calculations.

**7.52 Sample size calculations.** You are designing a study to test the null hypothesis that $\mu = 0$ versus the alternative that $\mu$ is positive. Assume that $\sigma$ is 10. Suppose that it would be important to be able to detect the alternative $\mu = 2$. Perform power calculations for a variety of sample sizes and determine how large a sample you would need to detect this alternative with power of at least 0.80.

**7.53 Determining the sample size.** Consider Example 7.9 (page 434). What is the minimum sample size needed for the power to be greater than 80% when $\mu = 1.0$?

# 7.2 Comparing Two Means

A nutritionist is interested in the effect of increased calcium on blood pressure. A psychologist wants to compare male and female college students' impressions of personality based on selected photographs. A bank wants to know which of two incentive plans will most increase the use of its credit cards. Two-sample problems such as these are among the most common situations encountered in statistical practice.

> **TWO-SAMPLE PROBLEMS**
> - The goal of inference is to compare the responses in two groups.
> - Each group is considered to be a sample from a distinct population.
> - The responses in each group are independent of those in the other group.

**LOOK BACK**
randomized comparative experiment, page 183

A two-sample problem can arise from a randomized comparative experiment that randomly divides the subjects into two groups and exposes each group to a different treatment. Comparing random samples separately selected from two populations is also a two-sample problem. Unlike the matched pairs designs studied earlier, there is no matching of the units in the two samples, and the two samples may be of different sizes. Inference procedures for two-sample data differ from those for matched pairs.

We can present two-sample data graphically by a back-to-back stemplot (for small samples) or by side-by-side boxplots (for larger samples). Now we will apply the ideas of formal inference in this setting. When both population distributions are symmetric, and especially when they are at least approximately Normal, a comparison of the mean responses in the two populations is most often the goal of inference.

We have two independent samples, from two distinct populations (such as subjects given a treatment and those given a placebo). The same variable is measured for both samples. We will call the variable $x_1$ in the first population and $x_2$ in the second because the variable may have different distributions in the two populations. Here is the notation that we will use to describe the two populations:

| Population | Variable | Mean | Standard deviation |
|:---:|:---:|:---:|:---:|
| 1 | $x_1$ | $\mu_1$ | $\sigma_1$ |
| 2 | $x_2$ | $\mu_2$ | $\sigma_2$ |

We want to compare the two population means, either by giving a confidence interval for $\mu_1 - \mu_2$ or by testing the hypothesis of no difference, $H_0: \mu_1 = \mu_2$.

Inference is based on two independent SRSs, one from each population. Here is the notation that describes the samples:

| Population | Sample size | Sample mean | Sample standard deviation |
|:---:|:---:|:---:|:---:|
| 1 | $n_1$ | $\bar{x}_1$ | $s_1$ |
| 2 | $n_2$ | $\bar{x}_2$ | $s_2$ |

Throughout this section, the subscripts 1 and 2 show the population to which a parameter or a sample statistic refers.

## The two-sample z statistic

The natural estimator of the difference $\mu_1 - \mu_2$ is the difference between the sample means, $\bar{x}_1 - \bar{x}_2$. If we are to base inference on this statistic, we must know its sampling distribution. First, the mean of the difference $\bar{x}_1 - \bar{x}_2$ is the difference of the means $\mu_1 - \mu_2$. This follows from the addition rule for means and the fact that the mean of any $\bar{x}$ is the same as the mean of the population. Because the samples are independent, their sample means $\bar{x}_1$ and $\bar{x}_2$ are independent random variables. The addition rule for variances says that the variance of the difference $\bar{x}_1 - \bar{x}_2$ is the sum of their variances, which is

**LOOK BACK**

addition rule for means, page 278

addition rule for variances, page 282

$$\frac{\sigma_1^2}{n_1} + \frac{\sigma_2^2}{n_2}$$

We now know the mean and variance of the distribution of $\bar{x}_1 - \bar{x}_2$ in terms of the parameters of the two populations. If the two population distributions are both Normal, then the distribution of $\bar{x}_1 - \bar{x}_2$ is also Normal. This is true because each sample mean alone is Normally distributed and because a difference of independent Normal random variables is also Normal.

**7.13 Heights of 10-year-old girls and boys.** A fourth-grade class has 12 girls and 8 boys. The children's heights are recorded on their 10th birthdays. What is the chance that the girls are taller than the boys? Of course, it is very unlikely that all of the girls are taller than all of the boys. We translate the question into the following: what is the probability that the mean height of the girls is greater than the mean height of the boys?

Based on information from the National Health and Nutrition Examination Survey,[17] we assume that the heights (in inches) of 10-year-old girls are $N(56.4, 2.7)$ and the heights of 10-year-old boys are $N(55.7, 3.8)$. The heights of the students in our class are assumed to be random samples from these populations. The two distributions are shown in Figure 7.12(a).

The difference $\bar{x}_1 - \bar{x}_2$ between the female and male mean heights varies in different random samples. The sampling distribution has mean

$$\mu_1 - \mu_2 = 56.4 - 55.7 = 0.7 \text{ inch}$$

and variance

$$\frac{\sigma_1^2}{n_1} + \frac{\sigma_2^2}{n_2} = \frac{2.7^2}{12} + \frac{3.8^2}{8}$$
$$= 2.41$$

The standard deviation of the difference in sample means is therefore $\sqrt{2.41} = 1.55$ inches.

If the heights vary Normally, the difference in sample means is also Normally distributed. The distribution of the difference in heights is shown in Figure 7.12(b). We standardize $\bar{x}_1 - \bar{x}_2$ by subtracting its mean (0.7) and dividing by its standard deviation (1.55). Therefore, the probability that the

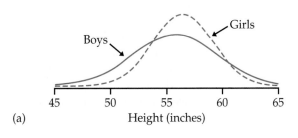

**FIGURE 7.12** Distributions for Example 7.13. (a) Distributions of heights of 10-year-old boys and girls. (b) Distribution of the difference between mean heights of 12 girls and 8 boys.

girls are taller than the boys is

$$P(\bar{x}_1 - \bar{x}_2 > 0) = P\left(\frac{(\bar{x}_1 - \bar{x}_2) - 0.7}{1.55} > \frac{0 - 0.7}{1.55}\right)$$

$$= P(Z > -0.45) = 0.67$$

Even though the population mean height of 10-year-old girls is greater than the population mean height of 10-year-old boys, the probability that the sample mean of the girls is greater than the sample mean of the boys in our class is only 67%. *Large samples are needed to see the effects of small differences.*

As Example 7.13 reminds us, any Normal random variable has the $N(0, 1)$ distribution when standardized. We have arrived at a new $z$ statistic.

---

### TWO-SAMPLE $z$ STATISTIC

Suppose that $\bar{x}_1$ is the mean of an SRS of size $n_1$ drawn from an $N(\mu_1, \sigma_1)$ population and that $\bar{x}_2$ is the mean of an independent SRS of size $n_2$ drawn from an $N(\mu_2, \sigma_2)$ population. Then the **two-sample $z$ statistic**

$$z = \frac{(\bar{x}_1 - \bar{x}_2) - (\mu_1 - \mu_2)}{\sqrt{\dfrac{\sigma_1^2}{n_1} + \dfrac{\sigma_2^2}{n_2}}}$$

has the standard Normal $N(0, 1)$ sampling distribution.

---

In the unlikely event that both population standard deviations are known, the two-sample $z$ statistic is the basis for inference about $\mu_1 - \mu_2$. Exact $z$ procedures are seldom used, however, because $\sigma_1$ and $\sigma_2$ are rarely known. In Chapter 6, we discussed the one-sample $z$ procedures in order to introduce the ideas of inference. Here we move directly to the more useful $t$ procedures.

## The two-sample $t$ procedures

Suppose now that the population standard deviations $\sigma_1$ and $\sigma_2$ are not known. We estimate them by the sample standard deviations $s_1$ and $s_2$ from our two samples. Following the pattern of the one-sample case, we substitute the standard errors for the standard deviations used in the two-sample $z$ statistic. The result is the *two-sample t statistic:*

$$t = \frac{(\bar{x}_1 - \bar{x}_2) - (\mu_1 - \mu_2)}{\sqrt{\dfrac{s_1^2}{n_1} + \dfrac{s_2^2}{n_2}}}$$

Unfortunately, this statistic does *not* have a $t$ distribution. A $t$ distribution replaces the $N(0, 1)$ distribution only when a single standard deviation ($\sigma$) in a $z$ statistic is replaced by its sample standard deviation ($s$). In this case, we replace two standard deviations ($\sigma_1$ and $\sigma_2$) by their estimates ($s_1$ and $s_2$), which does not produce a statistic having a $t$ distribution.

approximations for the
degrees of freedom

Nonetheless, we can approximate the distribution of the two-sample $t$ statistic by using the $t(k)$ distribution with an **approximation for the degrees of freedom $k$.** We use these approximations to find approximate values of $t^*$ for confidence intervals and to find approximate $P$-values for significance tests. Here are two approximations:

1. Use a value of $k$ that is calculated from the data. In general, it will not be a whole number.

2. Use $k$ equal to the smaller of $n_1 - 1$ and $n_2 - 1$.

Most statistical software uses the first option to approximate the $t(k)$ distribution for two-sample problems unless the user requests another method. Use of this approximation without software is a bit complicated; we will give the details later in this section. If you are not using software, the second approximation is preferred. This approximation is appealing because it is conservative.[18] Margins of error for confidence intervals are a bit larger than they need to be, so the true confidence level is larger than $C$. For significance testing, the true $P$-values are a bit smaller than those we obtain from the approximation; for tests at a fixed significance level, we are a little less likely to reject $H_0$ when it is true. In practice, the choice of approximation rarely makes a difference in our conclusion.

## The two-sample $t$ significance test

> ### THE TWO-SAMPLE $t$ SIGNIFICANCE TEST
>
> Suppose that an SRS of size $n_1$ is drawn from a Normal population with unknown mean $\mu_1$ and that an independent SRS of size $n_2$ is drawn from another Normal population with unknown mean $\mu_2$. To test the hypothesis $H_0: \mu_1 = \mu_2$, compute the **two-sample $t$ statistic**
>
> $$t = \frac{\bar{x}_1 - \bar{x}_2}{\sqrt{\dfrac{s_1^2}{n_1} + \dfrac{s_2^2}{n_2}}}$$
>
> and use $P$-values or critical values for the $t(k)$ distribution, where the degrees of freedom $k$ are either approximated by software or are the smaller of $n_1 - 1$ and $n_2 - 1$.

**EXAMPLE**

**7.14 Directed reading activities assessment.** An educator believes that new directed reading activities in the classroom will help elementary school pupils improve some aspects of their reading ability. She arranges for a third-grade class of 21 students to take part in these activities for an eight-week period. A control classroom of 23 third-graders follows the same curriculum without the activities. At the end of the eight weeks, all students are given a Degree of Reading Power (DRP) test, which measures the aspects of reading ability that the treatment is designed to improve. The data appear in Table 7.4.[19]

**TABLE 7.4**

DRP scores for third-graders

| Treatment Group | | | | Control Group | | | |
|---|---|---|---|---|---|---|---|
| 24 | 61 | 59 | 46 | 42 | 33 | 46 | 37 |
| 43 | 44 | 52 | 43 | 43 | 41 | 10 | 42 |
| 58 | 67 | 62 | 57 | 55 | 19 | 17 | 55 |
| 71 | 49 | 54 | | 26 | 54 | 60 | 28 |
| 43 | 53 | 57 | | 62 | 20 | 53 | 48 |
| 49 | 56 | 33 | | 37 | 85 | 42 | |

First examine the data:

```
       Control      Treatment
          970 | 1 |
          860 | 2 | 4
          773 | 3 | 3
      8632221 | 4 | 3334699
         5543 | 5 | 23467789
           20 | 6 | 127
              | 7 | 1
            5 | 8 |
```

A back-to-back stemplot suggests that there is a mild outlier in the control group but no deviation from Normality serious enough to forbid use of $t$ procedures. Separate Normal quantile plots for both groups (Figure 7.13) confirm that both are approximately Normal. The scores of the treatment group appear to be somewhat higher than those of the control group. The summary statistics are

| Group | $n$ | $\bar{x}$ | $s$ |
|---|---|---|---|
| Treatment | 21 | 51.48 | 11.01 |
| Control | 23 | 41.52 | 17.15 |

Because we hope to show that the treatment (Group 1) is better than the control (Group 2), the hypotheses are

$$H_0: \mu_1 = \mu_2$$

$$H_a: \mu_1 > \mu_2$$

The two-sample $t$ test statistic is

$$t = \frac{\bar{x}_1 - \bar{x}_2}{\sqrt{\dfrac{s_1^2}{n_1} + \dfrac{s_2^2}{n_2}}} = \frac{51.48 - 41.52}{\sqrt{\dfrac{11.01^2}{21} + \dfrac{17.15^2}{23}}}$$

$$= 2.31$$

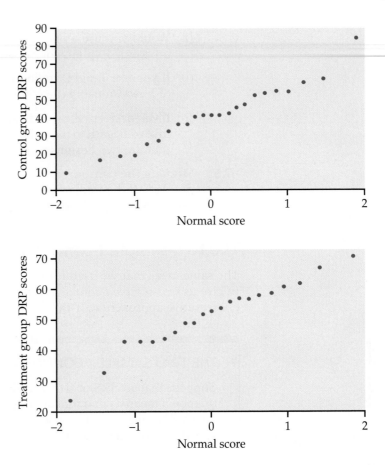

**FIGURE 7.13** Normal quantile plots of the DRP scores in Table 7.4.

The *P*-value for the one-sided test is $P(T \geq 2.31)$. Software gives the approximate *P*-value as 0.0132 and uses 37.9 as the degrees of freedom. For the second approximation, the degrees of freedom $k$ are equal to the smaller of

$$n_1 - 1 = 21 - 1 = 20 \quad \text{and} \quad n_2 - 1 = 23 - 1 = 22$$

Comparing 2.31 with the entries in Table D for 20 degrees of freedom, we see that *P* lies between 0.02 and 0.01. The data strongly suggest that directed reading activity improves the DRP score ($t = 2.31$, df $= 20$, $P < 0.02$).

Note that when we report a result such as this with $P < 0.02$, we imply that the result is *not* significant at the 0.01 level.

If your software gives *P*-values for only the two-sided alternative, $2P(T \geq |t|)$, you need to divide the reported value by 2 after checking that the means differ in the direction specified by the alternative hypothesis.

**df = 20**

| $p$ | 0.02 | 0.01 |
|-----|------|------|
| $t^*$ | 2.197 | 2.528 |

## USE YOUR KNOWLEDGE

**7.54 Comparison of two Web designs.** You want to compare the daily number of hits for two different Web designs that advertise your Internet business. You assign the next 50 days to either Design A or Design B, 25 days to each.

(a) Would you use a one-sided or two-sided significance test for this problem? Explain your choice.

(b) If you use Table D to find the critical value, what are the degrees of freedom using the second approximation?

(c) If you perform the significance test using $\alpha = 0.05$, how large (positive or negative) must the $t$ statistic be to reject the null hypothesis that the two designs result in the same average hits?

**7.55  More on the comparison of two Web designs.** Consider the previous problem. If the $t$ statistic for comparing the mean hits were 2.75, what $P$-value would you report? What would you conclude using $\alpha = 0.05$?

## The two-sample $t$ confidence interval

The same ideas that we used for the two-sample $t$ significance tests also apply to give us *two-sample t confidence intervals*. We can use either software or the conservative approach with Table D to approximate the value of $t^*$.

---

### THE TWO-SAMPLE $t$ CONFIDENCE INTERVAL

Suppose that an SRS of size $n_1$ is drawn from a Normal population with unknown mean $\mu_1$ and that an independent SRS of size $n_2$ is drawn from another Normal population with unknown mean $\mu_2$. The **confidence interval for $\mu_1 - \mu_2$** given by

$$(\bar{x}_1 - \bar{x}_2) \pm t^* \sqrt{\frac{s_1^2}{n_1} + \frac{s_2^2}{n_2}}$$

has confidence level at least $C$ no matter what the population standard deviations may be. Here, $t^*$ is the value for the $t(k)$ density curve with area $C$ between $-t^*$ and $t^*$. The value of the degrees of freedom $k$ is approximated by software or we use the smaller of $n_1 - 1$ and $n_2 - 1$.

---

To complete the analysis of the DRP scores we examined in Example 7.14, we need to describe the size of the treatment effect. We do this with a confidence interval for the difference between the treatment group and the control group means.

**EXAMPLE**

**7.15  How much improvement?**  We will find a 95% confidence interval for the mean improvement in the entire population of third-graders. The interval is

$$(\bar{x}_1 - \bar{x}_2) \pm t^* \sqrt{\frac{s_1^2}{n_1} + \frac{s_2^2}{n_2}} = (51.48 - 41.52) \pm t^* \sqrt{\frac{11.01^2}{21} + \frac{17.15^2}{23}}$$

$$= 9.96 \pm 4.31 t^*$$

Using software, the degrees of freedom are 37.9 and $t^* = 2.025$. This approximation gives

$$9.96 \pm (4.31 \times 2.025) = 9.96 \pm 8.72 = (1.2, 18.7)$$

The conservative approach uses the $t(20)$ distribution. Table D gives $t^* = 2.086$. With this approximation we have

$$9.96 \pm (4.31 \times 2.086) = 9.96 \pm 8.99 = (1.0, 18.9)$$

We can see that the conservative approach does, in fact, give a larger interval than the more accurate approximation used by software. However, the difference is pretty small.

We estimate the mean improvement to be about 10 points, but with a margin of error of almost 9 points with either method. Although we have good evidence of some improvement, the data do not allow a very precise estimate of the size of the average improvement.

The design of the study in Example 7.14 is not ideal. Random assignment of students was not possible in a school environment, so existing third-grade classes were used. The effect of the reading programs is therefore confounded with any other differences between the two classes. The classes were chosen to be as similar as possible—for example, in terms of the social and economic status of the students. Extensive pretesting showed that the two classes were on the average quite similar in reading ability at the beginning of the experiment. To avoid the effect of two different teachers, the researcher herself taught reading in both classes during the eight-week period of the experiment. We can therefore be somewhat confident that the two-sample test is detecting the effect of the treatment and not some other difference between the classes. This example is typical of many situations in which an experiment is carried out but randomization is not possible.

## USE YOUR KNOWLEDGE

**7.56 Two-sample $t$ confidence interval.** Assume $\bar{x}_1 = 100$, $\bar{x}_2 = 120$, $s_1 = 10$, $s_2 = 12$, $n_1 = 50$, and $n_2 = 50$. Find a 95% confidence interval for the difference in the corresponding values of $\mu$ using the second approximation for degrees of freedom. Does this interval include more or fewer values than a 99% confidence interval? Explain your answer.

**7.57 Another two-sample $t$ confidence interval.** Assume $\bar{x}_1 = 100$, $\bar{x}_2 = 120$, $s_1 = 10$, $s_2 = 12$, $n_1 = 10$, and $n_2 = 10$. Find a 95% confidence interval for the difference in the corresponding values of $\mu$ using the second approximation for degrees of freedom. Would you reject the null hypothesis that the population means are equal in favor of the two-sided alternative at significance level 0.05? Explain.

## Robustness of the two-sample procedures

The two-sample $t$ procedures are more robust than the one-sample $t$ methods. When the sizes of the two samples are equal and the distributions of the two populations being compared have similar shapes, probability values from the $t$ table are quite accurate for a broad range of distributions when the sample sizes are as small as $n_1 = n_2 = 5$.[20] When the two population distributions have different shapes, larger samples are needed. The guidelines for the use of one-sample $t$ procedures can be adapted to two-sample procedures by replacing "sample size" with the "sum of the sample sizes" $n_1 + n_2$. These guidelines are rather conservative, especially when the two samples are of equal size. *In planning a two-sample study, choose equal sample sizes if you can.* The two-sample $t$ procedures are most robust against non-Normality in this case, and the conservative probability values are most accurate.

Here is an example with moderately large sample sizes that are not equal. Even if the distributions are not Normal, we are confident that the sample means will be approximately Normal. The two-sample $t$ test is very robust in this case.

**7.16 Wheat prices.** The U.S. Department of Agriculture (USDA) uses sample surveys to produce important economic estimates. One pilot study estimated wheat prices in July and in September using independent samples of wheat producers in the two months. Here are the summary statistics, in dollars per bushel:[21]

| Month | $n$ | $\bar{x}$ | $s$ |
|---|---|---|---|
| September | 45 | $3.61 | $0.19 |
| July | 90 | $2.95 | $0.22 |

The September prices are higher on the average. But we have data from only a sample of producers each month. Can we conclude that national average prices in July and September are not the same? Or are these differences merely what we would expect to see due to random variation?

Because we did not specify a direction for the difference before looking at the data, we choose a two-sided alternative. The hypotheses are

$$H_0: \mu_1 = \mu_2$$

$$H_a: \mu_1 \neq \mu_2$$

Because the samples are moderately large, we can confidently use the $t$ procedures even though we lack the detailed data and so cannot verify the Normality condition.

The two-sample $t$ statistic is

$$t = \frac{\bar{x}_1 - \bar{x}_2}{\sqrt{\dfrac{s_1^2}{n_1} + \dfrac{s_2^2}{n_2}}} = \frac{3.61 - 2.95}{\sqrt{\dfrac{0.19^2}{45} + \dfrac{0.22^2}{90}}}$$

$$= 18.03$$

**df = 40**

| $p$ | 0.0005 |
|---|---|
| $t^*$ | 3.551 |

The conservative approach finds the *P*-value by comparing 18.03 to critical values for the $t(44)$ distribution because the smaller sample has 45 observations. We must double the table tail area $p$ because the alternative is two-sided. Table D does not have entries for 44 degrees of freedom. When this happens, we use the next smaller degrees of freedom. Our calculated value of $t$ is larger than the $p = 0.0005$ entry in the table. Doubling 0.0005, we conclude that the *P*-value is less than 0.001. The data give conclusive evidence that the mean wheat prices were higher in September than they were in July ($t = 18.03$, df $= 44$, $P < 0.001$).

In this example the exact *P*-value is very small because $t = 18$ says that the observed difference in means is 18 standard errors above the hypothesized difference of zero ($\mu_1 = \mu_2$). This is so unlikely that the probability is zero for all practical purposes. The difference in mean prices is not only highly significant but large enough (66 cents per bushel) to be important to producers.

In this and other examples, we can choose which population to label 1 and which to label 2. After inspecting the data, we chose September as Population 1 because this choice makes the $t$ statistic a positive number. This avoids any possible confusion from reporting a negative value for $t$. *Choosing the population labels is* **not** *the same as choosing a one-sided alternative after looking at the data.* Choosing hypotheses after seeing a result in the data is a violation of sound statistical practice.

## Inference for small samples

Small samples require special care. We do not have enough observations to examine the distribution shapes, and only extreme outliers stand out. The power of significance tests tends to be low, and the margins of error of confidence intervals tend to be large. Despite these difficulties, we can often draw important conclusions from studies with small sample sizes. If the size of an effect is as large as it was in the wheat price example, it should still be evident even if the $n$'s are small.

**EXAMPLE**

**7.17 More about wheat prices.** In the setting of Example 7.16, a quick survey collects prices from only 5 producers each month. The data are

| Month | Price of wheat ($/bushel) | | | | |
|---|---|---|---|---|---|
| September | $3.5900 | $3.6150 | $3.5950 | $3.5725 | $3.5825 |
| July | $2.9200 | $2.9675 | $2.9175 | $2.9250 | $2.9325 |

The prices are reported to the nearest quarter of a cent. First, examine the distributions with a back-to-back stemplot after rounding each price to the nearest cent.

is our guess at the standard error for the difference in the sample means. Therefore, if we wanted to assess a possible study in terms of the margin of error for the estimated difference, we would examine $t^*$ times this quantity.

If we do not assume that the standard deviations are equal, we need to guess both standard deviations and then combine these for our guess at the standard error:

$$\sqrt{\frac{\sigma_1^2}{n_1} + \frac{\sigma_2^2}{n_2}}$$

This guess is then used in the denominator of the noncentrality parameter. For the degrees of freedom, the conservative approximation is appropriate.

**EXAMPLE**

**7.23 Planning a new study of calcium versus placebo groups.** In Example 7.20 we examined the effect of calcium on blood pressure by comparing the means of a treatment group and a placebo group using a pooled two-sample $t$ test. The $P$-value was 0.059, failing to achieve the usual standard of 0.05 for statistical significance. Suppose that we wanted to plan a new study that would provide convincing evidence, say at the 0.01 level, with high probability. Let's examine a study design with 45 subjects in each group ($n_1 = n_2 = 45$). Based on our previous results we choose $\mu_1 - \mu_2 = 5$ as an alternative that we would like to be able to detect with $\alpha = 0.01$. For $\sigma$ we use 7.4, our pooled estimate from Example 7.20. The degrees of freedom are $n_1 + n_2 - 2 = 88$ and $t^* = 2.37$ for the significance test. The noncentrality parameter is

$$\delta = \frac{5}{7.4\sqrt{\dfrac{1}{45} + \dfrac{1}{45}}} = \frac{5}{1.56} = 3.21$$

Software gives the power as 0.7965, or 80%. The Normal approximation gives 0.7983, a very accurate result. With this choice of sample sizes we would expect the margin of error for a 95% confidence interval ($t^* = 1.99$) for the difference in means to be

$$t^* \times 7.4\sqrt{\frac{1}{45} + \frac{1}{45}} = 1.99 \times 1.56 = 3.1$$

With software it is very easy to examine the effects of variations on a study design. In the above example, we might want to examine the power for $\alpha = 0.05$ and the effects of reducing the sample sizes.

## USE YOUR KNOWLEDGE

**7.96 Power and $\mu_1 - \mu_2$.** If you repeat the calculation in Example 7.23 for other values of $\mu_1 - \mu_2$ that are larger than 5, would you expect the power to be higher or lower than 0.7965? Why?

**7.97 Power and the standard deviation.** If the true population standard deviation were 7.0 instead of the 7.4 hypothesized in Example 7.23, would the power for this new experiment be greater or smaller than 0.7965? Explain.

## SECTION 7.3   Summary

Inference procedures for comparing the standard deviations of two Normal populations are based on the **F statistic,** which is the ratio of sample variances:

$$F = \frac{s_1^2}{s_2^2}$$

If an SRS of size $n_1$ is drawn from the $x_1$ population and an independent SRS of size $n_2$ is drawn from the $x_2$ population, the $F$ statistic has the **F distribution** $F(n_1 - 1, n_2 - 1)$ if the two population standard deviations $\sigma_1$ and $\sigma_2$ are in fact equal.

The **F test for equality of standard deviations** tests $H_0 : \sigma_1 = \sigma_2$ versus $H_a : \sigma_1 \neq \sigma_2$ using the statistic

$$F = \frac{\text{larger } s^2}{\text{smaller } s^2}$$

and doubles the upper-tail probability to obtain the $P$-value.

The $t$ procedures are quite **robust** when the distributions are not Normal. The $F$ tests and other procedures for inference about the spread of one or more Normal distributions are so strongly affected by non-Normality that we do not recommend them for regular use.

The **power** of the pooled two-sample $t$ test is found by first computing the critical value for the significance test, the degrees of freedom, and the **noncentrality parameter** for the alternative of interest. These are used to find the power from the $t$ **distribution.** A Normal approximation works quite well. Calculating margins of error for various study designs and assumptions is an alternative procedure for evaluating designs.

## SECTION 7.3   Exercises

*For Exercise 7.95, see page 476; and for Exercises 7.96 and 7.97, see page 478.*

*In all exercises calling for use of the F test, assume that both population distributions are very close to Normal. The actual data are not always sufficiently Normal to justify use of the F test.*

**7.98   Comparison of standard deviations.** Here are some summary statistics from two independent samples from Normal distributions:

| Sample | $n$ | $s^2$ |
|--------|-----|-------|
| 1 | 10 | 3.1 |
| 2 | 16 | 9.3 |

You want to test the null hypothesis that the two population standard deviations are equal versus the two-sided alternative at the 5% significance level.

(a) Calculate the test statistic.

(b) Find the appropriate value from Table E that you need to perform the significance test.

(c) What do you conclude?

**7.99   Revisiting the cholesterol comparison.** Compare the standard deviations of total cholesterol in Exercise 7.61 (page 467). Give the test statistic, the degrees of freedom, and the $P$-value. Write a short summary of your analysis, including comments on the assumptions for the test.

**7.100 An HDL comparison.** HDL is also known as "good" cholesterol. Compare the standard deviations of HDL in Exercise 7.61 (page 467). Give the test statistic, the degrees of freedom, and the *P*-value. Write a short summary of your analysis, including comments on the assumptions for the test.

**7.101** CHALLENGE **Revisiting the multimedia evaluation study.** Mean scores on a knowledge test are compared for two groups of women in Exercise 7.63 (page 468). Compare the standard deviations using an *F* test. What do you conclude? Comment on the Normal assumption for these data. These standard deviations are so close that we are not particularly surprised at the result of the significance test. Assume that the sample standard deviation in the intervention is the value given in Exercise 7.63 (1.15). How large would the standard deviation in the control group need to be to reject the null hypothesis of equal standard deviations at the 5% level?

**7.102 Revisiting the self-control and food study.** Compare the standard deviations of the self-efficacy scores in Exercise 7.64 (page 468). Give the test statistic, the degrees of freedom, and the *P*-value. Write a short summary of your analysis, including comments on the assumptions for the test.

**7.103 Revisiting the dust exposure study.** The two-sample problem in Exercise 7.65 (page 468) compares drill and blast workers with outdoor concrete workers with respect to the total dust that they are exposed to in the workplace. Here it may be useful to know whether or not the standard deviations differ in the two groups. Perform the *F* test and summarize the results. Are you concerned about the assumptions here? Explain why or why not.

**7.104 More on the dust exposure study.** Exercise 7.66 (page 468) is similar to Exercise 7.65, but the response variable here is exposure to dust particles that can enter and stay in the lungs. Compare the standard deviations with a significance test and summarize the results. Be sure to comment on the assumptions.

**7.105 Revisiting the size of trees in the north and south.** The diameters of trees in the Wade Tract for random samples selected from the north and south portions of the tract are compared in Exercise 7.81 (page 471). Are there statistically significant differences in the standard deviations for these two parts of the tract? Perform the significance test and summarize the results. Does the Normal assumption appear reasonable for these data?

**7.106 Revisiting the size of trees in the east and west.** Tree diameters for the east and west halves of the Wade Tract are compared in Exercise 7.82 (page 471). Using the questions in the previous exercise as a guide, analyze these data.

**7.107 Revisiting the storage time study.** We studied the loss of vitamin C when bread is stored in Exercise 7.76 (page 470). Recall that two loaves were measured immediately after baking and another two loaves were measured after three days of storage. These are very small sample sizes.

(a) Use Table E to find the value that the ratio of variances would have to exceed for us to reject the null hypothesis (at the 5% level) that the standard deviations are equal. What does this suggest about the power of the test?

(b) Perform the test and state your conclusion.

**7.108 Planning a study to compare tree size.** In Exercise 7.81 (page 471) DBH data for longleaf pine trees in two parts of the Wade Tract are compared. Suppose that you are planning a similar study where you will measure the diameters of longleaf pine trees. Based on Exercise 7.81, you are willing to assume that the standard deviation is 20 cm. Suppose that a difference in mean DBH of 10 cm or more would be important to detect. You will use a *t* statistic and a two-sided alternative for the comparison.

(a) Find the power if you randomly sample 20 trees from each area to be compared.

(b) Repeat the calculations for 60 trees in each sample.

(c) If you had to choose between the 20 and 60 trees per sample, which would be acceptable? Give reasons for your answer.

**7.109** CHALLENGE **More on planning a study to compare tree size.** Refer to the previous exercise. Find the two standard deviations from Exercise 7.81. Do the same for the data in Exercise 7.82, which is a similar setting. These are somewhat smaller than the assumed value that you used in the previous exercise. Explain why it is generally a better idea to assume a standard deviation that is larger than you expect than one that is smaller. Repeat the power calculations for some other reasonable values of $\sigma$ and comment on the impact of the size of $\sigma$ for planning the new study.

7.110  **Planning a study to compare ad placement.** Refer to Exercise 7.80 (page 470), where we compared trustworthy ratings for ads from two different publications. Suppose that you are planning a similar study using two different publications that are not expected to show the differences seen when comparing the *Wall Street Journal* with the *National Enquirer*. You would like to detect a difference of 1.5 points using a two-sided

significance test with a 5% level of significance. Based on Exercise 7.80, it is reasonable to use 1.6 as the value of the standard deviation for planning purposes.

(a)  What is the power if you use sample sizes similar to those used in the previous study, for example, 65 for each publication?

(b)  Repeat the calculations for 100 in each group.

(c)  What sample size would you recommend for the new study?

## CHAPTER 7  Exercises

7.111  **LSAT scores.** The scores of four senior roommates on the Law School Admission Test (LSAT) are

$$158, \ 168, \ 143, \ 155$$

Find the mean, the standard deviation, and the standard error of the mean. Is it appropriate to calculate a confidence interval based on these data? Explain why or why not.

7.112  **Converting a two-sided *P*-value.** You use statistical software to perform a significance test of the null hypothesis that two means are equal. The software reports *P*-values for the two-sided alternative. Your alternative is that the first mean is greater than the second mean.

(a)  The software reports $t = 1.81$ with a *P*-value of 0.07. Would you reject $H_0$ with $\alpha = 0.05$? Explain your answer.

(b)  The software reports $t = -1.81$ with a *P*-value of 0.07. Would you reject $H_0$ with $\alpha = 0.05$? Explain your answer.

7.113  **Degrees of freedom and confidence interval width.** As the degrees of freedom increase, the *t* distributions get closer and closer to the $z$ ($N(0, 1)$) distribution. One way to see this is to look at how the value of $t^*$ for a 95% confidence interval changes with the degrees of freedom. Make a plot with degrees of freedom from 2 to 100 on the *x* axis and $t^*$ on the *y* axis. Draw a horizontal line on the plot corresponding to the value of $z^* = 1.96$. Summarize the main features of the plot.

7.114  **Degrees of freedom and $t^*$.** Refer to the previous exercise. Make a similar plot and summarize its features for the value of $t^*$ for a 90% confidence interval.

7.115  **Sample size and margin of error.** The margin of error for a confidence interval depends on the confidence level, the standard deviation, and the sample size. Fix the confidence level at 95% and the standard deviation at 1 to examine the effect of the sample size. Find the margin of error for sample sizes of 5 to 100 by 5s—that is, let $n = 5, 10, 15, \ldots, 100$. Plot the margins of error versus the sample size and summarize the relationship.

7.116  **More on sample size and margin of error.** Refer to the previous exercise. Make a similar plot and summarize its features for a 99% confidence interval.

7.117  **Alcohol consumption and body composition.** Individuals who consume large amounts of alcohol do not use the calories from this source as efficiently as calories from other sources. One study examined the effects of moderate alcohol consumption on body composition and the intake of other foods. Fourteen subjects participated in a crossover design where they either drank wine for the first 6 weeks and then abstained for the next 6 weeks or vice versa.[33] During the period when they drank wine, the subjects, on average, lost 0.4 kilograms (kg) of body weight; when they did not drink wine, they lost an average of 1.1 kg. The standard deviation of the difference between the weight lost under these two conditions is 8.6 kg. During the wine period, they consumed an average of 2589 calories; with no wine, the mean consumption was 2575. The standard deviation of the difference was 210.

(a)  Compute the differences in means and the standard errors for comparing body weight and caloric intake under the two experimental conditions.

(b) A report of the study indicated that there were no significant differences in these two outcome measures. Verify this result for each measure, giving the test statistic, degrees of freedom, and the *P*-value.

(c) One concern with studies such as this, with a small number of subjects, is that there may not be sufficient power to detect differences that are potentially important. Address this question by computing 95% confidence intervals for the two measures and discuss the information provided by the intervals.

(d) Here are some other characteristics of the study. The study periods lasted for 6 weeks. All subjects were males between the ages of 21 and 50 years who weighed between 68 and 91 kilograms (kg). They were all from the same city. During the wine period, subjects were told to consume two 135 milliliter (ml) servings of red wine and no other alcohol. The entire 6-week supply was given to each subject at the beginning of the period. During the other period, subjects were instructed to refrain from any use of alcohol. All subjects reported that they complied with these instructions except for three subjects, who said that they drank no more than 3 to 4 12-ounce bottles of beer during the no-alcohol period. Discuss how these factors could influence the interpretation of the results.

7.118 **Healthy bones study.** Healthy bones are continually being renewed by two processes. Through bone formation, new bone is built; through bone resorption, old bone is removed. If one or both of these processes are disturbed, by disease, aging, or space travel, for example, bone loss can be the result. Osteocalcin (OC) is a biochemical marker for bone formation: higher levels of bone formation are associated with higher levels of OC. A blood sample is used to measure OC, and it is much less expensive to obtain than direct measures of bone formation. The units are milligrams of OC per milliliter of blood (mg/ml). One study examined various biomarkers of bone turnover.[34] Here are the OC measurements on 31 healthy females aged 11 to 32 years who participated in this study:

| | | | | | | | |
|---|---|---|---|---|---|---|---|
| 68.9 | 56.3 | 54.6 | 31.2 | 36.4 | 31.4 | 52.8 | 38.4 |
| 35.7 | 76.5 | 44.4 | 40.2 | 77.9 | 54.6 | 9.9 | 20.6 |
| 20.0 | 17.2 | 24.2 | 20.9 | 17.9 | 19.7 | 15.9 | 20.8 |
| 8.1 | 19.3 | 16.9 | 10.1 | 47.7 | 30.2 | 17.2 | |

(a) Display the data with a stemplot or histogram and a boxplot. Describe the distribution.

(b) Find a 95% confidence interval for the mean OC. Comment on the suitability of using this procedure for these data.

7.119 **More on the healthy bones study.** Refer to the previous exercise. Tartrate resistant acid phosphatase (TRAP) is a biochemical marker for bone resorption that is also measured in blood. Here are the TRAP measurements, in units per liter (U/l), for the same 31 females:

| | | | | | | | |
|---|---|---|---|---|---|---|---|
| 19.4 | 25.5 | 19.0 | 9.0 | 19.1 | 14.6 | 25.2 | 14.6 |
| 28.8 | 14.9 | 10.7 | 5.9 | 23.7 | 19.0 | 6.9 | 8.1 |
| 9.5 | 6.3 | 10.1 | 10.5 | 9.0 | 8.8 | 8.2 | 10.3 |
| 3.3 | 10.1 | 9.5 | 8.1 | 18.6 | 14.4 | 9.6 | |

(a) Display the data with a stemplot or histogram and a boxplot. Describe the distribution.

(b) Find a 95% confidence interval for the mean TRAP. Comment on the suitability of using this procedure for these data.

7.120 **Transforming the data.** Refer to Exercise 7.118 and the OC data for 31 females. Variables that measure concentrations such as this often have distributions that are skewed to the right. For this reason it is common to work with the logarithms of the measured values. Here are the OC values transformed with the (natural) log:

| | | | | | | | |
|---|---|---|---|---|---|---|---|
| 4.23 | 4.03 | 4.00 | 3.44 | 3.59 | 3.45 | 3.97 | 3.65 |
| 3.58 | 4.34 | 3.79 | 3.69 | 4.36 | 4.00 | 2.29 | 3.03 |
| 3.00 | 2.84 | 3.19 | 3.04 | 2.88 | 2.98 | 2.77 | 3.03 |
| 2.09 | 2.96 | 2.83 | 2.31 | 3.86 | 3.41 | 2.84 | |

(a) Display the data with a stemplot and a boxplot. Describe the distribution.

(b) Find a 95% confidence interval for the mean OC. Comment on the suitability of using this procedure for these data.

(c) Transform the mean and the endpoints of the confidence interval back to the original scale, mg/ml. Compare this interval with the one you computed in Exercise 7.118.

7.121 **More on transforming the data.** Refer to Exercise 7.119 and the TRAP data for 31 females. Variables that measure concentrations such as this often have distributions that are skewed to the right. For this reason it is common to work with the logarithms of the measured values. Here are the TRAP values transformed with the (natural) log:

| | | | | | | | |
|---|---|---|---|---|---|---|---|
| 2.97 | 3.24 | 2.94 | 2.20 | 2.95 | 2.68 | 3.23 | 2.68 |
| 3.36 | 2.70 | 2.37 | 1.77 | 3.17 | 2.94 | 1.93 | 2.09 |
| 2.25 | 1.84 | 2.31 | 2.35 | 2.20 | 2.17 | 2.10 | 2.33 |
| 1.19 | 2.31 | 2.25 | 2.09 | 2.92 | 2.67 | 2.26 | |

(a) Display the data with a stemplot and a boxplot. Describe the distribution.

(b) Find a 95% confidence interval for the mean TRAP. Comment on the suitability of using this procedure for these data.

(c) Transform the mean and the endpoints of the confidence interval back to the original scale, U/l. Compare this interval with the one you computed in Exercise 7.119.

**7.122** ⚔ **Analysis of tree size using the complete data set.** The data used in Exercises 7.25 (page 442), 7.81, and 7.82 (page 471) were obtained by taking simple random samples from the 584 longleaf pine trees that were measured in the Wade Tract. The entire data set is given in the LONGLEAF data set. More details about this data set can be found in the Data Appendix at the back of the book. Find the 95% confidence interval for the mean DBH using the entire data set, and compare this interval with the one that you calculated in Exercise 7.25. Write a report about these data. Include comments on the effect of the sample size on the margin of error, the distribution of the data, the appropriateness of the Normality-based methods for this problem, and the generalizability of the results to other similar stands of longleaf pine or other kinds of trees in this area of the United States and other areas.

**7.123** ⚔ **More on the complete tree size data set.** Use the LONGLEAF data set to repeat the calculations that you performed in Exercises 7.81 and 7.82. Discuss the effect of the sample size on the results.

**7.124** ⚔ **Even more on the complete tree size data set.** The DBH measures in the LONGLEAF data set do not appear to be Normally distributed. Make a histogram of the data and a Normal quantile plot if you have the software available. Mark the mean and the median on the histogram. Now, transform the data using a logarithm. Does this make the distribution appear

to be Normal? Use the same graphical summaries with the mean and the median marked on the histogram. Write a summary of your conclusions, paying particular attention to the use of data such as these for inference using the methods based on Normal distributions.

**7.125 Competitive prices?** A retailer entered into an exclusive agreement with a supplier who guaranteed to provide all products at competitive prices. The retailer eventually began to purchase supplies from other vendors who offered better prices. The original supplier filed a legal action claiming violation of the agreement. In defense, the retailer had an audit performed on a random sample of invoices. For each audited invoice, all purchases made from other suppliers were examined and the prices were compared with those offered by the original supplier. For each invoice, the percent of purchases for which the alternate supplier offered a lower price than the original supplier was recorded.[35] Here are the data:

| | | | | | | | | | | |
|---|---|---|---|---|---|---|---|---|---|---|
| 0 | 100 | 0 | 100 | 33 | 34 | 100 | 48 | 78 | 100 | 77 | 100 | 38 |
| 68 | 100 | 79 | 100 | 100 | 100 | 100 | 100 | 100 | 89 | 100 | 100 |

Report the average of the percents with a 95% margin of error. Do the sample invoices suggest that the original supplier's prices are not competitive on the average?

**7.126 Weight-loss programs.** In a study of the effectiveness of weight-loss programs, 47 subjects who were at least 20% overweight took part in a group support program for 10 weeks. Private weighings determined each subject's weight at the beginning of the program and 6 months after the program's end. The matched pairs $t$ test was used to assess the significance of the average weight loss. The paper reporting the study said, "The subjects lost a significant amount of weight over time, $t(46) = 4.68$, $p < 0.01$." It is common to report the results of statistical tests in this abbreviated style.[36]

(a) Why was the matched pairs statistic appropriate?

(b) Explain to someone who knows no statistics but is interested in weight-loss programs what the practical conclusion is.

(c) The paper follows the tradition of reporting significance only at fixed levels such as $\alpha = 0.01$. In fact, the results are more significant than

"$p < 0.01$" suggests. What can you say about the $P$-value of the $t$ test?

**7.127** 🏔 **Do women perform better in school?** Some research suggests that women perform better than men in school but men score higher on standardized tests. Table 1.9 (page 29) presents data on a measure of school performance, grade point average (GPA), and a standardized test, IQ, for 78 seventh-grade students. Do these data lend further support to the previously found gender differences? Give graphical displays of the data and describe the distributions. Use significance tests and confidence intervals to examine this question, and prepare a short report summarizing your findings.

**7.128** 🏔 **Self-concept and school performance.** Refer to the previous exercise. Although self-concept in this study was measured on a scale with values in the data set ranging from 20 to 80, many prefer to think of this kind of variable as having only two possible values: low self-concept or high self-concept. Find the median of the self-concept scores in Table 1.9 and define those students with scores at or below the median to be low-self-concept students and those with scores above the median to be high-self-concept students. Do high-self-concept students have grade point averages that are different from low-self-concept students? What about IQ? Prepare a report addressing these questions. Be sure to include graphical and numerical summaries and confidence intervals, and state clearly the details of significance tests.

**7.129** **Behavior of pet owners.** On the morning of March 5, 1996, a train with 14 tankers of propane derailed near the center of the small Wisconsin town of Weyauwega. Six of the tankers were ruptured and burning when the 1700 residents were ordered to evacuate the town. Researchers study disasters like this so that effective relief efforts can be designed for future disasters. About half of the households with pets did not evacuate all of their pets. A study conducted after the derailment focused on problems associated with retrieval of the pets after the evacuation and characteristics of the pet owners. One of the scales measured "commitment to adult animals," and the people who evacuated all or some of their pets were compared with those who did not evacuate any of their pets. Higher scores indicate that the pet owner is more likely to take actions that benefit the pet.[37] Here are the data summaries:

| Group | $n$ | $\bar{x}$ | $s$ |
|---|---|---|---|
| Evacuated all or some pets | 116 | 7.95 | 3.62 |
| Did not evacuate any pets | 125 | 6.26 | 3.56 |

Analyze the data and prepare a short report describing the results.

**7.130** **Occupation and diet.** Do various occupational groups differ in their diets? A British study of this question compared 98 drivers and 83 conductors of London double-decker buses.[38] The conductors' jobs require more physical activity. The article reporting the study gives the data as "Mean daily consumption ($\pm$ se)." Some of the study results appear below:

|  | Drivers | Conductors |
|---|---|---|
| Total calories | $2821 \pm 44$ | $2844 \pm 48$ |
| Alcohol (grams) | $0.24 \pm 0.06$ | $0.39 \pm 0.11$ |

(a) What does "se" stand for? Give $\bar{x}$ and $s$ for each of the four sets of measurements.

(b) Is there significant evidence at the 5% level that conductors consume more calories per day than do drivers? Use the two-sample $t$ method to give a $P$-value, and then assess significance.

(c) How significant is the observed difference in mean alcohol consumption? Use two-sample $t$ methods to obtain the $P$-value.

(d) Give a 95% confidence interval for the mean daily alcohol consumption of London double-decker bus conductors.

(e) Give a 99% confidence interval for the difference in mean daily alcohol consumption between drivers and conductors.

**7.131** **Occupation and diet, continued (optional).** Use of the pooled two-sample $t$ test is justified in part (b) of the previous exercise. Explain why. Find the $P$-value for the pooled $t$ statistic, and compare with your result in the previous exercise.

**7.132** **Conditions for inference.** The report cited in Exercise 7.130 says that the distribution of alcohol consumption among the individuals studied is "grossly skew."

(a) Do you think that this skewness prevents the use of the two-sample $t$ test for equality of means? Explain your answer.

**(b)** (Optional) Do you think that the skewness of the distributions prevents the use of the $F$ test for equality of standard deviations? Explain your answer.

**7.133 More on conditions for inference.** Table 1.2 (page 10) gives literacy rates for men and women in 17 Islamic nations. Is it proper to apply the one-sample $t$ method to these data to give a 95% confidence interval for the mean literacy rate of Islamic men? Explain your answer.

**7.134** ⚠ **PCBs in fish.** Polychlorinated biphenyls (PCBs) are a collection of compounds that are no longer produced in the United States but are still found in the environment. Evidence suggests that they can cause harmful health effects when consumed. Because PCBs can accumulate in fish, efforts have been made to identify areas where fish contain excessive amounts so that recommendations concerning consumption limits can be made. There are over 200 types of PCBs. Data from the Environmental Protection Agency National Study of Residues in Lake Fish are given in the data set PCB. More details about this data set can be found in the Data Appendix. Various lakes in the United States were sampled and the amounts of PCBs in fish were measured. The variable PCB is the sum of the amounts of all PCBs found in the fish. The units are parts per billion (ppb).

**(a)** Use graphical and numerical summaries to describe the distribution of this variable. Include a histogram with the location of the mean and the median clearly marked.

**(b)** Do you think it is appropriate to use methods based on Normal distributions for these data? Explain why or why not.

**(c)** Find a 95% confidence interval for the mean. Will this interval contain approximately 95% of the observations in the data set? Explain your answer.

**(d)** Transform the PCB variable with a logarithm. Analyze the transformed data and summarize your results. Do you prefer to work with the raw data or with logs for this variable? Give reasons for your answer.

**(e)** Visit the Web site `http://epa.gov/ waterscience/fishstudy/` to find details about how the data were collected. Write a summary describing these details and discuss how the results from this study can be generalized to other settings.

**7.135** ⚠ **PCBs in fish, continued.** Refer to the previous exercise. Not all types of PCBs are equally harmful. A scale has been developed to convert the raw amount of each type of PCB to a toxic equivalent (TEQ). The PCB data set contains a variable TEQPCB that is the total TEQ from all PCBs found in each sample. Using the questions in the previous exercise, analyze these data and summarize the results.

**7.136** ⚠ **Inference using the complete CRP data set.** In Exercise 7.26 (page 442) you analyzed the C-reactive protein (CRP) scores for a random sample of 40 children who participated in a study in Papua New Guinea. Serum retinol for the same children was analyzed in Exercise 7.28. Data for all 90 children who participated in the study are given in the data set PNG, described in the Data Appendix. Researchers who analyzed these data along with data from several other countries were interested in whether or not infections (as indicated by high CRP values) were associated with lower levels of serum retinol. A child with a value of CRP greater than 5.0 mg/l is classified as recently infected. Those whose CRP is less than or equal to 5.0 mg/l are not. Compare the serum retinol levels of the infected and noninfected children. Include graphical and numerical summaries, comments on all assumptions, and details of your analyses. Write a short report summarizing your results.

**7.137** ⚠ **More on using the complete CRP data set.** Refer to the previous exercise. The researchers in this study also measured $\alpha$1-acid glycoprotein (AGP). This protein is similar to CRP in that it is an indicator of infection. However, it rises more slowly than CRP and reaches a maximum 2 to 3 days after an infection. The units for AGP are grams per liter (g/l), and any value greater than 1.0 g/l is an indication of infection. Analyze the data on AGP in the data set PNG and write a report summarizing your results.

**7.138** ⚠ **Male and female CS students (optional).** Is there a difference between the average SAT scores of males and females? The CSDATA data set, described in the Data Appendix, gives the Math (SATM) and Verbal (SATV) scores for a group of 224 computer science majors. The variable SEX indicates whether each individual is male or female.

**(a)** Compare the two distributions graphically, and then use the two-sample $t$ test to compare the average SATM scores of males and females.

Is it appropriate to use the pooled $t$ test for this comparison? Write a brief summary of your results and conclusions that refers to both versions of the $t$ test and to the $F$ test for equality of standard deviations. Also give a 95% confidence interval for the difference in the means.

(b) Answer part (a) for the SATV scores.

(c) The students in the CSDATA data set were all computer science majors who began college during a particular year. To what extent do you think that your results would generalize to (*i*) computer science students entering in different years, (*ii*) computer science majors at other colleges and universities, and (*iii*) college students in general?

**7.139  Different methods of teaching reading.** In the READING data set, described in the Data Appendix, the response variable Post3 is to be compared for three methods of teaching reading. The Basal method is the standard, or control, method, and the two new methods are DRTA and Strat. We can use the methods of this chapter to compare Basal with DRTA and Basal with Strat. Note that to make comparisons among three treatments it is more appropriate to use the procedures that we will learn in Chapter 12.

(a) Is the mean reading score with the DRTA method higher than that for the Basal method? Perform an analysis to answer this question, and summarize your results.

(b) Answer part (a) for the Strat method in place of DRTA.

**7.140  Sample size calculation (optional).** Example 7.13 (page 449) tells us that the mean height of 10-year-old girls is $N(56.4, 2.7)$ and for boys it is $N(55.7, 3.8)$. The null hypothesis that the mean heights of 10-year-old boys and girls are equal is clearly false. The difference in mean heights is $56.4 - 55.7 = 0.7$ inch. Small differences such as this can require large sample sizes to detect. To simplify our calculations, let's assume that the standard deviations are the same, say $\sigma = 3.2$, and that we will measure the heights of an equal number of girls and boys. How many would we need to measure to have a 90% chance of detecting the (true) alternative hypothesis?

**7.141  House prices.** How much more would you expect to pay for a home that has four

bedrooms than for a home that has three? Here are some data for West Lafayette, Indiana.[39] These are the asking prices (in dollars) that the owners have set for their homes.

Four-bedroom homes:

| | | | | | |
|---|---|---|---|---|---|
| 121,900 | 139,900 | 157,000 | 159,900 | 176,900 | 224,900 |
| 235,000 | 245,000 | 294,000 | | | |

Three-bedroom homes:

| | | | | | |
|---|---|---|---|---|---|
| 65,500 | 79,900 | 79,900 | 79,900 | 82,900 | 87,900 |
| 94,000 | 97,500 | 105,000 | 111,900 | 116,900 | 117,900 |
| 119,900 | 122,900 | 124,000 | 125,000 | 126,900 | 127,900 |
| 127,900 | 127,900 | 132,900 | 145,000 | 145,500 | 157,500 |
| 194,000 | 205,900 | 259,900 | 265,000 | | |

(a) Plot the asking prices for the two sets of homes and describe the two distributions.

(b) Test the null hypothesis that the mean asking prices for the two sets of homes are equal versus the two-sided alternative. Give the test statistic with degrees of freedom, the $P$-value, and your conclusion.

(c) Would you consider using a one-sided alternative for this analysis? Explain why or why not.

(d) Give a 95% confidence interval for the difference in mean asking prices.

(e) These data are not SRSs from a population. Give a justification for use of the two-sample $t$ procedures in this case.

**7.142  More on house prices.** Go to the Web site www.realtor.com and select two geographical areas of interest to you. You will compare the prices of similar types of homes in these two areas. State clearly how you define the areas and the type of homes. For example, you can use city names or zip codes to define the area and you can select single-family homes or condominiums. We view these homes as representative of the asking prices of homes for these areas at the time of your search. If the search gives a large number of homes, select a random sample. Be sure to explain exactly how you do this. Use the methods you have learned in this chapter to compare the asking prices. Be sure to include a graphical summary.

# Inference for Proportions

Does a new medicine reduce the chance of getting a cold? A randomized comparative experiment is often used to answer this question. This chapter describes procedures for statistical inference when the response variable is Yes/No.

## Introduction

Many statistical studies produce counts rather than measurements. For example, the data from an opinion poll that asks a sample of adults whether they approve of the conduct of the president in office are the counts of "Yes," "No," and "Don't know." In an experiment that compares the effectiveness of four cold prevention treatments, the data are the number of subjects given each treatment and the number of subjects in each treatment group who catch a cold during the next month. Similarly, in a survey on driving behavior, the proportions of men and women who admit to shouting, cursing, or making gestures to other drivers in the last year are compared using count data. This chapter, and the next, present procedures for statistical inference in these settings.

The parameters we want to do inference about are population proportions. Just as in the case of inference about population means, we may be concerned with a single population or with comparing two populations. Inference about proportions in these one-sample and two-sample settings is very similar to inference about means, which we discussed in Chapter 7.

We begin in Section 8.1 with inference about a single population proportion. The statistical model for a count is then the binomial distribution, which we studied in Section 5.1. Section 8.2 concerns methods for comparing two proportions. Binomial distributions again play an important role.

# 8.1 Inference for a Single Proportion

We want to estimate the proportion $p$ of some characteristic, such as approval of the president's conduct in office, among the members of a large population. We select a simple random sample (SRS) of size $n$ from the population and record the count $X$ of "successes" (such as "Yes" answers to a question about the president). We will use "success" to represent the characteristic of interest. The sample proportion of successes $\hat{p} = X/n$ estimates the unknown population proportion $p$. If the population is much larger than the sample (say, at least 20 times as large), the count $X$ has approximately the binomial distribution $B(n, p)$.[1] In statistical terms, we are concerned with inference about the probability $p$ of a success in the binomial setting.

If the sample size $n$ is very small, we must base tests and confidence intervals for $p$ on the binomial distributions. These are awkward to work with because of the discreteness of the binomial distributions.[2] But we know that when the sample is large, both the count $X$ and the sample proportion $\hat{p}$ are approximately Normal. We will consider only inference procedures based on the Normal approximation. These procedures are similar to those for inference about the mean of a Normal distribution.

**LOOK BACK**

Normal approximation for counts, page 323

## Large-sample confidence interval for a single proportion

The unknown population proportion $p$ is estimated by the sample proportion $\hat{p} = X/n$. If the sample size $n$ is sufficiently large, $\hat{p}$ has approximately the Normal distribution, with mean $\mu_{\hat{p}} = p$ and standard deviation $\sigma_{\hat{p}} = \sqrt{p(1-p)/n}$. This means that approximately 95% of the time $\hat{p}$ will be within $2\sqrt{p(1-p)/n}$ of the unknown population proportion $p$.

**LOOK BACK**

Normal approximation for proportions, page 323

standard error, page 418

Note that the standard deviation $\sigma_{\hat{p}}$ depends upon the unknown parameter $p$. To estimate this standard deviation using the data, we replace $p$ in the formula by the sample proportion $\hat{p}$. As we did in Chapter 7, we use the term *standard error* for the standard deviation of a statistic that is estimated from data. Here is a summary of the procedure.

---

### LARGE-SAMPLE CONFIDENCE INTERVAL FOR A POPULATION PROPORTION

Choose an SRS of size $n$ from a large population with unknown proportion $p$ of successes. The **sample proportion** is

$$\hat{p} = \frac{X}{n}$$

where $X$ is the number of successes. The **standard error of $\hat{p}$** is

$$SE_{\hat{p}} = \sqrt{\frac{\hat{p}(1-\hat{p})}{n}}$$

and the **margin of error** for confidence level $C$ is

$$m = z^* \text{SE}_{\hat{p}}$$

where $z^*$ is the value for the standard Normal density curve with area $C$ between $-z^*$ and $z^*$. An **approximate level $C$ confidence interval** for $p$ is

$$\hat{p} \pm m$$

Use this interval for 90%, 95%, or 99% confidence when the number of successes and the number of failures are both at least 15.

**EXAMPLE**

**8.1 Proportion of frequent binge drinkers.** Alcohol abuse has been described by college presidents as the number one problem on campus, and it is a major cause of death in young adults. How common is it? A survey of 13,819 students in U.S. four-year colleges collected information on drinking behavior and alcohol-related problems.[3]

The researchers defined "binge drinking" as having five or more drinks in a row for men and four or more drinks in a row for women. "Frequent binge drinking" was defined as binge drinking three or more times in the past two weeks. According to this definition, 3140 students were classified as frequent binge drinkers. The proportion of drinkers is

$$\hat{p} = \frac{3140}{13{,}819} = 0.227$$

To find a 95% confidence interval, first compute the standard error:

$$\text{SE}_{\hat{p}} = \sqrt{\frac{\hat{p}(1 - \hat{p})}{n}}$$

$$= \sqrt{\frac{(0.227)(1 - 0.227)}{13{,}819}}$$

$$= 0.00356$$

Approximately 95% of the time, $\hat{p}$ will be within two standard errors ($2 \times 0.00356 = 0.00712$) of the true $p$. From Table A or D we find the value of $z^*$ to be 1.960. So the confidence interval is

$$\hat{p} \pm z^* \text{SE}_{\hat{p}} = 0.227 \pm (1.960)(0.00356)$$

$$= 0.227 \pm 0.007$$

$$= (0.220, 0.234)$$

We estimate with 95% confidence that between 22.0% and 23.4% of college students are frequent binge drinkers. In other words, we estimate that 22.7% of college students are frequent binge drinkers, with a 95% confidence level margin of error of 0.7%.

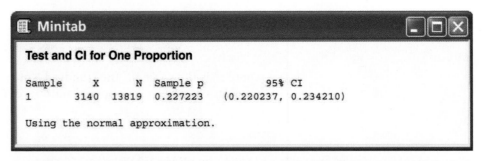

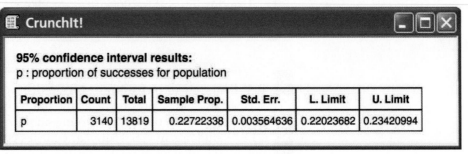

FIGURE 8.1 Minitab and CrunchIt! output for Example 8.1. By default, Minitab outputs an interval based on the binomial distribution. The large-sample confidence interval shown in the figure can be requested as an option.

Because the calculations for statistical inference for a single proportion are relatively straightforward, many software packages do not include them. Figure 8.1 gives output from Minitab and CrunchIt! for Example 8.1. As usual, the output reports more digits than are useful. *When you use software, be sure to think about how many digits are meaningful for your purposes.*

Remember that the margin of error in this confidence interval includes only random sampling error. There are other sources of error that are not taken into account. This survey used a design where the number of students sampled was proportional to the size of the college they attended; in this way we can treat the data as if we had an SRS. However, as is the case with many such surveys, we are forced to assume that the respondents provided accurate information. If the students did not answer the questions honestly, the results may be biased. Furthermore, we also have the typical problem of nonresponse. The response rate for this survey was 60%, a very good rate for surveys of this type. Do the students who did not respond have different drinking habits than those who did? If so, this is another source of bias.

We recommend the large-sample confidence interval for 90%, 95%, and 99% confidence whenever the number of successes and the number of failures are both at least 15. For smaller sample sizes, we recommend exact methods that use the binomial distribution. These are available as the default (for example, in Minitab and SAS) or as options in many statistical software packages and we do not cover them here. There is also an intermediate case between large samples and very small samples where a slight modification of the large-sample approach works quite well.[4] This method is called the "plus four" procedure and is described later.

## USE YOUR KNOWLEDGE

8.1 **Owning a cell phone.** In a 2004 survey of 1200 undergraduate students throughout the United States, 89% of the respondents said they owned a cell phone.[5] For 90% confidence, what is the margin of error?

8.2 **Importance of cell phone "features and functions."** In that same survey, one question asked what aspect was most important when buying a cell phone. "Features and functions" was the choice for 336 students. Give a 95% confidence interval for the proportion of U.S. students who find "features and functions" the most important aspect when buying a phone.

## BEYOND THE BASICS

### The Plus Four Confidence Interval for a Single Proportion

Computer studies reveal that confidence intervals based on the large-sample approach can be quite inaccurate when the number of successes and the number of failures are not at least 15. When this occurs, a simple adjustment to the confidence interval works very well in practice. The adjustment is based on assuming that the sample contains 4 additional observations, 2 of which are successes and 2 of which are failures. The estimator of the population proportion based on this *plus four* rule is

$$\tilde{p} = \frac{X + 2}{n + 4}$$

plus four estimate

This estimate was first suggested by Edwin Bidwell Wilson in 1927 and we call it the **plus four estimate.** The confidence interval is based on the $z$ statistic obtained by standardizing the plus four estimate $\tilde{p}$. Because $\tilde{p}$ is the sample proportion for our modified sample of size $n + 4$, it isn't surprising that the distribution of $\tilde{p}$ is close to the Normal distribution with mean $p$ and standard deviation $\sqrt{p(1 - p)/(n + 4)}$. To get a confidence interval, we estimate $p$ by $\tilde{p}$ in this standard deviation to get the standard error of $\tilde{p}$. Here is the final result.

---

**PLUS FOUR CONFIDENCE INTERVAL FOR A SINGLE PROPORTION**

Choose an SRS of size $n$ from a large population with unknown proportion $p$ of successes. The **plus four estimate of the population proportion** is

$$\tilde{p} = \frac{X + 2}{n + 4}$$

where $X$ is the number of successes. The **standard error of $\tilde{p}$** is

$$\text{SE}_{\tilde{p}} = \sqrt{\frac{\tilde{p}(1 - \tilde{p})}{n + 4}}$$

and the **margin of error** for confidence level $C$ is

$$m = z^* \text{SE}_{\tilde{p}}$$

where $z^*$ is the value for the standard Normal density curve with area $C$ between $-z^*$ and $z^*$. An **approximate level $C$ confidence interval** for

---

$p$ is

$$\tilde{p} \pm m$$

Use this interval for 90%, 95%, or 99% confidence whenever the sample size is at least $n = 10$.

**EXAMPLE**

**8.2 Percent of equol producers.** Research has shown that there are many health benefits associated with a diet that contains soy foods. Substances in soy called isoflavones are known to be responsible for these benefits. When soy foods are consumed, some subjects produce a chemical called equol, and it is thought that production of equol is a key factor in the health benefits of a soy diet. Unfortunately, not all people are equol producers; there appear to be two distinct subpopulations: equol producers and equol nonproducers.[6]

A nutrition researcher planning some bone health experiments would like to include some equol producers and some nonproducers among her subjects. A preliminary sample of 12 female subjects were measured, and 4 were found to be equol producers. We would like to estimate the proportion of equol producers in the population from which this researcher will draw her subjects.

The plus four estimate of the proportion of equol producers is

$$\tilde{p} = \frac{4+2}{12+4} = \frac{6}{16} = 0.375$$

For a 95% confidence interval, we use Table D to find $z^* = 1.96$. We first compute the standard error

$$SE_{\tilde{p}} = \sqrt{\frac{\tilde{p}(1 - \tilde{p})}{n + 4}}$$

$$= \sqrt{\frac{(0.375)(1 - 0.375)}{16}}$$

$$= 0.12103$$

and then the margin of error

$$m = z^*SE_{\tilde{p}}$$

$$= (1.96)(0.12103)$$

$$= 0.237$$

So the confidence interval is

$$\tilde{p} \pm m = 0.375 \pm 0.237$$

$$= (0.138, 0.612)$$

We estimate with 95% confidence that between 14% and 61% of women from this population are equol producers.

If the true proportion of equol users is near 14%, the lower limit of this interval, there may not be a sufficient number of equol producers in the study if subjects are tested only after they are enrolled in the experiment. It may be necessary to determine whether or not a potential subject is an equol producer. The study could then be designed to have the same number of equol producers and nonproducers.

## Significance test for a single proportion

**LOOK BACK**

**Normal approximation for proportions, page 323**

Recall that the sample proportion $\hat{p} = X/n$ is approximately Normal, with mean $\mu_{\hat{p}} = p$ and standard deviation $\sigma_{\hat{p}} = \sqrt{p(1-p)/n}$. For confidence intervals, we substitute $\hat{p}$ for $p$ in the last expression to obtain the standard error. When performing a significance test, however, the null hypothesis specifies a value for $p$, and we assume that this is the true value when calculating the $P$-value. Therefore, when we test $H_0 : p = p_0$, we substitute $p_0$ into the expression for $\sigma_{\hat{p}}$ and then standardize $\hat{p}$. Here are the details.

---

### LARGE-SAMPLE SIGNIFICANCE TEST FOR A POPULATION PROPORTION

Draw an SRS of size $n$ from a large population with unknown proportion $p$ of successes. To test the hypothesis $H_0 : p = p_0$, compute the **z statistic**

$$z = \frac{\hat{p} - p_0}{\sqrt{\dfrac{p_0(1 - p_0)}{n}}}$$

In terms of a standard Normal random variable $Z$, the approximate $P$-value for a test of $H_0$ against

$H_a : p > p_0$ is $P(Z \geq z)$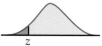

$H_a : p < p_0$ is $P(Z \leq z)$

$H_a : p \neq p_0$ is $2P(Z \geq |z|)$

---

**LOOK BACK**

**sign test for matched pairs, page 439**

We recommend the large-sample $z$ significance test as long as the expected number of successes, $np_0$, and the expected number of failures, $n(1 - p_0)$, are both at least 10. If this rule of thumb is not met, or if the population is less than 20 times as large as the sample, other procedures should be used. One such approach is to use the binomial distribution as we did with the sign test. Here is a large-sample example.

**8.3 Work stress.** According to the National Institute for Occupational Safety and Health,[7] job stress poses a major threat to the health of workers. A national survey of restaurant employees found that 75% said that work stress had a negative impact on their personal lives.[8] A sample of 100 employees of a restaurant chain finds that 68 answer "Yes" when asked, "Does work stress have a negative impact on your personal life?" Is this good reason to think that the proportion of all employees of this chain who would say "Yes" differs from the national proportion $p_0 = 0.75$?

To answer this question, we test

$$H_0: p = 0.75$$
$$H_a: p \neq 0.75$$

The expected numbers of "Yes" and "No" responses are $100 \times 0.75 = 75$ and $100 \times 0.25 = 25$. Both are greater than 10, so we can use the $z$ test. The test statistic is

$$z = \frac{\hat{p} - p_0}{\sqrt{\dfrac{p_0(1 - p_0)}{n}}}$$

$$= \frac{0.68 - 0.75}{\sqrt{\dfrac{(0.75)(0.25)}{100}}} = -1.62$$

From Table A we find $P(Z \leq -1.62) = 0.0526$. The $P$-value is the area in both tails, $P = 2 \times 0.0526 = 0.1052$. Figure 8.2 displays the $P$-value as an area under the standard Normal curve. We conclude that the chain restaurant data are compatible with the survey results ($\hat{p} = 0.68$, $z = -1.62$, $P = 0.11$).

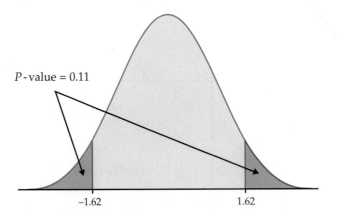

**FIGURE 8.2** The *P*-value for Example 8.3.

$P$-value = 0.11

−1.62    1.62

Figure 8.3 gives computer output from Minitab and CrunchIt! for this example. *Note that for some entries software gives many more digits than we need.* You should decide how many digits are important for your analysis. In general, we will round proportions to two digits, for example, 0.68, and nonsignificant *P*-values to two digits, for example, $P = 0.11$.

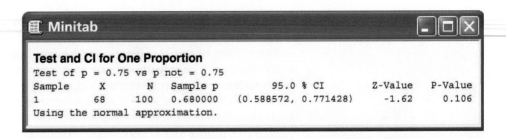

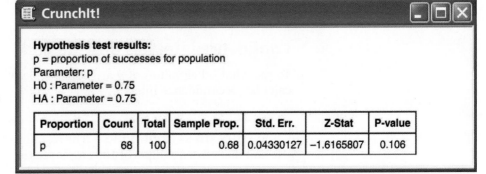

FIGURE 8.3 Minitab and CrunchIt! output for Example 8.3. By default, Minitab performs a test using the binomial distribution. The large-sample significance test shown in the figure can be requested as an option.

In this example we have arbitrarily chosen to associate the response "Yes" that work stress has a negative impact on the respondent's personal life with success and "No" with failure. Suppose we reversed the choice. If we observed that 68 respondents said "Yes," then the other $100 - 68 = 32$ people said "No." Let's repeat the significance test with "No" as the success outcome. The national comparison value for the significance test is now 25%, the proportion in the national survey who responded "No."

> **EXAMPLE**
>
> **8.4 Work stress, revisited.** A sample of 100 restaurant workers were asked whether or not work stress had a negative impact on their personal lives and 32 of them responded "No." A large national survey reported that 25% of workers reported a negative impact. We test the null hypothesis
>
> $$H_0 : p = 0.25$$
>
> against
>
> $$H_a : p \neq 0.25$$
>
> The test statistic is
>
> $$z = \frac{\hat{p} - 0.25}{\sqrt{\dfrac{(0.25)(0.75)}{100}}} = \frac{0.32 - 0.25}{\sqrt{\dfrac{(0.25)(0.75)}{100}}} = 1.62$$
>
> Using Table A, we find that $P = 0.11$.

When we interchanged "Yes" and "No" (or success and failure), we simply changed the sign of the test statistic $z$. The $P$-value remained the same. These facts are true in general. Our conclusion does not depend on an arbitrary choice of success and failure.

The results of our significance test have limited use in this example, as in many cases of inference about a single parameter. Of course, we do not expect the experience of the restaurant workers to be *exactly* the same as that of the workers in the national survey. *If the sample of restaurant workers is sufficiently large, we will have sufficient power to detect a very small difference. On the other hand, if our sample size is very small, we may be unable to detect differences that could be very important.* For these reasons we prefer to include a confidence interval as part of our analysis.

## Confidence intervals provide additional information

To see what other values of $p$ are compatible with the sample results, we will calculate a confidence interval.

**8.5 Work stress, continued.** The restaurant worker survey in Example 8.3 found that 68 of a sample of 100 employees agreed that work stress had a negative impact on their personal lives. That is, the sample size is $n = 100$ and the count of successes is $X = 68$. Because the number of successes and the number of failures are both at least 15, we will use the large-sample procedure to compute a 95% confidence interval. The sample proportion is

$$\hat{p} = \frac{X}{n}$$

$$= \frac{68}{100} = 0.68$$

The standard error is

$$SE_{\hat{p}} = \sqrt{\frac{\hat{p}(1 - \hat{p})}{n}}$$

$$= \sqrt{\frac{(0.68)(1 - 0.68)}{100}} = 0.0466$$

The $z$ critical value for 95% confidence is $z^* = 1.96$, so the margin of error is

$$m = z^* SE_{\hat{p}}$$

$$= (1.96)(0.0466)$$

$$= 0.091$$

The confidence interval is

$$\hat{p} \pm m = 0.68 \pm (1.96)(0.0466)$$

$$= 0.68 \pm 0.09$$

$$= (0.59, 0.77)$$

We are 95% confident that between 59% and 77% of the restaurant chain's employees feel that work stress is damaging their personal lives.

The confidence interval of Example 8.5 is much more informative than the significance test of Example 8.3. We have determined the values of $p$ that are consistent with the observed results. Note that the standard error used for the confidence interval is estimated from the data, whereas the denominator in the test statistic $z$ is based on the value assumed in the null hypothesis. A consequence of this fact is that the correspondence between the significance test result and the confidence interval is no longer exact. However, the correspondence is still very close. The confidence interval $(0.59, 0.77)$ gives an approximate range of $p_0$'s that would not be rejected by a test at the $\alpha = 0.05$ level of significance. We would not be surprised if the true proportion of restaurant workers who would say that work stress has a negative impact on their lives was as low as 60% or as high as 75%.

*We do not often use significance tests for a single proportion, because it is uncommon to have a situation where there is a precise $p_0$ that we want to test.* For physical experiments such as coin tossing or drawing cards from a well-shuffled deck, probability arguments lead to an ideal $p_0$. Even here, however, it can be argued, for example, that no real coin has a probability of heads *exactly* equal to 0.5. Data from past large samples can sometimes provide a $p_0$ for the null hypothesis of a significance test. In some types of epidemiology research, for example, "historical controls" from past studies serve as the benchmark for evaluating new treatments. Medical researchers argue about the validity of these approaches, because the past never quite resembles the present. In general, we prefer comparative studies whenever possible.

## USE YOUR KNOWLEDGE

8.3 **Working while enrolled in school.** A 1993 nationwide survey by the National Center for Education Statistics reports that 72% of all undergraduates work while enrolled in school.[9] You decide to test whether this percent is different at your university. In your random sample of 100 students, 77 said they were currently working.

(a) Give the null and alternative hypotheses for this study.

(b) Carry out the significance test. Report the test statistic and *P*-value.

(c) Does it appear that the percent of students working at your university is different at the $\alpha = 0.05$ level?

8.4 **Owning a cell phone, continued.** Refer to Exercise 8.1 (page 490). It was reported that cell phone ownership by undergraduate students in 2003 was 83%. Do the sample data in 2004 give good evidence that this percent has increased?

(a) Give the null and alternative hypotheses.

(b) Carry out the significance test. Report the test statistic and the *P*-value.

(c) State your conclusion using $\alpha = 0.05$.

## Choosing a sample size

**LOOK BACK**

choosing the sample
size, page 364

In Chapter 6, we showed how to choose the sample size $n$ to obtain a confidence interval with specified margin of error $m$ for a Normal mean. Because we are using a Normal approximation for inference about a population proportion, sample size selection proceeds in much the same way.

Recall that the margin of error for the large-sample confidence interval for a population proportion is

$$m = z^* \text{SE}_{\hat{p}} = z^* \sqrt{\frac{\hat{p}(1-\hat{p})}{n}}$$

Choosing a confidence level $C$ fixes the critical value $z^*$. The margin of error also depends on the value of $\hat{p}$ and the sample size $n$. Because we don't know the value of $\hat{p}$ until we gather the data, we must guess a value to use in the calculations. We will call the guessed value $p^*$. There are two common ways to get $p^*$:

1. Use the sample estimate from a pilot study or from similar studies done earlier.

2. Use $p^* = 0.5$. Because the margin of error is largest when $\hat{p} = 0.5$, this choice gives a sample size that is somewhat larger than we really need for the confidence level we choose. It is a safe choice no matter what the data later show.

Once we have chosen $p^*$ and the margin of error $m$ that we want, we can find the $n$ we need to achieve this margin of error. Here is the result.

---

### SAMPLE SIZE FOR DESIRED MARGIN OF ERROR

The level $C$ confidence interval for a proportion $p$ will have a margin of error approximately equal to a specified value $m$ when the sample size satisfies

$$n = \left(\frac{z^*}{m}\right)^2 p^*(1 - p^*)$$

Here $z^*$ is the critical value for confidence $C$, and $p^*$ is a guessed value for the proportion of successes in the future sample.

The margin of error will be less than or equal to $m$ if $p^*$ is chosen to be 0.5. The sample size required when $p^* = 0.5$ is

$$n = \frac{1}{4}\left(\frac{z^*}{m}\right)^2$$

---

The value of $n$ obtained by this method is not particularly sensitive to the choice of $p^*$ when $p^*$ is fairly close to 0.5. However, if the value of $p$ is likely to be smaller than about 0.3 or larger than about 0.7, use of $p^* = 0.5$ may result in a sample size that is much larger than needed.

**EXAMPLE**

**8.6 Planning a survey of students.** A large university is interested in assessing student satisfaction with the overall campus environment. The plan is to distribute a questionnaire to an SRS of students, but before proceeding, the university wants to determine how many students to sample. The questionnaire asks about a student's degree of satisfaction with various student services, each measured on a five-point scale. The university is interested in the proportion $p$ of students who are satisfied (that is, who choose either "satisfied" or "very satisfied," the two highest levels on the five-point scale).

The university wants to estimate $p$ with 95% confidence and a margin of error less than or equal to 3%, or 0.03. For planning purposes, they are willing to use $p^* = 0.5$. To find the sample size required,

$$n = \frac{1}{4}\left(\frac{z^*}{m}\right)^2 = \frac{1}{4}\left[\frac{1.96}{0.03}\right]^2 = 1067.1$$

Round up to get $n = 1068$. (Always round up. Rounding down would give a margin of error slightly greater than 0.03.)

Similarly, for a 2.5% margin of error we have (after rounding up)

$$n = \frac{1}{4}\left[\frac{1.96}{0.025}\right]^2 = 1537$$

and for a 2% margin of error,

$$n = \frac{1}{4}\left[\frac{1.96}{0.02}\right]^2 = 2401$$

News reports frequently describe the results of surveys with sample sizes between 1000 and 1500 and a margin of error of about 3%. These surveys generally use sampling procedures more complicated than simple random sampling, so the calculation of confidence intervals is more involved than what we have studied in this section. The calculations in Example 8.6 nonetheless show in principle how such surveys are planned.

In practice, many factors influence the choice of a sample size. The following example illustrates one set of factors.

**EXAMPLE**

**8.7 Assessing interest in Pilates classes.** The Division of Recreational Sports (Rec Sports) at a major university is responsible for offering comprehensive recreational programs, services, and facilities to the students. Rec Sports is continually examining its programs to determine how well it is meeting the needs of the students. Rec Sports is considering adding some new programs and would like to know how much interest there is in a new exercise program based on the Pilates method.[10] They will take a survey of undergraduate students. In the past, they emailed short surveys to all undergraduate students. The response rate obtained in this way was about 5%. This time they will send emails to a simple random sample of the students and will follow up with additional emails and eventually a phone call to get a higher response rate. Because of limited staff and the work involved with the follow-up, they would like to use a sample size of about 200. One of the questions they will ask is "Have you ever heard about the Pilates method of exercise?"

The primary purpose of the survey is to estimate various sample proportions for undergraduate students. Will the proposed sample size of $n = 200$ be adequate to provide Rec Sports with the needed information? To address this question, we calculate the margins of error of 95% confidence intervals for various values of $\hat{p}$.

**EXAMPLE**

**8.8 Margins of error.** In the Rec Sports survey, the margin of error of a 95% confidence interval for any value of $\hat{p}$ and $n = 200$ is

$$m = z^* SE_{\hat{p}}$$

$$= 1.96\sqrt{\frac{\hat{p}(1 - \hat{p})}{200}}$$

$$= 0.139\sqrt{\hat{p}(1 - \hat{p})}$$

The results for various values of $\hat{p}$ are

| $\hat{p}$ | $m$ | $\hat{p}$ | $m$ |
|------|-------|------|-------|
| 0.05 | 0.030 | 0.60 | 0.068 |
| 0.10 | 0.042 | 0.70 | 0.064 |
| 0.20 | 0.056 | 0.80 | 0.056 |
| 0.30 | 0.064 | 0.90 | 0.042 |
| 0.40 | 0.068 | 0.95 | 0.030 |
| 0.50 | 0.070 |      |       |

Rec Sports judged these margins of error to be acceptable, and they used a sample size of 200 in their survey.

The table in Example 8.8 illustrates two points. First, the margins of error for $\hat{p} = 0.05$ and $\hat{p} = 0.95$ are the same. The margins of error will always be the same for $\hat{p}$ and $1 - \hat{p}$. This is a direct consequence of the form of the confidence interval. Second, the margin of error varies only between 0.064 and 0.070 as $\hat{p}$ varies from 0.3 to 0.7, and the margin of error is greatest when $\hat{p} = 0.5$, as we claimed earlier. It is true in general that the margin of error will vary relatively little for values of $\hat{p}$ between 0.3 and 0.7. Therefore, when planning a study, it is not necessary to have a very precise guess for $p$. If $p^* = 0.5$ is used and the observed $\hat{p}$ is between 0.3 and 0.7, the actual interval will be a little shorter than needed but the difference will be small.

*Again it is important to emphasize that these calculations consider only the effects of sampling variability that are quantified in the margin of error.* Other sources of error, such as nonresponse and possible misinterpretation of questions, are not included in the table of margins of error for Example 8.8. Rec Sports is trying to minimize these kinds of errors. They did a pilot study using a small group of current users of their facilities to check the wording of the questions, and they devised a careful plan to follow up with the students who did not respond to the initial email.

## USE YOUR KNOWLEDGE

8.5   **Confidence level and sample size.** Refer to Example 8.6 (page 499). Suppose the university was interested in a 90% confidence interval with margin of error 0.03. Would the required sample size be smaller or larger than 1068 students? Verify this by performing the calculation.

8.6   **Calculating the sample size.** Refer to Exercise 8.3 (page 497). You plan to do a larger survey such that the 95% margin of error is no larger than 0.02. Using the results from the small survey of 100 students, what sample size would you use?

## SECTION 8.1   Summary

Inference about a population proportion $p$ from an SRS of size $n$ is based on the **sample proportion** $\hat{p} = X/n$. When $n$ is large, $\hat{p}$ has approximately the Normal distribution with mean $p$ and standard deviation $\sqrt{p(1-p)/n}$.

For large samples, the **margin of error for confidence level $C$** is

$$m = z^* \mathrm{SE}_{\hat{p}}$$

where $z^*$ is the value for the standard Normal density curve with area $C$ between $-z^*$ and $z^*$, and the **standard error of $\hat{p}$** is

$$\mathrm{SE}_{\hat{p}} = \sqrt{\frac{\hat{p}(1-\hat{p})}{n}}$$

The **level $C$ large-sample confidence interval** is

$$\hat{p} \pm m$$

We recommend using this interval for 90%, 95% and 99% confidence whenever the number of successes and the number of failures are both at least 15. When sample sizes are smaller, alternative procedures such as the **plus four estimate of the population proportion** are recommended.

The **sample size** required to obtain a confidence interval of approximate margin of error $m$ for a proportion is found from

$$n = \left(\frac{z^*}{m}\right)^2 p^*(1-p^*)$$

where $p^*$ is a guessed value for the proportion, and $z^*$ is the standard Normal critical value for the desired level of confidence. To ensure that the margin of error of the interval is less than or equal to $m$ no matter what $\hat{p}$ may be, use

$$n = \frac{1}{4}\left(\frac{z^*}{m}\right)^2$$

population having proportion $p_2$ of successes. The **plus four estimate of the difference in proportions** is

$$\tilde{D} = \tilde{p}_1 - \tilde{p}_2$$

where

$$\tilde{p}_1 = \frac{X_1 + 1}{n_1 + 2} \qquad \tilde{p}_2 = \frac{X_2 + 1}{n_2 + 2}$$

The **standard error of $\tilde{D}$** is

$$SE_{\tilde{D}} = \sqrt{\frac{\tilde{p}_1(1 - \tilde{p}_1)}{n_1 + 2} + \frac{\tilde{p}_2(1 - \tilde{p}_2)}{n_2 + 2}}$$

and the **margin of error** for confidence level $C$ is

$$m = z^* SE_{\tilde{D}}$$

where $z^*$ is the value for the standard Normal density curve with area $C$ between $-z^*$ and $z^*$. An **approximate level $C$ confidence interval** for $p_1 - p_2$ is

$$\tilde{D} \pm m$$

Use this method for 90%, 95%, or 99% confidence when both sample sizes are at least 5.

---

**EXAMPLE**

**8.10 Gender and sexual maturity.** In studies that look for a difference between genders, a major concern is whether or not apparent differences are due to other variables that are associated with gender. Because boys mature more slowly than girls, a study of adolescents that compares boys and girls of the same age may confuse a gender effect with an effect of sexual maturity. The "Tanner score" is a commonly used measure of sexual maturity.[22] Subjects are asked to determine their score by placing a mark next to a rough drawing of an individual at their level of sexual maturity. There are five different drawings, so the score is an integer between 1 and 5.

A pilot study included 12 girls and 12 boys from a population that will be used for a large experiment. Four of the boys and three of the girls had Tanner scores of 4 or 5, a high level of sexual maturity. Let's find a 95% confidence interval for the difference between the proportions of boys and girls who have high (4 or 5) Tanner scores in this population. The numbers of successes and failures in both groups are not all at least 10, so the large-sample approach is not recommended. On the other hand, the sample sizes are both at least 5, so the plus four method is appropriate.

The plus four estimate of the population proportion for boys is

$$\tilde{p}_1 = \frac{X_1 + 1}{n_1 + 2} = \frac{4 + 1}{12 + 2} = 0.3571$$

For girls, the estimate is

$$\tilde{p}_2 = \frac{X_2 + 1}{n_2 + 2} = \frac{3 + 1}{12 + 2} = 0.2857$$

Therefore, the estimate of the difference is

$$\tilde{D} = \tilde{p}_1 - \tilde{p}_2 = 0.3571 - 0.2857 = 0.071$$

The standard error of $\tilde{D}$ is

$$SE_{\tilde{D}} = \sqrt{\frac{\tilde{p}_1(1 - \tilde{p}_1)}{n_1 + 2} + \frac{\tilde{p}_2(1 - \tilde{p}_2)}{n_2 + 2}}$$

$$= \sqrt{\frac{(0.3571)(1 - 0.3571)}{12 + 2} + \frac{(0.2857)(1 - 0.2857)}{12 + 2}}$$

$$= 0.1760$$

For 95% confidence, $z^* = 1.96$ and the margin of error is

$$m = z^* SE_{\tilde{D}} = (1.96)(0.1760) = 0.345$$

The confidence interval is

$$\tilde{D} \pm m = 0.071 \pm 0.345 = (-0.274, 0.416)$$

With 95% confidence we can say that the difference in the proportions is between $-0.274$ and $0.416$. Alternatively, we can report that the difference in the proportions of boys and girls with high Tanner scores in this population is 7.1% with a 95% margin of error of 34.5%.

The very large margin of error in this example indicates that either boys or girls could be more sexually mature in this population and that the difference could be quite large. *Although the interval includes the possibility that there is no difference, corresponding to $p_1 = p_2$ or $p_1 - p_2 = 0$, we must be very cautious about concluding that there is **no** difference in the proportions.* With small sample sizes such as these, the data do not provide us with a lot of information for our inference. This fact is expressed quantitatively through the very large margin of error.

## Significance test for a difference in proportions

Although we prefer to compare two proportions by giving a confidence interval for the difference between the two population proportions, it is sometimes useful to test the null hypothesis that the two population proportions are the same.

We standardize $D = \hat{p}_1 - \hat{p}_2$ by subtracting its mean $p_1 - p_2$ and then dividing by its standard deviation

$$\sigma_D = \sqrt{\frac{p_1(1 - p_1)}{n_1} + \frac{p_2(1 - p_2)}{n_2}}$$

If $n_1$ and $n_2$ are large, the standardized difference is approximately $N(0, 1)$. For the large-sample confidence interval we used sample estimates in place of the unknown population values in the expression for $\sigma_D$. Although this approach would lead to a valid significance test, we instead adopt the more common practice of replacing the unknown $\sigma_D$ with an estimate that takes into account our null hypothesis $H_0: p_1 = p_2$. If these two proportions are equal, then we can view all of the data as coming from a single population. Let $p$ denote the common value of $p_1$ and $p_2$; then the standard deviation of $D = \hat{p}_1 - \hat{p}_2$ is

$$\sigma_D = \sqrt{\frac{p(1-p)}{n_1} + \frac{p(1-p)}{n_2}}$$

$$= \sqrt{p(1-p)\left(\frac{1}{n_1} + \frac{1}{n_2}\right)}$$

We estimate the common value of $p$ by the overall proportion of successes in the two samples:

$$\hat{p} = \frac{\text{number of successes in both samples}}{\text{number of observations in both samples}} = \frac{X_1 + X_2}{n_1 + n_2}$$

**pooled estimate of p**

This estimate of $p$ is called the **pooled estimate** because it combines, or pools, the information from both samples.

To estimate $\sigma_D$ under the null hypothesis, we substitute $\hat{p}$ for $p$ in the expression for $\sigma_D$. The result is a standard error for $D$ that assumes $H_0: p_1 = p_2$:

$$\text{SE}_{Dp} = \sqrt{\hat{p}(1-\hat{p})\left(\frac{1}{n_1} + \frac{1}{n_2}\right)}$$

The subscript on $\text{SE}_{Dp}$ reminds us that we pooled data from the two samples to construct the estimate.

---

### SIGNIFICANCE TEST FOR COMPARING TWO PROPORTIONS

To test the hypothesis

$$H_0: p_1 = p_2$$

compute the **z statistic**

$$z = \frac{\hat{p}_1 - \hat{p}_2}{\text{SE}_{Dp}}$$

where the **pooled standard error** is

$$\text{SE}_{Dp} = \sqrt{\hat{p}(1-\hat{p})\left(\frac{1}{n_1} + \frac{1}{n_2}\right)}$$

and where

$$\hat{p} = \frac{X_1 + X_2}{n_1 + n_2}$$

In terms of a standard Normal random variable $Z$, the $P$-value for a test of $H_0$ against

$H_a: p_1 > p_2$ is $P(Z \geq z)$

$H_a: p_1 < p_2$ is $P(Z \leq z)$

$H_a: p_1 \neq p_2$ is $2P(Z \geq |z|)$

This $z$ test is based on the Normal approximation to the binomial distribution. As a general rule, we will use it when the number of successes and the number of failures in each of the samples are at least 5.

**EXAMPLE**

**8.11 Gender and the proportion of frequent binge drinkers: the $z$ test.** Are men and women college students equally likely to be frequent binge drinkers? We examine the survey data in Example 8.9 (page 507) to answer this question. Here is the data summary:

| Population | $n$ | $X$ | $\hat{p} = X/n$ |
|---|---|---|---|
| 1 (men) | 5,348 | 1,392 | 0.260 |
| 2 (women) | 8,471 | 1,748 | 0.206 |
| Total | 13,819 | 3,140 | 0.227 |

The sample proportions are certainly quite different, but we will perform a significance test to see if the difference is large enough to lead us to believe that the population proportions are not equal. Formally, we test the hypotheses

$$H_0: p_1 = p_2$$
$$H_a: p_1 \neq p_2$$

The pooled estimate of the common value of $p$ is

$$\hat{p} = \frac{1392 + 1748}{5348 + 8471} = \frac{3140}{13,819} = 0.227$$

Note that this is the estimate on the bottom line of the data summary above.

drinking that we used for Example 8.1, survey results from both 1993 and 1999 are presented. Using the table below, test whether the proportions of frequent binge drinkers are different at the 5% level. Also construct a 95% confidence interval for the difference. Write a short summary of your results.

| Year | $n$ | $X$ |
| --- | --- | --- |
| 1993 | 14,995 | 2,973 |
| 1999 | 13,819 | 3,140 |

**8.47** **A comparison of the proportion of frequent binge drinkers, revisited.** Refer to Exercise 8.46. Redo the exercise in terms of the proportion of nonfrequent binge drinkers in each classification. Explain how you could have obtained these results from the calculations you did in Exercise 8.46.

**8.48** **Effects of reducing air pollution.** A study that evaluated the effects of a reduction in exposure to traffic-related air pollutants compared respiratory symptoms of 283 residents of an area with congested streets with 165 residents in a similar area where the congestion was removed because a bypass was constructed. The symptoms of the residents of both areas were evaluated at baseline and again a year after the bypass was completed.[24] For the residents of the congested streets, 17 reported that their symptoms of wheezing improved between baseline and one year later, while 35 of the residents of the bypass streets reported improvement.

(a) Find the two sample proportions.

(b) Report the difference in the proportions and the standard error of the difference.

(c) What are the appropriate null and alternative hypotheses for examining the question of interest? Be sure to explain your choice of the alternative hypothesis.

(d) Find the test statistic. Construct a sketch of the distribution of the test statistic under the assumption that the null hypothesis is true. Find the P-value and use your sketch to explain its meaning.

(e) Is no evidence of an effect the same as evidence that there is no effect? Use a 95% confidence interval to answer this question. Summarize your ideas in a way that could be understood by someone who has very little experience with statistics.

(f) The study was done in the United Kingdom. To what extent do you think that the results can be generalized to other circumstances?

**8.49** **Downloading music from the Internet.** A 2005 survey of Internet users reported that 22% downloaded music onto their computers. The filing of lawsuits by the recording industry may be a reason why this percent has decreased from the estimate of 29% from a survey taken two years before.[25] Assume that the sample sizes are both 1421. Using a significance test, evaluate whether or not there has been a change in the percent of Internet users who download music. Provide all details for the test and summarize your conclusion. Also report a 95% confidence interval for the difference in proportions and explain what information is provided in the interval that is not in the significance test results.

**8.50** **More on downloading music from the Internet.** Refer to the previous exercise. Suppose we are not exactly sure about the sizes of the samples. Redo the calculations for the significance test and the confidence interval under the following assumptions: (*i*) both sample sizes are 1000, (*ii*) both sample sizes are 1600, (*iii*) the first sample size is 1000 and the second is 1600. Summarize the effects of the sample sizes on the results.

**8.51** **Who gets stock options?** Different kinds of companies compensate their key employees in different ways. Established companies may pay higher salaries, while new companies may offer stock options that will be valuable if the company succeeds. Do high-tech companies tend to offer stock options more often than other companies? One study looked at a random sample of 200 companies. Of these, 91 were listed in the *Directory of Public High Technology Corporations* and 109 were not listed. Treat these two groups as SRSs of high-tech and non-high-tech companies. Seventy-three of the high-tech companies and 75 of the non-high-tech companies offered incentive stock options to key employees.[26]

(a) Give a 95% confidence interval for the difference in the proportions of the two types of companies that offer stock options.

(b) Compare the two groups of companies with a significance test.

(c) Summarize your analysis and conclusions.

**8.52** **Cheating during a test: 2002 versus 2004.** In Exercise 8.16, you examined the proportion of high school students who cheated on tests at least twice during the past year. Included in that study were the results for both 2002 and 2004. A reported 9054 out of 24,142 students said they cheated at least twice in 2004. A reported 5794 out of 12,121 students said they cheated at least twice in 2002. Give an estimate of the difference between these two proportions with a 90% confidence interval.

**8.53 Gender bias in textbooks.** To what extent do syntax textbooks, which analyze the structure of sentences, illustrate gender bias? A study of this question sampled sentences from 10 texts.[27] One part of the study examined the use of the words "girl," "boy," "man," and "woman." We will call the first two words *juvenile* and the last two *adult*. Is the proportion of female references that are juvenile (girl) equal to the proportion of male references that are juvenile (boy)? Here are data from one of the texts:

| Gender | n | X(juvenile) |
|--------|-----|-------------|
| Female | 60  | 48          |
| Male   | 132 | 52          |

(a) Find the proportion of juvenile references for females and its standard error. Do the same for the males.

(b) Give a 90% confidence interval for the difference and briefly summarize what the data show.

(c) Use a test of significance to examine whether the two proportions are equal.

**8.54** CHALLENGE **Bicycle accidents, alcohol, and gender.** In Exercise 8.28 (page 504) we examined the percent of fatally injured bicyclists tested for alcohol who tested positive. Here we examine the same data with respect to gender.

| Gender | n | X(tested positive) |
|--------|------|--------------------|
| Female | 191  | 27                 |
| Male   | 1520 | 515                |

(a) Summarize the data by giving the estimates of the two population proportions and a 95% confidence interval for their difference.

(b) The standard error $SE_D$ contains a contribution from each sample, $\hat{p}_1(1 - \hat{p}_1)/n_1$ and $\hat{p}_2(1 - \hat{p}_2)/n_2$. Which of these contributes the larger amount to the standard error of the difference? Explain why.

(c) Use a test of significance to examine whether the two proportions are equal.

**8.55 Pet ownership and gender: the significance test.** In Exercise 8.44 (page 517) we compared the proportion of pet owners who were women with the proportion of non–pet owners who were women in the Health ABC Study. Use a significance test to make the comparison and summarize the results of your analysis.

**8.56 Pet ownership and marital status: the significance test.** In Exercise 8.45 (page 517) we compared the proportion of pet owners who were married with the proportion of non–pet owners who were married in the Health ABC Study. Use a significance test to make the comparison and summarize the results of your analysis.

## CHAPTER 8 Exercises

**8.57 What's wrong?** For each of the following, explain what is wrong and why.

(a) A 90% confidence interval for the difference in two proportions includes errors due to nonresponse.

(b) A $z$ statistic is used to test the null hypothesis that $H_0 : \hat{p}_1 = \hat{p}_2$.

(c) If two sample proportions are equal, then the sample counts must be equal.

**8.58 Using a handheld phone while driving.** Refer to Exercise 8.14 (page 503). This same poll found that 58% of the respondents talked on a handheld phone while driving in the last year. Construct a 90% confidence interval for the proportion of U.S.

drivers who talked on a handheld phone while driving in the last year.

**8.59 Gender and using a handheld phone while driving.** Refer to the previous exercise. In this same report, this percent was broken down into 59% for men and 56% for women. Assuming that, among the 1048 respondents, there were an equal number of men and women, construct a 95% confidence interval for the difference in these proportions.

**8.60** CHALLENGE **Even more on downloading music from the Internet.** The following quotation is from a recent survey of Internet users. The sample size for the survey was 1371. Since 18% of those surveyed said they download music, the sample size for this subsample is 247.

Among current music downloaders, 38% say they are downloading less because of the RIAA suits.... About a third of current music downloaders say they use peer-to-peer networks.... 24% of them say they swap files using email and instant messaging; 20% download files from music-related Web sites like those run by music magazines or musician home-pages. And while online music services like iTunes are far from trumping the popularity of file-sharing networks, 17% of current music downloaders say they are using these paid services. Overall, 7% of Internet users say they have bought music at these new services at one time or another, including 3% who currently use paid services.[28]

(a) For each percent quoted, give the margin of error. You should express these in percents, as given in the quote.

(b) Rewrite the paragraph more concisely and include the margins of error.

(c) Pick either side A or side B below and give arguments in favor of the view that you select.

(A) The margins of error should be included because they are necessary for the reader to properly interpret the results.

(B) The margins of error interfere with the flow of the important ideas. It would be better to just report one margin of error and say that all of the others are no greater than this number.

If you choose view B, be sure to give the value of the margin of error that you report.

8.61    **Proportion of male heavy lottery players.** A study of state lotteries included a random digit dialing (RDD) survey conducted by the National Opinion Research Center (NORC). The survey asked 2406 adults about their lottery spending.[29] A total of 248 individuals were classified as "heavy" players. Of these, 152 were male. The study notes that 48.5% of U.S. adults are male. For this analysis, assume that the 248 heavy lottery players are a random sample of all heavy lottery players and that the margin of error for the 48.5% estimate of the percent of males in the U.S. adult population is so small that it can be neglected. Use a significance test to compare the proportion of males among heavy lottery players with the proportion of males in the U.S. adult population. Construct a 95% confidence interval for the proportion. Write a summary of what you have found. Be sure to comment on the possibility that some people may be reluctant to provide information about their lottery spending and how this might affect the results.

8.62    **Cell phone ownership: 2000 versus 2004.** Refer to Exercise 8.41 (page 517). The estimated proportion of undergraduates owning a phone in 2000 was 43%. We want to test whether the proportion of undergraduate cell phone owners has more than doubled in the last 4 years.

(a) Compute the quantity $\hat{p}_1 - 2\hat{p}_2$ where $\hat{p}_1$ is the 2004 estimate and $\hat{p}_2$ is the 2000 estimate.

(b) Using the rules for variances, compute the standard error of this estimate.

(c) Compute the $z$ statistic and $P$-value. What is your conclusion at the 5% level?

8.63    **More on the effects of reducing air pollution.** In Exercise 8.48 the effects of a reduction in air pollution on wheezing was examined by comparing the one-year change in symptoms in a group of residents who lived on congested streets with a group who lived in an area that had been congested but from which the congestion was removed when a bypass was built. The effect of the reduction in air pollution was assessed by comparing the proportions of residents in the two groups who reported that their wheezing symptoms improved. Here are some additional data from the same study:

| Symptom | Bypass | | Congested | |
|---|---|---|---|---|
| | $n$ | Improved | $n$ | Improved |
| Number of wheezing attacks | 282 | 45 | 163 | 21 |
| Wheezing disturbs sleep | 282 | 45 | 164 | 12 |
| Wheezing limits speech | 282 | 12 | 164 | 4 |
| Wheezing affects activities | 281 | 26 | 165 | 13 |
| Winter cough | 261 | 15 | 156 | 14 |
| Winter phlegm | 253 | 12 | 144 | 10 |
| Consulted doctor | 247 | 29 | 140 | 18 |

The table gives the number of subjects in each group and the number reporting improvement. So, for example, the proportion who reported improvement in the number of wheezing attacks was 21/163 in the congested group.

(a) The reported sample sizes vary from symptom to symptom. Give possible reasons for this and discuss the possible impact on the results.

(b) Calculate the difference in the proportions for each symptom. Make a table of symptoms ordered from highest to lowest based on these differences.

Include the estimates of the differences and the 95% confidence intervals in the table. Summarize your conclusions.

(c) Can you justify a one-sided alternative in this situation? Give reasons for your answer.

(d) Perform a significance test to compare the two groups for each of the symptoms. Summarize the results.

(e) Reanalyze the data using only the data from the bypass group. Give confidence intervals for the proportions that reported improved symptoms. Compare the conclusions that someone might make from these results with those you presented in part (b). Use your analyses of the data in this exercise to discuss the importance of a control group in studies such as this.

8.64 **"No Sweat" garment labels.** Following complaints about the working conditions in some apparel factories both in the United States and abroad, a joint government and industry commission recommended in 1998 that companies that monitor and enforce proper standards be allowed to display a "No Sweat" label on their products. Does the presence of these labels influence consumer behavior? A survey of U.S. residents aged 18 or older asked a series of questions about how likely they would be to purchase a garment under various conditions. For some conditions, it was stated that the garment had a "No Sweat" label; for others, there was no mention of such a label. On the basis of the responses, each person was classified as a "label user" or a "label nonuser."[30] There were 296 women surveyed. Of these, 63 were label users. On the other hand, 27 of 251 men were classified as users.

(a) Give a 95% confidence interval for the difference in the proportions.

(b) You would like to compare the women with the men. Set up appropriate hypotheses, and find the test statistic and the P-value. What do you conclude?

8.65 **Education of the customers.** To devise effective marketing strategies it is helpful to know the characteristics of your customers. A study compared demographic characteristics of people who use the Internet for travel arrangements and of people who do not.[31] Of 1132 Internet users, 643 had completed college. Among the 852 nonusers, 349 had completed college.

(a) Do users and nonusers differ significantly in the proportion of college graduates?

(b) Give a 95% confidence interval for the difference in the proportions.

8.66 **Income of the customers.** The study mentioned in the previous exercise also asked about income. Among Internet users, 493 reported income of less than $50,000 and 378 reported income of $50,000 or more. (Not everyone answered the income question.) The corresponding numbers for nonusers were 477 and 200. Perform a significance test to compare the incomes of users with nonusers and also give an estimate of the difference in proportions with a 95% margin of error.

8.67 **Nonresponse for the income question.** Refer to the previous two exercises. Give the total number of users and the total number of nonusers for the analysis of education. Do the same for the analysis of income. The difference is due to respondents who chose "Rather not say" for the income question. Give the proportions of "Rather not say" individuals for users and nonusers. Perform a significance test to compare these and give a 95% confidence interval for the difference. People are often reluctant to provide information about their income. Do you think that this amount of nonresponse for the income question is a serious limitation for this study?

8.68 **Improving the time to repair golf clubs.** The Ping Company makes custom-built golf clubs and competes in the $4 billion golf equipment industry. To improve its business processes, Ping decided to seek ISO 9001 certification.[32] As part of this process, a study of the time it took to repair golf clubs that were sent to the company by mail determined that 16% of orders were sent back to the customers in 5 days or less. Ping examined the processing of repair orders and made changes. Following the changes, 90% of orders were completed within 5 days. Assume that each of the estimated percents is based on a random sample of 200 orders.

(a) How many orders were completed in 5 days or less before the changes? Give a 95% confidence interval for the proportion of orders completed in this time.

(b) Do the same for orders after the changes.

(c) Give a 95% confidence interval for the improvement. Express this both for a difference in proportions and for a difference in percents.

8.69 **Parental pressure to succeed in school.** A Pew Research Center Poll used telephone interviews to ask American adults if parents are pushing to win when playing at an opponent's field or court, all other things being equal. Go to the Web site of your favorite sports team and find the proportion of wins for home games and the proportion of wins for away games. Now consider these games to be a random sample of the process that generates wins and losses. A complete analysis of data like these requires methods that are beyond what we have studied, but the methods discussed in this chapter will give us a reasonable approximation. Examine the home court advantage for your team and write a summary of your results. Be sure to comment on the effect of the sample size.

8.83    **Attitudes toward student loan debt.** The National Student Loan Survey asked the student loan borrowers in their sample about attitudes toward debt.[37] Here are some of the questions they asked, with the percent who responded in a particular way:

(a) "To what extent do you feel burdened by your student loan payments?" 55.5% said they felt burdened.

(b) "If you could begin again, taking into account your current experience, what would you borrow?" 54.4% said they would borrow less.

(c) "Since leaving school, my education loans have not caused me more financial hardship than I had anticipated at the time I took out the loans." 34.3% disagreed.

(d) "Making loan payments is unpleasant but I know that the benefits of education loans are worth it." 58.9% agreed.

(e) "I am satisfied that the education I invested in with my student loan(s) was worth the investment for career opportunities." 58.9% agreed.

(f) "I am satisfied that the education I invested in with my student loan(s) was worth the investment for personal growth." 71.5% agreed.

Assume that the sample size is 1280 for all of these questions. Compute a 95% confidence interval for each of the questions, and write a short report about what student loan borrowers think about their debt.

# Analysis of Two-Way Tables

There is growing evidence that early exposure to frightening movies is associated with lingering fright symptoms. Is this relationship different for boys and girls? Example 9.3 addresses this question.

## Introduction

We continue our study of methods for analyzing categorical data in this chapter. Inference about proportions in one-sample and two-sample settings was the focus of Chapter 8. We now study how to compare two or more populations when the response variable has two or more categories and how to test whether two categorical variables are independent. A single statistical test handles both of these cases.

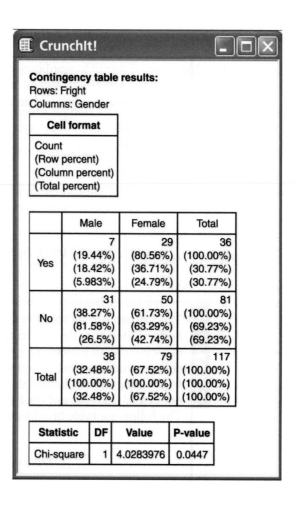

**Contingency table results:**
Rows: Fright
Columns: Gender

| Cell format |
| --- |
| Count |
| (Row percent) |
| (Column percent) |
| (Total percent) |

|  | Male | Female | Total |
| --- | --- | --- | --- |
| **Yes** | 7<br>(19.44%)<br>(18.42%)<br>(5.983%) | 29<br>(80.56%)<br>(36.71%)<br>(24.79%) | 36<br>(100.00%)<br>(30.77%)<br>(30.77%) |
| **No** | 31<br>(38.27%)<br>(81.58%)<br>(26.5%) | 50<br>(61.73%)<br>(63.29%)<br>(42.74%) | 81<br>(100.00%)<br>(69.23%)<br>(69.23%) |
| **Total** | 38<br>(32.48%)<br>(100.00%)<br>(32.48%) | 79<br>(67.52%)<br>(100.00%)<br>(67.52%) | 117<br>(100.00%)<br>(100.00%)<br>(100.00%) |

| Statistic | DF | Value | P-value |
| --- | --- | --- | --- |
| Chi-square | 1 | 4.0283976 | 0.0447 |

**FIGURE 9.1** CrunchIt! computer output for Example 9.3.

**EXAMPLE**

**9.4 Two-way table of ongoing fright symptoms and gender.** To compare the frequency of lingering fright symptoms across genders, we examine column percents. Here they are, rounded from the output for clarity:

**Column percents for gender**

|  | Gender | |
| --- | --- | --- |
| Ongoing fright symptoms | Male | Female |
| Yes | 18% | 37% |
| No | 82% | 63% |
| Total | 100% | 100% |

The "Total" row reminds us that 100% of the male and female students have been classified as having ongoing fright symptoms or not. (The sums sometimes differ slightly from 100% because of roundoff error.) The bar graph in Figure 9.2 compares the percents. The data reveal a clear relationship: 37% of the women have ongoing fright symptoms, as opposed to only 18% of the men.

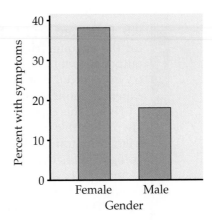

**FIGURE 9.2** Bar graph of the percents of male and female students with lingering fright symptoms, for Example 9.4.

The difference between the percents of students with lingering fears is reasonably large. A statistical test will tell us whether or not this difference can be plausibly attributed to chance. Specifically, if there is no association between gender and having ongoing fright symptoms, how likely is it that a sample would show a difference as large or larger than that displayed in Figure 9.2? In the remainder of this section we discuss the significance test to examine this question.

## The hypothesis: no association

The null hypothesis $H_0$ of interest in a two-way table is: There is *no association* between the row variable and the column variable. In Example 9.3, this null hypothesis says that gender and having ongoing fright symptoms are not related. The alternative hypothesis $H_a$ is that there is an association between these two variables. The alternative $H_a$ does not specify any particular direction for the association. For two-way tables in general, the alternative includes many different possibilities. Because it includes all sorts of possible associations, we cannot describe $H_a$ as either one-sided or two-sided.

In our example, the hypothesis $H_0$ that there is no association between gender and having ongoing fright symptoms is equivalent to the statement that the distributions of the ongoing fright symptoms variable are the same across the genders. For other two-way tables, where the columns correspond to independent samples from distinct populations, there are $c$ distributions for the row variable, one for each population. The null hypothesis then says that the $c$ distributions of the row variable are identical. The alternative hypothesis is that the distributions are not all the same.

## Expected cell counts

expected cell counts

To test the null hypothesis in $r \times c$ tables, we compare the observed cell counts with **expected cell counts** calculated under the assumption that the null hypothesis is true. A numerical summary of the comparison will be our test statistic.

**9.5 Expected counts from software.** The observed and expected counts for the ongoing fright symptoms example appear in the Minitab computer output shown in Figure 9.3. The expected counts are given as the second entry in each cell. For example, in the first cell the observed count is 7 and the expected count is 11.69.

How is this expected count obtained? Look at the percents in the right margin of the table in Figure 9.1. We see that 30.77% of all students had ongoing fright symptoms. If the null hypothesis of no relation between gender and ongoing fright is true, we expect this overall percent to apply to both men and women. In particular, we expect 30.77% of the men to have lingering fright symptoms. Since there are 38 men, the expected count is 30.77% of 38, or 11.69. The other expected counts are calculated in the same way.

```
Minitab

   Rows: Symptom    Columns: Gender

                1_Male   2_Female     All

   1_Yes             7         29      36
                 11.69      24.31   36.00

   2_No             31         50      81
                 26.31      54.69   81.00

   All              38         79     117
                 38.00      79.00  117.00

   Cell Contents:        Count
                         Expected count


   Pearson Chi-Square = 4.028, DF = 1, P-Value = 0.045
```

**FIGURE 9.3** Minitab computer output for Example 9.5.

The reasoning of Example 9.5 leads to a simple formula for calculating expected cell counts. To compute the expected count of men with ongoing fright symptoms, we multiplied the proportion of students with fright symptoms (36/117) by the number of men (38). From Figures 9.1 and 9.3 we see that the numbers 36 and 38 are the row and column totals for the cell of interest and that 117 is $n$, the total number of observations for the table. The expected cell count is therefore the product of the row and column totals divided by the table total.

$$\text{expected cell count} = \frac{\text{row total} \times \text{column total}}{n}$$

## The chi-square test

To test the $H_0$ that there is no association between the row and column classifications, we use a statistic that compares the entire set of observed counts with the set of expected counts. To compute this statistic,

- First, take the difference between each observed count and its corresponding expected count, and square these values so that they are all 0 or positive.

**LOOK BACK**
standardizing, page 61

- Since a large difference means less if it comes from a cell that is expected to have a large count, divide each squared difference by the expected count. This is a kind of standardization.

- Finally, sum over all cells.

The result is called the *chi-square statistic* $X^2$. The chi-square statistic was invented by the English statistician Karl Pearson (1857–1936) in 1900, for purposes slightly different from ours. It is the oldest inference procedure still used in its original form. With the work of Pearson and his contemporaries at the beginning of the last century, statistics first emerged as a separate discipline.

### CHI-SQUARE STATISTIC

The **chi-square statistic** is a measure of how much the observed cell counts in a two-way table diverge from the expected cell counts. The formula for the statistic is

$$X^2 = \sum \frac{(\text{observed count} - \text{expected count})^2}{\text{expected count}}$$

where "observed" represents an observed cell count, "expected" represents the expected count for the same cell, and the sum is over all $r \times c$ cells in the table.

If the expected counts and the observed counts are very different, a large value of $X^2$ will result. Large values of $X^2$ provide evidence against the null hypothesis. To obtain a *P*-value for the test, we need the sampling distribution of $X^2$ under the assumption that $H_0$ (no association between the row and column variables) is true. We once again use an approximation, related to the Normal approximation for binomial distributions. The result is a new distribution, the **chi-square distribution,** which we denote by $\chi^2$ ($\chi$ is the lowercase Greek letter chi).

**LOOK BACK**
**Normal approximation for counts, page 323**

chi-square distribution
$\chi^2$

Like the *t* distributions, the $\chi^2$ distributions form a family described by a single parameter, the degrees of freedom. We use $\chi^2(\text{df})$ to indicate a particular member of this family. Figure 9.4 displays the density curves of the $\chi^2(2)$ and $\chi^2(4)$ distributions. As the figure suggests, $\chi^2$ distributions take only positive

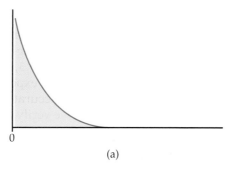
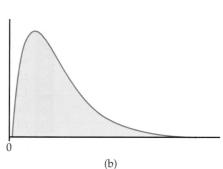

**FIGURE 9.4** (a) The $\chi^2(2)$ density curve. (b) The $\chi^2(4)$ density curve.

(a)          (b)

**9.9 Background music and wine sales: conditional distributions.**
When no music was played, there were 84 bottles of wine sold. Of these, 30 were French wine. Therefore, the column proportion for this cell is

$$\frac{30}{84} = 0.357$$

That is, 35.7% of the wine sold was French when no music was played. Similarly, 11 bottles of Italian wine were sold under this condition, and this is 13.1% of the sales:

$$\frac{11}{84} = 0.131$$

In all, we calculate nine percents. Here are the results:

**Column percents for wine and music**

| Wine | None | French | Italian | Total |
|------|------|--------|---------|-------|
| | | Music | | |
| French | 35.7 | 52.0 | 35.7 | 40.7 |
| Italian | 13.1 | 1.3 | 22.6 | 12.8 |
| Other | 51.2 | 46.7 | 41.7 | 46.5 |
| Total | 100.0 | 100.0 | 100.0 | 100.0 |

In addition to the conditional distributions of types of wine sold for each kind of music being played, the table also gives the marginal distribution of the types of wine sold. These percents appear in the rightmost column, labeled "Total."

The sum of the percents in each column should be 100, except for possible small roundoff errors. It is good practice to calculate each percent separately and then sum each column as a check. In this way we can find arithmetic errors that would not be uncovered if, for example, we calculated the column percent for the "Other" row by subtracting the sum of the percents for "French" and "Italian" from 100.

Figure 9.5 compares the distributions of types of wine sold for each of the three music conditions. There appears to be an association between the music played and the type of wine that customers buy. Sales of Italian wine are very low when French music is playing but are higher when Italian music or no music is playing. French wine is popular in this market, selling well under all music conditions but notably better when French music is playing.

Another way to look at these data is to examine the row percents. These fix a type of wine and compare its sales when different types of music are playing. Figure 9.6 displays these results. We see that more French wine is sold when French music is playing, and more Italian wine is sold when Italian music is playing. The negative effect of French music on sales of Italian wine is dramatic.

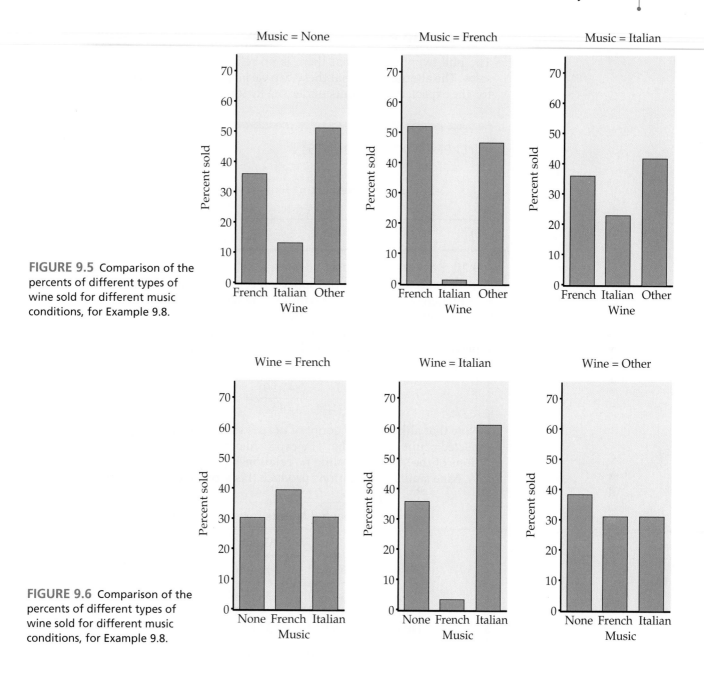

**FIGURE 9.5** Comparison of the percents of different types of wine sold for different music conditions, for Example 9.8.

**FIGURE 9.6** Comparison of the percents of different types of wine sold for different music conditions, for Example 9.8.

We observe a clear relationship between music type and wine sales for the 243 bottles sold during the study. The chi-square test assesses whether this observed association is statistically significant, that is, too strong to occur often just by chance. The test only confirms that there is some relationship. The percents we have compared describe the nature of the relationship. *The chi-square test does not in itself tell us what population our conclusion describes.* If the study was done in one market on a Saturday, the results may apply only to Saturday shoppers at this market. The researchers may invoke their understanding of consumer behavior to argue that their findings apply more generally, but that is beyond the scope of the statistical analysis.

## Computing expected cell counts

The null hypothesis is that there is no relationship between music and wine sales. The alternative is that these two variables are related. Here is the formula for the expected cell counts under the hypothesis of "no relationship."

---

**EXPECTED CELL COUNTS**

$$\text{expected count} = \frac{\text{row total} \times \text{column total}}{n}$$

---

**EXAMPLE**

**9.10 Background music and wine sales: expected cell counts.**   What is the expected count in the upper-left cell in the table of Example 9.8, bottles of French wine sold when no music is playing, under the null hypothesis that music and wine sales are independent?

The column total, the number of bottles of wine sold when no music is playing, is 84. The row total shows that 99 bottles of French wine were sold during the study. The total sales were 243. The expected cell count is therefore

$$\frac{(84)(99)}{243} = 34.222$$

Note that although any count of bottles sold must be a whole number, an expected count need not be. The expected count is the mean over many repetitions of the study, assuming no relationship.

Nine similar calculations produce this table of expected counts:

| | Expected counts for wine and music | | | |
|---|---|---|---|---|
| | Music | | | |
| Wine | None | French | Italian | Total |
| French | 34.222 | 30.556 | 34.222 | 99.000 |
| Italian | 10.716 | 9.568 | 10.716 | 31.000 |
| Other | 39.062 | 34.877 | 39.062 | 113.001 |
| Total | 84.000 | 75.001 | 84.000 | 243.001 |

We can check our work by adding the expected counts to obtain the row and column totals, as in the table. These should be the same as those in the table of observed counts except for small roundoff errors, such as 113.001 rather than 113 for the total number of bottles of other wine sold.

## The $X^2$ statistic and its $P$-value

The expected counts are all large, so we proceed with the chi-square test. We compare the table of observed counts with the table of expected counts using the $X^2$ statistic.[6] We must calculate the term for each cell, then sum over all nine cells. For French wine with no music, the observed count is 30 bottles and the expected count is 34.222. The contribution to the $X^2$ statistic for this cell is therefore

$$\frac{(30 - 34.222)^2}{34.222} = 0.5209$$

The $X^2$ statistic is the sum of nine such terms:

$$X^2 = \sum \frac{(\text{observed} - \text{expected})^2}{\text{expected}}$$

$$= \frac{(30 - 34.222)^2}{34.222} + \frac{(39 - 30.556)^2}{30.556} + \frac{(30 - 34.222)^2}{34.222}$$

$$+ \frac{(11 - 10.716)^2}{10.716} + \frac{(1 - 9.568)^2}{9.568} + \frac{(19 - 10.716)^2}{10.716}$$

$$+ \frac{(43 - 39.062)^2}{39.062} + \frac{(35 - 34.877)^2}{34.877} + \frac{(35 - 39.062)^2}{39.062}$$

$$= 0.5209 + 2.3337 + 0.5209 + 0.0075 + 7.6724 + 6.4038$$

$$+ 0.3971 + 0.0004 + 0.4223$$

$$= 18.28$$

Because there are $r = 3$ types of wine and $c = 3$ music conditions, the degrees of freedom for this statistic are

$$\text{df} = (r - 1)(c - 1) = (3 - 1)(3 - 1) = 4$$

**df = 4**

| $p$ | 0.0025 | 0.001 |
|-----|--------|-------|
| $\chi^2$ | 16.42 | 18.47 |

Under the null hypothesis that music and wine sales are independent, the test statistic $X^2$ has a $\chi^2(4)$ distribution. To obtain the $P$-value, look at the df = 4 row in Table F. The calculated value $X^2 = 18.28$ lies between the critical points for probabilities 0.0025 and 0.001. The $P$-value is therefore between 0.0025 and 0.001. Because the expected cell counts are all large, the $P$-value from Table F will be quite accurate. There is strong evidence ($X^2 = 18.28$, df = 4, $P < 0.0025$) that the type of music being played has an effect on wine sales.

The size and nature of the relationship between music and wine sales are described by row and column percents. These are displayed in Figures 9.5 and 9.6. Here is another way to look at the data: we see that just two of the nine terms that make up the chi-square sum contribute about 14 of the total $X^2 = 18.28$. Comparing the observed and expected counts in these two cells, we see that sales of Italian wine are much below expectation when French music is playing and much above expectation when Italian music is playing. We are led to a specific conclusion: sales of Italian wine are strongly affected by Italian and French music. Figure 9.6(b) displays this effect.

## Models for two-way tables

The chi-square test for the presence of a relationship between the two directions in a two-way table is valid for data produced from several different study designs. The precise statement of the null hypothesis "no relationship" in terms of population parameters is different for different designs. We now describe

## USE YOUR KNOWLEDGE

**9.5**  **Distribution of M&M colors.** M&M Mars Company has varied the mix of colors for M&M's Milk Chocolate Candies over the years. These changes in color blends are the result of consumer preference tests. Most recently, the color distribution is reported to be 13% brown, 14% yellow, 13% red, 20% orange, 24% blue, and 16% green.[10] You open up a 14-ounce bag of M&M's and find 61 brown, 59 yellow, 49 red, 77 orange, 141 blue, and 88 green. Use a goodness of fit test to examine how well this bag fits the percents stated by the M&M Mars Company.

## SECTION 9.3  Summary

The **chi-square goodness of fit test** is used to compare the sample distribution of a categorical variable from a population with a hypothesized distribution. The data for $n$ observations with $k$ possible outcomes are summarized as observed counts $n_1, n_2, \ldots, n_k$ in $k$ cells. The **null hypothesis** specifies probabilities $p_1, p_2, \ldots, p_k$ for the possible outcomes.

The analysis of these data is similar to the analyses of two-way tables discussed in Section 9.1. For each cell, the **expected count** is determined by multiplying the total number of observations $n$ by the specified probability $p_i$. The null hypothesis is tested by the usual **chi-square statistic**, which compares the observed counts, $n_i$, with the expected counts. Under the null hypothesis, $X^2$ has approximately the $\chi^2$ distribution with df $= k - 1$.

## CHAPTER 9   Exercises

*For Exercises 9.1 and 9.2, see pages 533 and 534; for Exercises 9.3 and 9.4, see page 544; and for Exercise 9.5, see page 548.*

**9.6**  **Why not use a chi-square test?** As part of the study on ongoing fright symptoms due to exposure to horror movies at a young age, the following table

was created based on the written responses from 119 students.[11] Explain why a chi-square test is not appropriate for this table.

Percent of students who reported each problem

| Movie or video | Type of Problem | | | |
|---|---|---|---|---|
| | Bedtime | | Waking | |
| | Short term | Enduring | Short term | Enduring |
| *Poltergeist* ($n = 29$) | 68 | 7 | 64 | 32 |
| *Jaws* ($n = 23$) | 39 | 4 | 83 | 43 |
| *Nightmare on Elm Street* ($n = 16$) | 69 | 13 | 37 | 31 |
| *Thriller* (music video) ($n = 16$) | 40 | 0 | 27 | 7 |
| *It* ($n = 24$) | 64 | 0 | 64 | 50 |
| *The Wizard of Oz* ($n = 12$) | 75 | 17 | 50 | 8 |
| *E.T.* ($n = 11$) | 55 | 0 | 64 | 27 |

**9.7**  **Age and time status of U.S. college students.** The Census Bureau provides estimates of numbers of people in the United States classified in various ways.[12] Let's look at college students. The following table gives us data to examine the relation between age and full-time or part-time status. The numbers in the table are expressed as thousands of U.S. college students.

U.S. college students by age and status: October 2004

| Age | Status | |
|---|---|---|
| | Full-time | Part-time |
| 15–19 | 3553 | 329 |
| 20–24 | 5710 | 1215 |
| 25–34 | 1825 | 1864 |
| 35 and over | 901 | 1983 |

(a) Give the joint distribution of age and status for this table.

(b) What is the marginal distribution of age? Display the results graphically.

(c) What is the marginal distribution of status? Display the results graphically.

(d) Compute the conditional distribution of age for each of the two status categories. Display the results graphically.

(e) Write a short paragraph describing the distributions and how they differ.

**9.8** **Time status versus gender for the 20–24 age category.** Refer to Exercise 9.7. The table below breaks down the 20–24 age category by gender.

| | Gender | | |
| Status | Male | Female | Total |
|---|---|---|---|
| Full-time | 2719 | 2991 | 5710 |
| Part-time | 535 | 680 | 1215 |
| Total | 3254 | 3671 | 6925 |

(a) Compute the marginal distribution for gender. Display the results graphically.

(b) Compute the conditional distribution of status for males and for females. Display the results graphically and comment on how these distributions differ.

(c) If you wanted to test the null hypothesis that there is no difference between these two conditional distributions, what would the expected cell counts be for the full-time status row of the table?

(d) Computer software gives $X^2 = 5.17$. Using Table F, give an appropriate bound for the $P$-value and state your conclusions at the 5% level.

**9.9** **Does using Rodham matter?** In April 2006, the Opinion Research Corporation conducted a telephone poll for CNN of 1012 adult Americans.[13] Half those polled were asked their opinion of Hillary Rodham Clinton. The other half were asked their opinion of Hillary Clinton. The table below summarizes the results. A chi-square test was used to determine if opinions differed based on the name.

| | Opinion | | | |
| Name | Favorable | Unfavorable | Never heard of | No opinion |
|---|---|---|---|---|
| Hillary Rodham Clinton | 50% | 42% | 2% | 6% |
| Hillary Clinton | 46% | 43% | 2% | 9% |

(a) Computer software gives $X^2 = 4.23$. Can we comfortably use the chi-square distribution to compute the $P$-value? Explain.

(b) What are the degrees of freedom for $X^2$?

(c) Give an appropriate bound for the $P$-value using Table F and state your conclusions.

**9.10** **Waking versus bedtime symptoms.** As part of the study on ongoing fright symptoms due to exposure to horror movies at a young age, the following table was presented to describe the lasting impact these movies have had during bedtime and waking life:

| | Waking symptoms | |
| Bedtime symptoms | Yes | No |
|---|---|---|
| Yes | 36 | 33 |
| No | 33 | 17 |

(a) What percent of the students have lasting waking-life symptoms?

(b) What percent of the students have both waking-life and bedtime symptoms?

(c) Test whether there is an association between waking-life and bedtime symptoms. State the null and alternative hypotheses, the $X^2$ statistic, and the $P$-value.

**9.11** **New treatment for cocaine addiction.** Cocaine addiction is difficult to overcome. Addicts have been reported to have a significant depletion of stimulating neurotransmitters and thus continue to take cocaine to avoid feelings of depression and anxiety. A 3-year study with 72 chronic cocaine users compared an antidepressant drug called desipramine with lithium and a placebo. (Lithium is a standard drug to treat cocaine addiction. A placebo is a substance containing no medication, used so that the effect of being in the study but not taking any drug can be seen.) One-third of the subjects, chosen at random, received each treatment.[14] Following are the results:

|  | Cocaine relapse? | |
|---|---|---|
| Treatment | Yes | No |
| Desipramine | 10 | 14 |
| Lithium | 18 | 6 |
| Placebo | 20 | 4 |

(a) Compare the effectiveness of the three treatments in preventing relapse using percents and a bar graph. Write a brief summary.

(b) Can we comfortably use the chi-square test to test the null hypothesis that there is no difference between treatments? Explain.

(c) Perform the significance test and summarize the results.

9.12 **Find the degrees of freedom and *P*-value.** For each of the following situations give the degrees of freedom and an appropriate bound on the *P*-value (give the exact value if you have software available) for the $X^2$ statistic for testing the null hypothesis of no association between the row and column variables.

(a) A 2 by 2 table with $X^2 = 1.25$.

(b) A 4 by 4 table with $X^2 = 18.34$.

(c) A 2 by 8 table with $X^2 = 24.21$.

(d) A 5 by 3 table with $X^2 = 12.17$.

9.13 **Can you construct the joint distribution from the marginal distributions?** Here are the row and column totals for a two-way table with two rows and two columns:

$$
\begin{array}{cc|c}
a & b & 50 \\
c & d & 150 \\
\hline
100 & 100 & 200
\end{array}
$$

Find *two different* sets of counts *a*, *b*, *c*, and *d* for the body of the table. This demonstrates that the relationship between two variables cannot be obtained solely from the two marginal distributions of the variables.

9.14 **Construct a table with no association.** Construct a $3 \times 2$ table of counts where there is no apparent association between the row and column variables.

9.15 **Gender versus motivation for volunteer service.** A study examined patterns and characteristics of volunteer-service for young people from high school through early adulthood.[15] Here

are some data that can be used to compare males and females on participation in unpaid volunteer service or community service and motivation for participation:

| | Participants | | | |
|---|---|---|---|---|
| | Motivation | | | |
| Gender | Strictly voluntary | Court-ordered | Other | Non-participants |
| Men | 31.9% | 2.1% | 6.3% | 59.7% |
| Women | 43.7% | 1.1% | 6.5% | 48.7% |

Note that the percents in each row sum to 100%.

(a) Graphically compare the volunteer-service profiles for men and women. Describe any differences that are striking.

(b) Find the proportion of men who volunteer. Do the same for women. Refer to the section on relative risk in Chapter 8 (page 515) and the discussion on page 535 of this chapter. Compute the relative risk of being a volunteer for females versus males. Write a clear sentence contrasting females and males using relative risk as your numerical summary.

9.16 **Gender versus motivation for volunteer service, continued.** Refer to the previous exercise. Recompute the table for volunteers only. To do this take the entries for each motivation and divide by the percent of volunteers. Do this separately for each gender. Verify that the percents sum to 100% for each gender. Give a graphical summary to compare the motivation of men and women who are volunteers. Compare this with your summary in part (a) of the previous exercise, and write a short paragraph describing similarities and differences in these two views of the data.

9.17 **Drinking status and class attendance.** As part of the 1999 College Alcohol Study, students who drank alcohol in the last year were asked if drinking ever resulted in missing a class.[16] The data are given in the following table:

| | Drinking status | | |
|---|---|---|---|
| Missed a class | Nonbinger | Occasional binger | Frequent binger |
| No | 4617 | 2047 | 1176 |
| Yes | 446 | 915 | 1959 |

(a) Summarize the results of this table graphically and numerically.

(b) What is the marginal distribution of drinking status? Display the results graphically.

(c) Compute the relative risk of missing a class for occasional bingers versus nonbingers and for frequent bingers versus nonbingers. Summarize these results.

(d) Perform the chi-square test for this two-way table. Give the test statistic, degrees of freedom, the P-value, and your conclusion.

9.18 **Sexual imagery in magazine ads.** In what ways do advertisers in magazines use sexual imagery to appeal to youth? One study classified each of 1509 full-page or larger ads as "not sexual" or "sexual," according to the amount and style of the dress of the male or female model in the ad. The ads were also classified according to the target readership of the magazine.[17] Here is the two-way table of counts:

| Model dress | Magazine readership | | | Total |
|---|---|---|---|---|
| | Women | Men | General interest | |
| Not sexual | 351 | 514 | 248 | 1113 |
| Sexual | 225 | 105 | 66 | 396 |
| Total | 576 | 619 | 314 | 1509 |

(a) Summarize the data numerically and graphically.

(b) Perform the significance test that compares the model dress for the three categories of magazine readership. Summarize the results of your test and give your conclusion.

(c) All of the ads were taken from the March, July, and November issues of six magazines in one year. Discuss this fact from the viewpoint of the validity of the significance test and the interpretation of the results.

9.19 **Intended readership of ads with sexual imagery.** The ads in the study described in the previous exercise were also classified according to the age group of the intended readership. Here is a summary of the data:

| Model dress | Magazine readership age group | |
|---|---|---|
| | Young adult | Mature adult |
| Not sexual | 72.3% | 76.1% |
| Sexual | 27.2% | 23.9% |
| Number of ads | 1006 | 503 |

Using parts (a) and (b) in the previous exercise as a guide, analyze these data and write a report summarizing your work.

9.20 **Air pollution from a steel mill.** One possible effect of air pollution is genetic damage. A study designed to examine this problem exposed one group of mice to air near a steel mill and another group to air in a rural area and compared the numbers of mutations in each group.[18] Here are the data for a mutation at the *Hm-2* gene locus:

| Mutation | Location | |
|---|---|---|
| | Steel mill air | Rural air |
| Yes | 30 | 23 |
| No | | |
| Total | 96 | 150 |

(a) Fill in the missing entries in the table.

(b) Summarize the data numerically and graphically.

(c) Is there evidence to conclude that the location is related to the occurrence of mutations? Perform the significance test and summarize the results.

9.21 **Dieting trends among male and female undergraduates.** A recent study of undergraduates looked at gender differences in dieting trends.[19] There were 181 women and 105 men who participated in the survey. The table below summarizes whether a student tried a low-fat diet or not by gender:

| Tried low-fat diet | Gender | |
|---|---|---|
| | Women | Men |
| Yes | 35 | 8 |
| No | | |

We will use the following example to explain the fundamentals of simple linear regression. Because regression calculations in practice are always done by statistical software, we will rely on computer output for the arithmetic. In the next section, we give an example that illustrates how to do the work with a calculator if software is unavailable.

**EXAMPLE**

**10.1 Relationship between speed driven and fuel efficiency.** Computers in some vehicles calculate various quantities related to the vehicle's performance. One of these is the fuel efficiency, or gas mileage, expressed as miles per gallon (mpg). Another is the average speed in miles per hour (mph). For one vehicle equipped in this way, mpg and mph were recorded each time the gas tank was filled, and the computer was then reset.[1] How does the speed at which the vehicle is driven affect the fuel efficiency? There are 234 observations available. We will work with a simple random sample of size 60.

Before starting our analysis, it is appropriate to consider the extent to which our results can reasonably be generalized. Because we have a simple random sample from a population of size 234, we are on firm ground in making inferences about this particular vehicle. However, as a practical matter, no one really cares about this particular vehicle. Our results are interesting only if they can be applied to other similar vehicles that are driven under similar conditions. Our statistical modeling for this data set is concerned about the process by which speed affects the fuel efficiency. Although we would not expect the parameters that describe the relationship between speed and fuel efficiency to be *exactly* the same for similar vehicles, we would expect to find qualitatively similar results.

In the statistical model for predicting fuel efficiency from speed, subpopulations are defined by the explanatory variable, speed. For a particular value of speed, say 30 mph, we can think about operating this vehicle repeatedly at this average speed. Variation in driving conditions and the behavior of the driver would be sources of variation that would give different values of mpg for this subpopulation.

**EXAMPLE**

**10.2 Graphical display of the fuel efficiency relationship.** We start our analysis with a graphical display of the data. Figure 10.3 is a plot of fuel efficiency versus speed for our sample of 60 observations. We use the variable names MPG and MPH. The least-squares regression line and a smooth function are also shown in the plot. Although there is a positive association between MPG and MPH, the fit is not linear. The smooth function shows us that the relationship levels off somewhat with increasing speed.

*Always start with a graphical display of the data.* There is no point in trying to do statistical inference if our data do not, at least approximately, meet the assumptions that are the foundation for our inference. At this point we need to make a choice. One possibility would be to confine our interest to speeds that are 30 mph or less, a region where it appears that a line would be a good fit to the data. Another possibility is to make some sort of transformation that will

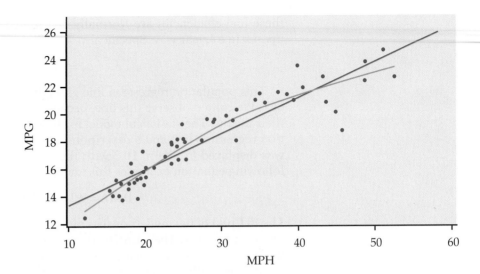

**FIGURE 10.3** Scatterplot of MPG versus MPH with a smooth function and the least-squares line, for Example 10.2.

make the relationship approximately linear for the entire set of data. We will choose the second option.

**10.3  Is this relationship linear?**   One type of function that looks similar to the smooth-function fit in Figure 10.3 is a logarithm. Therefore, we will examine the effect of transforming speed by taking the natural logarithm. The result is shown in Figure 10.4. In this plot the smooth function and the line are quite close. We are satisfied that the relationship between the log of speed and fuel efficiency is approximately linear for this set of data.

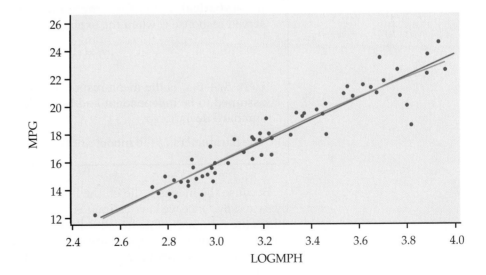

**FIGURE 10.4** Scatterplot of MPG versus logarithm of MPH with a smooth function and the least-squares line, for Example 10.3.

Now that we have an approximate linear relationship, we return to predicting fuel efficiency for different subpopulations, defined by the explanatory variable speed. Consider a particular value of speed, for example 30 mph, which in Figure 10.4 would be $x = \log(30) = 3.4$. Our statistical model assumes that these fuel efficiencies are Normally distributed with a mean $\mu_y$ that depends upon $x$ in a linear way. Specifically,

$$\mu_y = \beta_0 + \beta_1 x$$

This population regression line gives the mean fuel efficiency for all values of $x$. We cannot observe this line, because the observed responses $y$ vary about their means. The statistical model for linear regression consists of the population regression line and a description of the variation of $y$ about the line. This was displayed in Figure 10.2 with the line and the three Normal curves. The following equation expresses this idea in an equation:

$$\text{DATA} = \text{FIT} + \text{RESIDUAL}$$

The FIT part of the model consists of the subpopulation means, given by the expression $\beta_0 + \beta_1 x$. The RESIDUAL part represents deviations of the data from the line of population means. We assume that these deviations are Normally distributed with standard deviation $\sigma$. We use $\epsilon$ (the Greek letter epsilon) to stand for the RESIDUAL part of the statistical model. A response $y$ is the sum of its mean and a chance deviation $\epsilon$ from the mean. The deviations $\epsilon$ represent "noise," that is, variation in $y$ due to other causes that prevent the observed $(x, y)$-values from forming a perfectly straight line on the scatterplot.

---

### SIMPLE LINEAR REGRESSION MODEL

Given $n$ observations of the explanatory variable $x$ and the response variable $y$,

$$(x_1, y_1),\ (x_2, y_2), \ldots,\ (x_n, y_n)$$

the **statistical model for simple linear regression** states that the observed response $y_i$ when the explanatory variable takes the value $x_i$ is

$$y_i = \beta_0 + \beta_1 x_i + \epsilon_i$$

Here $\beta_0 + \beta_1 x_i$ is the mean response when $x = x_i$. The deviations $\epsilon_i$ are assumed to be independent and Normally distributed with mean 0 and standard deviation $\sigma$.

The parameters of the model are $\beta_0$, $\beta_1$, and $\sigma$.

---

Because the means $\mu_y$ lie on the line $\mu_y = \beta_0 + \beta_1 x$, they are all determined by $\beta_0$ and $\beta_1$. Once we have estimates of $\beta_0$ and $\beta_1$, the linear relationship determines the estimates of $\mu_y$ for all values of $x$. Linear regression allows us to do inference not only for subpopulations for which we have data but also for those corresponding to $x$'s not present in the data. We will learn how to do inference about

• the slope $\beta_1$ and the intercept $\beta_0$ of the population regression line,

• the mean response $\mu_y$ for a given value of $x$, and

• an individual future response $y$ for a given value of $x$.

## Estimating the regression parameters

**LOOK BACK**
least-squares
regression, page 112

The method of least squares presented in Chapter 2 fits a line to summarize a relationship between the observed values of an explanatory variable and a response variable. Now we want to use the least-squares line as a basis for inference about a population from which our observations are a sample. We can do this only when the statistical model just presented holds. In that setting, the slope $b_1$ and intercept $b_0$ of the least-squares line

$$\hat{y} = b_0 + b_1 x$$

estimate the slope $\beta_1$ and the intercept $\beta_0$ of the population regression line.

Using the formulas from Chapter 2, the slope of the least-squares line is

**LOOK BACK**
least-squares
equations, page 114
correlation, page 102
unbiased estimator,
page 217

$$b_1 = r\frac{s_y}{s_x}$$

and the intercept is

$$b_0 = \bar{y} - b_1\bar{x}$$

Here, $r$ is the correlation between $y$ and $x$, $s_y$ is the standard deviation of $y$, and $s_x$ is the standard deviation of $x$. Some algebra based on the rules for means of random variables (Section 4.4) shows that $b_0$ and $b_1$ are unbiased estimators of $\beta_0$ and $\beta_1$. Furthermore, $b_0$ and $b_1$ are Normally distributed with means $\beta_0$ and $\beta_1$ and standard deviations that can be estimated from the data. Normality of these sampling distributions is a consequence of the assumption that the $\epsilon_i$ are distributed Normally. A general form of the central limit theorem tells us that the distributions of $b_0$ and $b_1$ will still be approximately Normal even if the $\epsilon_i$ are not. On the other hand, outliers and influential observations can invalidate the results of inference for regression.

**LOOK BACK**
central limit theorem,
page 339

The predicted value of $y$ for a given value $x^*$ of $x$ is the point on the least-squares line $\hat{y} = b_0 + b_1 x^*$. This is an unbiased estimator of the mean response $\mu_y$ when $x = x^*$. The **residual** is

residual

$$e_i = \text{observed response} - \text{predicted response}$$

$$= y_i - \hat{y}_i$$

$$= y_i - b_0 - b_1 x_i$$

The residuals $e_i$ correspond to the model deviations $\epsilon_i$. The $e_i$ sum to 0, and the $\epsilon_i$ come from a population with mean 0.

The remaining parameter to be estimated is $\sigma$, which measures the variation of $y$ about the population regression line. Because this parameter is the standard deviation of the model deviations, it should come as no surprise that we use the residuals to estimate it. As usual, we work first with the variance and take the square root to obtain the standard deviation. For simple linear regression, the estimate of $\sigma^2$ is the average squared residual

$$s^2 = \frac{\sum e_i^2}{n-2}$$

$$= \frac{\sum (y_i - \hat{y}_i)^2}{n-2}$$

**LOOK BACK**

**sample variance,
page 40**

degrees of freedom

We average by dividing the sum by $n - 2$ in order to make $s^2$ an unbiased estimate of $\sigma^2$. The sample variance of $n$ observations uses the divisor $n - 1$ for this same reason. The quantity $n - 2$ is called the **degrees of freedom** for $s^2$. The estimate of $\sigma$ is given by

$$s = \sqrt{s^2}$$

We will use statistical software to calculate the regression for predicting fuel efficiency with the log of speed for Example 10.3. In entering the data, we chose the names LOGMPH for the log of speed and MPG for fuel efficiency. *It is good practice to use names, rather than just x and y, to remind yourself which data the output describes.*

**EXAMPLE**

**10.4 Statistical software output for fuel efficiency.**    Figure 10.5 gives the outputs for four commonly used statistical software packages and Excel. Other software will give similar information. The SPSS output reports estimates of our three parameters as $b_0 = -7.796$, $b_1 = 7.874$, and $s = 0.9995$. Be sure that you can find these entries in this output and the corresponding values in the other outputs.

The least-squares regression line is the straight line that is plotted in Figure 10.4. We would report it as

$$\widehat{\text{MPG}} = -7.80 + 7.87\text{LOGMPH}$$

with a model standard deviation of $s = 1.00$. Note that the number of digits provided varies with the software used and we have rounded off the values to three significant digits. *It is important to avoid cluttering up your report of the results of a statistical analysis with many digits that are not relevant.* Software often reports many more digits than are meaningful or useful.

The outputs contain other information that we will ignore for now. Computer outputs often give more information than we want or need. *The experienced user of statistical software learns to ignore the parts of the output that are not needed for the current problem.* This is done to reduce user frustration when a software package does not print out the particular statistics wanted for an analysis.

Now that we have fitted a line, we should examine the residuals for Normality and any remaining patterns in the data. We usually plot the residuals both against the case number (especially if this reflects the order in which the observations were collected) and against the explanatory variable. For this example, in place of case number, we prefer another variable that is similar but is recorded in a more useful scale. It is the total number of miles that the vehicle has been driven.

Figure 10.6 gives a plot of the residuals versus miles driven with a smooth-function fit. The smooth function suggests that the residuals increase slightly up to about 50,000 miles and then tend to decrease somewhat. With the data that we have for this example, it is difficult to decide if this effect is real or due to chance variation. It is not unreasonable to think that the vehicle performance decreases with age. Since the effect does not appear to be particularly large, we

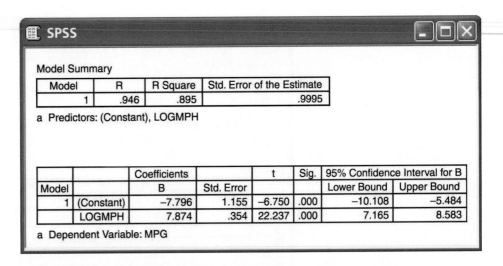

SPSS

**Model Summary**

| Model | R | R Square | Std. Error of the Estimate |
|---|---|---|---|
| 1 | .946 | .895 | .9995 |

a Predictors: (Constant), LOGMPH

| Model | | Coefficients | | t | Sig. | 95% Confidence Interval for B | |
|---|---|---|---|---|---|---|---|
| | | B | Std. Error | | | Lower Bound | Upper Bound |
| 1 | (Constant) | −7.796 | 1.155 | −6.750 | .000 | −10.108 | −5.484 |
| | LOGMPH | 7.874 | .354 | 22.237 | .000 | 7.165 | 8.583 |

a Dependent Variable: MPG

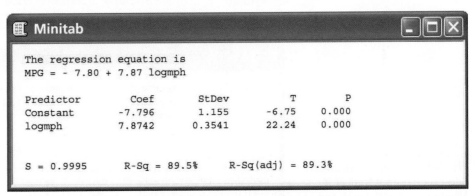

Minitab

```
The regression equation is
MPG = - 7.80 + 7.87 logmph

Predictor         Coef        StDev           T          P
Constant        -7.796        1.155        -6.75      0.000
logmph          7.8742       0.3541        22.24      0.000


S = 0.9995      R-Sq = 89.5%      R-Sq(adj) = 89.3%
```

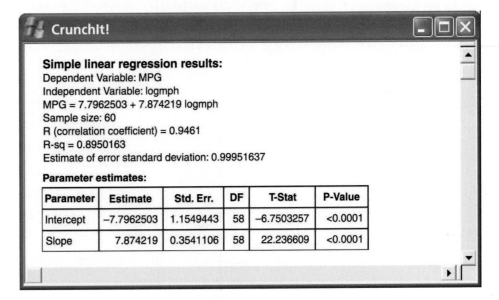

CrunchIt!

**Simple linear regression results:**
Dependent Variable: MPG
Independent Variable: logmph
MPG = 7.7962503 + 7.874219 logmph
Sample size: 60
R (correlation coefficient) = 0.9461
R-sq = 0.8950163
Estimate of error standard deviation: 0.99951637

**Parameter estimates:**

| Parameter | Estimate | Std. Err. | DF | T-Stat | P-Value |
|---|---|---|---|---|---|
| Intercept | −7.7962503 | 1.1549443 | 58 | −6.7503257 | <0.0001 |
| Slope | 7.874219 | 0.3541106 | 58 | 22.236609 | <0.0001 |

**FIGURE 10.5** Regression output from SPSS, Minitab, CrunchIt!, Excel, and SAS for the fuel efficiency example. *(continued)*

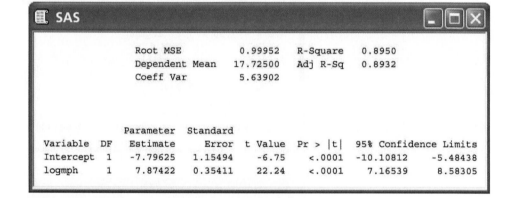

| | A | B | C | D | E | F | G |
|---|---|---|---|---|---|---|---|
| 1 | SUMMARY OUTPUT | | | | | | |
| 2 | | | | | | | |
| 3 | Regression Statistics | | | | | | |
| 4 | Multiple R | 0.946053015 | | | | | |
| 5 | R Square | 0.895016308 | | | | | |
| 6 | Adjusted R Square | 0.893206244 | | | | | |
| 7 | Standard Error | 0.999516364 | | | | | |
| 8 | Observations | 60 | | | | | |
| 9 | | | | | | | |
| 10 | ANOVA | | | | | | |
| 11 | | df | SS | MS | F | Significance F | |
| 12 | Regression | 1 | 493.9885883 | 493.9886 | 494.4668 | 4.50949E-30 | |
| 13 | Residual | 58 | 57.94391174 | 0.999033 | | | |
| 14 | Total | 59 | 551.9325 | | | | |
| 15 | | | | | | | |
| 16 | | Coefficients | Standard Error | tStet | P-value | Lower 95% | Upper 95% |
| 17 | intercept | –7.796250129 | 1.154944262 | –6.75033 | 7.69E-09 | –10.10812052 | –5.48437974 |
| 18 | logmph | 7.874219013 | 0.354110611 | 22.23661 | 4.51E-30 | 7.165390143 | 8583047883 |

**FIGURE 10.5** *(Continued)* Regression output from SPSS, Minitab, CrunchIt!, Excel, and SAS for the fuel efficiency example.

**SAS**

```
                    Root MSE           0.99952    R-Square    0.8950
                    Dependent Mean    17.72500    Adj R-Sq    0.8932
                    Coeff Var          5.63902

                   Parameter   Standard
Variable    DF     Estimate      Error    t Value   Pr > |t|    95% Confidence Limits
Intercept    1     -7.79625     1.15494    -6.75     <.0001     -10.10812     -5.48438
logmph       1      7.87422     0.35411    22.24     <.0001       7.16539      8.58305
```

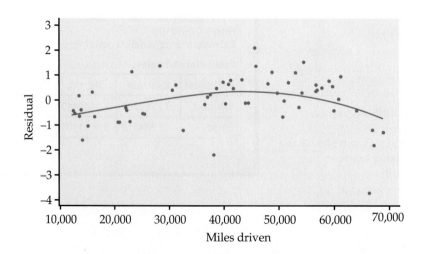

**FIGURE 10.6** Plot of residuals versus miles driven with a smooth function, for the fuel efficiency example.

will ignore it for the present analysis. With more data, however, it may be an interesting phenomenon to study.

The residuals are plotted versus the explanatory variable, log of mph (labeled LOGMPH), in Figure 10.7. No clear pattern is evident. There is one residual that is somewhat low, and we have seen it in all of our plots. Inspection of Figure 10.4 reveals that this observation does not appear to distort our least-squares regression line.

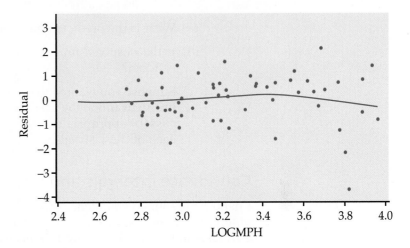

**FIGURE 10.7** Plot of residuals versus log of MPH with a smooth function, for the fuel efficiency example.

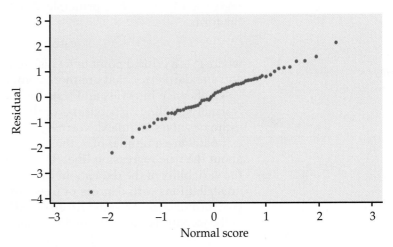

**FIGURE 10.8** Normal quantile plot of the residuals for the fuel efficiency example.

Finally, Figure 10.8 is a Normal quantile plot of the residuals. Because the plot looks fairly straight, we are confident that we do not have a serious violation of our assumption that the residuals are Normally distributed. Observe that the low outlier is also visible in this plot.

## USE YOUR KNOWLEDGE

**10.1** **Understanding a linear regression model.** Consider a linear regression model with $\mu_y = 40.5 - 2.5x$ and standard deviation $\sigma = 2.0$.

(a) What is the slope of the population regression line?

(b) Explain clearly what this slope says about the change in the mean of $y$ for a change in $x$.

(c) What is the subpopulation mean when $x = 10$?

(d) Between what 2 values would approximately 95% of the observed responses, $y$, fall when $x = 10$?

**10.2 More on speed's effect on fuel efficiency.** Refer to Example 10.4.

(a) What is the predicted mpg for the car when it averages 35 mph?

(b) If the observed mpg when $x = 35$ mph were 21.0, what is the residual?

(c) Suppose you wanted to use the estimated population regression line to examine the average mpg at 45, 55, 65, and 75 mph. Discuss the appropriateness of using the equation to predict mpg for each of these speeds.

## Confidence intervals and significance tests

Chapter 7 presented confidence intervals and significance tests for means and differences in means. In each case, inference rested on the standard errors of estimates and on $t$ distributions. Inference for the intercept and slope in a linear regression is similar in principle. For example, the confidence intervals have the form

$$\text{estimate} \pm t^* \text{SE}_{\text{estimate}}$$

where $t^*$ is a critical point of a $t$ distribution. It is the formulas for the estimate and standard error that are more complicated.

Confidence intervals and tests for the slope and intercept are based on the Normal sampling distributions of the estimates $b_1$ and $b_0$. Standardizing these estimates gives standard Normal $z$ statistics. The standard deviations of these estimates are multiples of $\sigma$, the model parameter that describes the variability about the true regression line. Because we do not know $\sigma$, we estimate it by $s$, the variability of the data about the least-squares line. When we do this, we get $t$ distributions with degrees of freedom $n - 2$, the degrees of freedom of $s$. We give formulas for the standard errors $\text{SE}_{b_1}$ and $\text{SE}_{b_0}$ in Section 10.2. For now we will concentrate on the basic ideas and let the computer do the computations.

---

CONFIDENCE INTERVALS AND SIGNIFICANCE TESTS FOR REGRESSION SLOPE AND INTERCEPT

A **level $C$ confidence interval for the intercept** $\beta_0$ is

$$b_0 \pm t^* \text{SE}_{b_0}$$

A **level $C$ confidence interval for the slope** $\beta_1$ is

$$b_1 \pm t^* \text{SE}_{b_1}$$

In these expressions $t^*$ is the value for the $t(n - 2)$ density curve with area $C$ between $-t^*$ and $t^*$.

To test the hypothesis $H_0: \beta_1 = 0$, compute the **test statistic**

$$t = \frac{b_1}{SE_{b_1}}$$

The **degrees of freedom** are $n - 2$. In terms of a random variable $T$ having the $t(n - 2)$ distribution, the $P$-value for a test of $H_0$ against

$H_a: \beta_1 > 0$ is $P(T \geq t)$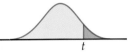

$H_a: \beta_1 < 0$ is $P(T \leq t)$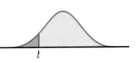

$H_a: \beta_1 \neq 0$ is $2P(T \geq |t|)$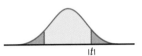

There is a similar significance test about the intercept $\beta_0$ that uses $SE_{b_0}$ and the $t(n - 2)$ distribution. Although computer outputs often include a test of $H_0: \beta_0 = 0$, this information usually has little practical value. From the equation for the population regression line, $\mu_y = \beta_0 + \beta_1 x$, we see that $\beta_0$ is the mean response corresponding to $x = 0$. In many practical situations, this subpopulation does not exist or is not interesting.

On the other hand, the test of $H_0: \beta_1 = 0$ is quite useful. When we substitute $\beta_1 = 0$ in the model, the $x$ term drops out and we are left with

$$\mu_y = \beta_0$$

This model says that the mean of $y$ does not vary with $x$. All of the $y$'s come from a single population with mean $\beta_0$, which we would estimate by $\bar{y}$. The hypothesis $H_0: \beta_1 = 0$ therefore says that there is no straight-line relationship between $y$ and $x$ and that linear regression of $y$ on $x$ is of no value for predicting $y$.

**EXAMPLE**

**10.5 Statistical software output, continued.** The computer outputs in Figure 10.5 for the fuel efficiency problem contain the information needed for inference about the regression slope and intercept. Let's look at the SPSS output. The column labeled Std. Error gives the standard errors of the estimates. The value of $SE_{b_1}$ appears on the line labeled with the variable name for the explanatory variable, LOGMPH. It is given as 0.354. In a summary we would report that the regression coefficient for the log of speed is 7.87 with a standard error of 0.35.

The $t$ statistic and $P$-value for the test of $H_0: \beta_1 = 0$ against the two-sided alternative $H_a: \beta_1 \neq 0$ appear in the columns labeled t and Sig. We can verify the $t$ calculation from the formula for the standardized estimate:

$$t = \frac{b_1}{SE_{b_1}} = \frac{7.874}{0.354} = 22.24$$

Some software will do these calculations directly if you input a value for the explanatory variable. Others will calculate the intervals for each value of $x$ in the data set. Creating a new data set with an additional observation with $x$ equal to the value of interest and $y$ missing will often work.

**EXAMPLE**

**10.8 Confidence interval for a speed of 30 mph.** Let's find the confidence interval for the mean response at 30 mph. We use $x = \log(30) = 3.4$ as the value for the explanatory variable. Our predicted fuel efficiency is

$$\widehat{MPG} = -7.80 + 7.87 \text{LOGMPH}$$

$$= -7.80 + (7.87)(3.4)$$

$$= 19.0$$

Software tells us that the 95% confidence interval for the mean response is 18.7 to 19.3 mpg.

If we operated this vehicle many times under similar conditions at an average speed of 30 mph, we would expect the fuel efficiency to be between 18.7 and 19.3 mpg. Note that many of the observations in Figure 10.9 lie outside the confidence bands. *These confidence intervals do not tell us what mileage to expect for a single observation at a particular average speed such as 30 mph.* We need a different kind of interval for this purpose.

## Prediction intervals

In the last example, we predicted the mean fuel efficiency when the average speed is 30 mph. Suppose we now want to predict a future observation of fuel efficiency when the vehicle is driven at 30 mph under similar conditions. Our best guess at the fuel efficiency is what we obtained before using the regression equation, that is, 19.0 mpg. The margin of error, on the other hand, is larger because it is harder to predict an individual value than to predict the mean.

The predicted response $y$ for an individual case with a specific value $x^*$ of the explanatory variable $x$ is

$$\hat{y} = b_0 + b_1 x^*$$

This is the same as the expression for $\hat{\mu}_y$. That is, the fitted line is used both to estimate the mean response when $x = x^*$ and to predict a single future response. We use the two notations $\hat{\mu}_y$ and $\hat{y}$ to remind ourselves of these two distinct uses.

prediction interval
A useful prediction should include a margin of error to indicate its accuracy. The interval used to predict a future observation is called a **prediction interval.** Although the response $y$ that is being predicted is a random variable, the interpretation of a prediction interval is similar to that for a confidence interval. Consider doing the following many times:

- Draw a sample of $n$ observations $(x_i, y_i)$ and then one additional observation $(x^*, y)$.

- Calculate the 95% prediction interval for $y$ when $x = x^*$ using the sample of size $n$.

Then 95% of the prediction intervals will contain the value of $y$ for the additional observation. In other words, the probability that this method produces an interval that contains the value of a future observation is 0.95.

The form of the prediction interval is very similar to that of the confidence interval for the mean response. The difference is that the standard error $SE_{\hat{y}}$ used in the prediction interval includes both the variability due to the fact that the least-squares line is not exactly equal to the true regression line *and* the variability of the future response variable $y$ around the subpopulation mean. (The formula for $SE_{\hat{y}}$ appears in Section 10.2.)

---

### PREDICTION INTERVAL FOR A FUTURE OBSERVATION

A **level $C$ prediction interval for a future observation** on the response variable $y$ from the subpopulation corresponding to $x^*$ is

$$\hat{y} \pm t^* SE_{\hat{y}}$$

where $t^*$ is the value for the $t(n-2)$ density curve with area $C$ between $-t^*$ and $t^*$.

---

Again, we use a graph to illustrate the results.

**EXAMPLE**

**10.9 Prediction intervals for fuel efficiency.** Figure 10.10 shows the upper and lower prediction limits, along with the data and the least-squares line. The 95% prediction limits are indicated by the dashed curves. Compare this figure with Figure 10.9, which shows the 95% confidence limits drawn to the same scale. The upper and lower limits of the prediction intervals are farther from the least-squares line than are the confidence limits. This results in most, but not all, of the observations in Figure 10.10 lying within the prediction bands.

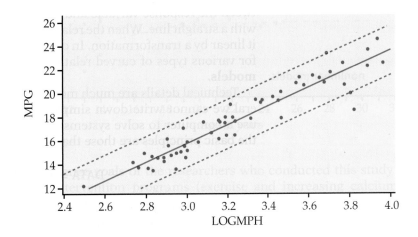

**FIGURE 10.10** The 95% prediction limits (dashed curves) for individual responses for the fuel efficiency example. Compare with Figure 10.9.

The overall deviation of any $y$ observation from the mean of the $y$'s is the sum of these two deviations:

$$(y_i - \bar{y}) = (\hat{y}_i - \bar{y}) + (y_i - \hat{y}_i)$$

In terms of deviations, this equation expresses the idea that DATA = FIT + RESIDUAL.

Several times we have measured variation by an average of squared deviations. If we square each of the three deviations above and then sum over all $n$ observations, it is an algebraic fact that the sums of squares add:

$$\sum (y_i - \bar{y})^2 = \sum (\hat{y}_i - \bar{y})^2 + \sum (y_i - \hat{y}_i)^2$$

We rewrite this equation as

$$\text{SST} = \text{SSM} + \text{SSE}$$

where

$$\text{SST} = \sum (y_i - \bar{y})^2$$

$$\text{SSM} = \sum (\hat{y}_i - \bar{y})^2$$

$$\text{SSE} = \sum (y_i - \hat{y}_i)^2$$

sum of squares    The SS in each abbreviation stands for **sum of squares,** and the T, M, and E stand for total, model, and error, respectively. ("Error" here stands for deviations from the line, which might better be called "residual" or "unexplained variation.") The total variation, as expressed by SST, is the sum of the variation due to the straight-line model (SSM) and the variation due to deviations from this model (SSE). This partition of the variation in the data between two sources is the heart of analysis of variance.

If $H_0: \beta_1 = 0$ were true, there would be no subpopulations and all of the $y$'s should be viewed as coming from a single population with mean $\mu_y$. The variation of the $y$'s would then be described by the sample variance

$$s_y^2 = \frac{\sum (y_i - \bar{y})^2}{n - 1}$$

The numerator in this expression is SST. The denominator is the total degrees of freedom, or simply DFT.

Just as the total sum of squares SST is the sum of SSM and SSE, the total degrees of freedom DFT is the sum of DFM and DFE, the degrees of freedom for the model and for the error:

$$\text{DFT} = \text{DFM} + \text{DFE}$$

The model has one explanatory variable $x$, so the degrees of freedom for this source are DFM = 1. Because DFT = $n - 1$, this leaves DFE = $n - 2$ as the degrees of freedom for error. For each source, the ratio of the sum of squares

mean square    to the degrees of freedom is called the **mean square,** or simply MS. The general formula for a mean square is

$$\text{MS} = \frac{\text{sum of squares}}{\text{degrees of freedom}}$$

Each mean square is an average squared deviation. MST is just $s_y^2$, the sample variance that we would calculate if all of the data came from a single population. MSE is also familiar to us:

$$\text{MSE} = s^2 = \frac{\sum (y_i - \hat{y}_i)^2}{n-2}$$

It is our estimate of $\sigma^2$, the variance about the population regression line.

---

### SUMS OF SQUARES, DEGREES OF FREEDOM, AND MEAN SQUARES

**Sums of squares** represent variation present in the responses. They are calculated by summing squared deviations. **Analysis of variance** partitions the total variation between two sources.

The sums of squares are related by the formula

$$\text{SST} = \text{SSM} + \text{SSE}$$

That is, the total variation is partitioned into two parts, one due to the model and one due to deviations from the model.

**Degrees of freedom** are associated with each sum of squares. They are related in the same way:

$$\text{DFT} = \text{DFM} + \text{DFE}$$

To calculate **mean squares,** use the formula

$$\text{MS} = \frac{\text{sum of squares}}{\text{degrees of freedom}}$$

---

interpretation of $r^2$

In Section 2.3 we noted that $r^2$ is the fraction of variation in the values of $y$ that is explained by the least-squares regression of $y$ on $x$. The sums of squares make this interpretation precise. Recall that $\text{SST} = \text{SSM} + \text{SSE}$. It is an algebraic fact that

$$r^2 = \frac{\text{SSM}}{\text{SST}} = \frac{\sum (\hat{y}_i - \bar{y})^2}{\sum (y_i - \bar{y})^2}$$

Because SST is the total variation in $y$ and SSM is the variation due to the regression of $y$ on $x$, this equation is the precise statement of the fact that $r^2$ is the fraction of variation in $y$ explained by $x$.

### The ANOVA F test

The null hypothesis $H_0: \beta_1 = 0$ that $y$ is not linearly related to $x$ can be tested by comparing MSM with MSE. The ANOVA test statistic is an **F statistic,**

F statistic

$$F = \frac{\text{MSM}}{\text{MSE}}$$

To find the 95% confidence interval we compute

$$\hat{\mu} \pm t^*\text{SE}_{\hat{\mu}} = 3.837 \pm (12.71)(3.158)$$
$$= 3.837 \pm 40.138$$
$$= 4 \pm 40$$

The interval is $-36$ to $44$ mg/kg/d of nitrogen.

Calculations for the prediction intervals are similar. The only difference is the use of the formula for $\text{SE}_{\hat{y}}$ in place of $\text{SE}_{\hat{\mu}}$.

Since the confidence interval for mean response includes the value 0, the corresponding intake 0.7 g/kg/d should be considered as a possible value for the intake requirement for this individual. Other intakes would also produce confidence intervals that would include the value of 0 for mean balance. Here is one method that is commonly used to determine a single value of the requirement for an individual.

**EXAMPLE**

**10.21 Estimating the protein requirement.** We define the estimated requirement for an individual to be the intake corresponding to zero balance using the fitted regression equation. To do this, we set the equation

$$\hat{\mu} = b_0 + b_1 x$$

equal to 0 and solve for the intake $x$. So,

$$x = -b_0/b_1$$
$$= -(-126.280)/185.882$$
$$= 0.68$$

The estimated protein requirement for this individual is 0.68 g/kg/d.

If we repeat these calculations using data collected on a large number of individuals, we can estimate the requirement distribution for a population. There are many interesting statistical issues related to this problem.[5]

## Inference for correlation

**LOOK BACK**
correlation, page 102

The correlation coefficient is a measure of the strength and direction of the linear association between two variables. Correlation does not require an explanatory-response relationship between the variables. We can consider the sample correlation $r$ as an estimate of the correlation in the population and base inference about the population correlation on $r$.

The correlation between the variables $x$ and $y$ when they are measured for every member of a population is the **population correlation.** As usual, we use Greek letters to represent population parameters. In this case $\rho$ (the Greek letter rho) is the population correlation. When $\rho = 0$, there is no linear association in the population. In the important case where the two variables $x$ and $y$ are both Normally distributed, the condition $\rho = 0$ is equivalent to the state-

population correlation $\rho$

ment that $x$ and $y$ are independent. That is, there is no association of any kind between $x$ and $y$. (Technically, the condition required is that $x$ and $y$ be **jointly Normal.** This means that the distribution of $x$ is Normal and also that the conditional distribution of $y$, given any fixed value of $x$, is Normal.) We therefore may wish to test the null hypothesis that a population correlation is 0.

**jointly Normal variables**

---

### TEST FOR A ZERO POPULATION CORRELATION

To test the hypothesis $H_0: \rho = 0$, compute the $t$ statistic:

$$t = \frac{r\sqrt{n-2}}{\sqrt{1-r^2}}$$

where $n$ is the sample size and $r$ is the sample correlation.

In terms of a random variable $T$ having the $t(n-2)$ distribution, the $P$-value for a test of $H_0$ against

$H_a: \rho > 0$ is $P(T \geq t)$

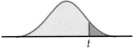

$H_a: \rho < 0$ is $P(T \leq t)$

$H_a: \rho \neq 0$ is $2P(T \geq |t|)$

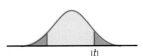

---

Most computer packages have routines for calculating correlations and some will provide the significance test for the null hypothesis that $\rho$ is zero.

**EXAMPLE**

**10.22 Correlation in the fuel efficiency study.** For the fuel efficiency example, the SPSS output appears in Figure 10.14. The sample correlation between fuel efficiency and the logarithm of speed is $r = 0.946$. SPSS calls this a Pearson correlation to distinguish it from other kinds of correlations that it can calculate. The $P$-value for a two-sided test of $H_0: \rho = 0$ is given as 0.000. This means that the actual $P$-value is less than 0.0005. We conclude that there is a nonzero correlation between MPG and LOGMPH.

If we wanted to test the one-sided alternative that the population correlation is negative, we divide the $P$-value in the output by 2, after checking that the sample coefficient is in fact negative.

If your software does not give the significance test, you can do the computations easily with a calculator.

| SPSS |
|---|

**Correlations**

|  |  |  |  |
|---|---|---|---|
| LOGMPH | Pearson Correlation | 1 | .946** |
|  | Sig. (2-tailed) | . | .000 |
|  | N | 60 | 60 |
| MPG | Pearson Correlation | .946** | 1 |
|  | Sig. (2-tailed) | .000 | . |
|  | N | 60 | 60 |

**. Correlation is significant at the 0.01 level (2-tailed).

FIGURE 10.14 Correlation output for Example 10.22.

**EXAMPLE**

**10.23 Correlation test using a calculator.** The correlation between MPG and LOGMPH is $r = 0.946$. Recall that $n = 60$. The $t$ statistic for testing the null hypothesis that the population correlation is zero is

$$t = \frac{r\sqrt{n-2}}{\sqrt{1-r^2}}$$

$$= \frac{0.946\sqrt{60-2}}{\sqrt{1-(0.946)^2}}$$

$$= 22.2$$

The degrees of freedom are $n - 2 = 58$. From Table D we conclude that $P < 0.0001$. This agrees with the SPSS output in Figure 10.14, where the $P$-value is given as 0.000. The data provide clear evidence that fuel efficiency and the log of speed are related.

There is a close connection between the significance test for a correlation and the test for the slope in a linear regression. Recall that

$$b_1 = r\frac{s_y}{s_x}$$

From this fact we see that if the slope is 0, so is the correlation, and vice versa. It should come as no surprise to learn that the procedures for testing $H_0: \beta_1 = 0$ and $H_0: \rho = 0$ are also closely related. In fact, the $t$ statistics for testing these hypotheses are numerically equal. That is,

$$\frac{b_1}{s_{b_1}} = \frac{r\sqrt{n-2}}{\sqrt{1-r^2}}$$

Check that this holds in both of our examples.

In our examples, the conclusion that there is a statistically significant correlation between the two variables would not come as a surprise to anyone familiar with the meaning of these variables. The significance test simply tells us whether or not there is evidence in the data to conclude that the population correlation is different from 0. The actual size of the correlation is of considerably more interest. We would therefore like to give a confidence interval for the population correlation. Unfortunately, most software packages do not perform this calculation. Because hand calculation of the confidence interval is very tedious, we do not give the method here.[6]

## USE YOUR KNOWLEDGE

10.5   **Research and development spending.** The National Science Foundation collects data on the research and development spending by universities and colleges in the United States.[7] Here are the data for the years 1999 to 2001 (using 1996 dollars):

| Year | 1999 | 2000 | 2001 |
|------|------|------|------|
| Spending (billions of dollars) | 26.4 | 28.0 | 29.7 |

Do the following by hand or with a calculator and verify your results with a software package.

(a) Make a scatterplot that shows the increase in research and development spending over time. Does the pattern suggest that the spending is increasing linearly over time?

(b) Find the equation of the least-squares regression line for predicting spending from year. Add this line to your scatterplot.

(c) For each of the three years, find the residual. Use these residuals to calculate the standard error $s$.

(d) Write the regression model for this setting. What are your estimates of the unknown parameters in this model?

(e) Compute a 95% confidence interval for the slope and summarize what this interval tells you about the increase in spending over time.

## SECTION 10.2   Summary

The **ANOVA table** for a linear regression gives the degrees of freedom, sum of squares, and mean squares for the model, error, and total sources of variation. The **ANOVA $F$ statistic** is the ratio MSM/MSE. Under $H_0: \beta_1 = 0$, this statistic has an $F(1, n - 2)$ distribution and is used to test $H_0$ versus the two-sided alternative.

The **square of the sample correlation** can be expressed as

$$r^2 = \frac{\text{SSM}}{\text{SST}}$$

and is interpreted as the proportion of the variability in the response variable $y$ that is explained by the explanatory variable $x$ in the linear regression.

The **standard errors** for $b_0$ and $b_1$ are

$$SE_{b_0} = s\sqrt{\frac{1}{n} + \frac{\bar{x}^2}{\sum(x_i - \bar{x})^2}}$$

$$SE_{b_1} = \frac{s}{\sqrt{\sum(x_i - \bar{x})^2}}$$

The **standard error** that we use for a confidence interval for the estimated mean response for the subpopulation corresponding to the value $x^*$ of the explanatory variable is

$$SE_{\hat{\mu}} = s\sqrt{\frac{1}{n} + \frac{(x^* - \bar{x})^2}{\sum(x_i - \bar{x})^2}}$$

The **standard error** that we use for a prediction interval for a future observation from the subpopulation corresponding to the value $x^*$ of the explanatory variable is

$$SE_{\hat{y}} = s\sqrt{1 + \frac{1}{n} + \frac{(x^* - \bar{x})^2}{\sum(x_i - \bar{x})^2}}$$

When the variables $y$ and $x$ are jointly Normal, the sample correlation is an estimate of the **population correlation** $\rho$. The test of $H_0: \rho = 0$ is based on the $t$ **statistic**

$$t = \frac{r\sqrt{n - 2}}{\sqrt{1 - r^2}}$$

which has a $t(n - 2)$ distribution under $H_0$. This test statistic is numerically identical to the $t$ statistic used to test $H_0: \beta_1 = 0$.

## CHAPTER 10   Exercises

*For Exercises 10.1 and 10.2, see pages 569 and 570; for Exercises 10.3 and 10.4, see page 576; and for Exercise 10.5, see page 593.*

10.6   **What's wrong?** For each of the following, explain what is wrong and why.

(a) The slope describes the change in $x$ for a change in $y$.

(b) The population regression line is $y = b_0 + b_1 x$.

(c) A 95% confidence interval for the mean response is the same width regardless of $x$.

10.7   **What's wrong?** For each of the following, explain what is wrong and why.

(a) The parameters of the simple linear regression model are $b_0$, $b_1$, and $s$.

(b) To test $H_0: b_1 = 0$, use a $t$ test.

(c) For a particular value of the explanatory variable $x$, the confidence interval for the mean response will be wider than the prediction interval for a future observation.

10.8   **95% confidence intervals for the slope.** Find a 95% confidence interval for the slope in each of the following settings:

(a) $n = 25$, $\hat{y} = 1.3 + 12.10x$, and $SE_{b_1} = 6.31$

(b) $n = 25$, $\hat{y} = 13.0 + 6.10x$, and $SE_{b_1} = 6.31$

(c) $n = 100$, $\hat{y} = 1.3 + 12.10x$, and $SE_{b_1} = 6.31$

10.9   **Significance test for the slope.** For each of the settings in the previous exercise, test the null hypothesis that the slope is zero versus the two-sided alternative.

**TABLE 10.1**

In-state tuition and fees (in dollars) for 32 public universities

| School | 2000 | 2005 | School | 2000 | 2005 | School | 2000 | 2005 |
|--------|------|------|--------|------|------|--------|------|------|
| Penn State | 7,018 | 11,508 | Virginia | 4,335 | 7,370 | Iowa State | 3,132 | 5,634 |
| Pittsburgh | 7,002 | 11,436 | Indiana | 4,405 | 7,112 | Oregon | 3,819 | 5,613 |
| Michigan | 6,926 | 9,798 | Cal-Santa Barbara | 3,832 | 6,997 | Iowa | 3,204 | 5,612 |
| Rutgers | 6,333 | 9,221 | Texas | 3,575 | 6,972 | Washington | 3,761 | 5,610 |
| Illinois | 4,994 | 8,634 | Cal-Irvine | 3,970 | 6,770 | Nebraska | 3,450 | 5,540 |
| Minnesota | 4,877 | 8,622 | Cal-San Diego | 3,848 | 6,685 | Kansas | 2,725 | 5,413 |
| Michigan State | 5,432 | 8,108 | Cal-Berkeley | 4,047 | 6,512 | Colorado | 3,188 | 5,372 |
| Ohio State | 4,383 | 8,082 | UCLA | 3,698 | 6,504 | North Carolina | 2,768 | 4,613 |
| Maryland | 5,136 | 7,821 | Purdue | 3,872 | 6,458 | Arizona | 2,348 | 4,498 |
| Cal-Davis | 4,072 | 7,457 | Wisconsin | 3,791 | 6,284 | Florida | 2,256 | 3,094 |
| Missouri | 4,726 | 7,415 | Buffalo | 4,715 | 6,068 | | | |

**10.10 Public university tuition: 2000 versus 2005.** Table 10.1 shows the in-state undergraduate tuition and required fees for 34 public universities in 2000 and 2005.[8]

(a) Plot the data with the 2000 tuition on the x-axis and describe the relationship. Are there any outliers or unusual values? Does a linear relationship between the tuition in 2000 and 2005 seem reasonable?

(b) Run the simple linear regression and state the least-squares regression line.

(c) Obtain the residuals and plot them versus the 2000 tuition amount. Is there anything unusual in the plot?

(d) Do the residuals appear to be approximately Normal? Explain.

(e) Give the null and alternative hypotheses for examining the relationship between 2000 and 2005 tuition amounts.

(f) Write down the test statistic and P-value for the hypotheses stated in part (e). State your conclusions.

**10.11 More on public university tuition.** Refer to Exercise 10.10.

(a) Construct a 95% confidence interval for the slope. What does this interval tell you about the percent increase in tuition between 2000 and 2005?

(b) The tuition at Stat U was $5000 in 2000. What is the predicted tuition in 2005?

(c) Find a 95% prediction interval for the 2005 tuition at Stat U and summarize the results.

**10.12 Are the two fuel efficiency measurements similar?** Refer to Exercise 7.24. In addition to the computer calculating mpg, the driver also recorded the mpg by dividing the miles driven by the amount of gallons at fill-up. The driver wants to determine if these calculations are different.

| Fill-up | 1 | 2 | 3 | 4 | 5 | 6 | 7 | 8 | 9 | 10 |
|---------|---|---|---|---|---|---|---|---|---|----|
| Computer | 41.5 | 50.7 | 36.6 | 37.3 | 34.2 | 45.0 | 48.0 | 43.2 | 47.7 | 42.2 |
| Driver | 36.5 | 44.2 | 37.2 | 35.6 | 30.5 | 40.5 | 40.0 | 41.0 | 42.8 | 39.2 |

| Fill-up | 11 | 12 | 13 | 14 | 15 | 16 | 17 | 18 | 19 | 20 |
|---------|----|----|----|----|----|----|----|----|----|----|
| Computer | 43.2 | 44.6 | 48.4 | 46.4 | 46.8 | 39.2 | 37.3 | 43.5 | 44.3 | 43.3 |
| Driver | 38.8 | 44.5 | 45.4 | 45.3 | 45.7 | 34.2 | 35.2 | 39.8 | 44.9 | 47.5 |

(a) Consider the driver's mpg calculations as the explanatory variable. Plot the data and describe the relationship. Are there any outliers or unusual values? Does a linear relationship seem reasonable?

(b) Run the simple linear regression and state the least-squares regression line.

(c) Summarize the results. Does it appear that the computer and driver calculations are the same? Explain.

**10.13 Beer and blood alcohol.** How well does the number of beers a student drinks predict his or her blood alcohol content? Sixteen student volunteers at Ohio State University drank a randomly assigned number of 12-ounce cans of beer. Thirty minutes later, a police officer measured their blood alcohol content (BAC). Here are the data:[9]

| Student | 1 | 2 | 3 | 4 | 5 | 6 | 7 | 8 |
|---------|---|---|---|---|---|---|---|---|
| Beers | 5 | 2 | 9 | 8 | 3 | 7 | 3 | 5 |
| BAC | 0.10 | 0.03 | 0.19 | 0.12 | 0.04 | 0.095 | 0.07 | 0.06 |

| Student | 9 | 10 | 11 | 12 | 13 | 14 | 15 | 16 |
|---------|---|----|----|----|----|----|----|----|
| Beers | 3 | 5 | 4 | 6 | 5 | 7 | 1 | 4 |
| BAC | 0.02 | 0.05 | 0.07 | 0.10 | 0.085 | 0.09 | 0.01 | 0.05 |

The students were equally divided between men and women and differed in weight and usual drinking habits. Because of this variation, many students don't believe that number of drinks predicts blood alcohol well.

(a) Make a scatterplot of the data. Find the equation of the least-squares regression line for predicting blood alcohol from number of beers and add this line to your plot. What is $r^2$ for these data? Briefly summarize what your data analysis shows.

(b) Is there significant evidence that drinking more beers increases blood alcohol on the average in the population of all students? State hypotheses, give a test statistic and $P$-value, and state your conclusion.

(c) Steve thinks he can drive legally 30 minutes after he drinks 5 beers. The legal limit is BAC = 0.08. Give a 90% prediction interval. Can he be confident he won't be arrested if he drives and is stopped?

10.14    **Predicting water quality.** The index of biotic integrity (IBI) is a measure of the water quality in streams. IBI and land use measures for a collection of streams in the Ozark Highland ecoregion of Arkansas were collected as part of a study.[10] Table 10.2 gives the data for IBI and the area of the watershed in square kilometers for streams in the original sample with area less than or equal to 70 km$^2$.

(a) Use numerical and graphical methods to describe the variable IBI. Do the same for area. Summarize your results.

(b) Plot the data and describe the relationship. Are there any outliers or unusual patterns?

(c) Give the statistical model for simple linear regression for this problem.

(d) State the null and alternative hypotheses for examining the relationship between IBI and area.

(e) Run the simple linear regression and summarize the results.

(f) Obtain the residuals and plot them versus area. Is there anything unusual in the plot?

(g) Do the residuals appear to be approximately Normal? Give reasons for your answer.

(h) Do the assumptions for the analysis of these data using the model you gave in part (c) appear to be reasonable? Explain your answer.

10.15    **More on predicting water quality.** The researchers who conducted the study described in the previous exercise also recorded the percent of the watershed area that was forest for each of the streams. The data are given in Table 10.3. Analyze these data using the questions in the previous exercise as a guide.

10.16    **Comparing the analyses.** In Exercises 10.14 and 10.15, you used two different explanatory variables to predict IBI. Summarize the two analyses and compare the results. If you had to choose between the two explanatory variables for predicting IBI, which one would you prefer? Give reasons for your answer.

10.17    **How an outlier can affect statistical significance.** Consider the data in Table 10.3 and the relationship between IBI and the percent of watershed area that was forest. The relationship between these two variables is almost significant at the .05 level. In this exercise you will demonstrate the potential effect of an outlier on statistical significance. Investigate what happens when you decrease the IBI to 0.0 for (1) an observation with 0% forest and (2) an observation with 100% forest.

---

**TABLE 10.2**

Watershed area and index of biotic integrity

| Area | IBI | Area | IBI | Area | IBI | Area | IBI | Area | IBI |
|------|-----|------|-----|------|-----|------|-----|------|-----|
| 21 | 47 | 29 | 61 | 31 | 39 | 32 | 59 | 34 | 72 |
| 34 | 76 | 49 | 85 | 52 | 89 | 2 | 74 | 70 | 89 |
| 6 | 33 | 28 | 46 | 21 | 32 | 59 | 80 | 69 | 80 |
| 47 | 78 | 8 | 53 | 8 | 43 | 58 | 88 | 54 | 84 |
| 10 | 62 | 57 | 55 | 18 | 29 | 19 | 29 | 39 | 54 |
| 49 | 78 | 9 | 71 | 5 | 55 | 14 | 58 | 9 | 71 |
| 23 | 33 | 31 | 59 | 18 | 81 | 16 | 71 | 21 | 75 |
| 32 | 64 | 10 | 41 | 26 | 82 | 9 | 60 | 54 | 84 |
| 12 | 83 | 21 | 82 | 27 | 82 | 23 | 86 | 26 | 79 |
| 16 | 67 | 26 | 56 | 26 | 85 | 28 | 91 | | |

**TABLE 10.3**

Percent forest and index of biotic integrity

| Forest | IBI | Forest | IBI | Forest | IBI | Forest | IBI | Forest | IBI |
|--------|-----|--------|-----|--------|-----|--------|-----|--------|-----|
| 0 | 47 | 0 | 61 | 0 | 39 | 0 | 59 | 0 | 72 |
| 0 | 76 | 3 | 85 | 3 | 89 | 7 | 74 | 8 | 89 |
| 9 | 33 | 10 | 46 | 10 | 32 | 11 | 80 | 14 | 80 |
| 17 | 78 | 17 | 53 | 18 | 43 | 21 | 88 | 22 | 84 |
| 25 | 62 | 31 | 55 | 32 | 29 | 33 | 29 | 33 | 54 |
| 33 | 78 | 39 | 71 | 41 | 55 | 43 | 58 | 43 | 71 |
| 47 | 33 | 49 | 59 | 49 | 81 | 52 | 71 | 52 | 75 |
| 59 | 64 | 63 | 41 | 68 | 82 | 75 | 60 | 79 | 84 |
| 79 | 83 | 80 | 82 | 86 | 82 | 89 | 86 | 90 | 79 |
| 95 | 67 | 95 | 56 | 100 | 85 | 100 | 91 | | |

Write a short summary of what you learn from this exercise.

**10.18 Predicting water quality for an area of 30 km².** Refer to Exercise 10.14.

(a) Find a 95% confidence interval for mean response corresponding to an area of 30 km².

(b) Find a 95% prediction interval for a future response.

(c) Write a short paragraph interpreting the meaning of the intervals in terms of Ozark Highland streams.

(d) Do you think that these results can be applied to other streams in Arkansas or in other states? Explain why or why not.

**10.19 Compare the predictions.** Case 21 in Table 10.2 and Table 10.3 corresponds to the same watershed area. For this case the area is 10 km² and the percent forest is 25%. A predicted index of biotic integrity based on area was computed in Exercise 10.14, while one based on percent forest was computed in Exercise 10.15. Compare these two estimates and explain why they differ. Use the idea of a prediction interval to interpret these results.

**10.20 U.S. versus overseas stock returns.** Returns on common stocks in the United States and overseas appear to be growing more closely correlated as economies become more interdependent. Suppose that the following population regression line connects the total annual returns (in percent) on two indexes of stock prices:

MEAN OVERSEAS RETURN

$$= 4.6 + 0.67 \times \text{U.S. RETURN}$$

(a) What is $\beta_0$ in this line? What does this number say about overseas returns when the U.S. market is flat (0% return)?

(b) What is $\beta_1$ in this line? What does this number say about the relationship between U.S. and overseas returns?

(c) We know that overseas returns will vary in years having the same return on U.S. common stocks. Write the regression model based on the population regression line given above. What part of this model allows overseas returns to vary when U.S. returns remain the same?

**10.21** CHALLENGE **Breaking strength of wood.** Exercise 2.144 (page 163) gives the modulus of elasticity (MOE) and the modulus of rupture (MOR) for 32 plywood specimens. Because measuring MOR involves breaking the wood but measuring MOE does not, we would like to predict the destructive test result, MOR, using the nondestructive test result, MOE.

(a) Describe the distribution of MOR using graphical and numerical summaries. Do the same for MOE.

(b) Make a plot of the two variables. Which should be plotted on the $x$ axis? Give a reason for your answer.

(c) Give the statistical model for this analysis, run the analysis, summarize the results, and write a short summary of your conclusions.

(d) Examine the assumptions needed for the analysis. Are you satisfied that there are no serious violations that would cause you to question the validity of your conclusions?

**10.22 Breaking strength of wood, continued.** Refer to the previous exercise. Consider an MOE of 2,000,000.

(a) Interpret the confidence interval for mean response and the prediction interval for a future observation for this value of MOE.

**TABLE 10.9**

SAT and ACT scores

| SAT | ACT | SAT | ACT | SAT | ACT | SAT | ACT |
|-----|-----|-----|-----|-----|-----|-----|-----|
| 1000 | 24 | 870 | 21 | 1090 | 25 | 800 | 21 |
| 1010 | 24 | 880 | 21 | 860 | 19 | 1040 | 24 |
| 920 | 17 | 850 | 22 | 740 | 16 | 840 | 17 |
| 840 | 19 | 780 | 22 | 500 | 10 | 1060 | 25 |
| 830 | 19 | 830 | 20 | 780 | 12 | 870 | 21 |
| 1440 | 32 | 1190 | 30 | 1120 | 27 | 1120 | 25 |
| 490 | 7 | 800 | 16 | 590 | 12 | 800 | 18 |
| 1050 | 23 | 830 | 16 | 990 | 24 | 960 | 27 |
| 870 | 18 | 890 | 23 | 700 | 16 | 880 | 21 |
| 970 | 21 | 880 | 24 | 930 | 22 | 1020 | 24 |
| 920 | 22 | 980 | 27 | 860 | 23 | 790 | 14 |
| 810 | 19 | 1030 | 23 | 420 | 21 | 620 | 18 |
| 1080 | 23 | 1220 | 30 | 800 | 20 | 1150 | 28 |
| 1000 | 19 | 1080 | 22 | 1140 | 24 | 970 | 20 |
| 1030 | 25 | 970 | 20 | 920 | 21 | 1060 | 24 |

(b) Find the least-squares regression line and draw it on your plot. Give the results of the significance test for the slope.

(c) What is the correlation between the two tests?

**10.54** ⚠ CHALLENGE **SAT versus ACT, continued.** Refer to the previous exercise. Find the predicted value of ACT for each observation in the data set.

(a) What is the mean of these predicted values? Compare it with the mean of the ACT scores.

(b) Compare the standard deviation of the predicted values with the standard deviation of the actual ACT scores. If least-squares regression is used to predict ACT scores for a large number of students such as these, the average predicted value will be accurate but the variability of the predicted scores will be too small.

(c) Find the SAT score for a student who is one standard deviation above the mean ($z = (x - \bar{x})/s = 1$). Find the predicted ACT score and standardize this score. (Use the means and standard deviations from this set of data for these calculations.)

(d) Repeat part (c) for a student whose SAT score is one standard deviation below the mean ($z = -1$).

(e) What do you conclude from parts (c) and (d)? Perform additional calculations for different $z$'s if needed.

**10.55** ⚠ CHALLENGE **Matching standardized scores.** Refer to the previous two exercises. An alternative to the least-squares method is based on matching standardized scores. Specifically, we set

$$\frac{(\hat{y} - \bar{y})}{s_y} = \frac{(x - \bar{x})}{s_x}$$

and solve for $y$. Let's use the notation $y = a_0 + a_1 x$ for this line. The slope is $a_1 = s_y/s_x$ and the intercept is $a_0 = \bar{y} - a_1\bar{x}$. Compare these expressions with the formulas for the least-squares slope and intercept (page 565).

(a) Using the data in Table 10.9, find the values of $a_0$ and $a_1$.

(b) Plot the data with the least-squares line and the new prediction line.

(c) Use the new line to find predicted ACT scores. Find the mean and the standard deviation of these scores. How do they compare with the mean and standard deviation of the ACT scores?

**10.56** **Length, width, and weight of perch.** Here are data for 12 perch caught in a lake in Finland:[20]

| Weight (grams) | Length (cm) | Width (cm) | Weight (grams) | Length (cm) | Width (cm) |
|---------------|-------------|------------|----------------|-------------|------------|
| 5.9 | 8.8 | 1.4 | 300.0 | 28.7 | 5.1 |
| 100.0 | 19.2 | 3.3 | 300.0 | 30.1 | 4.6 |
| 110.0 | 22.5 | 3.6 | 685.0 | 39.0 | 6.9 |
| 120.0 | 23.5 | 3.5 | 650.0 | 41.4 | 6.0 |
| 150.0 | 24.0 | 3.6 | 820.0 | 42.5 | 6.6 |
| 145.0 | 25.5 | 3.8 | 1000.0 | 46.6 | 7.6 |

In this exercise we will examine different models for predicting weight.

(a) Run the regression using length to predict weight. Do the same using width as the explanatory variable. Summarize the results. Be sure to include the value of $r^2$.

(b) Plot weight versus length and weight versus width. Include the least-squares lines in these plots. Do these relationships appear to be linear? Explain your answer.

10.57 **Transforming the perch data.** Refer to the previous exercise.

(a) Try to find a better model using a transformation of length. One possibility is to use the square. Make a plot and perform the regression analysis. Summarize the results.

(b) Do the same for width.

10.58 **Creating a new explanatory variable.** Refer to the previous two exercises.

(a) Create a new variable that is the product of length and width. Make a plot and run the regression using this new variable. Summarize the results.

(b) Write a short report summarizing and comparing the different regression analyses that you performed in this exercise and the previous two exercises.

10.59 **Index of biotic integrity.** Refer to the data on the index of biotic integrity and area in Exercise 10.14 (page 596) and the additional data on percent watershed area that was forest in Exercise 10.15. Find the correlations among these three variables, perform the test of statistical significance, and summarize the results. Which of these test results could have been obtained from the analyses that you performed in Exercises 10.14 and 10.15?

10.60 **Food neophobia.** Food neophobia is a personality trait associated with avoiding unfamiliar foods. In one study of 564 children who were 2 to 6 years of age, food neophobia and the frequency of consumption of different types of food were measured.[21] Here is a summary of the correlations:

| Type of food | Correlation |
|---|---|
| Vegetables | −0.27 |
| Fruit | −0.16 |
| Meat | −0.15 |
| Eggs | −0.08 |
| Sweet/fatty snacks | 0.04 |
| Starchy staples | −0.02 |

Perform the significance test for each correlation and write a summary about food neophobia and the consumption of different types of food.

10.61 **Personality traits and scores on the GRE.** A study reported correlations between several personality traits and scores on the Graduate Record Examination (GRE) for a sample of 342 test takers.[22] Here is a table of the correlations:

| | GRE score | | |
|---|---|---|---|
| Personality trait | Analytical | Quantitative | Verbal |
| Conscientiousness | −0.17 | −0.14 | −0.12 |
| Rationality | −0.06 | −0.03 | −0.08 |
| Ingenuity | −0.06 | −0.08 | −0.02 |
| Quickness | 0.21 | 0.15 | 0.26 |
| Creativity | 0.24 | 0.26 | 0.29 |
| Depth | 0.06 | 0.08 | 0.15 |

For each correlation, test the null hypothesis that the corresponding true correlation is zero. Reproduce the table and mark the correlations that have $P < 0.001$ with ***, those that have $P < 0.01$ with **, and those that have $P < 0.05$ with *. Some critics of standardized tests have suggested that the tests penalize students who are "deep thinkers" and those who are very creative. Others have suggested that students who work quickly do better on these tests. Write a summary of the results of your significance tests, taking into account these comments.

10.62 **Resting metabolic rate and exercise.** Metabolic rate, the rate at which the body consumes energy, is important in studies of weight gain, dieting, and exercise. The table below gives data on the lean body mass and resting metabolic rate for 12 women and 7 men who are subjects in a study of dieting. Lean body mass, given in kilograms, is a person's weight leaving out all fat. Metabolic rate is measured in calories burned per 24 hours, the same calories used to describe the energy content of foods. The researchers believe that lean body mass is an important influence on metabolic rate.

| Subject | Sex | Mass | Rate | Subject | Sex | Mass | Rate |
|---|---|---|---|---|---|---|---|
| 1 | M | 62.0 | 1792 | 11 | F | 40.3 | 1189 |
| 2 | M | 62.9 | 1666 | 12 | F | 33.1 | 913 |
| 3 | F | 36.1 | 995 | 13 | M | 51.9 | 1460 |
| 4 | F | 54.6 | 1425 | 14 | F | 42.4 | 1124 |
| 5 | F | 48.5 | 1396 | 15 | F | 34.5 | 1052 |
| 6 | F | 42.0 | 1418 | 16 | F | 51.1 | 1347 |
| 7 | M | 47.4 | 1362 | 17 | F | 41.2 | 1204 |
| 8 | F | 50.6 | 1502 | 18 | M | 51.9 | 1867 |
| 9 | F | 42.0 | 1256 | 19 | M | 46.9 | 1439 |
| 10 | M | 48.7 | 1614 | | | | |

$$\mu_y = \beta_0 + \beta_1 x$$

For any fixed value of $x$, the response $y$ varies Normally around this mean and has a standard deviation $\sigma$ that is the same for all values of $x$.

In the multiple regression setting, the response variable $y$ depends on $p$ explanatory variables, which we will denote by $x_1, x_2, \ldots, x_p$. The mean response depends on these explanatory variables according to a linear function

$$\mu_y = \beta_0 + \beta_1 x_1 + \beta_2 x_2 + \cdots + \beta_p x_p$$

Similar to simple linear regression, this expression is the population regression equation. We do not observe the mean response because the observed values of $y$ vary about their means. We can think of subpopulations of responses, each corresponding to a particular set of values for *all* of the explanatory variables $x_1, x_2, \ldots, x_p$. In each subpopulation, $y$ varies Normally with a mean given by the population regression equation. The regression model assumes that the standard deviation $\sigma$ of the responses is the same in all subpopulations.

**EXAMPLE**

**11.1 Predicting early success in college.** Our case study uses data collected at a large university on all first-year computer science majors in a particular year.[1] The purpose of the study was to attempt to predict success in the early university years. One measure of success was the cumulative grade point average (GPA) after three semesters. Among the explanatory variables recorded at the time the students enrolled in the university were average high school grades in mathematics (HSM), science (HSS), and English (HSE).

We will use high school grades to predict the response variable GPA. There are $p = 3$ explanatory variables: $x_1$ = HSM, $x_2$ = HSS, and $x_3$ = HSE. The high school grades are coded on a scale from 1 to 10, with 10 corresponding to A, 9 to A−, 8 to B+, and so on. These grades define the subpopulations. For example, the straight-C students are the subpopulation defined by HSM = 4, HSS = 4, and HSE = 4.

One possible multiple regression model for the subpopulation mean GPAs is

$$\mu_{\text{GPA}} = \beta_0 + \beta_1 \text{HSM} + \beta_2 \text{HSS} + \beta_3 \text{HSE}$$

For the straight-C subpopulation of students, the model gives the subpopulation mean as

$$\mu_{\text{GPA}} = \beta_0 + \beta_1 4 + \beta_2 4 + \beta_3 4$$

## Data for multiple regression

The data for a simple linear regression problem consist of observations $(x_i, y_i)$ of the two variables. Because there are several explanatory variables in multiple regression, the notation needed to describe the data is more elaborate. Each observation or case consists of a value for the response variable and for each of the explanatory variables. Call $x_{ij}$ the value of the $j$th explanatory variable for the $i$th case. The data are then

$$\text{Case 1: } (x_{11}, x_{12}, \ldots, x_{1p}, y_1)$$

$$\text{Case 2: } (x_{21}, x_{22}, \ldots, x_{2p}, y_2)$$

$$\vdots$$

$$\text{Case } n: (x_{n1}, x_{n2}, \ldots, x_{np}, y_n)$$

Here, $n$ is the number of cases and $p$ is the number of explanatory variables. Data are often entered into computer regression programs in this format. Each row is a case and each column corresponds to a different variable. The data for Example 11.1, with several additional explanatory variables, appear in this format in the CSDATA data set described in the Data Appendix.

## Multiple linear regression model

**LOOK BACK**
**DATA = FIT +**
**RESIDUAL, page 564**

We combine the population regression equation and assumptions about variation to construct the multiple linear regression model. The subpopulation means describe the FIT part of our statistical model. The RESIDUAL part represents the variation of observations about the means. We will use the same notation for the residual that we used in the simple linear regression model. The symbol $\epsilon$ represents the deviation of an individual observation from its subpopulation mean. We assume that these deviations are Normally distributed with mean 0 and an unknown standard deviation $\sigma$ that does not depend on the values of the $x$ variables. *These are assumptions that we can check by examining the residuals in the same way that we did for simple linear regression.*

---

MULTIPLE LINEAR REGRESSION MODEL

The **statistical model for multiple linear regression** is

$$y_i = \beta_0 + \beta_1 x_{i1} + \beta_2 x_{i2} + \cdots + \beta_p x_{ip} + \epsilon_i$$

for $i = 1, 2, \ldots, n$.

The **mean response** $\mu_y$ is a linear function of the explanatory variables:

$$\mu_y = \beta_0 + \beta_1 x_1 + \beta_2 x_2 + \cdots + \beta_p x_p$$

The **deviations** $\epsilon_i$ are independent and Normally distributed with mean 0 and standard deviation $\sigma$. In other words, they are an SRS from the $N(0, \sigma)$ distribution.

The parameters of the model are $\beta_0, \beta_1, \beta_2, \ldots, \beta_p$, and $\sigma$.

---

The assumption that the subpopulation means are related to the regression coefficients $\beta$ by the equation

$$\mu_y = \beta_0 + \beta_1 x_1 + \beta_2 x_2 + \cdots + \beta_p x_p$$

implies that we can estimate all subpopulation means from estimates of the $\beta$'s. To the extent that this equation is accurate, we have a useful tool for describing how the mean of $y$ varies with the collection of $x$'s.

correlations in Figure 11.3). The correlation between HSM and HSS is 0.58, and that between HSM and HSE is 0.45. Thus, when we have a regression model that contains all three high school grades as explanatory variables, there is considerable overlap of the predictive information contained in these variables. *The significance tests for individual regression coefficients assess the significance of each predictor variable assuming that all other predictors are included in the regression equation.* Given that we use a model with HSM and HSS as predictors, the coefficient of HSE is not statistically significant. Similarly, given that we have HSM and HSE in the model, HSS does not have a significant regression coefficient. HSM, however, adds significantly to our ability to predict GPA even after HSS and HSE are already in the model.

Unfortunately, we cannot conclude from this analysis that the *pair* of explanatory variables HSS and HSE contribute nothing significant to our model for predicting GPA once HSM is in the model. The impact of relations among the several explanatory variables on fitting models for the response is the most important new phenomenon encountered in moving from simple linear regression to multiple regression. We can only hint at the many complicated problems that arise.

## Residuals

As in simple linear regression, we should always examine the residuals as an aid to determining whether the multiple regression model is appropriate for the data. Because there are several explanatory variables, we must examine several residual plots. It is usual to plot the residuals versus the predicted values $\hat{y}$ and also versus each of the explanatory variables. Look for outliers, influential observations, evidence of a curved (rather than linear) relation, and anything else unusual. Again, we leave the task of making these plots as an exercise. The plots all appear to show more or less random noise around the center value of 0.

If the deviations $\epsilon$ in the model are Normally distributed, the residuals should be Normally distributed. Figure 11.5 presents a Normal quantile plot of the residuals. The distribution appears to be approximately Normal. There are many other specialized plots that help detect departures from the multiple regression model. Discussion of these, however, is more than we can undertake in this chapter.

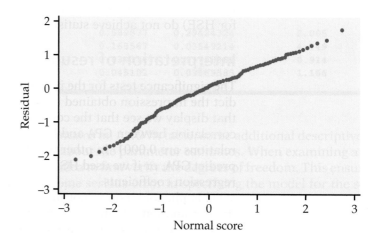

**FIGURE 11.5** Normal quantile plot of the residuals from the high school grades model. There are no important deviations from Normality.

**USE YOUR KNOWLEDGE**

11.6    **Residual plots for the CSDATA analysis.** The CSDATA data set can be found in the Data Appendix. Using a statistical package, fit the linear model with HSM and HSE as predictors and obtain the residuals and predicted values. Plot the residuals versus the predicted values, HSM, and HSE. Are the residuals more or less randomly dispersed around zero? Comment on any unusual patterns.

## Refining the model

Because the variable HSS has the largest *P*-value of the three explanatory variables (see Figure 11.4) and therefore appears to contribute the least to our explanation of GPA, we rerun the regression using only HSM and HSE as explanatory variables. The SAS output appears in Figure 11.6. The *F* statistic indicates that we can reject the null hypothesis that the regression coefficients for the two explanatory variables are both 0. The *P*-value is still 0.0001. The value of $R^2$ has dropped very slightly compared with our previous run, from 0.2046 to 0.2016. Thus, dropping HSS from the model resulted in the loss of very little explanatory power. The measure *s* of variation about the fitted equation (Root MSE in the printout) is nearly identical for the two regressions, another indication that we lose very little dropping HSS. The *t* statistics for the individual regression coefficients indicate that HSM is still clearly significant ($P = 0.0001$), while the statistic for HSE is larger than before (1.747 versus 1.166) and approaches the traditional 0.05 level of significance ($P = 0.082$).

**FIGURE 11.6** Multiple regression output for regression using HSM and HSE to predict GPA.

```
SAS                                                          _ □ X

Dependent Variable: GPA


                Analysis of Variance

                          Sum of          Mean
  Source          DF      Squares        Square     F Value    Prob>F

  Model            2      27.30349      13.65175     27.894    <.0001
  Error          221     108.15930       0.48941
  C Total        223     135.46279


      Root MSE        0.69958    R-Square      0.2016
      Dep Mean        2.63522    Adj R-sq      0.1943
      C.V.           26.54718


                    Parameter Estimates

                    Parameter      Standard     T for H0:
  Variable   DF     Estimate         Error     Parameter=0   Prob > |T|

  INTERCEP    1     0.624228      0.29172204      2.140         0.0335
  HSM         1     0.182654      0.03195581      5.716         0.0001
  HSE         1     0.060670      0.03472914      1.747         0.0820
```

*The following nine exercises use the CHEESE data set described in the Data Appendix.*

11.51   **Describing the explanatory variables.** For each of the four variables in the CHEESE data set, find the mean, median, standard deviation, and interquartile range. Display each distribution by means of a stemplot and use a Normal quantile plot to assess Normality of the data. Summarize your findings. Note that when doing regressions with these data, we do not assume that these distributions are Normal. Only the residuals from our model need to be (approximately) Normal. The careful study of each variable to be analyzed is nonetheless an important first step in any statistical analysis.

11.52   **Pairwise scatterplots of the explanatory variables.** Make a scatterplot for each pair of variables in the CHEESE data set (you will have six plots). Describe the relationships. Calculate the correlation for each pair of variables and report the *P*-value for the test of zero population correlation in each case.

11.53   **Simple linear regression model of Taste.** Perform a simple linear regression analysis using Taste as the response variable and Acetic as the explanatory variable. Be sure to examine the residuals carefully. Summarize your results. Include a plot of the data with the least-squares regression line. Plot the residuals versus each of the other two chemicals. Are any patterns evident? (The concentrations of the other chemicals are lurking variables for the simple linear regression.)

11.54   **Another simple linear regression model of Taste.** Repeat the analysis of Exercise 11.53 using Taste as the response variable and H2S as the explanatory variable.

11.55   **The final simple linear regression model of Taste.** Repeat the analysis of Exercise 11.53 using Taste as the response variable and Lactic as the explanatory variable.

11.56   **Comparing the simple linear regression models.** Compare the results of the regressions performed in the three previous exercises. Construct a table with values of the *F* statistic, its *P*-value, $R^2$, and the estimate *s* of the standard deviation for each model. Report the three regression equations. Why are the intercepts in these three equations different?

11.57   **Multiple regression model of Taste.** Carry out a multiple regression using Acetic and H2S to predict Taste. Summarize the results of your analysis. Compare the statistical significance of Acetic in this model with its significance in the model with Acetic alone as a predictor (Exercise 11.53). Which model do you prefer? Give a simple explanation for the fact that Acetic alone appears to be a good predictor of Taste, but with H2S in the model, it is not.

11.58   **Another multiple regression model of Taste.** Carry out a multiple regression using H2S and Lactic to predict Taste. Comparing the results of this analysis with the simple linear regressions using each of these explanatory variables alone, it is evident that a better result is obtained by using both predictors in a model. Support this statement with explicit information obtained from your analysis.

11.59   **The final multiple regression model of Taste.** Use the three explanatory variables Acetic, H2S, and Lactic in a multiple regression to predict Taste. Write a short summary of your results, including an examination of the residuals. Based on all of the regression analyses you have carried out on these data, which model do you prefer and why?

11.60   **Finding a multiple regression model on the Internet.** Search the Internet to find an example of the use of multiple regression. Give the setting of the example, describe the data, give the model, and summarize the results. Explain why the use of multiple regression in this setting was appropriate or inappropriate.

# One-Way Analysis of Variance

Which brand of tires lasts the longest under city driving conditions? The methods described in this chapter allow us to compare the average wear of each brand.

**12.1 Inference for One-Way Analysis of Variance**

**12.2 Comparing the Means**

## Introduction

Many of the most effective statistical studies are comparative. For example, we may wish to compare customer satisfaction of men and women using an online fantasy football site or compare the responses to various treatments in a clinical trial. We display these comparisons with back-to-back stemplots or side-by-side boxplots, and we measure them with five-number summaries or with means and standard deviations.

When only two groups are compared, Chapter 7 provides the tools we need to answer the question "Is the difference between groups statistically significant?" Two-sample *t* procedures compare the means of two Normal populations, and we saw that these procedures, unlike comparisons of spread, are sufficiently robust to be widely useful.

In this chapter, we will compare any number of means by techniques that generalize the two-sample *t* and share its robustness and usefulness. These methods will allow us to address comparisons such as

- Which of 4 advertising offers mailed to sample households produces the highest dollar sales?

- Which of 10 brands of automobile tires wears longest?

- How long do cancer patients live under each of 3 therapies for their lung cancer?

# 12.1 Inference for One-Way Analysis of Variance

**LOOK BACK**

comparing two means,
page 447

When comparing different populations or treatments, the data are subject to sampling variability. For example, we would not expect the same sales data if we mailed various advertising offers to a different sample of households. We therefore pose the question for inference in terms of the *mean* response. In Chapter 7 we met procedures for comparing the means of two populations. We are now ready to extend those methods to problems involving more than two populations. The statistical methodology for comparing several means is called **analysis of variance,** or simply **ANOVA.** In the sections that follow, we will examine the basic ideas and assumptions that are needed for ANOVA. Although the details differ, many of the concepts are similar to those discussed in the two-sample case.

ANOVA

one-way ANOVA

We will consider two ANOVA techniques. When there is only one way to classify the populations of interest, we use **one-way ANOVA** to analyze the data. For example, to compare the survival times for three different lung cancer therapies we use one-way ANOVA. This chapter presents the details for one-way ANOVA.

two-way ANOVA

In many other comparison studies, there is more than one way to classify the populations. For the advertising study, the company may also consider mailing the offers using two different envelope styles. Will each offer draw more sales on the average when sent in an attention-grabbing envelope? Analyzing the effect of advertising offer and envelope layout together requires **two-way ANOVA.** This technique will be discussed in Chapter 13. While adding yet more factors necessitates even higher-way ANOVA techniques, most of the new ideas in ANOVA with more than one factor already appear in two-way ANOVA.

## Data for one-way ANOVA

One-way analysis of variance is a statistical method for comparing several population means. We draw a simple random sample (SRS) from each population and use the data to test the null hypothesis that the population means are all equal. Consider the following two examples:

**EXAMPLE**

**12.1  Choosing the best magazine layout.**   A magazine publisher wants to compare three different layouts for a magazine that will be offered for sale at supermarket checkout lines. She is interested in whether there is a layout that better catches shoppers' attention and results in more sales. To investigate, she randomly assigns each of 60 stores to one of the three layouts and records the number of magazines that are sold in a one-week period.

**EXAMPLE**

**12.2  Average age of bookstore customers.**   How do five bookstores in the same city differ in the demographics of their customers? Are certain bookstores more popular among teenagers? Do upper-income shoppers tend to go to one store? A market researcher asks 50 customers of each store to respond to a questionnaire. Two variables of interest are the customer's age and income level.

These two examples are similar in that

- There is a single quantitative response variable measured on many units; the units are stores in the first example and customers in the second.

- The goal is to compare several populations: stores displaying three magazine layouts in the first example and customers of five bookstores in the second.

**LOOK BACK**
observation versus experiment, page 175

There is, however, an important difference. Example 12.1 describes an experiment in which stores are randomly assigned to layouts. Example 12.2 is an observational study in which customers are selected during a particular time period and not all agree to provide data. We will treat our samples of customers as random samples even though this is only approximately true.

In both examples, we will use ANOVA to compare the mean responses. The same ANOVA methods apply to data from random samples and to data from randomized experiments. *It is important to keep the data-production method in mind when interpreting the results. A strong case for causation is best made by a randomized experiment.*

## Comparing means

**LOOK BACK**
standard deviation of $\bar{x}$, page 338

The question we ask in ANOVA is "Do all groups have the same population mean?" We will often use the term *groups* for the populations to be compared in a one-way ANOVA. To answer this question we compare the sample means. Figure 12.1 displays the sample means for Example 12.1. It appears that Layout 2 has the highest average sales. But is the observed difference in sample means just the result of chance variation? We should not expect sample means to be equal, even if the population means are all identical.

The purpose of ANOVA is to assess whether the observed differences among sample means are *statistically significant*. Could a variation among the three sample means this large be plausibly due to chance, or is it good evidence for a difference among the population means? This question can't be answered from the sample means alone. Because the standard deviation of a sample mean $\bar{x}$ is the population standard deviation $\sigma$ divided by $\sqrt{n}$, the answer also depends upon both the variation within the groups of observations and the sizes of the samples.

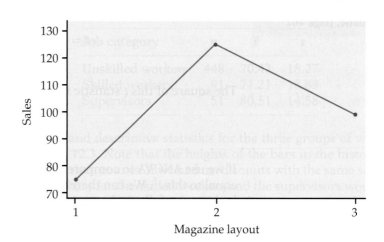

**FIGURE 12.1** Mean sales of magazines for three different magazine layouts.

$$DATA = FIT + RESIDUAL$$

view of statistical models. For one-way ANOVA, this corresponds to

$$x_{ij} = \mu_i + \epsilon_{ij}$$

We can think of these three terms as sources of variation. The ANOVA table separates the variation in the data into two parts: the part due to the fit and the remainder, which we call residual.

**EXAMPLE**

**12.8 ANOVA table for worker safety study.** The SPSS output in Figure 12.8 gives the sources of variation in the first column. Here, FIT is called Between Groups, RESIDUAL is called Within Groups, and DATA is the last entry, Total. Different software packages use different terms for these sources of variation but the basic concept is common to all. In place of FIT, some software packages use Between Groups, Model, or the name of the factor. Similarly, terms like Within Groups or Error are frequently used in place of RESIDUAL.

variation among groups

The Between Groups row in the table gives information related to the variation **among** group means. In writing ANOVA tables we will use the generic label "groups" or some other term that describes the factor being studied for this row.

variation within groups

The Within Groups row in the table gives information related to the variation **within** groups. We noted that the term "error" is frequently used for this source of variation, particularly for more general statistical models. This label is most appropriate for experiments in the physical sciences where the observations within a group differ because of measurement error. In business and the biological and social sciences, on the other hand, the within-group variation is often due to the fact that not all firms or plants or people are the same. This sort of variation is not due to errors and is better described as "residual" or "within-group" variation. Nevertheless, we will use the generic label "error" for this source of variation in writing ANOVA tables.

Finally, the Total row in the ANOVA table corresponds to the DATA term in our DATA = FIT + RESIDUAL framework. So, for analysis of variance,

$$DATA = FIT + RESIDUAL$$

translates into

$$Total = Between\ Groups + Within\ Groups$$

**LOOK BACK**

**sum of squares,
page 580**

The second column in the software output given in Figure 12.8 is labeled Sum of Squares. As you might expect, each sum of squares is a sum of squared deviations. We use SSG, SSE, and SST for the entries in this column, corresponding to groups, error, and total. Each sum of squares measures a different type of variation. SST measures variation of the data around the overall mean, $x_{ij} - \bar{x}$. Variation of the group means around the overall mean $\bar{x}_i - \bar{x}$ is measured by SSG. Finally, SSE measures variation of each observation around its group mean, $x_{ij} - \bar{x}_i$.

**EXAMPLE**

**12.9 ANOVA table for worker safety study, continued.** The Sum of Squares column in Figure 12.8 gives the values for the three sums of squares.

$$SST = 196391.4$$

$$SSG = \phantom{00}4662.2$$

$$SSE = 191729.2$$

Verify that $SST = SSG + SSE$.

This fact is true in general. The total variation is always equal to the among-group variation plus the within-group variation. Note that software output frequently gives many more digits than we need, as in this case. In this example it appears that most of the variation is coming from within groups.

**LOOK BACK**

degrees of freedom, page 42

Associated with each sum of squares is a quantity called the degrees of freedom. Because SST measures the variation of all $N$ observations around the overall mean, its degrees of freedom are $DFT = N - 1$. This is the same as the degrees of freedom for the ordinary sample variance with sample size $N$. Similarly, because SSG measures the variation of the $I$ sample means around the overall mean, its degrees of freedom are $DFG = I - 1$. Finally, SSE is the sum of squares of the deviations $x_{ij} - \bar{x}_i$. Here we have $N$ observations being compared with $I$ sample means, and $DFE = N - I$.

**EXAMPLE**

**12.10 Degrees of freedom for worker safety study.** In our worker safety example, we have $I = 3$ and $N = 590$. Therefore,

$$DFT = N - 1 = 590 - 1 = 589$$

$$DFG = I - 1 = 3 - 1 = 2$$

$$DFE = N - I = 590 - 3 = 587$$

These are the entries in the df column of Figure 12.8.

Note that the degrees of freedom add in the same way that the sums of squares add. That is, $DFT = DFG + DFE$.

**LOOK BACK**

mean square, page 581

For each source of variation, the mean square is the sum of squares divided by the degrees of freedom. You can verify this by doing the divisions for the values given on the output in Figure 12.8.

---

**SUMS OF SQUARES, DEGREES OF FREEDOM, AND MEAN SQUARES**

**Sums of squares** represent variation present in the data. They are calculated by summing squared deviations. In one-way ANOVA there are three **sources of variation:** groups, error, and total. The sums of squares are related by the formula

$$\text{SST} = \text{SSG} + \text{SSE}$$

Thus, the total variation is composed of two parts, one due to groups and one due to error.

**Degrees of freedom** are related to the deviations that are used in the sums of squares. The degrees of freedom are related in the same way as the sums of squares are:

$$\text{DFT} = \text{DFG} + \text{DFE}$$

To calculate each **mean square,** divide the corresponding sum of squares by its degrees of freedom.

We can use the error mean square to find $s_p$, the pooled estimate of the parameter $\sigma$ of our model. It is true in general that

$$s_p^2 = \text{MSE} = \frac{\text{SSE}}{\text{DFE}}$$

In other words, the error mean square is an estimate of the within-group variance, $\sigma^2$. The estimate of $\sigma$ is therefore the square root of this quantity. So,

$$s_p = \sqrt{\text{MSE}}$$

**EXAMPLE**

**12.11 MSE for worker safety study.** From the SPSS output in Figure 12.8 we see that the MSE is reported as 326.626. The pooled estimate of $\sigma$ is therefore

$$s_p = \sqrt{\text{MSE}}$$
$$= \sqrt{326.626} = 18.07$$

## The *F* test

If $H_0$ is true, there are no differences among the group means. The ratio MSG/MSE is a statistic that is approximately 1 if $H_0$ is true and tends to be larger if $H_a$ is true. This is the ANOVA $F$ statistic. In our example, MSG = 2331.116 and MSE = 326.626, so the ANOVA $F$ statistic is

$$F = \frac{\text{MSG}}{\text{MSE}} = \frac{2331.116}{326.626} = 7.137$$

When $H_0$ is true, the $F$ statistic has an $F$ distribution that depends upon two numbers: the *degrees of freedom for the numerator* and the *degrees of freedom for the denominator.* These degrees of freedom are those associated with the mean squares in the numerator and denominator of the $F$ statistic. For one-way ANOVA, the degrees of freedom for the numerator are DFG $= I - 1$, and the degrees of freedom for the denominator are DFE $= N - I$. We use the notation $F(I - 1, N - I)$ for this distribution.

APPLET

The *One-Way ANOVA* applet available on the Web site www.whfreeman.com/ips is an excellent way to see how the value of the $F$ statistic and the $P$-value depend upon the variability of the data within the groups and the differences between the means. See Exercises 12.18 and 12.19 for use of this applet.

---

**THE ANOVA $F$ TEST**

To test the null hypothesis in a one-way ANOVA, calculate the **$F$ statistic**

$$F = \frac{MSG}{MSE}$$

When $H_0$ is true, the $F$ statistic has the $F(I - 1, N - I)$ distribution. When $H_a$ is true, the $F$ statistic tends to be large. We reject $H_0$ in favor of $H_a$ if the $F$ statistic is sufficiently large.

The **$P$-value** of the $F$ test is the probability that a random variable having the $F(I - 1, N - I)$ distribution is greater than or equal to the calculated value of the $F$ statistic.

---

Tables of $F$ critical values are available for use when software does not give the $P$-value. Table E in the back of the book contains the $F$ critical values for probabilities $p = 0.100, 0.050, 0.025, 0.010,$ and $0.001$. For one-way ANOVA we use critical values from the table corresponding to $I - 1$ degrees of freedom in the numerator and $N - I$ degrees of freedom in the denominator.

**EXAMPLE**

**12.12 The ANOVA $F$ test for the worker safety study.** In the study of worker safety, we found $F = 7.14$. (Note that it is standard practice to round $F$ statistics to two places after the decimal point.) There were three populations, so the degrees of freedom in the numerator are DFG $= I - 1 = 2$. For this example the degrees of freedom in the denominator are DFE $= N - I = 590 - 3 = 587$. In Table E we first find the column corresponding to 2 degrees of freedom in the numerator. For the degrees of freedom in the denominator, we see that there are entries for 200 and 1000. These entries are very close. To be conservative we use critical values corresponding to 200 degrees of freedom in the denominator since these are slightly larger.

| $p$ | Critical value |
|---|---|
| 0.100 | 2.33 |
| 0.050 | 3.04 |
| 0.025 | 3.76 |
| 0.010 | 4.71 |
| 0.001 | 7.15 |

versus the alternative that the three population means are not all the same. We would report these results as $F(2,587) = 7.14$ with $P < 0.001$. (Note that we have given the degrees of freedom for the $F$ statistic in parentheses.) Because the $P$-value is very small, we conclude that the data provide clear evidence that the three population means are not all the same.

Having evidence that the three population means are not the same does not really tell us anything useful. We would really like our analysis to provide us with more specific information. The alternative hypothesis is true if

$$\mu_{UN} \neq \mu_{SK}$$

or if

$$\mu_{UN} \neq \mu_{SU}$$

or if

$$\mu_{SK} \neq \mu_{SU}$$

or if any combination of these statements is true. *When you reject the ANOVA null hypothesis, additional analyses are required to obtain useful results.*

Experts on safety in workplaces would suggest that supervisors face a very different safety environment than the other types of workers. Therefore, a reasonable question to ask is whether or not the mean of the supervisors is different from the mean of the others. We can take this question and translate it into a testable hypothesis.

**EXAMPLE**

**12.15 An additional comparison of interest.** To compare the supervisors with the other two groups of workers we construct the following null hypothesis:

$$H_{01}: \frac{1}{2}(\mu_{UN} + \mu_{SK}) = \mu_{SU}$$

We could use the two-sided alternative

$$H_{a1}: \frac{1}{2}(\mu_{UN} + \mu_{SK}) \neq \mu_{SU}$$

but we could also argue that the one-sided alternative

$$H_{a1}: \frac{1}{2}(\mu_{UN} + \mu_{SK}) < \mu_{SU}$$

is appropriate for this problem because we expect the unskilled workers and the skilled workers to have a work environment that is less safe than the supervisors' work environment.

In the example above we used $H_{01}$ and $H_{a1}$ to designate the null and alternative hypotheses. The reason for this is that there is a natural additional set of hypotheses that we should examine for this example. We use $H_{02}$ and $H_{a2}$ for these hypotheses.

EXAMPLE

**12.16 Another comparison of interest.** Do the data provide any evidence to support a conclusion that the unskilled workers and the skilled workers have different mean SCI scores? We translate this question into the following null and alternative hypotheses:

$$H_{02}: \mu_{\text{UN}} = \mu_{\text{SK}}$$

$$H_{a2}: \mu_{\text{UN}} \neq \mu_{\text{SK}}$$

Each of $H_{01}$ and $H_{02}$ says that a combination of population means is 0. These combinations of means are called contrasts because the coefficients sum to zero. We use $\psi$, the Greek letter psi, for contrasts among population means. For comparing the supervisors with the other two groups of workers, we have

$$\psi_1 = -\frac{1}{2}(\mu_{\text{UN}} + \mu_{\text{SK}}) + \mu_{\text{SU}}$$

$$= (-0.5)\mu_{\text{UN}} + (-0.5)\mu_{\text{SK}} + (1)\mu_{\text{SU}}$$

and for comparing the unskilled workers with the skilled workers

$$\psi_2 = (1)\mu_{\text{UN}} + (-1)\mu_{\text{SK}}$$

In each case, the value of the contrast is 0 when $H_0$ is true. *Note that we have chosen to define the contrasts so that they will be positive when the alternative of interest (what we expect) is true. Whenever possible, this is a good idea because it makes some computations easier.*

sample contrast

A contrast expresses an effect in the population as a combination of population means. To estimate the contrast, form the corresponding **sample contrast** by using sample means in place of population means. Under the ANOVA assumptions, a sample contrast is a linear combination of independent Normal variables and therefore has a Normal distribution. We can obtain the standard error of a contrast by using the rules for variances. Inference is based on $t$ statistics. Here are the details.

**LOOK BACK**

**rules for variances, page 282**

---

### CONTRASTS

A **contrast** is a combination of population means of the form

$$\psi = \sum a_i \mu_i$$

where the coefficients $a_i$ sum to 0. The corresponding **sample contrast** is

$$c = \sum a_i \bar{x}_i$$

The **standard error of $c$** is

$$\text{SE}_c = s_p \sqrt{\sum \frac{a_i^2}{n_i}}$$

**LOOK BACK**
pooled two-sample *t*
procedures, page 462

We performed the first calculation when we analyzed the contrast $\psi_2 = \mu_1 - \mu_2$ in the previous section. These *t* statistics are very similar to the pooled two-sample *t* statistic for comparing two population means. The difference is that we now have more than two populations, so each statistic uses the pooled estimator $s_p$ from all groups rather than the pooled estimator from just the two groups being compared. This additional information about the common $\sigma$ increases the power of the tests. The degrees of freedom for all of these statistics are DFE = 587, those associated with $s_p$.

Because we do not have any specific ordering of the means in mind as an alternative to equality, we must use a two-sided approach to the problem of deciding which pairs of means are significantly different.

---

### MULTIPLE COMPARISONS

To perform a **multiple-comparisons procedure,** compute *t* **statistics** for all pairs of means using the formula

$$t_{ij} = \frac{\bar{x}_i - \bar{x}_j}{s_p\sqrt{\dfrac{1}{n_i} + \dfrac{1}{n_j}}}$$

If

$$|t_{ij}| \geq t^{**}$$

we declare that the population means $\mu_i$ and $\mu_j$ are different. Otherwise, we conclude that the data do not distinguish between them. The value of $t^{**}$ depends upon which multiple-comparisons procedure we choose.

---

One obvious choice for $t^{**}$ is the upper $\alpha/2$ critical value for the $t(\text{DFE})$ distribution. This choice simply carries out as many separate significance tests of fixed level $\alpha$ as there are pairs of means to be compared. The procedure based on this choice is called the **least-significant differences method,** or simply LSD.

**LSD method**

*LSD has some undesirable properties, particularly if the number of means being compared is large.* Suppose, for example, that there are $I = 20$ groups and we use LSD with $\alpha = 0.05$. There are 190 different pairs of means. If we perform 190 *t* tests, each with an error rate of 5%, our overall error rate will be unacceptably large. We expect about 5% of the 190 to be significant even if the corresponding population means are the same. Since 5% of 190 is 9.5, we expect 9 or 10 false rejections.

The LSD procedure fixes the probability of a false rejection for each single pair of means being compared. It does not control the overall probability of *some* false rejection among all pairs. Other choices of $t^{**}$ control possible errors in other ways. The choice of $t^{**}$ is therefore a complex problem, and a detailed discussion of it is beyond the scope of this text. Many choices for $t^{**}$ are used in practice. One major statistical package allows selection from a list of over a dozen choices.

Bonferroni method

We will discuss only one of these, called the **Bonferroni method.** Use of this procedure with $\alpha = 0.05$, for example, guarantees that the probability of *any* false rejection among all comparisons made is no greater than 0.05. This is much stronger protection than controlling the probability of a false rejection at 0.05 for *each separate* comparison.

**EXAMPLE**

**12.23 Applying the Bonferroni method.** We apply the Bonferroni multiple-comparisons procedure with $\alpha = 0.05$ to the data from the worker safety study. The value of $t^{**}$ for this procedure (from software or special tables) is 2.13. Of the statistics $t_{12} = -0.38$, $t_{13} = -3.78$, and $t_{23} = -2.94$ calculated in the beginning of this section, only $t_{13}$ and $t_{23}$ are significant. These two statistics compare supervisors with each of the other two groups.

Of course, we prefer to use software for the calculations.

**EXAMPLE**

**12.24 Interpreting software output.** The output generated by SPSS for Bonferroni comparisons appears in Figure 12.10. The software uses an asterisk to indicate that the difference in a pair of means is statistically significant. These results agree with the calculations that we performed in Examples 12.22 and 12.23. Note that each comparison is given twice in the output.

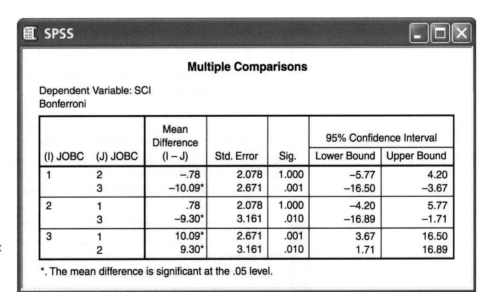

**Multiple Comparisons**

Dependent Variable: SCI
Bonferroni

| (I) JOBC | (J) JOBC | Mean Difference (I – J) | Std. Error | Sig. | 95% Confidence Interval Lower Bound | Upper Bound |
|---|---|---|---|---|---|---|
| 1 | 2 | –.78 | 2.078 | 1.000 | –5.77 | 4.20 |
|   | 3 | –10.09* | 2.671 | .001 | –16.50 | –3.67 |
| 2 | 1 | .78 | 2.078 | 1.000 | –4.20 | 5.77 |
|   | 3 | –9.30* | 3.161 | .010 | –16.89 | –1.71 |
| 3 | 1 | 10.09* | 2.671 | .001 | 3.67 | 16.50 |
|   | 2 | 9.30* | 3.161 | .010 | 1.71 | 16.89 |

\*. The mean difference is significant at the .05 level.

**FIGURE 12.10** Software output giving the multiple-comparisons analysis for the worker safety example.

The data in the worker safety study provided a clear result: the supervisors have the highest mean SCI score, and we are unable to see a difference between the unskilled workers and the skilled workers. Unfortunately, this type of clarity does not always emerge from a multiple-comparisons analysis. For example, with three groups, we can (*a*) fail to detect a difference between Groups 1 and

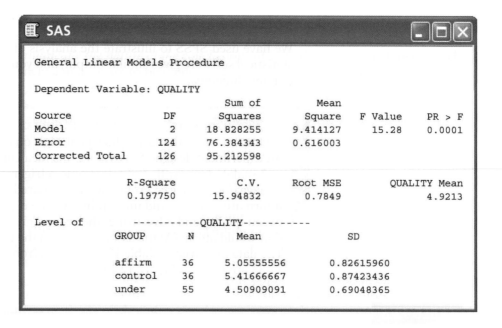

**FIGURE 12.11** SAS, Excel, Minitab, and TI-83 output for the advertising study in Example 12.26. *(continued)*

## Power*

Recall that the power of a test is the probability of rejecting $H_0$ when $H_a$ is in fact true. Power measures how likely a test is to detect a specific alternative. When planning a study in which ANOVA will be used for the analysis, it is important to perform power calculations to check that the sample sizes are adequate to detect differences among means that are judged to be important. Power calculations also help evaluate and interpret the results of studies in

---

*This section is optional.

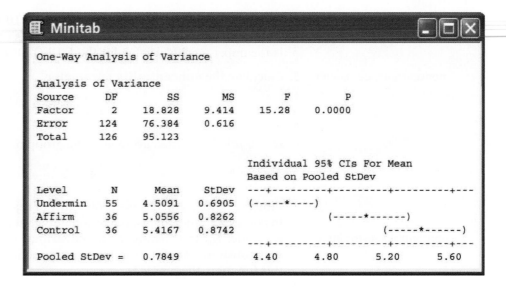

**FIGURE 12.11** *(Continued)*
SAS, Excel, Minitab, and TI-83 output for the advertising study in Example 12.26.

which $H_0$ was not rejected. We sometimes find that the power of the test was so low against reasonable alternatives that there was little chance of obtaining a significant $F$.

In Chapter 7 we found the power for the two-sample $t$ test. One-way ANOVA is a generalization of the two-sample $t$ test, so it is not surprising that the procedure for calculating power is quite similar. Here are the steps that are needed:

**LOOK BACK**
power, page 433

1. Specify

   (a) an alternative ($H_a$) that you consider important; that is, values for the true population means $\mu_1, \mu_2, \ldots, \mu_I$;

   (b) sample sizes $n_1, n_2, \ldots, n_I$; usually these will all be equal to the common value $n$;

   (c) a level of significance $\alpha$, usually equal to 0.05; and

   (d) a guess at the standard deviation $\sigma$.

2. Use the degrees of freedom DFG $= I - 1$ and DFE $= N - I$ to find the critical value that will lead to the rejection of $H_0$. This value, which we denote by $F^*$, is the upper $\alpha$ critical value for the $F$(DFG, DFE) distribution.

**noncentrality parameter**    3. Calculate the **noncentrality parameter**[4]

$$\lambda = \frac{\sum n_i (\mu_i - \overline{\mu})^2}{\sigma^2}$$

where $\overline{\mu}$ is a weighted average of the group means

$$\overline{\mu} = \sum \frac{n_i}{N} \mu_i$$

4. Find the power, which is the probability of rejecting $H_0$ when the alternative hypothesis is true, that is, the probability that the observed $F$ is greater than $F^*$. Under $H_a$, the $F$ statistic has a distribution known as the **noncentral F distribution.** SAS, for example, has a function for this distribution. Using this function, the power is

**noncentral F distribution**

$$\text{Power} = 1 - \text{PROBF}(F^*, \text{DFG}, \text{DFE}, \lambda)$$

Note that, if the $n_i$ are all equal to the common value $n$, then $\overline{\mu}$ is the ordinary average of the $\mu_i$ and

$$\lambda = \frac{n \sum (\mu_i - \overline{\mu})^2}{\sigma^2}$$

If the means are all equal (the ANOVA $H_0$), then $\lambda = 0$. The noncentrality parameter measures how unequal the given set of means is. Large $\lambda$ points to an alternative far from $H_0$, and we expect the ANOVA $F$ test to have high power. Software makes calculation of the power quite easy, but tables and charts are also available.

**EXAMPLE**

**12.27 Power of a reading comprehension study.**    Suppose that a study on reading comprehension for three different teaching methods has 10 students in each group. How likely is this study to detect differences in the mean responses that would be viewed as important? A previous study performed in a different setting found sample means of 41, 47, and 44, and the pooled standard deviation was 7. Based on these results, we will use $\mu_1 = 41$, $\mu_2 = 47$, $\mu_3 = 44$, and $\sigma = 7$ in a calculation of power. The $n_i$ are equal, so $\overline{\mu}$ is simply the average of the $\mu_i$:

$$\overline{\mu} = \frac{41 + 47 + 44}{3} = 44$$

The noncentrality parameter is therefore

$$\begin{aligned}
\lambda &= \frac{n \sum (\mu_i - \overline{\mu})^2}{\sigma^2} \\
&= \frac{(10)[(41 - 44)^2 + (47 - 44)^2 + (44 - 44)^2]}{49} \\
&= \frac{(10)(18)}{49} = 3.67
\end{aligned}$$

Because there are three groups with 10 observations per group, DFG = 2 and DFE = 27. The critical value for $\alpha = 0.05$ is $F^* = 3.35$. The power is therefore

$$1 - \text{PROBF}(3.35, 2, 27, 3.67) = 0.3486$$

The chance that we reject the ANOVA $H_0$ at the 5% significance level is only about 35%.

If the assumed values of the $\mu_i$ in this example describe differences among the groups that the experimenter wants to detect, then we would want to use more than 10 subjects per group. Although $H_0$ is assumed to be false, the chance of rejecting it is only about 35%. This chance can be increased to acceptable levels by increasing the sample sizes.

**EXAMPLE**

**12.28 Changing the sample size.**   To decide on an appropriate sample size for the experiment described in the previous example, we repeat the power calculation for different values of $n$, the number of subjects in each group. Here are the results:

| $n$ | DFG | DFE | $F^*$ | $\lambda$ | Power |
|-----|-----|-----|-------|-----------|-------|
| 20  | 2   | 57  | 3.16  | 7.35      | 0.65  |
| 30  | 2   | 87  | 3.10  | 11.02     | 0.84  |
| 40  | 2   | 117 | 3.07  | 14.69     | 0.93  |
| 50  | 2   | 147 | 3.06  | 18.37     | 0.97  |
| 100 | 2   | 297 | 3.03  | 36.73     | $\approx 1$ |

With $n = 40$, the experimenters have a 93% chance of rejecting $H_0$ with $\alpha = 0.05$ and thereby demonstrating that the groups have different means. In the long run, 93 out of every 100 such experiments would reject $H_0$ at the $\alpha = 0.05$ level of significance. Using 50 subjects per group increases the chance of finding significance to 97%. With 100 subjects per group, the experimenters are virtually certain to reject $H_0$. The exact power for $n = 100$ is 0.99989. In most real-life situations the additional cost of increasing the sample size from 50 to 100 subjects per group would not be justified by the relatively small increase in the chance of obtaining statistically significant results.

## SECTION 12.2   Summary

**One-way analysis of variance (ANOVA)** is used to compare several population means based on independent SRSs from each population. The populations are assumed to be Normal with possibly different means and the same standard deviation.

To do an analysis of variance, first compute sample means and standard deviations for all groups. Side-by-side boxplots give an overview of the data. Examine Normal quantile plots (either for each group separately or for the residuals) to detect outliers or extreme deviations from Normality. Compute the ratio of the largest to the smallest sample standard deviation. If this ratio is less than 2 and the Normal quantile plots are satisfactory, ANOVA can be performed.

The **null hypothesis** is that the population means are *all* equal. The **alternative hypothesis** is true if there are *any* differences among the population means.

ANOVA is based on separating the total variation observed in the data into two parts: variation **among group means** and variation **within groups.** If the variation among groups is large relative to the variation within groups, we have evidence against the null hypothesis.

An **analysis of variance table** organizes the ANOVA calculations. **Degrees of freedom, sums of squares, and mean squares** appear in the table. The *F* **statistic** and its *P*-**value** are used to test the null hypothesis.

Specific questions formulated before examination of the data can be expressed as **contrasts.** Tests and confidence intervals for contrasts provide answers to these questions.

If no specific questions are formulated before examination of the data and the null hypothesis of equality of population means is rejected, **multiple-comparisons** methods are used to assess the statistical significance of the differences between pairs of means.

The **power** of the *F* test depends upon the sample sizes, the variation among population means, and the within-group standard deviation.

## CHAPTER 12   Exercises

*For Exercises 12.1 and 12.2, see pages 643 and 644; for Exercises 12.3 and 12.4, see pages 647 and 648; for Exercises 12.5 and 12.6, see page 655; and for Exercises 12.7 and 12.8, see page 665.*

**12.9   Describing the ANOVA model.** For each of the following situations, identify the response variable and the populations to be compared, and give $I$, the $n_i$, and $N$.

(a) A poultry farmer is interested in reducing the cholesterol level in his marketable eggs. He wants to compare two different cholesterol-lowering drugs added to the hen's standard diet as well as an all-vegetarian diet. He assigns 25 of his hens to each of the three treatments.

(b) A researcher is interested in students' opinions regarding an additional annual fee to support non-income-producing varsity sports. Students were asked to rate their acceptance of this fee on a five-point scale. She received 94 responses, of which 31 were from students who attend varsity football or basketball games only, 18 were from students who also attend other varsity competitions, and 45 who did not attend any varsity games.

(c) A professor wants to evaluate the effectiveness of his teaching assistants. In one class period, the 42 students were randomly divided into three equal-sized groups, and each group was taught power calculations from one of the assistants. At the beginning of the next class, each student took a quiz on power calculations, and these scores were compared.

**12.10   Describing the ANOVA model, continued.** For each of the following situations, identify the response variable and the populations to be compared, and give $I$, the $n_i$, and $N$.

(a) A developer of a virtual-reality (VR) teaching tool for the deaf wants to compare the effectiveness of different navigation methods. A total of 40 children were available for the experiment, of which equal numbers were randomly assigned to use a joystick, wand, dancemat, or gesture-based pinch gloves. The time (in seconds) to complete a designed VR path is recorded for each child.

(b) To study the effects of pesticides on birds, an experimenter randomly (and equally) allocated 65 chicks to five diets (a control and four with a different pesticide included). After a month, the calcium content (milligrams) in a 1-centimeter length of bone from each chick was measured.

(c) A university sandwich shop wants to compare the effects of providing free food with a sandwich order on sales. The experiment will be conducted from 11:00 A.M. to 2:00 P.M. for the next 20 weekdays. On each day, customers will be offered one of the following: a free drink, free chips, a free cookie, or nothing. Each option will be offered 5 times.

**12.11** **Determining the degrees of freedom.** Refer to Exercise 12.9. For each situation, give the following:

(a) Degrees of freedom for the model, for error, and for the total.

(b) Null and alternative hypotheses.

(c) Numerator and denominator degrees of freedom for the $F$ statistic.

**12.12** **Determining the degrees of freedom, continued.** Refer to Exercise 12.10. For each situation, give the following:

(a) Degrees of freedom for the model, for error, and for the total.

(b) Null and alternative hypotheses.

(c) Numerator and denominator degrees of freedom for the $F$ statistic.

**12.13** **Data collection and the interpretation of results.** Refer to Exercise 12.9. For each situation, discuss the method of obtaining the data and how this will affect the extent to which the results can be generalized.

**12.14** **Data collection, continued.** Refer to Exercise 12.10. For each situation, discuss the method of obtaining the data and how this will affect the extent to which the results can be generalized.

**12.15** **A one-way ANOVA example.** A study compared 4 groups with 8 observations per group. An $F$ statistic of 3.33 was reported.

(a) Give the degrees of freedom for this statistic and the entries from Table E that correspond to this distribution.

(b) Sketch a picture of this $F$ distribution with the information from the table included.

(c) Based on the table information, how would you report the $P$-value?

(d) Can you conclude that all pairs of means are different? Explain your answer.

**12.16** **Calculating the ANOVA $F$ test $P$-value.** For each of the following situations, find the degrees of

freedom for the $F$ statistic and then use Table E to approximate the $P$-value.

(a) Seven groups are being compared with 5 observations per group. The value of the $F$ statistic is 2.31.

(b) Five groups are being compared with 11 observations per group. The value of the $F$ statistic is 2.83.

(c) Six groups are being compared using 66 total observations. The value of the $F$ statistic is 4.08.

**12.17** **Calculating the ANOVA $F$ test $P$-value, continued.** For each of the following situations, find the $F$ statistic and the degrees of freedom. Then draw a sketch of the distribution under the null hypothesis and shade in the portion corresponding to the $P$-value. State how you would report the $P$-value.

(a) Compare 5 groups with 9 observations per group, MSE = 50, and MSG = 127.

(b) Compare 4 groups with 7 observations per group, SSG = 40, and SSE = 153.

**12.18** **The effect of increased variation within groups.** The *One-Way ANOVA* applet lets you see how the $F$ statistic and the $P$-value depend on the variability of the data within groups and the differences among the means.

(a) The black dots are at the means of the three groups. Move these up and down until you get a configuration that gives a $P$-value of about 0.01. What is the value of the $F$ statistic?

(b) Now increase the variation within the groups by dragging the mark on the pooled standard error scale to the right. Describe what happens to the $F$ statistic and the $P$-value. Explain why this happens.

**12.19** **The effect of increased variation between groups.** Set the pooled standard error for the *One-Way ANOVA* applet at a middle value. Drag the black dots so that they are approximately equal.

(a) What is the $F$ statistic? Give its $P$-value.

(b) Drag the mean of the second group up and the mean of the third group down. Describe the effect on the $F$ statistic and its $P$-value. Explain why they change in this way.

**12.20** **Calculating the pooled standard deviation.** An experiment was run to compare four groups. The sample sizes were 25, 28, 150, and 21, and the corresponding estimated standard deviations were 42, 38, 20, and 45.

(a) Is it reasonable to use the assumption of equal standard deviations when we analyze these data? Give a reason for your answer.

(b) Give the values of the variances for the four groups.

(c) Find the pooled variance.

(d) What is the value of the pooled standard deviation?

(e) Explain why your answer in part (d) is much closer to the standard deviation for the third group than to any of the other standard deviations.

**12.21 Sleep deprivation and reaction times.** Sleep deprivation experienced by physicians during residency training and the possible negative consequences are of concern to many in the health care community. One study of 33 resident anesthesiologists compared their changes from baseline in reaction times on four tasks.[5] Under baseline conditions, the physicians reported getting an average of 7.04 hours of sleep. While on duty, however, the average was 1.66 hours. For each of the tasks the researchers reported a statistically significant increase in the reaction time when the residents were working in a state of sleep deprivation.

(a) If each task is analyzed separately as the researchers did in their report, what is the appropriate statistical method to use? Explain your answer.

(b) Is it appropriate to use a one-way ANOVA with $I = 4$ to analyze these data? Explain why or why not.

**12.22** CHALLENGE **The two-sample $t$ test and one-way ANOVA.** Refer to the LDL level data in Exercise 7.61 (page 467). Find the two-sample pooled $t$ statistic for comparing men with women. Then formulate the problem as an ANOVA and report the results of this analysis. Verify that $F = t^2$.

**12.23 The importance of recreational sports to college satisfaction.** The National Intramural-Recreational Sports Association (NIRSA) performed a survey to look at the value of recreational sports on college campuses.[6] One of the questions asked each student to rate the importance of recreational sports to college satisfaction and success. Responses were on

a 10-point scale with 1 indicating total lack of importance and 10 indicating very high importance. The following table summarizes these results:

| Class | $n$ | Mean score |
|---|---|---|
| Freshman | 724 | 7.6 |
| Sophomore | 536 | 7.6 |
| Junior | 593 | 7.5 |
| Senior | 437 | 7.3 |

(a) To compare the mean scores across classes, what are the degrees of freedom for the ANOVA $F$ statistic?

(b) The MSG = 11.806. If $s_p = 2.16$, what is the $F$ statistic?

(c) Give an approximate (from a table) or exact (from software) $P$-value. What do you conclude?

**12.24 Restaurant ambience and consumer behavior.** There have been numerous studies investigating the effects of restaurant ambience on consumer behavior. A recent study investigated the effects of musical genre on consumer spending.[7] At a single high-end restaurant in England over a 3-week period, there were a total of 141 participants; 49 of them were subjected to background pop music (for example, Britney Spears, Culture Club, and Ricky Martin) while dining, 44 to background classical music (for example, Vivaldi, Handel, and Strauss), and 48 to no background music. For each participant, the total food bill, adjusted for time spent dining, was recorded. The following table summarizes the means and standard deviations:

| Background music | Mean bill | $n$ | $s$ |
|---|---|---|---|
| Classical | 24.130 | 44 | 2.243 |
| Pop | 21.912 | 49 | 2.627 |
| None | 21.697 | 48 | 3.332 |
| Total | 22.531 | 141 | 2.969 |

(a) Plot the means versus the type of background music. Does there appear to be a difference in spending?

(b) Is it reasonable to assume that the variances are equal? Explain.

(c) The $F$ statistic is 10.62. Give the degrees of freedom and either an approximate (from a table) or an exact (from software) $P$-value. What do you conclude?

(d) Refer back to part (a). Without doing any formal analysis, describe the pattern in the means that is likely responsible for your conclusion in part (c).

(e) To what extent do you think the results of this study can be generalized to other settings? Give reasons for your answer.

**12.25 The effects of two stimulant drugs.** An experimenter was interested in investigating the effects of two stimulant drugs (labeled A and B). She divided 20 rats equally into 5 groups (placebo, Drug A low, Drug A high, Drug B low, and Drug B high) and, 20 minutes after injection of the drug, recorded each rat's activity level (higher score is more active). The following table summarizes the results:

| Treatment | $\bar{x}$ | $s^2$ |
|---|---|---|
| Placebo | 14.00 | 8.00 |
| Low A | 15.25 | 12.25 |
| High A | 18.25 | 12.25 |
| Low B | 16.75 | 6.25 |
| High B | 22.50 | 11.00 |

(a) Plot the means versus the type of treatment. Does there appear to be a difference in the activity level? Explain.

(b) Is it reasonable to assume that the variances are equal? Explain your answer, and if reasonable, compute $s_p$.

(c) Give the degrees of freedom for the $F$ statistic.

(d) The $F$ statistic is 4.35. Find the associated $P$-value and state your conclusions.

**12.26** **Exam accommodations and end-of-term grades.** The Americans with Disabilities Act (ADA) requires that students with learning disabilities (LD) and/or attention deficit disorder (ADD) be given certain accommodations when taking examinations. One study designed to assess the effects of these accommodations examined the relationship between end-of-term grades and the number of accommodations given.[8] The researchers reported the mean grades with sample sizes and standard deviations versus the number of accommodations in a table similar to this:

| Accommodations | Mean grade | $n$ | $s$ |
|---|---|---|---|
| 0 | 2.7894 | 160 | 0.85035 |
| 1 | 2.8605 | 38 | 0.83068 |
| 2 | 2.5757 | 37 | 0.82745 |
| 3 | 2.6286 | 7 | 1.03072 |
| 4 | 2.4667 | 3 | 1.66233 |
| Total | 2.7596 | 245 | 0.85701 |

(a) Plot the means versus the number of accommodations. Is there a pattern evident?

(b) A large number of digits are reported for the means and the standard deviations. Do you think that all of these are necessary? Give reasons for your answer and describe how you would report these results.

(c) Should we pool to obtain an estimate of an assumed standard deviation for these data? Explain your answer and give the pooled estimate if your answer is Yes.

(d) The small numbers of observations with 3 or 4 accommodations lead to estimates that are highly variable in these groups compared with the other groups. Inclusion of groups with relatively few observations in an ANOVA can also lead to low power. We could eliminate these two levels from the analysis or we could combine them with the 37 observations in the group above to form a new group with 2 or more accommodations. Which of these options would you prefer? Give reasons for your answer.

(e) The 245 grades reported in the table were from a sample of 61 students who completed three, four, or five courses during a spring term at one college and who were qualified to receive accommodations. Students in the sample were self-identified, in the sense that they had to request qualification. Even when qualified, some students choose not to request accommodations for some or all of their courses. Based on these facts, would you advise that ANOVA methods be used for these data? Explain your answer. (The authors did not present the results of an ANOVA in their publication.)

(f) To what extent do you think the results of this study can be generalized to other settings? Give reasons for your answer.

(g) Most reasonable approaches to the analysis of these data would conclude that the data fail to provide evidence that the number of accommodations is related to the mean grades. Does this imply that the accommodations are not

| *H. caribaea* red | | | | | | | |
|---|---|---|---|---|---|---|---|
| 41.90 | 42.01 | 41.93 | 43.09 | 41.47 | 41.69 | 39.78 | 40.57 |
| 39.63 | 42.18 | 40.66 | 37.87 | 39.16 | 37.40 | 38.20 | 38.07 |
| 38.10 | 37.97 | 38.79 | 38.23 | 38.87 | 37.78 | 38.01 | |

| *H. caribaea* yellow | | | | | | | |
|---|---|---|---|---|---|---|---|
| 36.78 | 37.02 | 36.52 | 36.11 | 36.03 | 35.45 | 38.13 | 37.1 |
| 35.17 | 36.82 | 36.66 | 35.68 | 36.03 | 34.57 | 34.63 | |

Do a complete analysis that includes description of the data and a significance test to compare the mean lengths of the flowers for the three species.

12.36  **Air quality in poultry-processing plants, continued.** Refer to Exercise 12.34. There is not sufficient information to examine the distributions in detail, but it is not unreasonable to expect count data such as these to be skewed. Reanalyze the data after taking logs of the CFU counts. Summarize your work and compare the results you have found here with what you obtained in Exercise 12.34.

12.37  **Taking the log of the response variable.** The distributions of the flower lengths in Exercise 12.35 are somewhat skewed. Take logs of the lengths and reanalyze the data. Write a summary of your results and include a comparison with the results you found in Exercise 12.35.

12.38  **Do poets die young?** According to William Butler Yeats, "She is the Gaelic muse, for she gives inspiration to those she persecutes. The Gaelic poets die young, for she is restless, and will not let them remain long on earth." One study designed to investigate this issue examined the age at death for writers from different cultures and genders.[13] Three categories of writers examined were novelists, poets, and nonfiction writers. The ages at death for female writers in these categories from North America are given in Table 12.2. Most of the writers are from the United States, but Canadian and Mexican writers are also included.

(a) Use graphical and numerical methods to describe the data.

(b) Examine the assumptions necessary for ANOVA. Summarize your findings.

(c) Run the ANOVA and report the results.

(d) Use a contrast to compare the poets with the two other types of writers. Do you think that the quotation from Yeats justifies the use of a one-sided alternative for examining this contrast? Explain your answer.

(e) Use another contrast to compare the novelists with the nonfiction writers. Explain your choice for an alternative hypothesis for this contrast.

(f) Use a multiple-comparisons procedure to compare the three means. How do the conclusions from this approach compare with those using the contrasts?

12.39  **Do isoflavones increase bone mineral density?** Kudzu is a plant that was imported to the United States from Japan and now covers over seven million acres in the South. The plant contains chemicals called isoflavones that have been shown to have beneficial effects on bones. One study used three groups of rats to compare a control group with rats that were fed either a low dose or a high dose of isoflavones from kudzu.[14] One of the outcomes examined was the bone mineral density

**TABLE 12.2**

Age at death for women writers

| Type | Age at death | | | | | | | | | | | | | | |
|---|---|---|---|---|---|---|---|---|---|---|---|---|---|---|---|
| Novels | 57 | 90 | 67 | 56 | 90 | 72 | 56 | 90 | 80 | 74 | 73 | 86 | 53 | 72 | 86 |
| ($n = 67$) | 82 | 74 | 60 | 79 | 80 | 79 | 77 | 64 | 72 | 88 | 75 | 79 | 74 | 85 | 71 |
| | 78 | 57 | 54 | 50 | 59 | 72 | 60 | 77 | 50 | 49 | 73 | 39 | 73 | 61 | 90 |
| | 77 | 57 | 72 | 82 | 54 | 62 | 74 | 65 | 83 | 86 | 73 | 79 | 63 | 72 | 85 |
| | 91 | 77 | 66 | 75 | 90 | 35 | 86 | | | | | | | | |
| Poems | 88 | 69 | 78 | 68 | 72 | 60 | 50 | 47 | 74 | 36 | 87 | 55 | 68 | 75 | 78 |
| ($n = 32$) | 85 | 69 | 38 | 58 | 51 | 72 | 58 | 84 | 30 | 79 | 90 | 66 | 45 | 70 | 48 |
| | 31 | 43 | | | | | | | | | | | | | |
| Nonfiction | 74 | 86 | 87 | 68 | 76 | 73 | 63 | 78 | 83 | 86 | 40 | 75 | 90 | 47 | 91 |
| ($n = 24$) | 94 | 61 | 83 | 75 | 89 | 77 | 86 | 66 | 97 | | | | | | |

in the femur (in grams per square centimeter). Here are the data:

| Treatment | Bone mineral density (g/cm²) | | | | | |
|---|---|---|---|---|---|---|
| Control | 0.228 | 0.207 | 0.234 | 0.220 | 0.217 | 0.228 |
| | 0.209 | 0.221 | 0.204 | 0.220 | 0.203 | 0.219 |
| | 0.218 | 0.245 | 0.210 | | | |
| Low dose | 0.211 | 0.220 | 0.211 | 0.233 | 0.219 | 0.233 |
| | 0.226 | 0.228 | 0.216 | 0.225 | 0.200 | 0.208 |
| | 0.198 | 0.208 | 0.203 | | | |
| High dose | 0.250 | 0.237 | 0.217 | 0.206 | 0.247 | 0.228 |
| | 0.245 | 0.232 | 0.267 | 0.261 | 0.221 | 0.219 |
| | 0.232 | 0.209 | 0.255 | | | |

(a) Use graphical and numerical methods to describe the data.

(b) Examine the assumptions necessary for ANOVA. Summarize your findings.

(c) Run the ANOVA and report the results.

(d) Use a multiple-comparisons method to compare the three groups.

(e) Write a short report explaining the effect of kudzu isoflavones on the femur of the rat.

12.40 **A consumer price promotion study.** If a supermarket product is offered at a reduced price frequently, do customers expect the price of the product to be lower in the future? This question was examined by researchers in a study conducted on students enrolled in an introductory management course at a large midwestern university. For 10 weeks 160 subjects received information about the products. The treatment conditions corresponded to the number of promotions (1, 3, 5, or 7) that were described during this 10-week period. Students were randomly assigned to four groups.[15] Table 12.3 gives the data.

(a) Make a Normal quantile plot for the data in each of the four treatment groups. Summarize the information in the plots and draw a conclusion regarding the Normality of these data.

(b) Summarize the data with a table containing the sample size, mean, standard deviation, and standard error for each group.

(c) Is the assumption of equal standard deviations reasonable here? Explain why or why not.

(d) Run the one-way ANOVA. Give the hypotheses tested, the test statistic with degrees of freedom, and the $P$-value. Summarize your conclusion.

12.41 **A consumer price promotion study, continued.** Refer to the previous exercise. Use the Bonferroni or another multiple-comparisons procedure to compare the group means. Summarize the results and support your conclusions with a graph of the means.

12.42 **Do piano lessons improve the spatial-temporal reasoning of preschool children?** The data in Table 12.4 contain the change in spatial-

| TABLE 12.3 | | | | | | | | | | |
|---|---|---|---|---|---|---|---|---|---|---|
| Price promotion data | | | | | | | | | | |
| Number of promotions | Expected price (dollars) | | | | | | | | | |
| 1 | 3.78 | 3.82 | 4.18 | 4.46 | 4.31 | 4.56 | 4.36 | 4.54 | 3.89 | 4.13 |
| | 3.97 | 4.38 | 3.98 | 3.91 | 4.34 | 4.24 | 4.22 | 4.32 | 3.96 | 4.73 |
| | 3.62 | 4.27 | 4.79 | 4.58 | 4.46 | 4.18 | 4.40 | 4.36 | 4.37 | 4.23 |
| | 4.06 | 3.86 | 4.26 | 4.33 | 4.10 | 3.94 | 3.97 | 4.60 | 4.50 | 4.00 |
| 3 | 4.12 | 3.91 | 3.96 | 4.22 | 3.88 | 4.14 | 4.17 | 4.07 | 4.16 | 4.12 |
| | 3.84 | 4.01 | 4.42 | 4.01 | 3.84 | 3.95 | 4.26 | 3.95 | 4.30 | 4.33 |
| | 4.17 | 3.97 | 4.32 | 3.87 | 3.91 | 4.21 | 3.86 | 4.14 | 3.93 | 4.08 |
| | 4.07 | 4.08 | 3.95 | 3.92 | 4.36 | 4.05 | 3.96 | 4.29 | 3.60 | 4.11 |
| 5 | 3.32 | 3.86 | 4.15 | 3.65 | 3.71 | 3.78 | 3.93 | 3.73 | 3.71 | 4.10 |
| | 3.69 | 3.83 | 3.58 | 4.08 | 3.99 | 3.72 | 4.41 | 4.12 | 3.73 | 3.56 |
| | 3.25 | 3.76 | 3.56 | 3.48 | 3.47 | 3.58 | 3.76 | 3.57 | 3.87 | 3.92 |
| | 3.39 | 3.54 | 3.86 | 3.77 | 4.37 | 3.77 | 3.81 | 3.71 | 3.58 | 3.69 |
| 7 | 3.45 | 3.64 | 3.37 | 3.27 | 3.58 | 4.01 | 3.67 | 3.74 | 3.50 | 3.60 |
| | 3.97 | 3.57 | 3.50 | 3.81 | 3.55 | 3.08 | 3.78 | 3.86 | 3.29 | 3.77 |
| | 3.25 | 3.07 | 3.21 | 3.55 | 3.23 | 2.97 | 3.86 | 3.14 | 3.43 | 3.84 |
| | 3.65 | 3.45 | 3.73 | 3.12 | 3.82 | 3.70 | 3.46 | 3.73 | 3.79 | 3.94 |

# 13.1 The Two-Way ANOVA Model

We begin with a discussion of the advantages of the two-way ANOVA design and illustrate these with some examples. Then we discuss the model and the assumptions.

## Advantages of two-way ANOVA

In one-way ANOVA, we classify populations according to one categorical variable, or factor. In the two-way ANOVA model, there are two factors, each with its own number of levels. When we are interested in the effects of two factors, a two-way design offers great advantages over several single-factor studies. Several examples will illustrate these advantages.

> **EXAMPLE**
>
> **13.1 Design 1: Choosing the best magazine layout and cover.** In Example 12.1, a magazine publisher wants to compare three different magazine layouts. To do this, she plans to randomly assign the three design layouts equally among 60 supermarkets. The number of magazines sold during a one-week period is the outcome variable.
>
> Now suppose a second experiment is planned for the following week to compare four different covers for the magazine. A similar experimental design will be used, with the four covers randomly assigned among the same 60 supermarkets.

Here is a picture of the design of the first experiment with the sample sizes:

| Layout | $n$ |
|--------|-----|
| 1 | 20 |
| 2 | 20 |
| 3 | 20 |
| Total | 60 |

And this represents the second experiment:

| Cover | $n$ |
|-------|-----|
| 1 | 15 |
| 2 | 15 |
| 3 | 15 |
| 4 | 15 |
| Total | 60 |

In the first experiment 20 stores were assigned to each level of the factor for a total of 60 stores. In the second experiment 15 stores were assigned to each level of the factor for a total of 60 stores. The total amount of time for the two experiments is two weeks. Each experiment will be analyzed using one-way ANOVA. The factor in the first experiment is magazine layout with three levels, and the factor in the second experiment is magazine cover with four levels. Let's now consider combining the two experiments into one.

**EXAMPLE**

**13.2 Design 2: Choosing the best magazine layout and cover.** Suppose we use a two-way approach for the magazine design problem. There are two factors, layout and cover. Since layout has three levels and cover has four levels, this is a $3 \times 4$ design. This gives a total of 12 possible combinations of layout and cover. With a total of 60 stores, we could assign each combination of layout and cover to 5 stores. The number of magazines sold during a one-week period is the outcome variable.

Here is a picture of the two-way design with the sample sizes:

| Layout | Cover | | | | Total |
|---|---|---|---|---|---|
| | 1 | 2 | 3 | 4 | |
| 1 | 5 | 5 | 5 | 5 | 20 |
| 2 | 5 | 5 | 5 | 5 | 20 |
| 3 | 5 | 5 | 5 | 5 | 20 |
| Total | 15 | 15 | 15 | 15 | 60 |

cell

Each combination of the factors in a two-way design corresponds to a **cell.** The $3 \times 4$ ANOVA for the magazine experiment has twelve cells, each corresponding to a particular combination of layout and cover.

With the two-way design for layout, notice that we have 20 stores assigned to each level, the same as what we had for the one-way experiment for layout alone. Similarly, there are 15 stores assigned to each level of cover. Thus, the two-way design gives us the same amount of information for estimating the sales for each level of each factor as we had with the two one-way designs. The difference is that we can collect all of the information in only one week. By combining the two factors into one experiment, we have increased our efficiency by reducing the amount of data to be collected by half.

**EXAMPLE**

**13.3 Can increased palm oil consumption reduce malaria?** Malaria is a serious health problem causing an estimated 2.7 million deaths per year, mostly in Africa.[1] Some research suggests that vitamin A can reduce episodes of malaria in young children. Red palm oil is a good source of vitamin A and is readily available in Nigeria, a country where malaria accounts for about 30% of the deaths of young children. Can an increase in the consumption of red palm oil reduce the occurrence and severity of malaria in this region?[2]

**2.** We can reduce the residual variation in a model by including a second factor thought to influence the response.

**3.** We can investigate interactions between factors.

These considerations also apply to study designs with more than two factors. We will be content to explore only the two-way case. The choice of sampling or experimental design is fundamental to any statistical study. *Factors and levels must be carefully selected by an individual or team who understands both the statistical models and the issues that the study will address.*

## The two-way ANOVA model

When discussing two-way models in general, we will use the labels A and B for the two factors. For particular examples and when using statistical software, it is better to use meaningful names for these categorical variables. Thus, in Example 13.2 we would say that the factors are layout and cover, and in Example 13.4 we would say the factors are dosage and gender.

The numbers of levels of the factors are often used to describe the model. Again using our earlier examples, we would say Example 13.2 represents a $3 \times 4$ ANOVA and Example 13.4 illustrates a $3 \times 2$ ANOVA. In general, Factor A will have $I$ levels and Factor B will have $J$ levels. Therefore, we call the general two-way problem an $I \times J$ ANOVA.

In a two-way design every level of A appears in combination with every level of B, so that $I \times J$ groups are compared. The sample size for level $i$ of Factor A and level $j$ of Factor B is $n_{ij}$.[3] The total number of observations is

$$N = \sum n_{ij}$$

---

### ASSUMPTIONS FOR TWO-WAY ANOVA

We have independent SRSs of size $n_{ij}$ from each of $I \times J$ Normal populations. The population means $\mu_{ij}$ may differ, but all populations have the same standard deviation $\sigma$. The $\mu_{ij}$ and $\sigma$ are unknown parameters.

Let $x_{ijk}$ represent the $k$th observation from the population having Factor A at level $i$ and Factor B at level $j$. The statistical model is

$$x_{ijk} = \mu_{ij} + \epsilon_{ijk}$$

for $i = 1, \ldots, I$ and $j = 1, \ldots, J$ and $k = 1, \ldots, n_{ij}$, where the deviations $\epsilon_{ijk}$ are from an $N(0, \sigma)$ distribution.

---

**LOOK BACK**
**one-way model,**
**page 645**

Similar to the one-way model, the FIT part is the group means $\mu_{ij}$, and the RESIDUAL part is the deviations $\epsilon_{ijk}$ of the individual observations from their group means. To estimate a group mean $\mu_{ij}$ we use the sample mean of the observations in the samples from this group:

$$\bar{x}_{ij} = \frac{1}{n_{ij}} \sum_k x_{ijk}$$

The $k$ below the $\sum$ means that we sum the $n_{ij}$ observations that belong to the $(i, j)$th sample.

The RESIDUAL part of the model contains the unknown $\sigma$. We calculate the sample variances for each SRS and pool these to estimate $\sigma^2$:

$$s_p^2 = \frac{\sum(n_{ij} - 1)s_{ij}^2}{\sum(n_{ij} - 1)}$$

Just as in one-way ANOVA, the numerator in this fraction is SSE and the denominator is DFE. Also, DFE is the total number of observations minus the number of groups. That is, DFE $= N - IJ$. The estimator of $\sigma$ is $s_p$.

## Main effects and interactions

In this section we will further explore the FIT part of the two-way ANOVA, which is represented in the model by the population means $\mu_{ij}$. The two-way design gives some structure to the set of means $\mu_{ij}$.

So far, because we have independent samples from each of $I \times J$ groups, we have presented the problem as a one-way ANOVA with $IJ$ groups. Each population mean $\mu_{ij}$ is estimated by the corresponding sample mean $\bar{x}_{ij}$, and we can calculate sums of squares and degrees of freedom as in one-way ANOVA. In accordance with the conventions used by many computer software packages, we use the term *model* when discussing the sums of squares and degrees of freedom calculated as in one-way ANOVA with $IJ$ groups. Thus, SSM is a model sum of squares constructed from deviations of the form $\bar{x}_{ij} - \bar{x}$, where $\bar{x}$ is the average of all of the observations and $\bar{x}_{ij}$ is the mean of the $(i, j)$th group. Similarly, DFM is simply $IJ - 1$.

In two-way ANOVA, the terms SSM and DFM can be further broken down into terms corresponding to a main effect for A, a main effect for B, and an AB interaction. Each of SSM and DFM is then a sum of terms:

$$\text{SSM} = \text{SSA} + \text{SSB} + \text{SSAB}$$

and

$$\text{DFM} = \text{DFA} + \text{DFB} + \text{DFAB}$$

The term SSA represents variation among the means for the different levels of Factor A. Because there are $I$ such means, DFA $= I - 1$ degrees of freedom. Similarly, SSB represents variation among the means for the different levels of Factor B, with DFB $= J - 1$.

Interactions are a bit more involved. We can see that SSAB, which is SSM $-$ SSA $-$ SSB, represents the variation in the model that is not accounted for by the main effects. By subtraction we see that its degrees of freedom are

$$\text{DFAB} = (IJ - 1) - (I - 1) - (J - 1)$$
$$= (I - 1)(J - 1)$$

There are many kinds of interactions. The easiest way to study them is through examples.

If the effect being tested is zero, the calculated $F$ statistic has an $F$ distribution with numerator degrees of freedom corresponding to the effect and denominator degrees of freedom equal to DFE. Large values of the $F$ statistic lead to rejection of the null hypothesis. The $P$-value is the probability that a random variable having the corresponding $F$ distribution is greater than or equal to the calculated value.

The following example illustrates how to do a two-way ANOVA. As with the one-way ANOVA, we focus our attention on interpretation of the computer output.

**EXAMPLE**

**13.11  A study of cardiovascular risk factors.**   A study of cardiovascular risk factors compared runners who averaged at least 15 miles per week with a control group described as "generally sedentary." Both men and women were included in the study.[7] The design is a 2 × 2 ANOVA with the factors group and gender. There were 200 subjects in each of the four combinations. One of the variables measured was the heart rate after 6 minutes of exercise on a treadmill. SAS computer analysis produced the outputs in Figure 13.5 and Figure 13.6.

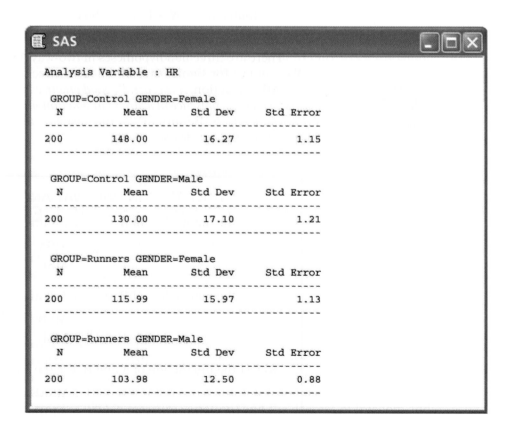

**FIGURE 13.5** Summary statistics for heart rates in the four groups of a 2 × 2 ANOVA, for Example 13.11.

SAS

```
General Linear Models Procedure
Dependent Variable: HR

                                Sum of          Mean
Source              DF         Squares        Square    F Value    Pr > F
Model                3       215256.09      71752.03     296.35    0.0001
Error              796       192729.83        242.12
Corrected Total    799       407985.92

              R-square             C.V.      Root MSE              HR Mean
              0.527607         12.49924        15.560               124.49

Source              DF       Type I SS   Mean Square    F Value    Pr > F
GROUP                1       168432.08     168432.08     695.65    0.0001
GENDER               1        45030.00      45030.00     185.98    0.0001
GROUP*GENDER         1         1794.01       1794.01       7.41    0.0066
```

**FIGURE 13.6** Two-way ANOVA output for heart rates, for Example 13.11.

We begin with the usual preliminary examination. From Figure 13.5 we see that the ratio of the largest to the smallest standard deviation is less than 2. Therefore, we are not concerned about violating the assumption of equal population standard deviations. Normal quantile plots (not shown) do not reveal any outliers, and the data appear to be reasonably Normal.

The ANOVA table at the top of the output in Figure 13.6 is in effect a one-way ANOVA with four groups: female control, female runner, male control, and male runner. In this analysis Model has 3 degrees of freedom, and Error has 796 degrees of freedom. The $F$ test and its associated $P$-value for this analysis refer to the hypothesis that all four groups have the same population mean. We are interested in the main effects and interaction, so we ignore this test.

The sums of squares for the group and gender main effects and the group-by-gender interaction appear at the bottom of Figure 13.6 under the heading Type I SS. These sum to the sum of squares for Model. Similarly, the degrees of freedom for these sums of squares sum to the degrees of freedom for Model. Two-way ANOVA splits the variation among the means (expressed by the Model sum of squares) into three parts that reflect the two-way layout.

Because the degrees of freedom are all 1 for the main effects and the interaction, the mean squares are the same as the sums of squares. The $F$ statistics for the three effects appear in the column labeled F Value, and the $P$-values are under the heading Pr > F. For the group main effect, we verify the calculation of $F$ as follows:

$$F = \frac{\text{MSG}}{\text{MSE}} = \frac{168,432}{242.12} = 695.65$$

All three effects are statistically significant. The group effect has the largest $F$, followed by the gender effect and then the group-by-gender interaction. To interpret these results, we examine the plot of means with bars indicating one standard error in Figure 13.7. Note that the standard errors are quite small due to the large sample sizes. The significance of the main effect for group is due to the fact that the controls have higher average heart rates than the runners for both genders. This is the largest effect evident in the plot.

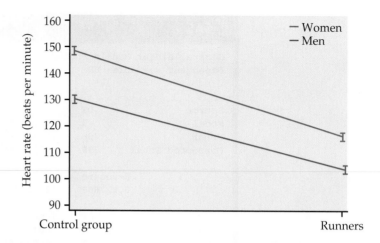

**FIGURE 13.7** Plot of the group means with standard errors for heart rates in the 2 × 2 ANOVA, for Example 13.11.

The significance of the main effect for gender is due to the fact that the females have higher heart rates than the men in both groups. The differences are not as large as those for the group effect, and this is reflected in the smaller value of the $F$ statistic.

The analysis indicates that a complete description of the average heart rates requires consideration of the interaction in addition to the main effects. The two lines in the plot are not parallel. This interaction can be described in two ways. The female-male difference in average heart rates is greater for the controls than for the runners. Alternatively, the difference in average heart rates between controls and runners is greater for women than for men. As the plot suggests, the interaction is not large. It is statistically significant because there were 800 subjects in the study.

Two-way ANOVA output for other software is similar to that given by SAS. Figure 13.8 gives the analysis of the heart rate data using Excel and Minitab.

## SECTION 13.2    Summary

**Two-way analysis of variance** is used to compare population means when populations are classified according to two factors.

ANOVA assumes that the populations are Normal with possibly different means and the same standard deviation and that independent SRSs are drawn from each population.

As with one-way ANOVA, preliminary analysis includes examination of means, standard deviations, and Normal quantile plots. **Marginal means** are calculated by taking averages of the cell means across rows and columns. Pooling is used to estimate the within-group variance.

ANOVA separates the total variation into parts for the **model** and **error.** The model variation is separated into parts for each of the **main effects** and the **interaction.**

The calculations are organized into an **ANOVA table.** $F$ statistics and $P$-values are used to test hypotheses about the main effects and the interaction.

Careful inspection of the means is necessary to interpret significant main effects and interactions. Plots are a useful aid.

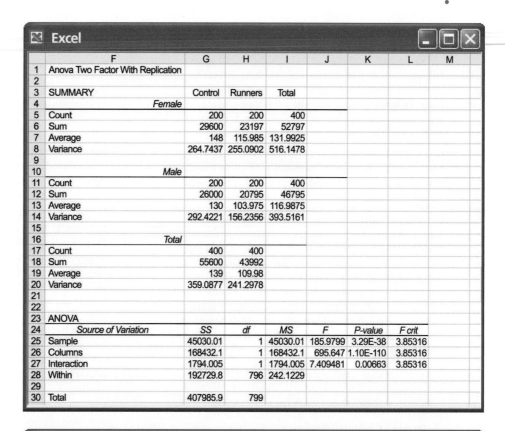

Excel

| | F | G | H | I | J | K | L | M |
|---|---|---|---|---|---|---|---|---|
| 1 | Anova Two Factor With Replication | | | | | | | |
| 2 | | | | | | | | |
| 3 | SUMMARY | | Control | Runners | Total | | | |
| 4 | Female | | | | | | | |
| 5 | Count | | 200 | 200 | 400 | | | |
| 6 | Sum | | 29600 | 23197 | 52797 | | | |
| 7 | Average | | 148 | 115.985 | 131.9925 | | | |
| 8 | Variance | | 264.7437 | 255.0902 | 516.1478 | | | |
| 9 | | | | | | | | |
| 10 | Male | | | | | | | |
| 11 | Count | | 200 | 200 | 400 | | | |
| 12 | Sum | | 26000 | 20795 | 46795 | | | |
| 13 | Average | | 130 | 103.975 | 116.9875 | | | |
| 14 | Variance | | 292.4221 | 156.2356 | 393.5161 | | | |
| 15 | | | | | | | | |
| 16 | Total | | | | | | | |
| 17 | Count | | 400 | 400 | | | | |
| 18 | Sum | | 55600 | 43992 | | | | |
| 19 | Average | | 139 | 109.98 | | | | |
| 20 | Variance | | 359.0877 | 241.2978 | | | | |
| 21 | | | | | | | | |
| 22 | | | | | | | | |
| 23 | ANOVA | | | | | | | |
| 24 | Source of Variation | | SS | df | MS | F | P-value | F crit |
| 25 | Sample | | 45030.01 | 1 | 45030.01 | 185.9799 | 3.29E-38 | 3.85316 |
| 26 | Columns | | 168432.1 | 1 | 168432.1 | 695.647 | 1.10E-110 | 3.85316 |
| 27 | Interaction | | 1794.005 | 1 | 1794.005 | 7.409481 | 0.00663 | 3.85316 |
| 28 | Within | | 192729.8 | 796 | 242.1229 | | | |
| 29 | | | | | | | | |
| 30 | Total | | 407985.9 | 799 | | | | |

Minitab

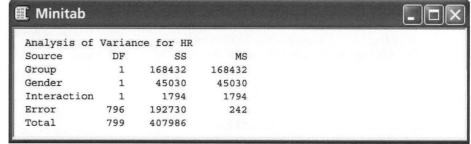

```
Analysis of Variance for HR
Source        DF        SS        MS
Group          1    168432    168432
Gender         1     45030     45030
Interaction    1      1794      1794
Error        796    192730       242
Total        799    407986
```

**FIGURE 13.8** Excel and Minitab two-way ANOVA output for the heart rate study, for Example 13.11.

# CHAPTER 13 Exercises

*For Exercises 13.1 and 13.2, see page 694.*

**13.3** **Describing a two-way ANOVA model.** A $2 \times 3$ ANOVA was run with 6 observations per cell.

(a) Give the degrees of freedom for the $F$ statistic that is used to test for interaction in this analysis and the entries from Table E that correspond to this distribution.

(b) Sketch a picture of this distribution with the information from the table included.

(c) The calculated value of the $F$ statistic is 2.73. How would you report the $P$-value?

(d) Would you expect a plot of the means to look parallel? Explain your answer.

**13.4** **Determining the critical value of $F$.** For each of the following situations, state how large the $F$ statistic needs to be for rejection of the null hypothesis at the 5% level. Sketch each distribution and indicate the region where you would reject.

(a) The main effect for the first factor in a $3 \times 5$ ANOVA with 3 observations per cell.

(b) The interaction in a $3 \times 3$ ANOVA with 3 observations per cell.

(c) The interaction in a 2 × 2 ANOVA with 51 observations per cell.

**13.5  Identifying the factors of a two-way ANOVA model.** For each of the following situations, identify both factors and the response variable. Also, state the number of levels for each factor ($I$ and $J$) and the total number of observations ($N$).

(a) A child psychologist is interested in studying how a child's percent of pretend play differs with gender and age (4, 8, and 12 months). There are 11 infants assigned to each cell of the experiment.

(b) Brewers malt is produced from germinating barley. A homebrewer wants to determine the best conditions to germinate the barley. A total of 30 lots of barley seed were equally and randomly assigned to 10 germination conditions. The conditions are combinations of the week after harvest (1, 3, 6, 9, or 12 weeks) and the amount of water used in the process (4 or 8 milliliters). The percent of seeds germinating is the outcome variable.

(c) A virologist wants to compare the effects of two different media (A and B) and three different incubation times (12, 18, and 24 hours) on the growth of the Ebola virus. She plans on doing four replicates of each combination.

**13.6  Determining the degrees of freedom.** For each part in Exercise 13.5, outline the ANOVA table, giving the sources of variation and the degrees of freedom.

**13.7  The effects of proximity and visibility on food intake.** A recent study investigated the influence that proximity and visibility of food have on food intake.[8] A total of 40 secretaries from the University of Illinois participated in the study. A candy dish full of individually wrapped chocolates was placed either at the desk of the participant or at a location 2 meters from the participant. The candy dish was either a clear (candy visible) or opaque (candy not visible) covered bowl. After a week, the researchers noted not only the number of candies consumed per day but also the self-reported number of candies consumed by each participant. The table at the top of the next column summarizes the mean difference between these two values (reported minus actual).

(a) Make a plot of the means and describe the patterns that you see. Does the plot suggest an interaction between visibility and proximity?

(b) This study actually took 4 weeks, with each participant being observed at each treatment

|          | Visibility | |
| Proximity | Clear | Opaque |
| --- | --- | --- |
| Proximate | −1.2 | −0.8 |
| Less proximate | 0.5 | 0.4 |

combination in a random order. Explain why a "repeated-measures" design like this may be beneficial.

**13.8  Hypotension and endurance exercise.** In sedentary individuals, low blood pressure (hypotension) often occurs after a single bout of aerobic exercise and lasts nearly two hours. This can cause dizziness, light-headedness, and possibly fainting upon standing. It is thought that endurance exercise training can reduce the degree of postexercise hypotension. To test this, researchers studied 16 endurance-trained and 16 sedentary men and women.[9] The following table summarizes the postexercise systolic arterial pressure (mmHg) after 60 minutes of upright cycling:

| Group | $n$ | $\bar{x}$ | Std. error |
| --- | --- | --- | --- |
| Women, sedentary | 8 | 100.7 | 3.4 |
| Women, endurance | 8 | 105.3 | 3.6 |
| Men, sedentary | 8 | 114.2 | 3.8 |
| Men, endurance | 8 | 110.2 | 2.3 |

(a) Make a plot similar to Figure 13.4 with the systolic blood pressure on the $y$ axis and training level on the $x$ axis. Describe the pattern you see.

(b) From the table, one can show that SSA = 677.12, SSB = 0.72, SSAB = 147.92, and SSE = 2478 where A is the gender effect and B is the training level. Construct the ANOVA table with $F$ statistics and degrees of freedom, and state your conclusions regarding main effects and interaction.

(c) The researchers also measured the before-exercise systolic blood pressure of the participants and looked at a model that incorporated both the pre- and postexercise values. Explain why it is likely beneficial to incorporate both measurements in the study.

**13.9  Evaluation of an intervention program.** The National Crime Victimization Survey estimates

**TABLE 13.1**

Safety behaviors of abused women

| Behavior | Intervention group (%) | | | Control group (%) | | |
|---|---|---|---|---|---|---|
| | Baseline | 3 months | 6 months | Baseline | 3 months | 6 months |
| Hide money | 68.0 | 60.0 | 62.7 | 60.0 | 37.8 | 35.1 |
| Hide extra keys | 52.7 | 76.0 | 68.9 | 53.3 | 33.8 | 39.2 |
| Abuse code to alert family | 30.7 | 74.7 | 60.0 | 22.7 | 27.0 | 43.2 |
| Hide extra clothing | 37.3 | 73.6 | 52.7 | 42.7 | 32.9 | 27.0 |
| Asked neighbors to call police | 49.3 | 73.0 | 66.2 | 32.0 | 45.9 | 40.5 |
| Know Social Security number | 93.2 | 93.2 | 100.0 | 89.3 | 93.2 | 98.6 |
| Keep rent, utility receipts | 75.3 | 95.5 | 89.4 | 70.3 | 84.7 | 80.9 |
| Keep birth certificates | 84.0 | 90.7 | 93.3 | 77.3 | 90.4 | 93.2 |
| Keep driver's license | 93.3 | 93.3 | 97.3 | 94.7 | 95.9 | 98.6 |
| Keep telephone numbers | 96.0 | 98.7 | 100.0 | 90.7 | 97.3 | 100.0 |
| Removed weapons | 50.0 | 70.6 | 38.5 | 40.7 | 23.8 | 5.9 |
| Keep bank account numbers | 81.0 | 94.3 | 96.2 | 76.2 | 85.5 | 94.4 |
| Keep insurance policy number | 70.9 | 90.4 | 89.7 | 68.3 | 84.2 | 94.8 |
| Keep marriage license | 71.1 | 92.3 | 84.6 | 63.3 | 73.2 | 80.0 |
| Hide valuable jewelry | 78.7 | 84.5 | 83.9 | 74.0 | 75.0 | 80.3 |

that there were over 400,000 violent crimes committed against women by their intimate partner that resulted in physical injury. An intervention study designed to increase safety behaviors of abused women compared the effectiveness of six telephone intervention sessions with a control group of abused women who received standard care. Fifteen different safety behaviors were examined.[10] One of the variables analyzed was the total number of behaviors (out of 15) that each woman performed. Here is a summary of the means of this variable at baseline (just before the first telephone call) and at follow-up 3 and 6 months later:

| | Time | | |
|---|---|---|---|
| Group | Baseline | 3 months | 6 months |
| Intervention | 10.4 | 12.5 | 11.9 |
| Control | 9.6 | 9.9 | 10.4 |

(a) Find the marginal means. Are they useful for understanding the results of this study?

(b) Plot the means. Do you think there is an interaction? Describe the meaning of an interaction for this study.

(*Note:* This exercise is from a repeated-measures design, and the data are not particularly Normal because they are counts with values from 1 to 15. Although we cannot use the methods in this

chapter for statistical inference in this setting, the example does illustrate ideas about interactions.)

13.10 ⚠ CHALLENGE **More on the assessment of an intervention program.** Refer to the previous exercise. Table 13.1 gives the percents of women who responded that they performed each of the 15 safety behaviors studied.

(a) Summarize these data graphically. Do you think that your graphical display is more effective than Table 13.1 for describing the results of this study? Explain why or why not.

(b) Note any particular patterns in the data that would be important to someone who would use these results to design future intervention programs for abused women.

(c) The study was conducted "at a family violence unit of a large urban District Attorney's Office that serves an ethnically diverse population of three million citizens." To what extent do you think that this fact limits the conclusions that can be drawn?

13.11 **The acceptability of lying.** Lying is a common component of all human relationships. To investigate the acceptability of lying under various scenarios, researchers questioned 229 high school students from a West Coast public high school and 261 college students from a state university in the Midwest.[11] As part of the questioning, participants were asked to read a vignette in which the protagonist lies to his or her parents and to evaluate the acceptability of lying on a 4-point scale (1 = totally unacceptable,

Some of the computer exercises in the text refer to 16 relatively large data sets that are on the CD that accompanies this text. The CD also contains data for many other exercises and examples.

Background information for each of the 16 data sets is presented below. For most, the first five cases are given here.

## 1   BIOMARKERS

Text Reference:
Exercises 7.118–7.121,
10.26–10.29,
and 11.34–11.39

Healthy bones are continually being renewed by two processes. Through bone formation, new bone is built; through bone resorption, old bone is removed. If one or both of these processes is disturbed, by disease, aging, or space travel, for example, bone loss can be the result. The variables VO+ and VO− measure bone formation and bone resorption, respectively. Osteocalcin (OC) is a biochemical marker for bone formation: higher levels of bone formation are associated with higher levels of OC. A blood sample is used to measure OC, and it is much less expensive to obtain than direct measures of bone formation. The units are milligrams of OC per milliliter of blood (mg/ml). Similarly, tartrate resistant acid phosphatase (TRAP) is a biochemical marker for bone resorption that is also measured in blood. It is measured in units per liter (U/l). These variables were measured in a study of 31 healthy women aged 11 to 32 years. The results were published in C. M. Weaver et al., "Quantification of biochemical markers of bone turnover by kinetic measures of bone formation and resorption in young healthy females," *Journal of Bone and Mineral Research*, 12 (1997), pp. 1714–1720. Variables with the first letter "L" are the logarithms of the measured variables. The data were provided by Linda McCabe. The first five cases are given in the table below.

| VO+ | VO− | OC | LOC | TRAP | LTRAP | LVO+ | LVO− |
|-----|-----|-----|-----|------|-------|------|------|
| 1606 | 903 | 68.9 | 4.233 | 19.4 | 2.965 | 7.382 | 6.806 |
| 2240 | 1761 | 56.3 | 4.031 | 25.5 | 3.239 | 7.714 | 7.474 |
| 2221 | 1486 | 54.6 | 4.000 | 19.0 | 2.944 | 7.706 | 7.304 |
| 896 | 1116 | 31.2 | 3.440 | 9.0 | 2.197 | 6.798 | 7.018 |
| 2545 | 2236 | 36.4 | 3.595 | 19.1 | 2.950 | 7.842 | 7.712 |

## 2   BRFSS

Text Reference:
Exercises 2.136, 2.137, and 2.166

With support from the Centers for Disease Control and Prevention (CDC), the Behavioral Risk Factor Surveillance System (BRFSS) conducts the world's largest, ongoing telephone survey of health conditions and risk behaviors in the United States. The prevalence of various health risk factors by state is summarized on the CDC Web site, www. cdc.gov/brfss. The data set BRFSS contains data on 29 demographic characteristics and risk factors for each state. The demographic characteristics are age (percents aged 18 to 24, 25 to 34, 35 to 44, 45 to 54, 55 to 64, and 65 or over), education (less than high school, high school or GED, some post–high school, college), income (less than $15,000, $15,000 to $25,000, $25,000 to $35,000, $35,000 to $50,000, $50,000 or more), and percent female. Risk factors are body mass index (BMI is weight in kilograms divided by the square of height in meters; the classifications are less than 25, 25 to 30, and

30 or more; 18.5 to 24.9 is considered normal, 25 to 29.9 is overweight, and 30 or over is obese), alcohol consumption (at least one drink within the last 30 days, heavy is more than two drinks per day for men and more than one drink per day for women, binge is five or more drinks [four for women] on one occasion during the past 30 days), physical exercise (at least 10 minutes at a time during a usual week), fruits and vegetables (eat at least five servings per day), physical activity (30 or more minutes of moderate physical activity five or more days per week or vigorous physical activity for 20 or more minutes three or more days per week), and smoking (every day, some days, former smoker, never smoked).

## 3 CHEESE

Text Reference:
Examples 14.5 and 14.9;
Exercises 11.51–11.59
and 14.36–14.38

As cheddar cheese matures, many chemical processes take place. The taste of matured cheese is related to the concentration of several chemicals in the final product. In a study of cheddar cheese from the LaTrobe Valley of Victoria, Australia, samples of cheese were analyzed for their chemical composition and were subjected to taste tests.

Data for one type of cheese-manufacturing process appear below. The variable "Case" is used to number the observations from 1 to 30. "Taste" is the response variable of interest. The taste scores were obtained by combining the scores from several tasters.

Three of the chemicals whose concentrations were measured were acetic acid, hydrogen sulfide, and lactic acid. For acetic acid and hydrogen sulfide (natural) log transformations were taken. Thus, the explanatory variables are the transformed concentrations of acetic acid ("Acetic") and hydrogen sulfide ("H2S") and the untransformed concentration of lactic acid ("Lactic"). These data are based on experiments performed by G. T. Lloyd and E. H. Ramshaw of the CSIRO Division of Food Research, Victoria, Australia. Some results of the statistical analyses of these data are given in G. P. McCabe, L. McCabe, and A. Miller, "Analysis of taste and chemical composition of cheddar cheese, 1982–83 experiments," CSIRO Division of Mathematics and Statistics Consulting Report VT85/6; and in I. Barlow et al., "Correlations and changes in flavour and chemical parameters of cheddar cheeses during maturation," *Australian Journal of Dairy Technology*, 44 (1989), pp. 7–18. The table below gives the data for the first five cases.

| Case | Taste | Acetic | H2S | Lactic |
|------|-------|--------|-------|--------|
| 01 | 12.3 | 4.543 | 3.135 | 0.86 |
| 02 | 20.9 | 5.159 | 5.043 | 1.53 |
| 03 | 39.0 | 5.366 | 5.438 | 1.57 |
| 04 | 47.9 | 5.759 | 7.496 | 1.81 |
| 05 | 5.6 | 4.663 | 3.807 | 0.99 |

## 4 CSDATA

Text Reference:
Example 11.1;
Exercises 1.173, 2.155,
3.93, 3.94, 7.138,
11.5, 11.6, and
14.39–14.42

The computer science department of a large university was interested in understanding why a large proportion of their first-year students failed to graduate as computer science majors. An examination of records from the registrar indicated that most of the attrition occurred during the first three semesters. Therefore, they decided to study all first-year students entering their program in a particular year and to follow their progress for the first three semesters.

The variables studied included the grade point average after three semesters and a collection of variables that would be available as students entered their program. These included scores on standardized tests such as the SATs and high school grades in various subjects. The individuals who conducted the study were also interested in examining differences between men and women in this program. Therefore, sex was included as a variable.

Data on 224 students who began study as computer science majors in a particular year were analyzed. A few exceptional cases were excluded, such as students who did not have complete data available on the variables of interest (a few students were admitted who did not take the SATs). Data for the first five students appear below. There are eight variables for each student. OBS is a variable used to identify the student. The data files kept by the registrar identified students by social security number, but for this study they were simply given a number from 1 to 224. The grade point average after three semesters is the variable GPA. This university uses a four-point scale, with A corresponding to 4, B to 3, C to 2, etc. A straight-A student has a 4.00 GPA.

The high school grades included in the data set are the variables HSM, HSS, and HSE. These correspond to average high school grades in math, science, and English. High schools use different grading systems (some high schools have a grade higher than A for honors courses), so the university's task in constructing these variables is not easy. The researchers were willing to accept the university's judgment and used its values. High school grades were recorded on a scale from 1 to 10, with 10 corresponding to A, 9 to A−, 8 to B+, etc.

The SAT scores are SATM and SATV, corresponding to the Mathematics and Verbal parts of the SAT. Gender was recorded as 1 for men and 2 for women. This is an arbitrary code. For software packages that can use alphanumeric variables (that is, values do not have to be numbers), it is more convenient to use M and F or Men and Women as values for the sex variable. With this kind of user-friendly capability, you do not have to remember who are the 1s and who are the 2s.

Results of the study are reported in P. F. Campbell and G. P. McCabe, "Predicting the success of freshmen in a computer science major," *Communications of the ACM*, 27 (1984), pp. 1108–1113. The table below gives data for the first five students.

| OBS | GPA | HSM | HSS | HSE | SATM | SATV | SEX |
|-----|-----|-----|-----|-----|------|------|-----|
| 001 | 3.32 | 10 | 10 | 10 | 670 | 600 | 1 |
| 002 | 2.26 | 6 | 8 | 5 | 700 | 640 | 1 |
| 003 | 2.35 | 8 | 6 | 8 | 640 | 530 | 1 |
| 004 | 2.08 | 9 | 10 | 7 | 670 | 600 | 1 |
| 005 | 3.38 | 8 | 9 | 8 | 540 | 580 | 1 |

## 5   DANDRUFF

Text Reference:
Exercises 12.54–12.57

The DANDRUFF data set is based on W. L. Billhimer et al., "Results of a clinical trial comparing 1% pyrithione zinc and 2% ketoconazole shampoos," *Cosmetic Dermatology*, 9 (1996), pp. 34–39. The study reported in this paper is a clinical trial that compared three treatments for dandruff and a placebo. The treatments were 1% pyrithione zinc shampoo (PyrI), the same shampoo but with instructions to shampoo two times (PyrII), 2% ketoconazole shampoo (Keto), and a placebo shampoo (Placebo). After six weeks of treatment, eight sections of the scalp were examined and given a score that measured the amount of scalp flaking on a 0 to 10 scale. The response variable was the sum of these eight scores. An analysis of the baseline flaking measure indicated that randomization of patients to treatments was successful in that no differences were found between the groups. At baseline there were 112 subjects in each of the three treatment groups and 28 subjects in the Placebo group. During the clinical trial 3 dropped out from the PyrII group and 6 from the Keto group. No patients dropped out of the other two groups.

## 14 READING

Text Reference:
Exercises 7.139, 12.61, and 12.62

Jim Baumann and Leah Jones, Purdue University College of Education, conducted a study to compare three methods of teaching reading comprehension. The 66 students who participated in the study were randomly assigned to the methods (22 to each). The standard practice of comparing new methods with a traditional one was used in this study. The traditional method is called Basal and the two innovative methods are called DRTA and Strat.

In the data set the variable Subject is used to identify the individual students. The values are 1 to 66. The method of instruction is indicated by the variable Group, with values B, D, and S, corresponding to Basal, DRTA, and Strat. Two pretests and three posttests were given to all students. These are the variables Pre1, Pre2, Post1, Post2, and Post3. Data for the first five subjects are given below.

| Subject | Group | Pre1 | Pre2 | Post1 | Post2 | Post3 |
|---------|-------|------|------|-------|-------|-------|
| 01 | B | 4 | 3 | 5 | 4 | 41 |
| 02 | B | 6 | 5 | 9 | 5 | 41 |
| 03 | B | 9 | 4 | 5 | 3 | 43 |
| 04 | B | 12 | 6 | 8 | 5 | 46 |
| 05 | B | 16 | 5 | 10 | 9 | 46 |

## 15 RUNNERS

Text Reference:
Example 13.11;
Exercises 1.81 and 1.143

A study of cardiovascular risk factors compared runners who averaged at least 15 miles per week with a control group described as "generally sedentary." Both men and women were included in the study. The data set was constructed based on information provided in P. D. Wood et al., "Plasma lipoprotein distributions in male and female runners," in P. Milvey (ed.), *The Marathon: Physiological, Medical, Epidemiological, and Psychological Studies,* New York Academy of Sciences, 1977.

The study design is a 2 × 2 ANOVA with the factors group and gender. There were 200 subjects in each of the four combinations. The variables are Id, a numeric subject identifier; Group, with values "Control" and "Runners"; Gender, with values Female and Male; and HeartRate, heart rate (beats per minute) after the subject ran for 6 minutes on a treadmill. Here are the data for the first five subjects:

| Id | Group | Gender | Beats |
|----|-------|--------|-------|
| 1 | Control | Female | 159 |
| 2 | Control | Female | 183 |
| 3 | Control | Female | 140 |
| 4 | Control | Female | 140 |
| 5 | Control | Female | 125 |

## 16 WORKERS

Text Reference:
Exercises 1.82–1.84,
1.172

Each March, the Bureau of Labor Statistics carries out an Annual Demographic Supplement to its monthly Current Population Survey. The data set WORKERS contains data about 71,076 people from one of these surveys. We included all people between the ages of 25 and 64 who have worked but whose main work experience is not in agriculture. Moreover, we combined the 16 levels of education in the BLS survey to form 6 levels.

There are five variables in the data set. Age is age in years. Education is the highest level of education a person has reached, with the following values: 1 = did not reach high school; 2 = some high school but no high school diploma; 3 = high school diploma; 4 = some college but no bachelor's degree (this includes people with an associate degree); 5 = bachelor's degree; 6 = postgraduate degree (master's, professional, or doctorate). Sex is coded as 1 = male and 2 = female. Total income is income from all sources. Note that income can be less than zero in some cases. Job class is a categorization of the person's main work experience, with 5 = private sector (outside households); 6 = government; 7 = self-employed. Here are the first five cases:

| Age | Education | Sex | Total income | Job class |
|-----|-----------|-----|--------------|-----------|
| 25 | 2 | 2 | 7,234 | 5 |
| 25 | 5 | 1 | 37,413 | 5 |
| 25 | 4 | 2 | 29,500 | 5 |
| 25 | 3 | 2 | 13,500 | 5 |
| 25 | 4 | 1 | 17,660 | 6 |

The first individual is a 25-year-old female who did not graduate from high school, works in the private sector, and had $7234 of income.

An Excel copy of the WORKERS data set is posted on the companion Web site, www.whfreeman.com/ips6e.

## TABLE E

### F critical values (continued)

| | | | | | Degrees of freedom in the numerator | | | | | |
|---|---|---|---|---|---|---|---|---|---|---|---|
| | | *p* | 1 | 2 | 3 | 4 | 5 | 6 | 7 | 8 | 9 |
| | 8 | .100 | 3.46 | 3.11 | 2.92 | 2.81 | 2.73 | 2.67 | 2.62 | 2.59 | 2.56 |
| | | .050 | 5.32 | 4.46 | 4.07 | 3.84 | 3.69 | 3.58 | 3.50 | 3.44 | 3.39 |
| | | .025 | 7.57 | 6.06 | 5.42 | 5.05 | 4.82 | 4.65 | 4.53 | 4.43 | 4.36 |
| | | .010 | 11.26 | 8.65 | 7.59 | 7.01 | 6.63 | 6.37 | 6.18 | 6.03 | 5.91 |
| | | .001 | 25.41 | 18.49 | 15.83 | 14.39 | 13.48 | 12.86 | 12.40 | 12.05 | 11.77 |
| | 9 | .100 | 3.36 | 3.01 | 2.81 | 2.69 | 2.61 | 2.55 | 2.51 | 2.47 | 2.44 |
| | | .050 | 5.12 | 4.26 | 3.86 | 3.63 | 3.48 | 3.37 | 3.29 | 3.23 | 3.18 |
| | | .025 | 7.21 | 5.71 | 5.08 | 4.72 | 4.48 | 4.32 | 4.20 | 4.10 | 4.03 |
| | | .010 | 10.56 | 8.02 | 6.99 | 6.42 | 6.06 | 5.80 | 5.61 | 5.47 | 5.35 |
| | | .001 | 22.86 | 16.39 | 13.90 | 12.56 | 11.71 | 11.13 | 10.70 | 10.37 | 10.11 |
| | 10 | .100 | 3.29 | 2.92 | 2.73 | 2.61 | 2.52 | 2.46 | 2.41 | 2.38 | 2.35 |
| | | .050 | 4.96 | 4.10 | 3.71 | 3.48 | 3.33 | 3.22 | 3.14 | 3.07 | 3.02 |
| | | .025 | 6.94 | 5.46 | 4.83 | 4.47 | 4.24 | 4.07 | 3.95 | 3.85 | 3.78 |
| | | .010 | 10.04 | 7.56 | 6.55 | 5.99 | 5.64 | 5.39 | 5.20 | 5.06 | 4.94 |
| | | .001 | 21.04 | 14.91 | 12.55 | 11.28 | 10.48 | 9.93 | 9.52 | 9.20 | 8.96 |
| | 11 | .100 | 3.23 | 2.86 | 2.66 | 2.54 | 2.45 | 2.39 | 2.34 | 2.30 | 2.27 |
| | | .050 | 4.84 | 3.98 | 3.59 | 3.36 | 3.20 | 3.09 | 3.01 | 2.95 | 2.90 |
| | | .025 | 6.72 | 5.26 | 4.63 | 4.28 | 4.04 | 3.88 | 3.76 | 3.66 | 3.59 |
| | | .010 | 9.65 | 7.21 | 6.22 | 5.67 | 5.32 | 5.07 | 4.89 | 4.74 | 4.63 |
| | | .001 | 19.69 | 13.81 | 11.56 | 10.35 | 9.58 | 9.05 | 8.66 | 8.35 | 8.12 |
| | 12 | .100 | 3.18 | 2.81 | 2.61 | 2.48 | 2.39 | 2.33 | 2.28 | 2.24 | 2.21 |
| | | .050 | 4.75 | 3.89 | 3.49 | 3.26 | 3.11 | 3.00 | 2.91 | 2.85 | 2.80 |
| | | .025 | 6.55 | 5.10 | 4.47 | 4.12 | 3.89 | 3.73 | 3.61 | 3.51 | 3.44 |
| | | .010 | 9.33 | 6.93 | 5.95 | 5.41 | 5.06 | 4.82 | 4.64 | 4.50 | 4.39 |
| | | .001 | 18.64 | 12.97 | 10.80 | 9.63 | 8.89 | 8.38 | 8.00 | 7.71 | 7.48 |
| | 13 | .100 | 3.14 | 2.76 | 2.56 | 2.43 | 2.35 | 2.28 | 2.23 | 2.20 | 2.16 |
| | | .050 | 4.67 | 3.81 | 3.41 | 3.18 | 3.03 | 2.92 | 2.83 | 2.77 | 2.71 |
| | | .025 | 6.41 | 4.97 | 4.35 | 4.00 | 3.77 | 3.60 | 3.48 | 3.39 | 3.31 |
| | | .010 | 9.07 | 6.70 | 5.74 | 5.21 | 4.86 | 4.62 | 4.44 | 4.30 | 4.19 |
| | | .001 | 17.82 | 12.31 | 10.21 | 9.07 | 8.35 | 7.86 | 7.49 | 7.21 | 6.98 |
| | 14 | .100 | 3.10 | 2.73 | 2.52 | 2.39 | 2.31 | 2.24 | 2.19 | 2.15 | 2.12 |
| | | .050 | 4.60 | 3.74 | 3.34 | 3.11 | 2.96 | 2.85 | 2.76 | 2.70 | 2.65 |
| | | .025 | 6.30 | 4.86 | 4.24 | 3.89 | 3.66 | 3.50 | 3.38 | 3.29 | 3.21 |
| | | .010 | 8.86 | 6.51 | 5.56 | 5.04 | 4.69 | 4.46 | 4.28 | 4.14 | 4.03 |
| | | .001 | 17.14 | 11.78 | 9.73 | 8.62 | 7.92 | 7.44 | 7.08 | 6.80 | 6.58 |
| | 15 | .100 | 3.07 | 2.70 | 2.49 | 2.36 | 2.27 | 2.21 | 2.16 | 2.12 | 2.09 |
| | | .050 | 4.54 | 3.68 | 3.29 | 3.06 | 2.90 | 2.79 | 2.71 | 2.64 | 2.59 |
| | | .025 | 6.20 | 4.77 | 4.15 | 3.80 | 3.58 | 3.41 | 3.29 | 3.20 | 3.12 |
| | | .010 | 8.68 | 6.36 | 5.42 | 4.89 | 4.56 | 4.32 | 4.14 | 4.00 | 3.89 |
| | | .001 | 16.59 | 11.34 | 9.34 | 8.25 | 7.57 | 7.09 | 6.74 | 6.47 | 6.26 |
| | 16 | .100 | 3.05 | 2.67 | 2.46 | 2.33 | 2.24 | 2.18 | 2.13 | 2.09 | 2.06 |
| | | .050 | 4.49 | 3.63 | 3.24 | 3.01 | 2.85 | 2.74 | 2.66 | 2.59 | 2.54 |
| | | .025 | 6.12 | 4.69 | 4.08 | 3.73 | 3.50 | 3.34 | 3.22 | 3.12 | 3.05 |
| | | .010 | 8.53 | 6.23 | 5.29 | 4.77 | 4.44 | 4.20 | 4.03 | 3.89 | 3.78 |
| | | .001 | 16.12 | 10.97 | 9.01 | 7.94 | 7.27 | 6.80 | 6.46 | 6.19 | 5.98 |
| | 17 | .100 | 3.03 | 2.64 | 2.44 | 2.31 | 2.22 | 2.15 | 2.10 | 2.06 | 2.03 |
| | | .050 | 4.45 | 3.59 | 3.20 | 2.96 | 2.81 | 2.70 | 2.61 | 2.55 | 2.49 |
| | | .025 | 6.04 | 4.62 | 4.01 | 3.66 | 3.44 | 3.28 | 3.16 | 3.06 | 2.98 |
| | | .010 | 8.40 | 6.11 | 5.19 | 4.67 | 4.34 | 4.10 | 3.93 | 3.79 | 3.68 |
| | | .001 | 15.72 | 10.66 | 8.73 | 7.68 | 7.02 | 6.56 | 6.22 | 5.96 | 5.75 |

*Degrees of freedom in the denominator* (row label along left margin)

## TABLE E

### F critical values (continued)

| | Degrees of freedom in the numerator | | | | | | | | | | |
|---|---|---|---|---|---|---|---|---|---|---|---|
| 10 | 12 | 15 | 20 | 25 | 30 | 40 | 50 | 60 | 120 | 1000 |
| 2.54 | 2.50 | 2.46 | 2.42 | 2.40 | 2.38 | 2.36 | 2.35 | 2.34 | 2.32 | 2.30 |
| 3.35 | 3.28 | 3.22 | 3.15 | 3.11 | 3.08 | 3.04 | 3.02 | 3.01 | 2.97 | 2.93 |
| 4.30 | 4.20 | 4.10 | 4.00 | 3.94 | 3.89 | 3.84 | 3.81 | 3.78 | 3.73 | 3.68 |
| 5.81 | 5.67 | 5.52 | 5.36 | 5.26 | 5.20 | 5.12 | 5.07 | 5.03 | 4.95 | 4.87 |
| 11.54 | 11.19 | 10.84 | 10.48 | 10.26 | 10.11 | 9.92 | 9.80 | 9.73 | 9.53 | 9.36 |
| 2.42 | 2.38 | 2.34 | 2.30 | 2.27 | 2.25 | 2.23 | 2.22 | 2.21 | 2.18 | 2.16 |
| 3.14 | 3.07 | 3.01 | 2.94 | 2.89 | 2.86 | 2.83 | 2.80 | 2.79 | 2.75 | 2.71 |
| 3.96 | 3.87 | 3.77 | 3.67 | 3.60 | 3.56 | 3.51 | 3.47 | 3.45 | 3.39 | 3.34 |
| 5.26 | 5.11 | 4.96 | 4.81 | 4.71 | 4.65 | 4.57 | 4.52 | 4.48 | 4.40 | 4.32 |
| 9.89 | 9.57 | 9.24 | 8.90 | 8.69 | 8.55 | 8.37 | 8.26 | 8.19 | 8.00 | 7.84 |
| 2.32 | 2.28 | 2.24 | 2.20 | 2.17 | 2.16 | 2.13 | 2.12 | 2.11 | 2.08 | 2.06 |
| 2.98 | 2.91 | 2.85 | 2.77 | 2.73 | 2.70 | 2.66 | 2.64 | 2.62 | 2.58 | 2.54 |
| 3.72 | 3.62 | 3.52 | 3.42 | 3.35 | 3.31 | 3.26 | 3.22 | 3.20 | 3.14 | 3.09 |
| 4.85 | 4.71 | 4.56 | 4.41 | 4.31 | 4.25 | 4.17 | 4.12 | 4.08 | 4.00 | 3.92 |
| 8.75 | 8.45 | 8.13 | 7.80 | 7.60 | 7.47 | 7.30 | 7.19 | 7.12 | 6.94 | 6.78 |
| 2.25 | 2.21 | 2.17 | 2.12 | 2.10 | 2.08 | 2.05 | 2.04 | 2.03 | 2.00 | 1.98 |
| 2.85 | 2.79 | 2.72 | 2.65 | 2.60 | 2.57 | 2.53 | 2.51 | 2.49 | 2.45 | 2.41 |
| 3.53 | 3.43 | 3.33 | 3.23 | 3.16 | 3.12 | 3.06 | 3.03 | 3.00 | 2.94 | 2.89 |
| 4.54 | 4.40 | 4.25 | 4.10 | 4.01 | 3.94 | 3.86 | 3.81 | 3.78 | 3.69 | 3.61 |
| 7.92 | 7.63 | 7.32 | 7.01 | 6.81 | 6.68 | 6.52 | 6.42 | 6.35 | 6.18 | 6.02 |
| 2.19 | 2.15 | 2.10 | 2.06 | 2.03 | 2.01 | 1.99 | 1.97 | 1.96 | 1.93 | 1.91 |
| 2.75 | 2.69 | 2.62 | 2.54 | 2.50 | 2.47 | 2.43 | 2.40 | 2.38 | 2.34 | 2.30 |
| 3.37 | 3.28 | 3.18 | 3.07 | 3.01 | 2.96 | 2.91 | 2.87 | 2.85 | 2.79 | 2.73 |
| 4.30 | 4.16 | 4.01 | 3.86 | 3.76 | 3.70 | 3.62 | 3.57 | 3.54 | 3.45 | 3.37 |
| 7.29 | 7.00 | 6.71 | 6.40 | 6.22 | 6.09 | 5.93 | 5.83 | 5.76 | 5.59 | 5.44 |
| 2.14 | 2.10 | 2.05 | 2.01 | 1.98 | 1.96 | 1.93 | 1.92 | 1.90 | 1.88 | 1.85 |
| 2.67 | 2.60 | 2.53 | 2.46 | 2.41 | 2.38 | 2.34 | 2.31 | 2.30 | 2.25 | 2.21 |
| 3.25 | 3.15 | 3.05 | 2.95 | 2.88 | 2.84 | 2.78 | 2.74 | 2.72 | 2.66 | 2.60 |
| 4.10 | 3.96 | 3.82 | 3.66 | 3.57 | 3.51 | 3.43 | 3.38 | 3.34 | 3.25 | 3.18 |
| 6.80 | 6.52 | 6.23 | 5.93 | 5.75 | 5.63 | 5.47 | 5.37 | 5.30 | 5.14 | 4.99 |
| 2.10 | 2.05 | 2.01 | 1.96 | 1.93 | 1.91 | 1.89 | 1.87 | 1.86 | 1.83 | 1.80 |
| 2.60 | 2.53 | 2.46 | 2.39 | 2.34 | 2.31 | 2.27 | 2.24 | 2.22 | 2.18 | 2.14 |
| 3.15 | 3.05 | 2.95 | 2.84 | 2.78 | 2.73 | 2.67 | 2.64 | 2.61 | 2.55 | 2.50 |
| 3.94 | 3.80 | 3.66 | 3.51 | 3.41 | 3.35 | 3.27 | 3.22 | 3.18 | 3.09 | 3.02 |
| 6.40 | 6.13 | 5.85 | 5.56 | 5.38 | 5.25 | 5.10 | 5.00 | 4.94 | 4.77 | 4.62 |
| 2.06 | 2.02 | 1.97 | 1.92 | 1.89 | 1.87 | 1.85 | 1.83 | 1.82 | 1.79 | 1.76 |
| 2.54 | 2.48 | 2.40 | 2.33 | 2.28 | 2.25 | 2.20 | 2.18 | 2.16 | 2.11 | 2.07 |
| 3.06 | 2.96 | 2.86 | 2.76 | 2.69 | 2.64 | 2.59 | 2.55 | 2.52 | 2.46 | 2.40 |
| 3.80 | 3.67 | 3.52 | 3.37 | 3.28 | 3.21 | 3.13 | 3.08 | 3.05 | 2.96 | 2.88 |
| 6.08 | 5.81 | 5.54 | 5.25 | 5.07 | 4.95 | 4.80 | 4.70 | 4.64 | 4.47 | 4.33 |
| 2.03 | 1.99 | 1.94 | 1.89 | 1.86 | 1.84 | 1.81 | 1.79 | 1.78 | 1.75 | 1.72 |
| 2.49 | 2.42 | 2.35 | 2.28 | 2.23 | 2.19 | 2.15 | 2.12 | 2.11 | 2.06 | 2.02 |
| 2.99 | 2.89 | 2.79 | 2.68 | 2.61 | 2.57 | 2.51 | 2.47 | 2.45 | 2.38 | 2.32 |
| 3.69 | 3.55 | 3.41 | 3.26 | 3.16 | 3.10 | 3.02 | 2.97 | 2.93 | 2.84 | 2.76 |
| 5.81 | 5.55 | 5.27 | 4.99 | 4.82 | 4.70 | 4.54 | 4.45 | 4.39 | 4.23 | 4.08 |
| 2.00 | 1.96 | 1.91 | 1.86 | 1.83 | 1.81 | 1.78 | 1.76 | 1.75 | 1.72 | 1.69 |
| 2.45 | 2.38 | 2.31 | 2.23 | 2.18 | 2.15 | 2.10 | 2.08 | 2.06 | 2.01 | 1.97 |
| 2.92 | 2.82 | 2.72 | 2.62 | 2.55 | 2.50 | 2.44 | 2.41 | 2.38 | 2.32 | 2.26 |
| 3.59 | 3.46 | 3.31 | 3.16 | 3.07 | 3.00 | 2.92 | 2.87 | 2.83 | 2.75 | 2.66 |
| 5.58 | 5.32 | 5.05 | 4.78 | 4.60 | 4.48 | 4.33 | 4.24 | 4.18 | 4.02 | 3.87 |

*(Continued)*

**2.71** No; the scatterplot shows little or no association, and regression explains only 1.3% of the variation in stock return.

**2.73** For metabolic rate on body mass, the slope is 26.9 cal/day per kg. For body mass on metabolic rate, the slope is 0.0278 kg per cal/day.

**2.75** $\hat{y} = 33.67 + 0.54x$. The predicted height is 69.85 inches.

**2.77** (a) $\bar{x} = 95$ and $s_x = 53.3854$ minutes; $\bar{y} = 12.6611$ and $s_y = 8.4967$ cm; $r = 0.9958$ (no units). (b) $\bar{y} = 32.1591$ and $s_y = 21.5816$ inches; $r$ is unchanged. (c) 0.4025 inches/minute.

**2.79** $r = 0.40$.

**2.81** (b) The slope is 0.000051; the plot suggests a horizontal line (slope 0). (c) Storing the oil doesn't help, as the total toxin level does not change over time.

**2.83** The sum is 0.01.

**2.85** (a) The plot is curved (low at the beginning and end of the year, high in the middle). (b) $\hat{y} = 39.392 + 1.4832x$; it does not fit well. (c) Residuals are negative for January through March and October through December, and positive from April to September. (d) A similar pattern would be expected in any city that is subject to seasonal temperature variation. (e) Seasons in the Southern Hemisphere are reversed.

**2.87** (b) No, because the pattern is not linear. (c) The sum is 0.01. The first two and last four residuals are negative, and those in the middle are positive.

**2.89** The variation of individual data increases the scatter, thus decreasing the strength of the relationship.

**2.91** For example, a student who in the past might have received a grade of B (and a lower SAT score) now receives an A (but has a lower SAT score than an A student in the past).

**2.93** $r = 0.08795$ and $b = 0.000811$ kg/cal.

**2.95** (a) Player 7's point is influential. (b) The first line omits Player 7.

**2.97** (b) $\hat{y} = 6.47 + 1.01x$. (c) The largest residuals are the Porsche Boxster (2.365) and Lamborghini Murcielago (−2.545). (d) The Insight pulls the line toward its point.

**2.99** Without the Insight, $\hat{y} = 4.87 + 1.11x$. For city mileages between 10 and 30 MPG, the difference in predicted highway mileage (with or without the Insight) is no more than 1.4 MPG, so the Insight is not very influential; it falls near the line suggested by the other points.

**2.101** (a) Drawing the "best line" by eye is a very inaccurate process.

**2.103** The plot should show a positive association when either group of points is viewed separately and should show a large number of bachelor's degree economists in business and graduate degree economists in academia.

**2.105** 1684 are binge drinkers; 8232 are not.

**2.107** $8232/17{,}096 \doteq 0.482$.

**2.109** $1630/7180 \doteq 0.227$.

**2.111** (a) About 3,388,000. (b) 0.207, 0.024; 0.320, 0.071; 0.104, 0.104; 0.046, 0.125. (c) 0.230, 0.391, 0.208, 0.171. (d) 0.677, 0.323.

**2.113** Full-time: 0.305, 0.472, 0.154, 0.069. Part-time: 0.073, 0.220, 0.321, 0.386.

**2.115** Two-year FT: 0.479, 0.521. Two-year PT: 0.458, 0.542. Four-year FT: 0.466, 0.534. Four-year PT: 0.394, 0.606. Graduate school: 0.455, 0.545. Vocational school: 0.539, 0.461.

**2.117** Start by setting $a$ equal to any number from 0 to 200.

**2.119** (a) 51.1%. (b) Small, 41.7%; medium, 51.7%; large, 60.0%. (d) Small, 39.8%; medium, 33.0%; large, 27.3%.

**2.121** Success (nonrelapse) rates were 58.3% (desipramine), 25.0% (lithium), and 16.7% (placebo).

**2.123** Age is one lurking variable: married men are generally older.

**2.125** No; self-confidence and improving fitness could be a common response to some other personality trait, or high self-confidence could make a person more likely to join the exercise program.

**2.127** Students with music experience may have other advantages (wealthier parents, better school systems, etc.).

**2.129** The diagram should show that either chemical exposure or time standing up or both or neither affect miscarriages.

**2.131** Spending more time watching TV means that less time is spent on other activities; this may suggest lurking variables.

**2.133** (a) Given two groups of the same age, where one group walks and the other does not, the walkers are half as likely to die in (say) the next year. (b) Men who choose to walk might also choose (or have chosen, earlier in life) other habits and behaviors that reduce mortality.

**2.135** A school that accepts weaker students but graduates a higher-than-expected number of them would have a positive residual, while a school with a stronger incoming class but a lower-than-expected graduation rate would have a negative residual. It seems reasonable to measure school quality by how much benefit students receive from attending the school.

**2.137** (a) The scatterplot shows a moderate positive association. (b) The regression line ($y = 1.1353x + 4.5503$) fits the overall trend. (c) For example, a state whose point falls above the line has a higher percent of college graduates than we would expect based on the percent who eat 5 servings of fruits and vegetables. (d) No; association is not evidence of causation.

**2.141** These results support the idea (the slope is negative), but the relationship is only moderately strong ($r^2 = 0.34$).

**2.143 (a)** One possible measure is mean response: 106.2 spikes/second for pure tones, 176.6 spikes/second for monkey calls. **(b)** $\hat{y} = 93.9 + 0.778x$. The third point has the largest residual. The first point is an outlier in the $x$ direction. **(c)** The correlation drops only slightly (from 0.6386 to 0.6101) when the third point is removed; it drops more drastically (to 0.4793) without the first point. **(d)** Without the first point, the line is $\hat{y} = 101 + 0.693x$; without the third point, it is $\hat{y} = 98.4 + 0.679x$.

**2.145** Based on the quantile plot, the distribution is close to Normal.

**2.147 (a)** Lines appear to fit the data well; there do not appear to be any outliers or influential points. **(b)** 18.9 ft$^3$ before, 15.7 ft$^3$ after. **(c)** 770.4 ft$^3$ before, 634.8 ft$^3$ after. **(d)** About \$50.44.

**2.149 (a)** $\hat{y} = 259.58 - 19.464x$; the relationship appears to be curved. **(b)** Either $\hat{y} = 5.9732 - 0.2184x$ or $\hat{y} = 2.5941 - 0.09486x$; the relationship appears to be linear.

**2.151** It is now more common for these stocks to rise and fall together.

**2.153** Number of firefighters and amount of damage are common responses to the seriousness of the fire.

**2.155** $\hat{y} = 1.28 + 0.00227x$, $r = 0.252$, and $r^2 = 6.3\%$. By itself, SATM does not give reliable predictions of GPA.

**2.157** A notably higher percent of women are "strictly voluntary" participants.

**2.159 (a)** Males: 490 admitted, 310 not. Females: 400 admitted, 300 not. **(b)** Males: 61.25% admitted. Females: 57.14% admitted. **(c)** Business school: 66.67% of males, 66.67% of females. Law school: 45% of males, 50% of females. **(d)** Most male applicants apply to the business school, where admission is easier. More women apply to the law school, which is more selective.

**2.161** First- and second-year: A has 8.3% small classes; B has 17.1% small. Upper-level: A has 77.5% small; B has 83.3% small.

**2.163** There is some suggestion that sexual ads are more common in magazines intended for young-adult readers, but the difference in percents is fairly small.

**2.165 (a)** Wagering on collegiate sports appears to be more common in Division II, and even more in Division III. **(b)** Even with smaller sample sizes (1000 or more), the estimates should be fairly accurate (barring dishonest responses). **(c)** Our conclusion might not hold for the true percents.

# CHAPTER 3

**3.1** Any group of friends is unlikely to include a representative cross section of all students.

**3.3** A computer programmer (and his friends) are not representative of all young people.

**3.7** This is an observational study. Explanatory variable: cell phone usage. Response variable: presence/absence of brain cancer.

**3.9** An experiment: each subject is (presumably randomly) assigned to a treatment group. Explanatory variable: teaching method. Response variable: change in each student's test score.

**3.11** Experimental units: food samples. Treatments: radiation exposure. Response variable: lipid oxidation. Factor: radiation exposure. Levels: nine different levels of radiation. It is likely that different lipids react to radiation differently.

**3.13** Those who volunteer to use the software may be better students (or worse).

**3.17 (a)** Students in the front rows have a different classroom experience from those in the back. (And if they chose their own seats, those who choose seats in the front may be different from those who choose back seats.) **(b)** There is no control group. **(c)** It is hard to compare different classes (zoology and botany) in different semesters.

**3.19** Those evaluating the exams should not know which teaching approach was used, and the students should not be told that they are being taught using the new (or old) method.

**3.21** Possible response variables include increase in weight or height, number of leaves, etc.

**3.23** Experimental units: pine tree seedlings. Factor: amount of light. Treatments: full light, or shaded to 5% of normal. Response variable: dry weight at end of study.

**3.25** Subjects: adults from selected households. Factors: level of identification and offer of survey results. Six treatments: interviewer's name/university name/both names, with or without results. Response variable: whether or not the interview is completed.

**3.27** Assign 9 subjects to each treatment. The first three groups are 03, 22, 29, 26, 01, 12, 11, 31, 21; 32, 30, 09, 23, 07, 27, 20, 06, 33; 05, 16, 28, 10, 18, 13, 25, 19, 04.

**3.29 (a)** Randomly assign 7 rats to each group. **(b)** Group 1 includes rats 16, 04, 21, 19, 07, 10, and 13. Group 2 is 15, 05, 09, 08, 18, 03, and 01.

**3.31** Assign 6 schools to each treatment group. Choose 16, 21, 06, 12, 02, 04 for Group 1; 14, 15, 23, 11, 09, 03 for Group 2; 07, 24, 17, 22, 01, 13 for Group 3; and the rest for Group 4.

**3.33 (a)** There are three factors (roller type, dyeing cycle time, and temperature), yielding eight treatments and requiring 24 fabric specimens.

**3.35 (a)** Population = 1 to 150, sample size 25, then click "Reset" and "Sample." **(b)** Without resetting, click "Sample" again.

(df = 12.1) or $16,452 to $114,344 (df = 8). **(e)** It seems that these houses should be a fair representation of three- and four-bedroom houses in West Lafayette.

## CHAPTER 8

**8.1** 0.01486.

**8.3** **(a)** $H_0: p = 0.72$; $H_a: p \neq 0.72$. **(b)** $z \doteq 1.11$, $P = 0.2670$. **(c)** No.

**8.5** A smaller sample is needed for 90% confidence; $n = 752$.

**8.7** **(a)** Yes. **(b)** Yes. **(c)** No. **(d)** No. **(e)** No.

**8.9** **(a)** Margin of error equals $z^*$ times standard error. **(b)** Use Normal distributions for proportions. **(c)** $H_0$ should refer to $p$, not $\hat{p}$.

**8.11** **(a)** $\hat{p} = 0.6341$, 0.6214 to 0.6467. This interval was found using a procedure that includes the correct proportion 95% of the time. **(b)** We do not know if those who did respond can reliably represent those who did not.

**8.13** **(a)** ±0.001321. **(b)** Other sources of error are much more significant than sampling error.

**8.15** **(a)** 0.3506 to 0.4094. **(b)** Yes; some respondents might not admit to such behavior.

**8.17** **(a)** $\hat{p} = 0.3275$; 0.3008 to 0.3541. **(b)** Speakers and listeners probably perceive sermon length differently.

**8.19** 0.1304 to 0.1696.

**8.21** **(a)** No. **(b)** Yes. **(c)** Yes. **(d)** No.

**8.23** 0.6345 to 0.7455.

**8.25** 0.2180 to 0.2510.

**8.27** 0.8230 to 0.9370.

**8.29** **(a)** $z = 1.34$, $P = 0.1802$. **(b)** 0.4969 to 0.5165.

**8.31** $n = 171$ or 172.

**8.33** The sample sizes are 35, 62, 81, 93, 97, 93, 81, 62, and 35; take $n = 97$.

**8.35** $\hat{p}_m - \hat{p}_w = 0.1214$; the interval is −0.0060 to 0.2488.

**8.37** $z \doteq 1.86$, $P = 0.0629$.

**8.39** **(a)** Yes. **(b)** No. **(c)** No. **(d)** Yes. **(e)** No.

**8.41** $z \doteq 4.24$, $P < 0.0001$. Confidence interval: 0.0323 to 0.0877.

**8.43** $z = 20.18$, so $P$ is tiny. Confidence interval: 0.1962 to 0.2377. Nonresponse error could render this interval and test result meaningless.

**8.45** −0.0017 to 0.0897.

**8.47** $z \doteq 6.01$, $P < 0.0001$; confidence interval is 0.0195 to 0.0384 (all the same as in Exercise 8.46).

**8.49** $z = 4.28$ and $P < 0.0001$. Confidence interval: 0.0381 to 0.1019.

**8.51** **(a)** −0.0053 to 0.2335. **(b)** $z \doteq 1.83$, $P = 0.0336$. **(c)** We have fairly strong evidence that high-tech

companies are more likely to offer stock options, but the difference in proportions could be very small or as large as 23%.

**8.53** **(a)** $\hat{p}_f = 0.8$, SE $\doteq 0.05164$; $\hat{p}_m = 0.3939$, SE $\doteq 0.04253$. **(b)** 0.2960 to 0.5161.

**8.55** $z = 2.10$, $P = 0.0360$.

**8.57** **(a)** Confidence intervals account for only sampling error. **(b)** $H_0$ should refer to $p_1$ and $p_2$. **(c)** Only if $n_1 = n_2$.

**8.59** −0.0298 to 0.0898.

**8.61** $\hat{p} = 0.6129$, $z = 4.03$, $P < 0.0001$; confidence interval is 0.5523 to 0.6735.

**8.63** **(a)** People have different symptoms; for example, not all who wheeze consult a doctor. **(b)** Sleep: 0.0864, 0.0280 to 0.1448. Number: 0.0307, −0.0361 to 0.0976. Speech: 0.0182, −0.0152 to 0.0515. Activities: 0.0137, −0.0395 to 0.0670. Doctor: −0.0112, −0.0796 to 0.0573. Phlegm: −0.0220, −0.0711 to 0.0271. Cough: −0.0323, −0.0853 to 0.0207. **(c)** It is reasonable to expect that the bypass proportions would be higher. **(d)** In the same order: $z = 2.64$, $P = 0.0042$; $z = 0.88$, $P = 0.1897$; $z = 0.99$, $P = 0.1600$; $z = 0.50$, $P = 0.3100$; $z = -0.32$, $P = 0.6267$; $z = -0.92$, $P = 0.8217$; $z = -1.25$, $P = 0.8950$. **(e)** 95% confidence interval for sleep improvement: 0.1168 to 0.2023. Part (b) showed improvement relative to control group, which is a better measure of the effect of the bypass.

**8.65** **(a)** $z = 6.98$, $P < 0.0001$. **(b)** 0.1145 to 0.2022.

**8.67** Education: 1132 users, 852 nonusers. Income: 871 users, 677 nonusers. For users, $\hat{p}_1 = 0.2306$; for nonusers, $\hat{p}_2 = 0.2054$. $z = 1.34$, $P = 0.1802$; −0.0114 to 0.0617. The lack of response about income makes the conclusions for Exercise 8.66 suspect.

**8.69** The margin of error is ±2.8%.

**8.71** $z = 8.95$, $P < 0.0001$; 0.3720 to 0.5613.

**8.73** All $\hat{p}$-values are greater than 0.5. Texts 3, 7, and 8 have (respectively) $z = 0.82$, $P = 0.4122$; $z = 3.02$, $P = 0.0025$; and $z = 2.10$, $P = 0.0357$. For the other texts, $z \geq 4.64$ and $P < 0.00005$.

**8.77** $z$: 0.90, 1.01, 1.27, 1.42, 2.84, 3.18, 4.49. $P$: 0.3681, 0.3125, 0.2041, 0.1556, 0.0045, 0.0015, 0.0000.

**8.79** **(a)** $n = 342$. **(b)** $n = (z^*/m)^2/2$.

**8.81** **(a)** $p_0 = 0.7911$. **(b)** $\hat{p} = 0.3897$, $z = -29.1$, $P$ is tiny. **(c)** $\hat{p}_1 = 0.3897$, $\hat{p}_2 = 0.7930$, $z = -29.2$, $P$ is tiny.

**8.83** **(a)** 0.5278 to 0.5822. **(b)** 0.5167 to 0.5713. **(c)** 0.3170 to 0.3690. **(d)** 0.5620 to 0.6160. **(e)** 0.5620 to 0.6160. **(f)** 0.6903 to 0.7397.

## CHAPTER 9

**9.1** **(a)** Given Explanatory = 1: 37.5% Yes, 62.5% No. Given Explanatory = 2: 47.5% Yes, 52.5% No. **(c)** When Explanatory = 2, "Yes" and "No" are nearly evenly split.

**9.3 (a)** $0.10 < P < 0.15$. **(b)** $0.01 < P < 0.02$.
**(c)** $0.025 < P < 0.05$. **(d)** $0.025 < P < 0.05$.

**9.5** $X^2 \doteq 15.2$, df $= 5$, $0.005 < P < 0.01$.

**9.7 (a)** 0.2044, 0.0189; 0.3285, 0.0699; 0.1050, 0.1072;
0.0518, 0.1141. **(b)** 0.2234, 0.3984, 0.2123, 0.1659.
**(c)** 0.6898, 0.3102. **(d)** Full-time students: 0.2964, 0.4763,
0.1522, 0.0752. Part-time students: 0.0610, 0.2254, 0.3458,
0.3678.

**9.9 (a)** Yes, this seems to satisfy the assumptions.
**(b)** df $= 3$. **(c)** $0.20 < P < 0.25$.

**9.11 (a)** Success (nonrelapse) rates were 58.3%
(desipramine), 25.0% (lithium), and 16.7% (placebo).
**(b)** Yes; this seems to satisfy the assumptions.
**(c)** $X^2 = 10.5$, df $= 2$, $P = 0.005$.

**9.13** Start by setting $a$ equal to any number from 0 to 50.

**9.15 (a)** A notably higher percent of women are "strictly
voluntary" participants. **(b)** 40.3% of men and 51.3% of
women are participants; the relative risk is 1.27.

**9.17 (a)** For example, among nonbingers, only 8.8% have
missed class, while 30.9% of occasional and 62.5% of
frequent bingers have missed class. **(b)** 45.37% of subjects
were nonbingers, 26.54% were occasional bingers, and
28.09% were frequent bingers. **(c)** Occasional versus
nonbingers: 3.5068. Frequent versus nonbingers: 7.0937.
**(d)** $X^2 \doteq 2672$, df $= 2$, $P$ is tiny.

**9.19 (b)** $X^2 = 2.591$, df $= 1$, $P = 0.108$.

**9.21 (a)** 146 women/No, 97 men/No. **(b)** For example,
19.34% of women, versus 7.62% of men, have tried low-fat
diets. **(c)** $X^2 = 7.143$, df $= 1$, $P = 0.008$.

**9.23 (a)** $X^2 = 76.7$, df $= 2$, $P < 0.0001$. **(b)** Even with much
smaller numbers of students, $P$ is still very small. **(c)** Our
conclusion might not hold for the true percents. **(d)** Lack
of independence could cause the estimated percents to be
too large or too small.

**9.25** $X^2 = 12.0$, df $= 1$, $P = 0.001$. The smallest expected
count is 6, so the test is valid.

**9.27** $X^2 = 23.1$, df $= 4$, $P < 0.0005$. Dog owners have less
education, and cat owners more, than we would expect
if there were no relationship between pet ownership and
educational level.

**9.29** The missing entries are 202, 64, 38, 33. $X^2 = 50.5$,
df $= 9$, $P < 0.0005$. The largest contributions to $X^2$
come from chemistry/engineering, physics/engineering,
and biology/liberal arts (more than expected), and
biology/engineering and chemistry/liberal arts (less than
expected).

**9.31** $X^2 = 3.955$, df $= 4$, $P = 0.413$.

**9.33 (a)** Cats: $X^2 = 6.611$, df $= 2$, $P = 0.037$. Dogs:
$X^2 = 26.939$, df $= 2$, $P < 0.0005$. **(b)** Dogs from pet stores
are less likely to go to a shelter, while "other source" dogs
are more likely to go. **(c)** The control group data should be
reasonably like an SRS.

**9.35** $X^2 = 43.487$, df $= 12$, $P < 0.0005$. Science has a
large proportion of low-scoring students, while liberal
arts/education has a large proportion of high-scoring
students.

**9.37** $X^2 = 852.433$, df $= 1$, $P < 0.0005$.

**9.39 (a)** $X^2 = 2.506$, df $= 2$, $P = 0.286$. **(b)** Divide each
echinacea count by 337 and each placebo count by 370.
**(c)** The only significant results are for rash ($z = 2.74$,
$P = 0.0061$), drowsiness ($z = 2.09$, $P = 0.0366$), and
other ($z = 2.09$, $P = 0.0366$). A $10 \times 2$ table would not be
appropriate, because each URI could have multiple adverse
events. **(d)** All results are unfavorable to echinacea, so we
are not concerned with having detected "false-positives."
**(e)** We do not have independent observations, but we
would expect the dependence to have the same effect on
both groups, so our conclusions should be fairly reliable.

**9.41** $X^2 = 3.781$, df $= 3$, $P = 0.2861$.

## CHAPTER 10

**10.1 (a)** $-2.5$. **(b)** When $x$ increases by 1, $\mu_y$ decreases by
2.5. **(c)** 15.5. **(d)** 11.5 and 19.5.

**10.3 (a)** An increase of 7.16 to 8.58 mpg. **(b)** A decrease of
7.16 to 8.58 mpg. **(c)** An increase of 3.58 to 4.29 mpg.

**10.5 (a)** The plot suggests a linear increase. **(b)** $\hat{y} =$
$-3271.9667 + 1.65x$. **(c)** Residuals: 0.01667, $-0.03333$,
0.01667. $s \doteq 0.04082$. **(d)** Given $x$ (the year), spending
comes from an $N(\mu_y, \sigma)$ distribution, where $\mu_y = \beta_0 + \beta_1 x$.
Estimates: $b_0 \doteq -3271.9667$, $b_1 \doteq 1.65$, $s \doteq 0.04082$.
**(e)** With 95% confidence, R&D spending increases from
1.283 to 2.017 billion dollars per year.

**10.7 (a)** $\beta_0$, $\beta_1$, and $\sigma$. **(b)** $H_0$ should refer to $\beta_1$. **(c)** The
confidence interval will be narrower than the prediction
interval.

**10.9 (a)** $t = 1.92$, $P = 0.0677$. **(b)** $t = 0.97$, $P = 0.3437$.
**(c)** $t = 1.92$, $P = 0.0581$.

**10.11 (a)** 1.2095 to 1.5765; a $1 difference in tuition in
2000 changes 2005 tuition by between $1.21 and $1.58,
so we estimate that tuition increased by 21% to 58%.
**(b)** $8024. **(c)** $6717 to $9331.

**10.13 (a)** $\hat{y} = -0.0127 + 0.0180x$, $r^2 \doteq 80.0\%$. **(b)** $H_0$:
$\beta_1 = 0$; $H_a$: $\beta_1 \neq 0$; $t = 7.48$, $P < 0.0001$. **(c)** The predicted
mean is 0.07712; the interval is 0.06808 to 0.08616.

**10.15 (a)** $x$ (percent forested) is right-skewed;
$\bar{x} = 39.3878\%$, $s_x = 32.2043\%$. $y$ (IBI) is left-skewed;
$\bar{y} = 65.9388$, $s_y = 18.2796$. **(b)** A weak positive association,
with more scatter in $y$ for small $x$. **(c)** $y_i = \beta_0 + \beta_1 x_i + \epsilon_i$,
$i = 1, 2, \ldots, 49$; $\epsilon_i$ are independent $N(0, \sigma)$ variables. **(d)** $H_0$:
$\beta_1 = 0$; $H_a$: $\beta_1 \neq 0$. **(e)** $\widehat{IBI} = 59.9 + 0.153\,\text{Area}$; $s = 17.79$.
For testing the hypotheses in (d), $t = 1.92$ and $P = 0.061$.
**(f)** Residual plot shows a slight curve. **(g)** Residuals are
left-skewed.

## CHAPTER 13

**p. 683**  © John Van Hasselt/CORBIS
**p. 685**  © John Van Hasselt/CORBIS
**p. 686**  Professor Pietro M. Motta/Photo Researchers
**p. 691**  Joel Rafkin/PhotoEdit
**p. 692**  © Banana Stock

## CHAPTER 14

**p. 711**  [p. 14-1]  Index Stock/JupiterImages
**p. 717**  [p. 14-7]  BrandX Pictures/photolibrary
**p. 720**  [p. 14-10]  Rod Planck/Photo Researchers

## CHAPTER 15

**p. 733**  [p. 15-1]  Jim Wark/Airphoto
**p. 735**  [p. 15-3]  Nigel Cattlin/Photo Researchers

**p. 743**  [p. 15-11]  © Jeff Greenberg/PhotoEdit
**p. 749**  [p. 15-17]  © lizabeth Crews/The Image Works
**p. 754**  [p. 15-22]  iconsportsmedia.com

## CHAPTER 16

**p. 16-1**  Greg Epperson/Index Stock Imagery/Jupiter Images
**p. 16-13**  © Brad Perks Lightscapes/Alamy
**p. 16-20**  AP Photo/Mark Humphrey
**p. 16-49**  click/MorgueFile

## CHAPTER 17

**p. 17-1**  © A. T. Willett/Alamy
**p. 17-4**  Michael Rosenfeld/Getty
**p. 17-7**  Darryl Leniuk/Masterfile
**p. 17-2**  © Jeff Greenberg/The Image Works

Table entry for z is the area under the standard normal curve to the left of z.

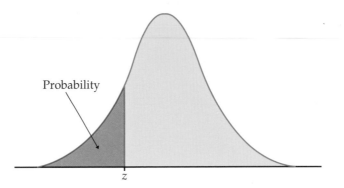

Probability

z

## TABLE A

### Standard normal probabilities

| z | .00 | .01 | .02 | .03 | .04 | .05 | .06 | .07 | .08 | .09 |
|---|-----|-----|-----|-----|-----|-----|-----|-----|-----|-----|
| −3.4 | .0003 | .0003 | .0003 | .0003 | .0003 | .0003 | .0003 | .0003 | .0003 | .0002 |
| −3.3 | .0005 | .0005 | .0005 | .0004 | .0004 | .0004 | .0004 | .0004 | .0004 | .0003 |
| −3.2 | .0007 | .0007 | .0006 | .0006 | .0006 | .0006 | .0006 | .0005 | .0005 | .0005 |
| −3.1 | .0010 | .0009 | .0009 | .0009 | .0008 | .0008 | .0008 | .0008 | .0007 | .0007 |
| −3.0 | .0013 | .0013 | .0013 | .0012 | .0012 | .0011 | .0011 | .0011 | .0010 | .0010 |
| −2.9 | .0019 | .0018 | .0018 | .0017 | .0016 | .0016 | .0015 | .0015 | .0014 | .0014 |
| −2.8 | .0026 | .0025 | .0024 | .0023 | .0023 | .0022 | .0021 | .0021 | .0020 | .0019 |
| −2.7 | .0035 | .0034 | .0033 | .0032 | .0031 | .0030 | .0029 | .0028 | .0027 | .0026 |
| −2.6 | .0047 | .0045 | .0044 | .0043 | .0041 | .0040 | .0039 | .0038 | .0037 | .0036 |
| −2.5 | .0062 | .0060 | .0059 | .0057 | .0055 | .0054 | .0052 | .0051 | .0049 | .0048 |
| −2.4 | .0082 | .0080 | .0078 | .0075 | .0073 | .0071 | .0069 | .0068 | .0066 | .0064 |
| −2.3 | .0107 | .0104 | .0102 | .0099 | .0096 | .0094 | .0091 | .0089 | .0087 | .0084 |
| −2.2 | .0139 | .0136 | .0132 | .0129 | .0125 | .0122 | .0119 | .0116 | .0113 | .0110 |
| −2.1 | .0179 | .0174 | .0170 | .0166 | .0162 | .0158 | .0154 | .0150 | .0146 | .0143 |
| −2.0 | .0228 | .0222 | .0217 | .0212 | .0207 | .0202 | .0197 | .0192 | .0188 | .0183 |
| −1.9 | .0287 | .0281 | .0274 | .0268 | .0262 | .0256 | .0250 | .0244 | .0239 | .0233 |
| −1.8 | .0359 | .0351 | .0344 | .0336 | .0329 | .0322 | .0314 | .0307 | .0301 | .0294 |
| −1.7 | .0446 | .0436 | .0427 | .0418 | .0409 | .0401 | .0392 | .0384 | .0375 | .0367 |
| −1.6 | .0548 | .0537 | .0526 | .0516 | .0505 | .0495 | .0485 | .0475 | .0465 | .0455 |
| −1.5 | .0668 | .0655 | .0643 | .0630 | .0618 | .0606 | .0594 | .0582 | .0571 | .0559 |
| −1.4 | .0808 | .0793 | .0778 | .0764 | .0749 | .0735 | .0721 | .0708 | .0694 | .0681 |
| −1.3 | .0968 | .0951 | .0934 | .0918 | .0901 | .0885 | .0869 | .0853 | .0838 | .0823 |
| −1.2 | .1151 | .1131 | .1112 | .1093 | .1075 | .1056 | .1038 | .1020 | .1003 | .0985 |
| −1.1 | .1357 | .1335 | .1314 | .1292 | .1271 | .1251 | .1230 | .1210 | .1190 | .1170 |
| −1.0 | .1587 | .1562 | .1539 | .1515 | .1492 | .1469 | .1446 | .1423 | .1401 | .1379 |
| −0.9 | .1841 | .1814 | .1788 | .1762 | .1736 | .1711 | .1685 | .1660 | .1635 | .1611 |
| −0.8 | .2119 | .2090 | .2061 | .2033 | .2005 | .1977 | .1949 | .1922 | .1894 | .1867 |
| −0.7 | .2420 | .2389 | .2358 | .2327 | .2296 | .2266 | .2236 | .2206 | .2177 | .2148 |
| −0.6 | .2743 | .2709 | .2676 | .2643 | .2611 | .2578 | .2546 | .2514 | .2483 | .2451 |
| −0.5 | .3085 | .3050 | .3015 | .2981 | .2946 | .2912 | .2877 | .2843 | .2810 | .2776 |
| −0.4 | .3446 | .3409 | .3372 | .3336 | .3300 | .3264 | .3228 | .3192 | .3156 | .3121 |
| −0.3 | .3821 | .3783 | .3745 | .3707 | .3669 | .3632 | .3594 | .3557 | .3520 | .3483 |
| −0.2 | .4207 | .4168 | .4129 | .4090 | .4052 | .4013 | .3974 | .3936 | .3897 | .3859 |
| −0.1 | .4602 | .4562 | .4522 | .4483 | .4443 | .4404 | .4364 | .4325 | .4286 | .4247 |
| 0.0 | .5000 | .4960 | .4920 | .4880 | .4840 | .4801 | .4761 | .4721 | .4681 | .4641 |

Table entry for z is the area under the standard normal curve to the left of z.

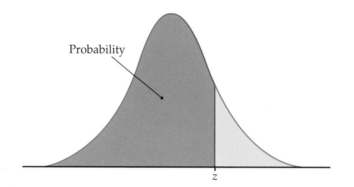
Probability

## TABLE A

### Standard normal probabilities (continued)

| z | .00 | .01 | .02 | .03 | .04 | .05 | .06 | .07 | .08 | .09 |
|---|-----|-----|-----|-----|-----|-----|-----|-----|-----|-----|
| 0.0 | .5000 | .5040 | .5080 | .5120 | .5160 | .5199 | .5239 | .5279 | .5319 | .5359 |
| 0.1 | .5398 | .5438 | .5478 | .5517 | .5557 | .5596 | .5636 | .5675 | .5714 | .5753 |
| 0.2 | .5793 | .5832 | .5871 | .5910 | .5948 | .5987 | .6026 | .6064 | .6103 | .6141 |
| 0.3 | .6179 | .6217 | .6255 | .6293 | .6331 | .6368 | .6406 | .6443 | .6480 | .6517 |
| 0.4 | .6554 | .6591 | .6628 | .6664 | .6700 | .6736 | .6772 | .6808 | .6844 | .6879 |
| 0.5 | .6915 | .6950 | .6985 | .7019 | .7054 | .7088 | .7123 | .7157 | .7190 | .7224 |
| 0.6 | .7257 | .7291 | .7324 | .7357 | .7389 | .7422 | .7454 | .7486 | .7517 | .7549 |
| 0.7 | .7580 | .7611 | .7642 | .7673 | .7704 | .7734 | .7764 | .7794 | .7823 | .7852 |
| 0.8 | .7881 | .7910 | .7939 | .7967 | .7995 | .8023 | .8051 | .8078 | .8106 | .8133 |
| 0.9 | .8159 | .8186 | .8212 | .8238 | .8264 | .8289 | .8315 | .8340 | .8365 | .8389 |
| 1.0 | .8413 | .8438 | .8461 | .8485 | .8508 | .8531 | .8554 | .8577 | .8599 | .8621 |
| 1.1 | .8643 | .8665 | .8686 | .8708 | .8729 | .8749 | .8770 | .8790 | .8810 | .8830 |
| 1.2 | .8849 | .8869 | .8888 | .8907 | .8925 | .8944 | .8962 | .8980 | .8997 | .9015 |
| 1.3 | .9032 | .9049 | .9066 | .9082 | .9099 | .9115 | .9131 | .9147 | .9162 | .9177 |
| 1.4 | .9192 | .9207 | .9222 | .9236 | .9251 | .9265 | .9279 | .9292 | .9306 | .9319 |
| 1.5 | .9332 | .9345 | .9357 | .9370 | .9382 | .9394 | .9406 | .9418 | .9429 | .9441 |
| 1.6 | .9452 | .9463 | .9474 | .9484 | .9495 | .9505 | .9515 | .9525 | .9535 | .9545 |
| 1.7 | .9554 | .9564 | .9573 | .9582 | .9591 | .9599 | .9608 | .9616 | .9625 | .9633 |
| 1.8 | .9641 | .9649 | .9656 | .9664 | .9671 | .9678 | .9686 | .9693 | .9699 | .9706 |
| 1.9 | .9713 | .9719 | .9726 | .9732 | .9738 | .9744 | .9750 | .9756 | .9761 | .9767 |
| 2.0 | .9772 | .9778 | .9783 | .9788 | .9793 | .9798 | .9803 | .9808 | .9812 | .9817 |
| 2.1 | .9821 | .9826 | .9830 | .9834 | .9838 | .9842 | .9846 | .9850 | .9854 | .9857 |
| 2.2 | .9861 | .9864 | .9868 | .9871 | .9875 | .9878 | .9881 | .9884 | .9887 | .9890 |
| 2.3 | .9893 | .9896 | .9898 | .9901 | .9904 | .9906 | .9909 | .9911 | .9913 | .9916 |
| 2.4 | .9918 | .9920 | .9922 | .9925 | .9927 | .9929 | .9931 | .9932 | .9934 | .9936 |
| 2.5 | .9938 | .9940 | .9941 | .9943 | .9945 | .9946 | .9948 | .9949 | .9951 | .9952 |
| 2.6 | .9953 | .9955 | .9956 | .9957 | .9959 | .9960 | .9961 | .9962 | .9963 | .9964 |
| 2.7 | .9965 | .9966 | .9967 | .9968 | .9969 | .9970 | .9971 | .9972 | .9973 | .9974 |
| 2.8 | .9974 | .9975 | .9976 | .9977 | .9977 | .9978 | .9979 | .9979 | .9980 | .9981 |
| 2.9 | .9981 | .9982 | .9982 | .9983 | .9984 | .9984 | .9985 | .9985 | .9986 | .9986 |
| 3.0 | .9987 | .9987 | .9987 | .9988 | .9988 | .9989 | .9989 | .9989 | .9990 | .9990 |
| 3.1 | .9990 | .9991 | .9991 | .9991 | .9992 | .9992 | .9992 | .9992 | .9993 | .9993 |
| 3.2 | .9993 | .9993 | .9994 | .9994 | .9994 | .9994 | .9994 | .9995 | .9995 | .9995 |
| 3.3 | .9995 | .9995 | .9995 | .9996 | .9996 | .9996 | .9996 | .9996 | .9996 | .9997 |
| 3.4 | .9997 | .9997 | .9997 | .9997 | .9997 | .9997 | .9997 | .9997 | .9997 | .9998 |